MASS SPECTROMETRY of POLYMERS

MASS SPECTROMETRY of POLYMERS

Edited by

Giorgio Montaudo
Robert P. Lattimer

CRC Press
Taylor & Francis Group
Boca Raton London New York

CRC Press is an imprint of the
Taylor & Francis Group, an **informa** business

CRC Press
Taylor & Francis Group
6000 Broken Sound Parkway NW, Suite 300
Boca Raton, FL 33487-2742

© 2002 by Taylor & Francis Group, LLC
CRC Press is an imprint of Taylor & Francis Group, an Informa business

First issued in paperback 2019

No claim to original U.S. Government works

ISBN-13: 978-0-367-45509-5 (pbk)
ISBN-13: 978-0-8493-3127-5 (hbk)

Visit the Taylor & Francis Web site at
http://www.taylorandfrancis.com

and the CRC Press Web site at
http://www.crcpress.com

Library of Congress Card Number 2001037684

Library of Congress Cataloging-in-Publication Data

Montaudo, Giorgio.
  Mass spectrometry of polymers / Giorgio Montaudo, Robert Lattimer.
       p.   cm.
  Includes bibliographical references and index.
  ISBN 0-8493-3127-7 (alk. paper)
  1.Polymers--Analysis. 2. Mass spectrometry. I. Lattimer, Robert (Robert P.) II. Title.

QD139.P6 M66 2001
547′.7046—dc21   2001037684

Preface

Mass spectrometry involves the study of ions in the vapor phase. This analytical method has a number of features and advantages that make it an extremely valuable tool for the identification and structural elucidation of organic molecules—including synthetic polymers:

(i) The amount of sample needed is small; for direct analysis, a microgram or less of material is normally sufficient.

(ii) The molar mass of the material can be obtained directly by measuring the mass of the molecular ion or a "quasimolecular ion" containing the intact molecule.

(iii) Molecular structures can be elucidated by examining molar masses, ion fragmentation patterns, and atomic compositions determined by mass spectrometry.

(iv) Mixtures can be analyzed by using "soft" desorption/ionization methods and hyphenated techniques (such as GC/MS, LC/MS, and MS/MS).

Mass spectrometric (MS) methods are routinely used to characterize a wide variety of biopolymers, such as proteins, polysaccharides, and nucleic acids. Nevertheless, despite its advantages, mass spectrometry has been underutilized in the past for studying synthetic polymer systems. It is fair to say that, until recently, polymer scientists have been rather unfamiliar with the advances made in the field of mass spectrometry.

However, mass spectrometry in recent years has rapidly become an indispensable tool in polymer analysis, and modern MS today complements in many ways the structural data provided by NMR and IR methods. Contemporary MS of polymers is emerging as a revolutionary discipline. It is capable of changing the analytical protocols established for years for the molecular and structural analysis of macromolecules.

Some of the most significant applications of modern MS to synthetic polymers are (a) chemical structure and end-group analysis, (b) direct measurement of molar mass and molar mass distribution, (c) copolymer composition and sequence distribution, and (d) detection and identification of impurities and additives in polymeric materials.

In view of the recent developments in this area, a book such as *Mass Spectrometry of Polymers* appears opportune. Even more, in our opinion there is an acute need for a state-of-the-art book that summarizes the progress recently made. No books currently exist that deal systematically with the whole subject. Therefore we present here an effort to summarize the current status of the use of mass spectrometry in polymer characterization.

The Distinctiveness of MS

A basic question one might ask is "why pursue mass spectral techniques for analysis of higher-molar mass polymers?"[1] After all, a number of "classical" methods are available that have proved very successful at analyzing polymers (e.g., gel permeation chromatography, vapor pressure osmometry, laser light scattering, magnetic resonance, infrared and ultraviolet/visible spectroscopies). In light of this success, what does mass spectrometry have to offer?

It turns out that there are important reasons to pursue polymer MS developments other than scientific curiosity and desire for methodological improvements.[1] Classical techniques, for example, are always averaging methods; i.e., they measure the average properties of a mixture of oligomers and thus do not examine individual molecules. Furthermore, classical techniques do not normally yield information on the different types of oligomers that may be present, nor do they distinguish and identify impurities and additives in polymer samples. Copolymers and blends will often not be distinguished as to polymer type. Finally, most classical methods do not provide absolute, direct molar-mass distributions for polymers; instead they rely on calibrations made using accepted standards. Mass spectrometry clearly has great potential to examine individual oligomers/components in polymeric systems, and this can add much information to complement and extend the "classical" methods.

Historical Background

In order to analyze any material by mass spectrometry, the sample must first be vaporized (or desorbed) and ionized in the instrument's vacuum system. Since polymers are generally nonvolatile, many mass spectral methods have involved *degradation* of the polymeric material prior to analysis of the more volatile fragments. Two traditional methods to examine polymers have been flash-pyrolysis GC/MS and direct pyrolysis in the ion source of the instrument.

In recent years, however, there has been a marked tendency toward the use of *direct* MS techniques. While a continued effort to introduce mass spectrometry as a major technique for the structural analysis of polymers has been made over the past three decades, MS analysis did not have a great impact upon the polymer community until the past five years or so. During this period outstanding progress has been made in the application of MS to some crucial problems involving the characterization of synthetic polymers.

Developments in two general areas have spurred this progress. Sector and quadrupole *mass analyzers*, the traditional methods of separation of ions in mass spectrometry, have recently been complemented by the development of powerful Fourier transform (FT-MS) and time-of-flight (TOF-MS) instruments. The TOF analyzers are particularly well-suited for detecting higher molar-mass species present in polymers.

Parallel to this progress, new *ionization methods* have been developed that are based on the direct desorption of ions from polymer surfaces. With the introduction of "desorption/ionization" techniques, it has become possible to eject large molecules into the gas phase directly from the sample surface, and thereby mass spectra of intact polymer molecules have been produced. Much progress to date has been made using matrix-assisted laser desorption/ionization (MALDI-MS), which is capable of generating quasimolecular ions in the range of 10^6 Daltons (Da) and beyond.

A brief list of ionization methods is given in Table 1. (One may quibble a bit about the dates given in the table, but we believe these are more or less accurate.) Up until about 1970, the only ionization method in common use was electron impact (EI). Field ionization (FI) was developed in the 1950s, but it was never very popular, and chemical ionization (CI) was just getting started. These three methods (EI, CI, FI) depend upon vaporization of the sample by heating, which pretty much limits polymer applications to small, stable oligomers or to polymer degradation products (formed by pyrolysis or other methods). Field desorption (FD-MS), invented in 1969, was the first "desorption/ionization" method. FD- and FI-MS are often very useful (particularly for analysis of less polar polymers), but they have never been in widespread use.

TABLE 1

History of Ionization Methods

Electron impact (EI) 1918
Field ionization (FI) 1954
Chemical ionization (CI) 1968
Field desorption (FD) 1969
Desorption chemical ionization (DCI) 1973
^{252}Cf plasma desorption (PD) 1974
Laser desorption (LD) 1975
Static secondary ion mass spectrometry (SSIMS) 1976
Atmospheric pressure chemical ionization (APCI) 1976
Thermospray (TSP) 1978
Electrohydrodynamic ionization (EH) 1978
Fast atom bombardment (FAB) 1982
Potassium ionization of desorbed species (KIDS) 1984
Electrospray ionization (ESI) 1984
Multiphoton ionization (MPI) 1987
Matrix-assisted laser desorption/ionization (MALDI) 1988

The 1970s and 1980s saw the advent of several new "soft" desorption/ionization methods, many of which are now well-established in analytical mass spectrometry. The term "desorption/ionization" refers to a method in which the desorption (vaporization) and ionization steps occur essentially simultaneously. MALDI and several other techniques listed in Table 1 have important applications in polymer analysis.

One reason for the underutilization of mass spectrometry in polymer analysis lies in the historical development. Magnetic resonance (NMR), infrared (IR), and ultraviolet/visible (UV/vis) spectroscopies have a long history in polymer analysis, while mass spectrometry is a relative newcomer. NMR, IR, and UV/vis techniques of course have the advantage that the polymer does not need to be vaporized prior to analysis. Thus these techniques gained a strong following in the polymer community long before mass spectrometric techniques were developed that could analyze intact macromolecules. In fact, mass spectrometry obtained a rather dubious reputation among many polymer scientists; this skepticism toward polymer MS continued even into the 1990s.

The well-known polymer analyst Jack Koenig, in his widely-read book *Spectroscopy of Polymers* (1992) said: "The majority of the spectroscopic techniques, such as UV and visible or mass spectroscopy, do not meet the specifications of the spectroscopic probe [for polymers]."[2] Koenig's rather skeptical opinion of mass spectrometry for polymer analysis was typical of the viewpoint of many scientists prior to the mid-1990s.

Fortunately, the use of mass spectrometry for polymer analysis took on a new dimension at the turn of the century. Figure 1 lists the number of polymer mass spectrometry publications in the CAplus (Chemical Abstracts) database over the years 1965–2000. Up until the mid-1990s there was a steady—but not dramatic—increase in the number of articles. Starting in 1995, however, there has been a marked increase in the number of polymer mass spectrometry reports in the literature. Also the number of symposia and conferences devoted to the subject has grown considerably in the last few years.

The major reason for this increase has been the use of MALDI-MS for numerous polymer applications. MALDI is by no means the only mass spectral method that is useful for polymer analysis, but it has provided the impetus to get polymer people interested in what mass spectrometry can do.

We find it encouraging that Koenig has included a chapter on mass spectrometry in the second edition of his book (1999).[3] At the end of the Mass Spectrometry chapter, Koenig makes these concluding remarks: "Modern MS, particularly with the advent of MALDI, is finally causing polymer chemists to be interested in MS as a structural analysis tool. . . . I expect that in the future MS will join IR and NMR as regular techniques used by polymer chemists."[3]

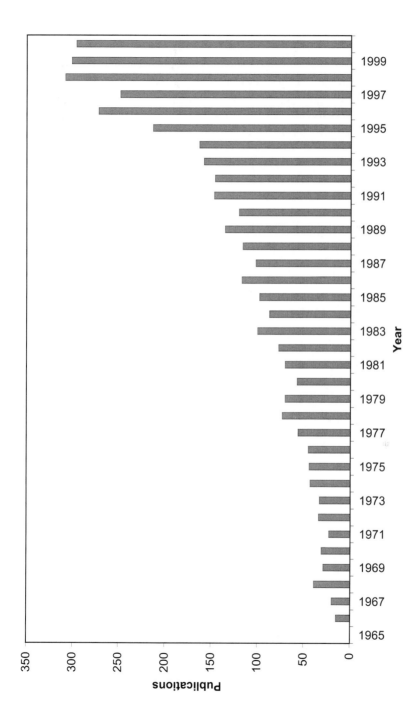

FIGURE 1
Polymer mass spectrometry publications.

Book Organization and Scope

The book consists of two introductory chapters followed by nine chapters on applications. Since it is relatively new to polymer science, mass spectrometry needs to be introduced in some detail, and this is done in Chapter 1. On the other hand, many analytical chemists will need an introduction to polymer characterization methods, and this is done in Chapter 2. The rest of the chapters cover in detail the most relevant applications of mass spectrometry to the analysis of polymers.

Because of the low volatility of polymeric materials, many mass spectral methods for polymers have involved pyrolysis (or thermal degradation), and this topic is covered in Chapter 3 (pyrolysis-GC/MS), Chapter 5 (direct pyrolysis-MS), and Chapter 6 (pyrolysis-FI/FD-MS). Chemical degradation methods are discussed in connection with fast atom bombardment analysis (Chapter 7).

For synthetic polymers, the most popular desorption/ionization method has been matrix-assisted laser desorption/ionization (MALDI-MS, Chapter 10). Several other techniques have important applications in polymer analysis. The more widely used methods are covered in this book: electrospray (Chapter 4), field ionization/desorption (Chapter 6), fast atom bombardment (Chapter 7), secondary ion mass spectrometry (Chapter 8), and laser desorption (Chapters 9 and 11).

The present book is designed to be practical in nature. That is, the individual chapters are not intended to be exhaustive reviews in a particular field. Instead, they introduce the subject and describe typical applications in a tutorial manner, with pertinent references from the literature. We trust that the book will be useful to both novices and experienced practitioners in polymer MS.

G. Montaudo
Catania, Italy

R. P. Lattimer
Brecksville, Ohio

References

1. Schulten, H.-R. and Lattimer, R. P., Applications of Mass Spectrometry to Polymers, *Mass Spectrom. Rev.*, 3, 231, 1984.
2. Koenig, J. L., *Spectroscopy of Polymers*, American Chemical Society, Washington, DC, 1992.
3. Koenig, J. L., *Spectroscopy of Polymers: Second Edition*, Elsevier, Amsterdam, 1999.

The Editors

Robert Lattimer, B.S., Ph.D., is a Senior Research Associate at Noveon, Inc. (formerly a division of the BF Goodrich Co.) in Brecksville, Ohio. He has been supervisor of mass spectrometry since 1974. Dr. Lattimer has a B.S. in chemistry from the University of Missouri and a Ph.D. in physical chemistry from the University of Kansas. He was a postdoctoral associate at the University of Michigan prior to coming to BF Goodrich/Noveon.

Dr. Lattimer is an internationally recognized authority in the analytical characterization and degradation of polymeric materials. His research interests include mechanisms of crosslinking and pyrolysis of polymers, and the mass spectrometric analysis of polymeric systems. He is Editor of the *Journal of Analytical and Applied Pyrolysis* and a past Associate Editor of *Rubber Chemistry and Technology*. Dr. Lattimer is past Chairman of the Gordon Research Conference on Analytical Pyrolysis, and he received the ACS Rubber Division's Sparks-Thomas Award in 1990. He has won two Rubber Division Best Paper Awards, as well as three Honorable Mentions.

Dr. Lattimer is a member of the American Chemical Society and its Rubber, Polymer, and Analytical Divisions. He is a past Councilor and Chairman of the Akron Section ACS. He is a member and past Vice President of the American Society for Mass Spectrometry.

Dr. Lattimer lives in Hudson, Ohio, with his wife Mary and two sons, Scott and Paul.

Giorgio Montaudo, Ph.D. is a Professor of industrial chemistry at the Department of Chemistry, University of Catania, Italy and Director of the Institute for Chemistry & Technology of Polymeric Materials of the National Council of Research of Italy, Catania. Dr. Montaudo received a Ph.D. in chemistry from the University of Catania. He was a postdoctoral associate at the Polytechnic Institute of Brooklyn (1966) and at the University of Michigan (1967-68 and 1971) and he was a Humboldt Foundation Fellow, 1973 at Mainz University. Dr. Montaudo has been active in the field of the synthesis, degradation, and characterization of polymeric materials. A major section of his activity has been dedicated to develop mass spectrometry of polymers as analytical and structural tools for the analysis of polymers. He is the author of more than 300 publications in international journals and chapters in books.

Dr. Montaudo serves on the Editorial Board of *Journal of Analytical & Applied Pyrolysis; Macromolecules; Macromolecular Chemistry & Physics; Polymer International; Polymer Degradation & Stability;* and *European Mass Spectrometry.* He is a past member of the Editorial Board of *Journal of Polymer*

Science, and *Trends in Polymer Science.* He received the Award of the Italian Chemical Industry, Milan 1990. His participation in over 120 international invited lectures includes: Charles M. McKnight Lecture, April 1998, The University of Akron; Visiting Professor, May-July 1980, Mainz University; Visiting Professor, March-September 1988, University of Cincinnati; Visiting Professor, September-November 1995, Universitè Pierre & Marie Curie Paris.

Dr. Montaudo lives in Catania, Italy, with his wife Paola. He has a son, Maurizio, and a daughter, Matilde.

Contributors

Mattanjah S. de Vries   University of California, Santa Barbara, California

David M. Hercules   Vanderbilt University, Nashville, Tennessee

Heinrich E. Hunziker   University of California, Santa Barbara, California

Robert Lattimer   Noveon, Inc., Brecksville, Ohio

Giorgio Montaudo   University of Catania, Catania, Italy

Maurizio S. Montaudo   Istituto per la Chimica e la Tecnologia dei Materiali Polimerici, Consiglio Nazionale delle Ricerche, Catania, Italy

Hajime Ohtani   Nagoya University, Nagoya, Japan

Salvador J. Pastor   University of Arkansas, Fayetteville, Arkansas

Michael J. Polce   The University of Akron, Akron, Ohio

Laszlo Prokai   Univeristy of Florida, Gainesville, Florida

Concetto Puglisi   Istituto per la Chimica e la Tecnologia dei Materiali Polimerici, Consiglio Nazionale delle Ricerche, Catania, Italy

Filippo Samperi   Istituto per la Chimica e la Tecnologia dei Materiali Polimerici, Consiglio Nazionale delle Ricerche, Catania, Italy

Shin Tsuge   Nagoya University, Nagoya, Japan

Chrys Wesdemiotis   The University of Akron, Akron, Ohio

Charles L. Wilkins   University of Arkansas, Fayetteville, Arkansas

Contents

1

Introduction to Mass Spectrometry of Polymers

Michael J. Polce and Chrys Wesdemiotis

CONTENTS

0-8493-3127-7/02/$0.00+$1.50
© 2002 by CRC Press LLC

1.1   Introduction

Mass spectral analyses involve the formation of gaseous ions from an analyte (M) and subsequent measurement of the mass-to-charge ratio (m/z) of these ions.[1] Depending on the ionization method used, the sample is converted to molecular or quasimolecular ions and their fragments. Molecular ions are generally radical cations (M^+), formed by electron removal from M; electron addition to yield M^- is used occasionally for electronegative samples.[2,3] Quasimolecular ions may be either positive or negative and arise by adding to M, or subtracting from it, an ion; common examples include $[M + H]^+$, $[M - H]^-$, $[M + Na]^+$, and $[M + Cl]^-$. "Soft" ionization methods generate predominantly molecular or quasimolecular ions, whereas "hard" ionization methods also yield fragment ions.[1-3] The mass spectrometer separates the ions generated upon ionization according to their mass-to-charge ratio (or a related property) to give a graph of ion abundance vs. m/z. Mixtures are often preseparated by gas or liquid chromatography, so that a mass spectrum can be obtained for each individual component to thereby facilitate sample characterization.[2,3]

The exact m/z value of the molecular or quasimolecular ion reveals the ion's elemental composition and, thus, allows for the compositional analysis of the sample under study.[1] If the molecular ions are unstable and decompose completely, the resulting fragmentation patterns can be used as a fingerprint for the identification of the sample.[1] Fragment ions also provide important information about the primary structure (i.e., connectivity or sequence) of the sample molecules.[1-3] With soft ionization methods that produce little or no fragments, fragmentation can be induced by employing tandem mass spectrometry (MS/MS).[4,5]

Mass spectrometry methods have experienced a steadily increasing use in polymer analyses[6] due to their high *sensitivity* ($<10^{-15}$ mol suffice for analysis), *selectivity* (minor components can be analyzed within a mixture), *specificity* (exact mass and fragmentation patterns serve as particularly specific compositional characteristics), and *speed* (data acquisition possible within seconds). As mentioned, the analysis of a polymer (or any other sample) by mass spectrometry presupposes that the polymer can at least partly be converted to gas-phase ions. This chapter briefly reviews the ionization methods and instrumentation available today for the characterization of synthetic macromolecules.

1.2 Ionization Methods

There are three major methods for the preparation of gaseous ions. (i) Volatile materials are generally ionized by interaction of their vapors with electrons, ions, or strong electric fields. (ii) Strong electric fields can also ionize non-volatile materials. In addition, ions from nonvolatile and thermally labile compounds can be desorbed into the gas phase via bombardment of the appropriately prepared sample with fast atoms, ions, or laser photons and via rapid heating. (iii) Alternatively, liquid solutions of the analyte may directly be converted to gas phase ions via spray techniques. Method (i) can only be applied to monomers and low-mass oligomers or in conjunction with degradation methods (principally pyrolysis). Methods (ii) and (iii) on the other hand are amenable to intact polymers. The ensuing sections describe the specific properties of these ionization methods.

1.2.1 Ionization of Volatile Materials

1.2.1.1 Electron Ionization (EI)

In this method, the sample is thermally vaporized and approximately 10^{-5} Torr of its vapors enter the ion source volume where they are ionized by collision with an electron beam of (typically) 70 eV kinetic energy. Electron ionization can produce intact molecular radical cations, $M^{+\cdot}$, by ejection of an electron from the sample molecules (Eq. 1.1).[1,7] This process has a yield of ~0.01% and deposits a wide distribution of internal energies to the newly formed molecular ions; as a result, many $M^{+\cdot}$ are formed excited enough to yield a number of fragment ions (Eq. 1.2) via competitive (F_1^+, F_2^+, F_3^+) and consecutive (f_a^+, f_b^+, f_c^+) decompositions.

$$M + e^- \rightarrow M^{+\cdot} + 2e^- \tag{1.1}$$

$$\nearrow F_1^+ \rightarrow f_a^+ \rightarrow \tag{1.2a}$$

$$M^{+\cdot} \rightarrow F_2^+ \rightarrow f_b^+ \rightarrow \tag{1.2b}$$

$$\searrow F_3^+ \rightarrow f_c^+ \rightarrow \tag{1.2c}$$

The EI mass spectrum that results is comprised of the molecular ion and all fragment ions; the degree of fragmentation can be reduced by lowering the electron energy to ≤ 15 eV.[1,7] Figure 1.1 shows the EI mass spectra of the photolysis products of poly(ethylene) and poly(propylene).[8] Each spectrum shows the molecular ions of several hydrocarbon subunits (*m/z* values

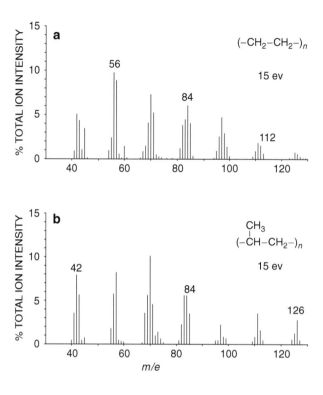

FIGURE 1.1
EI mass spectra using 15 eV ionizing electrons of the laser pyrolysis products of (a) poly(ethylene) and (b) poly(propylene). (Reprinted from Ref. 8 with permission of John Wiley & Sons)

marked) as well as their fragment ions; their distinctive fragmentation patterns help identify the composition of the original polyolefin.

From functionalized polymers or copolymers, complex mixtures of several monomers, small oligomers, and other products may arise upon degradation. In such cases, it is advantageous to use GC/MS, which makes it possible to obtain mass spectra of the single-mixture constituents. The mass spectra identify the individual components, while the total ion chromatograms reconstructed from the spectra reveal quantitative compositional information about the polymer, for example, the proportion of oligomers in a random or block copolymer. GC/MS of pyrolyzed polymers is covered in considerable detail in Chapter 3.

1.2.1.2   Chemical Ionization (CI)

In chemical ionization, gaseous analyte molecules are ionized by ion-molecule reactions with reagent ions, formed by electron ionization from the appropriate reagent gas.[9] The CI ion source is similar to the EI source, but is operated at a higher pressure (0.1–2 Torr). The chemical ionization process is illustrated for a proton transfer reaction, which is the most common ionization mode.[2,3,7,9] The sample and a large excess ($\sim 10^3$ fold) of the reagent gas (RH) are introduced simultaneously into the source. The reagent molecules are ionized by electron impact and react with other reagent molecules to form reactant ions, RH_2^+ (Eq. 1.3), which protonate the sample (Eq 1.4).

$$RH^{+\cdot} + RH \rightarrow RH_2^+ + R^\cdot \quad \text{(reagent ion formation)} \qquad (1.3)$$

$$RH_2^+ + M \rightarrow RH + MH^+ \quad \text{(proton transfer)} \qquad (1.4)$$

$$RH_2^+ + M \rightarrow [M + RH_2]^+ \quad \text{(electrophilic addition)} \qquad (1.5)$$

Typical protonation reagents are CH_5^+, $(CH_3)_3C^+$, and NH_4^+. Proton transfer proceeds at the collision rate (every encounter has 100% efficiency) with exothermic reactions, i.e., when the proton affinity (PA) of M is larger than the PA of RH.[9] The reaction exothermicity (ΔPA) ends up as internal energy of MH^+, which thus can be controlled by the choice of RH. When ΔPA is small, which is true for reagents of high proton affinity (such as NH_3), the internal energy of MH^+ is low and little (if any) fragmentation takes place. In contrast, when ΔPA is large, an appreciable fraction of MH^+ undergoes fragmentation. Endothermic proton transfer is usually not observed; in such a case, electrophilic addition (Eq. 1.5) is much more likely.[7,9] The large source pressure ensures that RH_2^+ is thermalized (to avoid endothermic reactions) and that M is ionized by a chemical reaction (Eqs. 1.4 or 1.5) and not by electron ionization.

$$Ar^{+\cdot} + M \rightarrow Ar + M^{+\cdot} \quad \text{(charge exchange)} \qquad (1.6)$$

$$H_3^+ + M \rightarrow 2H_2 + [M - H]^+ \quad \text{(anion abstraction)} \qquad (1.7)$$

$$CH_3O^- + M \rightarrow CH_3OH + [M - H]^- \quad \text{(cation abstraction)} \qquad (1.8)$$

$$Cl^- + M \rightarrow [M + Cl]^- \quad \text{(nucleophilic addition)} \qquad (1.9)$$

Depending on the chemical properties of the analyte, reactions other than proton transfer and electrophilic addition can be used to produce analyte molecular or quasimolecular ions. Equations 1.6 through 1.9 exemplify these alternatives with specific reactant ions, which are particularly effective for the given reactions.[7,9] Overall, negative chemical ionization (Eqs. 1.8 and 1.9) is used less frequently than positive chemical ionization (Eqs. 1.4 through 1.7).

CI can be used for the analysis of pyrolytic or photolytic degradation products with or without online chromatographic separation (see Chapters 4 and 5). A variant, namely desorption chemical ionization (DCI) is applicable to intact low-mass polymers as well. In DCI, the sample is not vaporized *before*

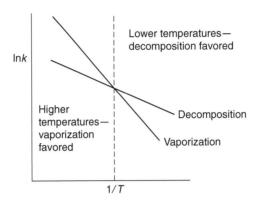

FIGURE 1.2
Dependence on temperature of the rate constants of decomposition and vaporization. (Reprinted from Ref. 10 with permission of the American Chemical Society)

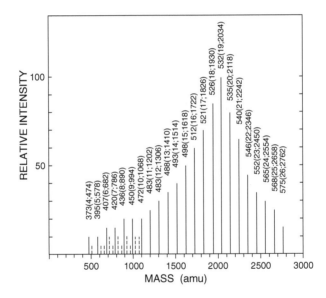

FIGURE 1.3
Partial DCI spectrum of poly(styrene) using argon as the reagent gas (Eq. 1.6). The solid lines are n-mer molecular ions, and the dashed lines are fragment ions. The numbers not in parentheses are the evaporation temperatures in K. The first and second numbers in parentheses are the number of monomer units and the monoisotopic mass, respectively. (Reprinted from Ref. 11 with permission of the American Chemical Society)

entering the CI source but is rapidly heated *inside* the source. Rapid heating enhances the probability of sample evaporation vis-à-vis sample decomposition (cf. Figure 1.2);[10] once the sample is in the gaseous state, it is immediately ionized by the surrounding CI reagent ions. A DCI application is illustrated in Figure 1.3, which reproduces the spectrum of a poly(styrene), acquired by rapid evaporation of the polymer from an electrically heated rhenium filament. DCI can be combined with K^+ ionization to form $[M + K]^+$ adducts; this approach, termed "potassium ion ionization of desorbed species" (KIDS).

1.2.1.3 Field Ionization (FI)

In FI, gaseous analyte molecules (M) approach a surface of high curvature that is maintained at a high positive potential, giving rise to a strong electric field near the surface (of the order of 10^7 V/cm). Under the influence of the field, quantum tunneling of a valence electron from M to the anode surface can take place in about 10^{-12} s, creating $M^{+\cdot}$. $[M + H]^+$ may also form with polar analytes by hydrogen abstraction from or near the anode.[2,3] Molecular ions produced via FI possess lower internal energies than those produced via EI and, thus, fragment less. This is documented in Figure 1.4 by the EI vs. FI spectra of poly(ethylene).

The residence times of an ion in the FI and EI sources are approximately 10^{-12} and 10^{-6} s, respectively. The smaller residence time upon FI eliminates or reduces the extent of rearrangements; as a result, isomers that produce very similar EI spectra may be distinguishable by their FI spectra.[7]

1.2.2 Desorption/Ionization Methods

1.2.2.1 Field Desorption (FD)

FD and FI have the same ionization mechanism. In FD, the sample is not vaporized into the gaseous state but deposited directly onto the surface carrying the strong field (called emitter). Under the strong fields used, no heating or only mild heating of the emitter is needed to desorb $M^{+\cdot}$ or $[M + H]^+$. Metal salts may be added to the sample to form other types of quasi-molecular ions, such as $[M + Na]^+$ or $[M + K]^+$.[2,3,6a,7] Field desorption leads to less excited ions than FI and often gives molecular or quasimolecular ions only, facilitating compositional analyses.[6a,7] FD has been successfully applied to polymers with molecular weights up to ca. 10,000 Da;[12,13] an example is shown in Figure 1.5.[14] The method is particularly useful for hydrocarbon polymers with no functional groups which even today are hard to ionize by any other methods (see Chapter 6).

1.2.2.2 Secondary Ion Mass Spectrometry (SIMS)

This method has traditionally been used for the elemental analysis of surfaces ("dynamic" SIMS). Organic materials can be subjected to SIMS, too, by depositing them as a thin film on a metal (or other) foil, occasionally together with a salt.[15,16] The sample is bombarded by a *primary ion* beam (e.g., Ar^+ or Cs^+), which leads to the sputtering of *secondary ions* from the surface. The latter can be $M^{+\cdot}$, M^{-}, $[M + Ag]^+$ (if a silver surface is used), or $[M + alkali]^+$

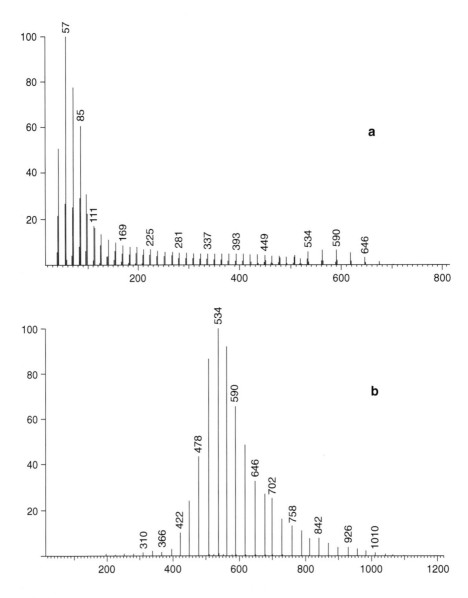

FIGURE 1.4
(a) Electron ionization (70 eV) and (b) field ionization mass spectra of poly(ethylene) 630. (Courtesy of Dr. Robert P. Lattimer, BF Goodrich Company)

(if the sample is doped with an alkali metal ion salt).[15–17] This SIMS technique is often referred to as "static" or "organic" SIMS and, as a high-energy process, normally causes extensive fragmentation.[17] The structural insight rendered by SIMS is discussed in detail in Chapter 8.

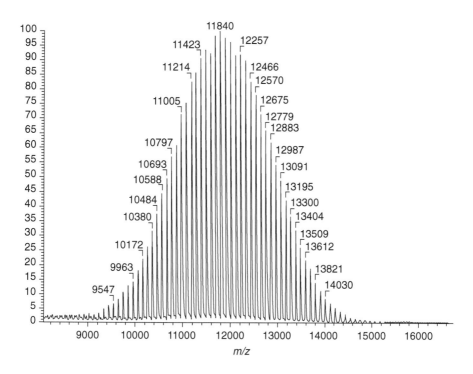

FIGURE 1.5
Field desorption mass spectrum of poly(styrene) 12500. (Reprinted from Ref. 14 with permission of John Wiley & Sons)

1.2.2.3 Fast Atom Bombardment (FAB) and Liquid Secondary Ion Mass Spectrometry (LSIMS)

FAB[18] and LSIMS[19] are conceptually identical with static SIMS. Now, the sample is mixed with a viscous liquid of low volatility, such as glycerol, thioglycerol, 3-nitrobenzylalcohol, or diethanolamine. A droplet of the mixture is bombarded by a fast (keV) beam of ions (LSIMS) or atoms (FAB), producing ions characteristic of the matrix and the analyte, as shown in Figure 1.6.[20] The analyte ions usually are $[M + H]^+$, $[M - H]^-$, or attachment ions of M with added or adventitious alkali metal ions. It is believed that these ions are formed by ion-molecule reactions in the selvedge region (gas phase region just above the liquid surface of the droplet being bombarded). Ions that are preformed in solution, such as quaternary ammonium cations and salt cluster ions, may be directly desorbed into the gas phase.

The liquid matrix provides continuous surface renewal, so that intense primary beams can be used to produce intense and long-lasting spectra. Further, the ion source is at ambient temperature, preventing the thermal degradation of labile compounds. FAB and liquid SIMS are, however, limited to polar polymers that are miscible with the polar liquid matrices necessary

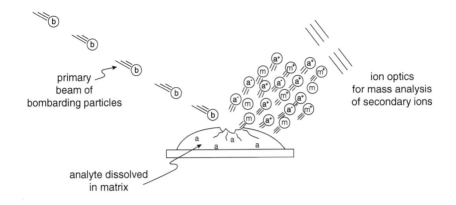

FIGURE 1.6
Bombardment of an analyte sample (a) dissolved in a liquid matrix (m) by a primary beam of atoms or ions (b) to produce sample ions that are characteristic of the analyte. (Reprinted from Ref. 20 with permission of John Wiley & Sons)

for these ionization methods (see Chapter 7). FAB and LSIMS have extensively been applied to low-molecular-weight polyglycols and related compounds (<5,000 Da).[21–23] Figure 1.7 shows the mass spectrum of a poly(ethylene glycol) with added NaBr; quasimolecular ions ($[M + Na]^+$) and fragments from H_2O loss can readily be identified. Many other fragments appear at low m/z where matrix ions and matrix cluster ions can also contribute; for this reason, fragmentation of FAB and LSIMS generated ions is often sought through MS/MS experiments.[22,23]

1.2.2.4   *Matrix-Assisted Laser Desorption Ionization (MALDI)*

MALDI[24] is the newest and most promising desorption method for synthetic macromolecules.[25] The polymer is dissolved in the appropriate solvent and mixed with a solution of the matrix to achieve a molar ratio of analyte to matrix of 1:100–1:50,000. A solution of an auxiliary ionization agent (e.g., a metal ion salt) may be added and a small droplet (≤1 μL) of the resulting mixture are loaded onto a target surface (Figure 1.8).[26] As the solvent evaporates, a solid solution of the sample (and the auxiliary agent) in the matrix is obtained, which is bombarded by laser light. The matrix must have a strong absorption at the wavelength emitted by the laser; normally pulsed UV (N_2, 337 nm) and IR (CO_2, 10.6 μm) lasers are employed. Upon irradiation of the crystalline sample mixture, intact protonated, deprotonated, or metal ion attached molecules are desorbed for m/z analysis.[24–27]

MALDI is extremely sensitive, with the total amount of sample deposited onto the target being in the pico- to femtomole range. Polymers up to about 10^6 Da can be ionized by this method (Figure 1.9).[28] Up to approximately 50,000 Da, singly charged ions are formed exclusively or predominantly, while at higher molecular weights multiply charged ions are usually coproduced in considerable abundance. The high dilution of the analyte in the

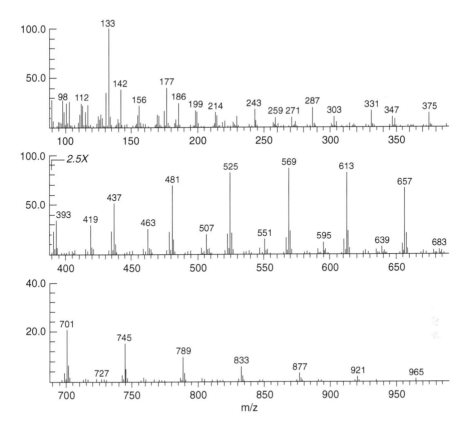

FIGURE 1.7
FAB mass spectrum of poly(ethylene glycol) 600. (Reprinted from Ref. 21 with permission of Elsevier Science)

matrix prohibits analyte-analyte interactions, which could lead to the formation of analyte clusters, thereby complicating molecular-weight assignments.[7]

The MALDI matrices are usually organic compounds. In UV-MALDI, which is most widely used for synthetic polymers,[25] the matrix is an aromatic organic compound carrying oxo, hydroxyl, and/or carboxyl groups; commonly selected matrices are 2,5-dihydroxybenzoic acid (DHB), 2-(4-hydroxyphenylazo)-benzoic acid (HABA), α-cyano-4-hydroxycinnamic acid (αCHCA), trans-3-indoleacrylic acid (IAA), dithranol, and all-trans retinoic acid (Figure 1.10). The macromolecules are not energized directly upon irradiation; the light is rather absorbed by the matrix which is ionized and dissociated. This process breaks down the crystalline structure of the matrix, changing it to a super-compressed gas, in which charge transfer reactions with the analyte molecules can take place (mainly H^+ or metal ion transfer).[27] As the gas expands, it transports entrapped analyte ions and molecules from the surface into the gas phase where, at the selvedge, further charge transfer reactions to neutral analyte molecules are possible. Collisions within the expanding gas

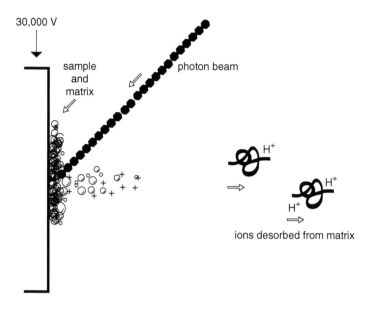

FIGURE 1.8
MALDI source. (Reprinted from Ref. 26 with permission of Academic Press)

("matrix plume") dissipate most of the internal energy of the analyte ions formed. The sequence of these desorption/ionization events is schematically summarized in Figure 1.11.[27]

MALDI is today the ionization method of choice for the analysis of the compositions, end groups, and molecular weight distributions of intact synthetic polymers. The promise and limitations (particularly in reproducing actual molecular weight distributions) of MALDI, which have been the subject of vigorous debate in the literature, are presented in more detail in Chapter 10. Here, MALDI's capabilities are exemplified by Figure 1.12, which shows the mass spectrum of a poly(ethylene glycol) that was derivatized with the drug acetaminophen.[29] The exact m/z values of the $[M + Na]^+$ ions observed confirm that the polyglycol carries the drug labels at both ends, as depicted below.

Further, only one distribution is observed, consistent with the absence of mono- or underivatized PEG.

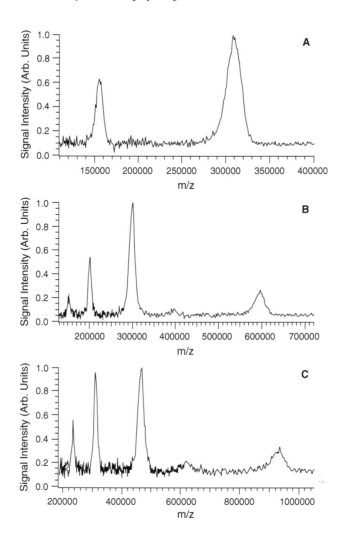

FIGURE 1.9
MALDI mass spectra of poly(styrene)s with nominal molecular weight of (a) 330,000, (b) 600,000, and (c) 900,000. (Reprinted from Ref. 28 with permission of the American Chemical Society)

1.2.3 Spray Ionization Methods

1.2.3.1 Thermospray (TSP)

In TSP,[30] a solution of the sample and an auxiliary electrolyte (usually ammonium acetate) passes through a heated capillary to enter the pumped ion chamber (1–10 Torr). A supersonic beam of charged droplets emerges at the entrance of the chamber, which is heated to aid the desolvation of the droplets. TSP produces essentially equal numbers of positively and negatively charged droplets; the droplet charge is determined by the electrolyte ion

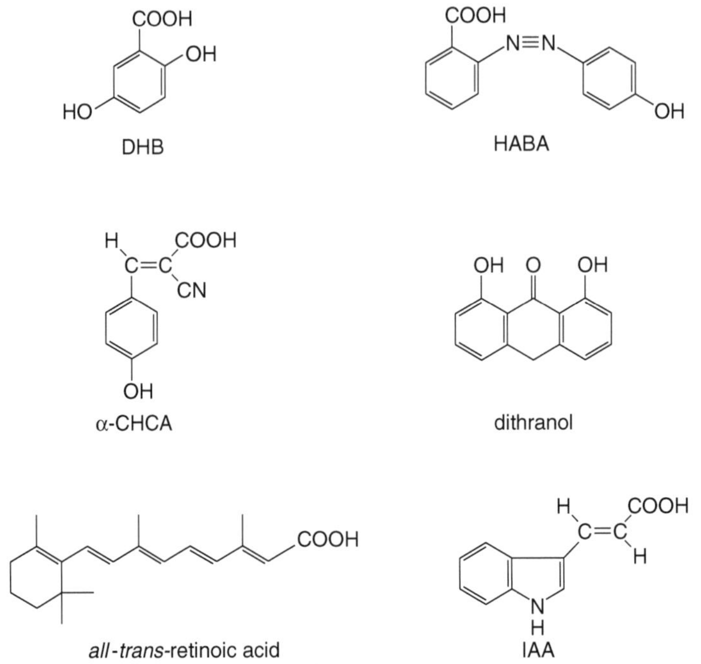

FIGURE 1.10
Common MALDI matrices used for synthetic polymers.

contained statistically in excess. Analyte molecules (M) and electrolyte ions (NH_4^+, CH_3COO^-) are evaporated from these droplets as their size decreases; subsequent ion-molecule reactions between M and the vaporized electrolyte ions yield analyte-indicative quasimolecular ions, such as $[M + H]^+$, $[M - H]^-$, or $[M + NH_4]^+$. Ionic analytes eliminate the need for added electrolyte. Figure 1.13 illustrates the thermospray spectrum of a poly(ethylene glycol) mixture;[31] the most abundant ions correspond to ammoniated and protonated molecules. Although no significant fragmentation is evident in Figure 1.13, TSP of biomolecules generally yields more fragments than other spray methods. TSP was primarily developed as an LC/MS interface and has mainly been used in the biological and pharmaceutical areas;[2,3,7,30,31] applications to synthetic polymers are scarce.

1.2.3.2 Electrospray Ionization (ESI)

ESI[32–36] is closely related to TSP. Now, a strong electric field is applied to the capillary carrying the analyte solution and the spray is produced at atmospheric pressure (cf. Figure 1.14).[3] Typically, the potential difference between the capillary and the 0.3–2 cm distant counter electrode is 3–6 kV. Spraying under these conditions produces highly charged droplets whose charge is determined by the polarity of the field applied to the capillary. Desolvation

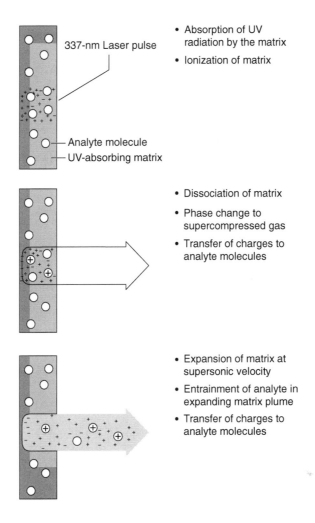

FIGURE 1.11
Schematic of the MALDI desorption process. (Reprinted from Ref. 27 with permission of the American Chemical Society)

of the droplets is aided by a counter-current flow of warm nitrogen gas (Figure 1.14). Other source designs replace or augment the effects of the counter-flow gas by making the droplets and solvated ions pass through a long heated metal capillary inserted between atmosphere and the first pumping stage of the mass spectrometer ("transfer capillary"). Residual solvent molecules are removed in the collisionally activated dissociation (CAD) region, located between the exit of the heated capillary and the entrance to the mass spectrometer.[2,3,7,32–36]

The flow rate through the electrically charged ESI capillary is kept in the range of 1–10 μL/min by a syringe pump or the LC system. The electrospray process (depicted in Figure 1.14) may be assisted by the coaxial flow of a

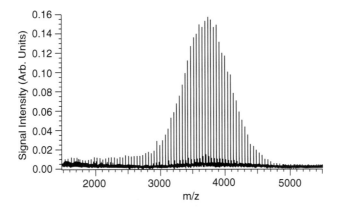

FIGURE 1.12
MALDI mass spectrum of poly(ethylene glycol) bis(acetaminophen) using HABA as matrix and Na^+ cationization; $M_n = 3{,}520$; $D = 1.014$. (Reprinted from Ref. 29 with permission of the American Chemical Society)

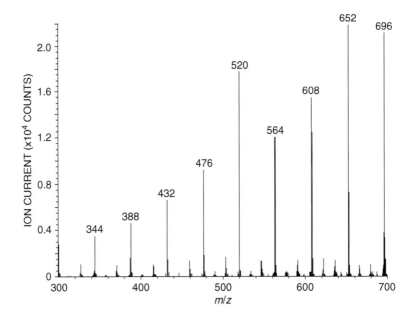

FIGURE 1.13
TSP mass spectrum of an equimolar mixture of four poly(ethylene glycols), viz. PEG 300, 600, 1000, and 1450; the m/z range 300–700 is displayed. The major ions correspond to $[M + NH_4]^+$ adducts. (Reprinted from Ref. 31 with permission of the American Chemical Society)

nebulizing gas. This variant, which is called ionspray, can sustain flow rates of ≥200 μL/min that are more compatible with LC/MS interfacing.[2,3,7]

ESI generally produces *multiply* charged quasimolecular ions except for analytes with molecular weights <1,000 Da. A continuous distribution of

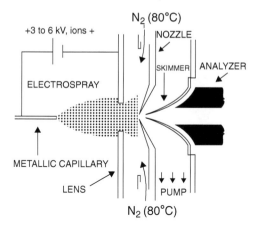

FIGURE 1.14
Schematic diagram of an ESI source. (Reprinted from Ref. 3 with permission of John Wiley & Sons)

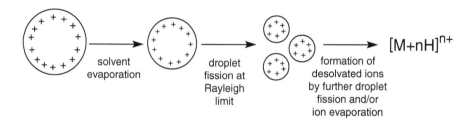

FIGURE 1.15
The formation of ions from droplets upon ESI. (Reprinted (modified) from Ref. 35 with permission of John Wiley & Sons)

charge states is usually observed, with the most abundant charge state being proportional to the molecular weight; for many compounds, the quasimolecular ions appear in the m/z 500–2,000 range. Basic and acidic samples readily ionize to give a series of $[M + nH]^{n+}$ cations and $[M - nH]^{n-}$ anions, respectively. For molecules without such ionizable sites, which is true for most synthetic polymers, salts are added to the solution being electrosprayed to obtain adducts with metal (e.g., Na^+, K^+), NH_4^+, or other ions.[36]

How the droplets produced upon ESI are transformed to the ions observed in the mass spectrum is not yet well understood.[2,3,7,32–36] As the solvent in the droplets evaporates, the droplets shrink and the ions contained in them accumulate at the surface to minimize coulombic repulsion between the charges. This process can continue until the Rayleigh instability limit is reached, at which the droplets disintegrate ("explode") into smaller droplets (Figure 1.15) that also shrink by solvent loss. Sequential subdivision through *coulombic explosion at the Rayleigh limit* may be repeated until the

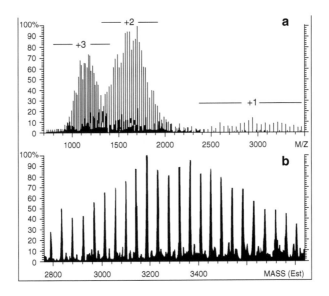

FIGURE 1.16
(a) ESI mass spectrum of poly(ethylene glycol) 3400. (b) Transformation of the m/z to a mass scale (see text). (Reprinted from Ref. 37 with permission of Elsevier Science)

droplet contains one analyte species. Another possibility is that the microdroplets arising from division at the Rayleigh limit may have so much charge accumulated on them that bare or solvated quasimolecular ions can be evaporated from their surface (Figure 1.15); this *ion evaporation* (or *ion desorption*) mechanism resembles the field desorption process discussed in Section 1.2.2.1.

ESI of synthetic polymers produces a series of variously charged ions per n-mer. Figure 1.16 shows the spectrum of PEG 3400 which, under the conditions used, gives rise to $[M + nNa]^{n+}$ ions in the +1, +2, and +3 charge states.[37] The mass spectrum can be deconvoluted to a distribution of oligomers (Figure 1.16b) by determining the total ion abundance due to each oligomer (obtained by summing the different charge state peaks of the same oligomer). This transformation is cumbersome (if possible at all) when the different charge states are unresolved, which is true for larger polymers. For this reason, ESI polymer applications have practically been limited to low-molecular-weight oligomers or dendrimers (see Chapter 4). The latter are essentially monodisperse, as is the vast majority of biomolecules, for which ESI is indispensable.[38] Interpretable ESI spectra from larger synthetic polymers can only be obtained at mass spectrometric resolutions that separate the isotopic clusters of the individual charge states. For a polymer of ~20,000 Da average molecular weight, a resolution of ≥50,000 would be needed and can be achieved with Fourier transform ion cyclotron resonance mass spectrometry (vide infra).[39]

1.3 Mass Analyzers

Mass analyzers disperse ions in space or time according to their mass-to-charge ratios (m/z). Certain analyzers separate the ions simultaneously, while others are scanned to transmit to the detector a narrow m/z range at a given time. Important features of a mass analyzer are its upper mass limit, transmission, resolving power, mass accuracy, dynamic range, and operating pressure. Table 1.1 gives the characteristics of the devices used most frequently.[40]

Depending on the resolution of the mass analyzer in the mass range of interest, a polymeric ion is either resolved into its individual isotopes or observed as an unresolved peak at the average m/z value of all isotopes. Monoisotopic m/z ratios are preferred because they provide a higher mass accuracy. Moreover, resolved isotopic patterns clearly reveal the presence (in the polymer) of elements with unique isotopic distributions (such as bromine, silver, or iron), thus supplying superior compositional insight.[1]

TABLE 1.1

Characteristics of Mass Analyzers[40]

Transmission (for ions separated in space) is defined as the ratio between the ion fluxes exiting and entering the mass analyzer. It decreases in the order: time-of-flight >> quadrupole > sectors

Upper m/z limit:
time-of-flight ($>10^6$) > quadrupole ion trap (10^5) > quadrupole (10^4) ≈ sectors ≈ ICR trap

Resolution is defined as the ratio between m/z value and the peak width in m/z units. It is mass-dependent. At m/z 1,000, it follows the order: ICR trap (10^6) > double sector (10^5)[a] > time-of-flight (10^3–10^4) ≈ quadrupole ion trap > quadrupole (10^3) ≈ magnetic sector > electric sector ($<10^2$)[b]

Mass accuracy (at m/z 1,000):
sectors (<5 ppm) > ICR trap (<10 ppm) > time-of-flight (0.01–0.1%) > quadrupole (0.1%) ≈ quadrupole ion trap

Dynamic range is the concentration range over which the measured signal abundance varies linearly with concentration. It is expressed as the ratio between the largest and lowest concentration and decreases in the order: sectors (10^7) > quadrupole (10^5) > quadrupole ion trap ≈ ICR trap ≈ time-of-flight

Operating pressure in Torr:
quadrupole ion trap (10^{-3}) > quadrupole (10^{-5}) > sectors (10^{-7}) ≈ time-of-flight > ICR trap (10^{-9})

---

[a] Either *EB* or *BE*.

[b] Can only separate ions that have different kinetic energies, as is true for the fragments formed from a fast-moving ion. The poor resolution is due to the release of internal energy into kinetic energy during fragmentation.

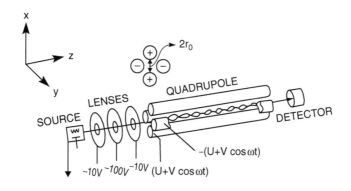

FIGURE 1.17
A simple quadrupole mass spectrometer consisting of an ion source, focusing lenses, a quadru-
pole mass filter, and an ion detector. (Reprinted (modified) from Ref. 3 with permission of John
Wiley & Sons)

1.3.1 Scanning Mass Analyzers

1.3.1.1   Quadrupole Mass Filter

Quadrupole mass analyzers consist of four parallel circular or hyperbolic
rods (Figure 1.17).[2,3,7,41] Each pair of opposite rods are electrically connected
and supplied voltages of the same magnitude but different polarity. The
voltage applied to each pair consists of a direct current (DC), U, and a radio-
frequency (rf) component, $V \cos \omega t$. Typical values are several hundred volts
for U, several thousand volts for V, and megahertz for ω. Since the total
potential of each rod is $+(U + V \cos \omega t)$ or $-(U + V \cos \omega t)$, the rf field period-
ically alternates the rods' polarity.[3,7]

Ions are accelerated along the z-axis (~5–20 eV) before entering the space
between the quadrupole rods where they experience the combined field
resulting from the rod potentials. A cation is drawn toward the negative pole
and vice versa; if the rod potential changes sign before the ion discharges,
the ion changes direction, thus oscillating through the rods. Whether an ion
succeeds passing through the rods or discharges on them is controlled by
the DC and rf voltages, as shown in Figure 1.18 for ions of three different
masses;[3,41] in this figure, the ion of mass m_2 has a stable trajectory (i.e., it can
be transmitted through the quadrupole) at the U and V values lying within
the dotted curve. Any other U and V values lead to unstable trajectories (i.e.,
to discharge at the rods).

The quadrupole is scanned with the U/V ratio kept constant, i.e., along the
scan line of Figure 1.18, which allows the successive separation of different
masses by sequentially bringing each of them into stable trajectories through
the rods. If U = 0 (no DC), all ions above a certain mass limit (determined by
V) have stable trajectories through the quadrupole; such an rf-only quadrupole
is therefore ideally suitable as a collision cell in a quadrupole-based tandem
mass spectrometer (vide infra).[41]

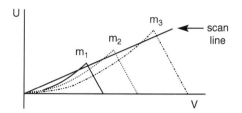

FIGURE 1.18

Stability conditions, expressed in terms of U vs. V plots, for three ions with masses $m_1 < m_2 < m_3$. The U and V values leading to stable trajectories through the quadrupole lie within the area defined by the V axis and the solid lines (m_1), dotted lines (m_2), or bullet lines (m_3). The quadrupole is scanned along the scan line, i.e., with the ratio U/V kept constant; the slope of this line determines the resolution. Gradually increasing U and V successively brings ions m_1, m_2, and m_3 through stable trajectories. If $U = 0$, the scan line overlaps with the V axis; the resolution is zero, and all ions pass the quadrupole. (Adapted from Ref. 3 with permission of John Wiley & Sons)

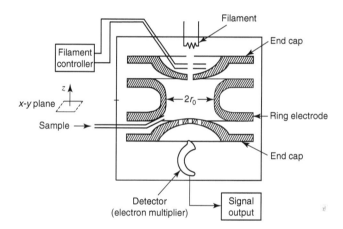

FIGURE 1.19

Schematic diagram of the quadrupole ion trap. (Reprinted from Ref. 2 with permission of Cambridge University Press)

1.3.1.2 Quadrupole Ion Trap

The ion trap can be viewed as a three-dimensional quadrupole and consists of a ring electrode and two end caps (Figure 1.19).[2,3,7,41] Ions are stored inside the trap by biasing the ring electrode with an rf voltage of low amplitude and grounding the end caps. Repulsive forces between the trapped ions increase the velocities and amplitudes of their motion, which could lead to their ejection from the trap. This is prevented by introducing a He bath gas in the trap (10^{-3} Torr), so that the ions are collisionally cooled and drift toward the trap center.

The ions may be created within the trap, for example by a short pulse of electrons, or injected into the trap from an external desorption or electrospray source.[42] A mass spectrum is subsequently obtained by scanning the rf potential, so that ions of increasing m/z successively develop unstable trajectories and escape the trap to strike an external ion detector (Figure 1.19).

By applying a DC voltage to the ring electrode and ramping the rf voltage, it is possible to isolate a given ion in the trap for the acquisition of its tandem mass spectrum. MS/MS is then accomplished by "tickling" the isolated ion with a supplementary alternating voltage of appropriate frequency, applied to the end caps. The newly formed fragment ions are finally detected by scanning the rf voltage, as explained above.[7,41] Note that the MS/MS steps are separated in time but not in space, in contrast to beam instruments (vide infra). By successively repeating the isolation/reaction/detection steps, sequential MS/MS experiments (i.e., MS[n]) can be performed on ions originating from a single ionization event.[43]

1.3.1.3   Magnetic and Electric Sectors

These analyzers function with fast ion beams (keV range). The kinetic energy (E_k) of an ion accelerated by a potential V (2–10 kV) upon exiting the ion source is given by Eq. 1.10, where e and v are the electron charge and the ion velocity, respectively. When the moving ion enters a magnetic sector, whose field (B) is perpendicular to the ion's velocity (Figure 1.20),[7] it is forced into a circular trajectory of radius r, as given by Eq. 1.11; based on this equation, a magnetic sector deflects according to momentum. Combining Eqs. 1.10 and 1.11 gives the mass analysis equation of the magnetic sector (Eq. 1.12). Usually, r is set by the flight tube and slits, and B is scanned to sequentially transmit ions of different m/z values to the collector (Figure 1.20).

$$E_k = zeV = mv^2/2 \tag{1.10}$$

$$mv^2/r = Bzev \Rightarrow mv = Bzer \tag{1.11}$$

$$m/z = B^2r^2e/2V \tag{1.12}$$

$$r = (2mE_k)^{0.5}/zeB \tag{1.13}$$

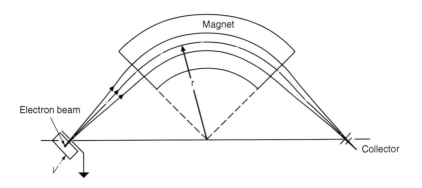

FIGURE 1.20
Schematic diagram of a mass spectrometer using a magnetic sector for m/z analysis. (Reprinted from Ref. 7 with permission of John Wiley & Sons)

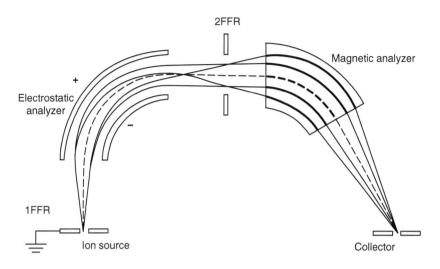

2FFR

Magnetic analyzer

Electrostatic
analyzer

1FFR

Ion source

Collector

FIGURE 1.21
Schematic diagram of an electric and a magnetic sector as they are arranged in a forward geometry double-focusing mass spectrometer (Nier-Johnson geometry). (Reprinted from Ref. 7 with permission of John Wiley & Sons)

Rearrangement of Eq. 1.12 to Eq. 1.13 shows that ions with a given mass and charge will pass the magnet at the same field B, only if their kinetic energies, E_k, are the same. Unfortunately, most ionization methods produce ions with a kinetic energy distribution, which degrades resolution by spreading over a wider range the value of B necessary for the separation of ions with a common m/z ratio. This problem can be compensated for by an electrostatic analyzer (also called electric sector); such a sector consists of two parallel cylindrical electrodes of opposite polarity with a homogeneous field E between them (Figure 1.21).[7] This field causes an ion entering it at right angles to follow a circular path of radius R, so that the electrostatic (zeE) and centrifugal (mv^2/R) forces on the ion are balanced (Eq. 1.14). According to Eqs. 1.14 and 1.10, ions accelerated to the same kinetic energy (E_k) and traversing the same electric field (E) have the same trajectory (R) through the electric sector independent of their m/z value. Thus, the electric sector is a kinetic energy analyzer and can be used to energy-focus the ion beams sent to (or coming from) a magnetic sector. The result is a double-focusing mass spectrometer of superior resolution (10^5 possible), as compared to an instrument comprised of a single magnet. The magnet (B) follows or precedes the electrostatic analyzer (E) in forward (EB, cf. Figure 1.21) or reverse (BE) geometry sector instruments.

$$zeE = mv^2/R = 2E_k/R \Rightarrow R = 2E_k/zeE \qquad (1.14)$$

If the ion kinetic energy changes due to dissociation after acceleration, the electric sector can also be used as a mass analyzer. A scan of the sector field, E, separates the fragment ions based on their kinetic energies (Eq. 1.14), which are proportional to (Eq. 1.10) and, thus, identify their m/z values. This capability allows one to perform tandem mass spectrometry experiments with a double-focusing *EB* or *BE* instrument (vide infra).

1.3.2   Nonscanning Mass Analyzers

1.3.2.1   *Time-of-Flight (TOF) Analyzers*

For m/z separation with a TOF analyzer,[27] the ions produced in the ion source are first accelerated through a potential V to acquire kinetic energies in the keV range (cf. Eq. 1.10). Then, the ions traverse ("fly") a distance d to reach the detector after a time t which is measured.[2,3,7,27] The flight time depends on the ion velocity v (Eq. 1.15) which in turn is dependent on m/z (Eqs. 1.10 and 1.16). Overall, the time taken by an ion to reach the detector is proportional to the square root of the ion's m/z value. Based on this relationship, the larger ions (higher masses) take longer times to arrive at the detector. To avoid overlaps with simultaneously arriving smaller ions produced later, the ion source must be pulsed.

$$t = d/v \tag{1.15}$$

$$v = (2zeV/m)^{0.5} \Rightarrow t = (m/z)^{0.5}d/(2eV)^{0.5} \tag{1.16}$$

A simple, linear TOF analyzer (Figure 1.22a) suffers from poor resolution due to the spread in velocity of ions of the same mass. This spread results partly from the fact that the ions are formed with some initial (varying) kinetic energy and in different regions of the source; as a result, they are accelerated to different final kinetic energies. Additionally, ions formed at different times and different locations would travel slightly different paths. The initial kinetic energy, spatial, and temporal distributions of the ions,[27] which degrade the resolution, can be minimized substantially by *time-lag focusing* ("*delayed extraction*")[44] and the use of a *reflectron* (Figure 1.22b).[27]

Time-lag focusing introduces a delay time between ion formation and acceleration. The time lag causes ions drifting toward (or away from) the detector to acquire a lower (higher) kinetic energy upon acceleration ("extraction") than other ions of the same mass, so that they all reach the detector at the same time. The reflectron is an ion mirror with an electric field which retards and reflects the entering ions (Figure 1.22b). Faster ions (i.e., those with higher kinetic energies) penetrate deeper into the reflectron and take more time to turn around; as a result, they reach the reflectron detector at the same time as the slower ions of the same mass, which spend less time inside the reflectron.

A reflectron TOF analyzer also permits tandem mass spectrometry studies. For this, an ion gate is needed to select the desired precursor ion among those produced in the source. A deflector electrode can serve as the gate, which is pulsed off at a selected time to allow the passage of a specific precursor ion (Figure 1.23). Post-source decay (PSD)[45] of this ion between the gate and the

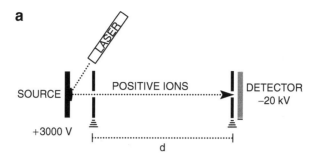

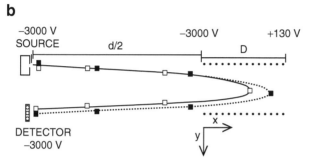

FIGURE 1.22
Schematic diagram of a (a) linear and (b) reflectron time-of-flight mass spectrometer. The signs
■ and □ represent ions of the same m/z ratio but with different kinetic energies. The ions enter
and exit the reflectron with the same kinetic energies. The ions with lower kinetic energy (□)
spent less time inside the reflectron and are reflected earlier. With properly adjusted parameters,
slower and faster ions reach the detector simultaneously. (Reprinted from Ref. 3 with permission
of John Wiley & Sons)

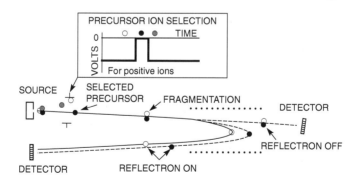

FIGURE 1.23
TOF mass spectrometer equipped with an ion gate (deflection electrode) and a reflectron. Pulsing
the ion gate at the correct time allows one to select the desired precursor ion (●). Fragments
formed before reaching the reflectron are dispersed inside the reflectron based on their kinetic
energies (which are proportional to their m/z ratios). (Reprinted from Ref. 3 with permission of
John Wiley & Sons)

reflectron lenses generates fragment ions with lower kinetic energies, which penetrate less into the reflectron and, hence, arrive earlier at the detector (Figure 1.23). To adequately resolve the fragment ions, PSD spectra are obtained in segments by stepping the reflectron voltages; for an ion at m/z 1,000–2,000, 6–8 segments are combined to assess the fragments formed over the entire m/z range.

Due to its high transmission and fast response, the TOF analyzer can easily be interfaced with pulsed ionization methods, such as MALDI, where limited ion currents are produced. With continuous, intense ion beams (as in ESI), the highest sensitivity and resolution are obtained using orthogonal extraction of the ions into the TOF tube; for this, the ions are first carried into a trapping device from which they are later extracted (accelerated) into a TOF drift tube in the orthogonal direction.[27,46] The essentially unlimited mass range of the TOF mass analyzer, combined with the improved resolution achievable with a reflectron and time-lag focusing ($\sim10^3$–10^4),[27] make TOF mass spectrometers ideal for the compositional analysis of synthetic and biological macromolecules with MALDI.

1.3.2.2   Fourier-Transform Ion Cyclotron Resonance (FTICR)

In FTICR (or FTMS) analyzers,[47,48] the ions are trapped inside a cell by crossed magnetic and electric fields (Figure 1.24); in this respect, such an analyzer is related with the quadrupole ion trap (vide supra). The FTICR cell is located inside a superconducting magnet with field B (usually 3–10 tesla). The magnetic field causes ions formed in (or injected into) the cell to move in circular trajectories perpendicular to the B axis with an angular cyclotron frequency ω_c (Eq. 1.17).[3,7] The ions also move along the B axis; to prevent them from escaping axially from the cell, a small DC field is applied

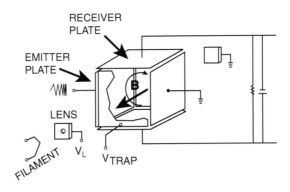

FIGURE 1.24
Schematic diagram of the ion cyclotron resonance mass spectrometer. (Reprinted from Ref. 3 with permission of John Wiley & Sons)

on the trapping plates (Figure 1.24).

$$\omega_c = zeB/m \Rightarrow m/z = eB/\omega_c \qquad (1.17)$$

Ions with the cyclotron frequency ω_c trapped in the cell can be excited by applying an oscillating electric field of the same frequency on the transmitter plates (normal to the direction of B, cf. Figure 1.24).[3] The energy transferred increases the radius of the ions' orbital motion without affecting ω_c; the excited ions move coherently, i.e., as a packet, closer to the receiver plates (cf. Figure 1.24), inducing image currents onto them which are amplified and measured.[3,7]

To obtain a mass spectrum over the desired m/z interval, *all* ions within this interval are excited simultaneously by a rapid frequency sweep of the voltage on the transmitter plates. The excitation pulse increases the orbital radii of all ions *and* puts ions of the same m/z ratio in phase. The orbiting ions create a complex wave signal in the circuit connecting the receiver plates, which is monitored over time as the coherent motion of the ions is destroyed by collisions (Figure 1.25).[3] Fourier-transformation of this time-domain signal furnishes the individual cyclotron frequencies and, hence, the m/z values (Eq. 1.17) of the ions (Figure 1.25).

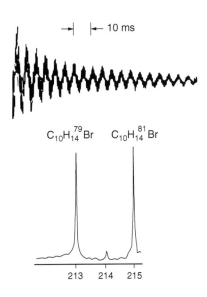

FIGURE 1.25
Free induction decay of the image current generated on the receiver plates by $C_{10}H_{14}Br^+$ ions that were excited at their resonance frequencies (top). A Fourier transform converts this time-domain signal into the frequency-domain, i.e., into ion intensity vs. frequency, from which the mass spectrum (i.e., ion intensity vs. m/z) is obtained via Eq. 1.17. (Reprinted from Ref. 3 with permission of John Wiley & Sons)

The mass resolution of FTICR analyzers is large, because frequencies can be selected and measured very accurately. The resolution depends on the observation time of the decaying time-domain signal, which in turn is linked with the frequency of ion-molecule collisions in the cell. For high resolution, a very low cell pressure ($\leq 10^{-9}$ Torr) is necessary. The resolution decreases with mass due to the inverse relationship between ω_c and m/z (Eq. 1.17). Practically, FTICR is used in the $m/z \leq 3,000$ range, although ions up to m/z 20,000 have been observed.[49] FTICR is readily interfaced with MALDI and ESI and has, therefore, been increasingly useful for the analysis of high-mass polymers (see Chapter 9 by Wilkins) and biomolecules.[49]

By careful selection of the resonance frequencies, a specific ion can be isolated in the cell with high resolution for the study of its dissociations or ion-molecule reactions. The products of such MS/MS experiments can be analyzed with high resolution by the same procedure used to measure normal MS spectra (see above). Moreover, since the various MS/MS steps are separated in time, they can be repeated sequentially several times (MS^n), as was the case with the ion-trap mass spectrometer (vide supra).

1.4 Detectors

The detector converts ions of a given m/z value into a measurable electrical signal whose intensity is proportional to the corresponding ion current. With beam instruments (sectors, quadrupoles, or TOF analyzers) and the quadrupole ion trap, the ions are first separated according to their m/z value before detection, usually by an electron multiplier or a photon multiplier. The operation of these most common detectors is briefly outlined below.

With FTICR, all ions produced (or isolated) in the ICR cell are detected simultaneously, as has been explained in Section 1.3.2.2. The time domain signal, i.e., the image current produced at the receiver plates by the orbiting ions (cf. Figures 1.24 and 1.25), can readily be amplified to yield a measurable signal; moreover, the ions are not destroyed during the detection process and, thus, can be re-excited and remeasured if an improvement of the signal to noise ratio is needed.[47,48]

1.4.1 Electron Multipliers and Related Devices

The electron multiplier consists either of a series of discrete dynodes or of a continuous channel of dynodes.[3,7,40,50] Figure 1.26 illustrates the continuous-dynode type, also called "channeltron," which is the most widely used electron multiplier in modern instruments. A high negative potential is applied to the channeltron entrance, while the opposite end (anode) is usually grounded. Secondary electrons produced at the entrance by impinging ions or electrons

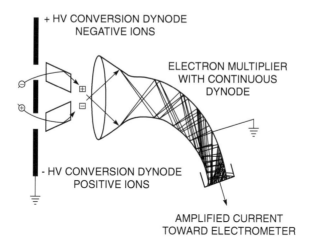

+ HV CONVERSION DYNODE
NEGATIVE IONS

ELECTRON MULTIPLIER
WITH CONTINUOUS
DYNODE

- HV CONVERSION DYNODE
POSITIVE IONS

AMPLIFIED CURRENT
TOWARD ELECTROMETER

FIGURE 1.26

Channeltron electron multiplier with conversion dynodes for positive and negative ions; $\oplus$ and $\ominus$ represent the positive or negative ions, respectively, that strike the conversion dynodes. (Reprinted from Ref. 3 with permission of John Wiley & Sons)

experience a cascade of collisions with the walls, ejecting more and more electrons as they are accelerated down the channel (Figure 1.26). Typically 10–20 stages of amplification take place until the anode is reached, where conventional amplifiers are connected in series before the signal readout.

Mass-analyzed ions may strike the electron multiplier directly, ejecting secondary electrons that initiate the cascading emission of additional electrons down the channeltron. Alternatively, the ions may first collide with conversion dynodes, situated in front of the channeltron, as shown in Figure 1.26. Positive ions hit the negative dynode, generating small negative ions and electrons; negative ions strike the positive dynode, generating small positive ions. These particles are then accelerated into the channeltron to start the amplification process described above. The gain, i.e., the number of secondary particles emitted per incoming ion, lies in the range of 10^6–10^7.

Miniature cylindrical or curved channeltrons with diameters and lengths of a few μm can be easily manufactured. In detectors known as *microchannel plates* or *array detectors*, a large number of these miniature electron multipliers is assembled on a flat plate (Figure 1.27). The amplification factor (gain) in each channel is ca. 10^5 and can be increased to 10^8 by using two or more interconnected parallel plates. Array detectors can measure both the position and the intensity of an ion beam, which makes them particularly useful for sector instruments that disperse *and* transmit simultaneously a range of m/z values. Microchannel arrays are also used in TOF mass spectrometers, where the ions can have considerable spatial distributions in the area perpendicular to the ion beam. Often, a conversion dynode is placed in front of the microchannel plates to augment the detection sensitivity of macromolecular ions ($m/z > 30,000$); this combination has been termed "high-mass" detector.

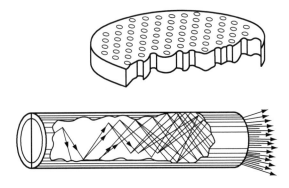

FIGURE 1.27
Cross section of a microchannel plate (top) and electron multiplication within a microchannel (bottom). (Reproduced from Ref. 3 with permission of John Wiley & Sons)

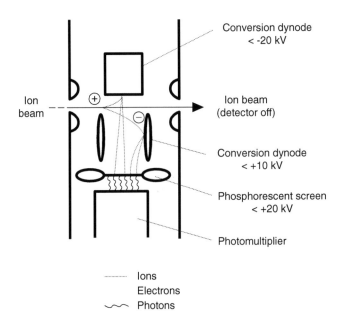

FIGURE 1.28
Photon multiplier detector. (Reproduced from Ref. 3 with permission of John Wiley & Sons)

1.4.2 Photon Multipliers

These detectors consist of conversion dynodes, a scintillator (normally a phosphorescent screen), and a photomultiplier tube (Figure 1.28).[3] Cations and anions are accelerated to the negative and positive conversion dynodes, respectively. The ensuing ion-surface collisions lead to the generation of secondary electrons which strike the phosphorescent screen where they are converted to photons. The latter are detected by the photomultiplier. Overall, the ion beam reaching the detector is amplified ca. 10^4–10^5 times. In contrast to electrons, photons can readily exit the vacuum system of the mass spectrometer through a glass window. For this reason, the photomultiplier is generally installed *outside* the vacuum system, which substantially increases its lifetime vis-à-vis that of the electron multiplier.

A microchannel plate can also be interfaced with a scintillator and photomultiplier; here, the secondary electrons exiting the plate channels first strike the scintillator for electron-to-photon conversion; the photons generated in this process are subsequently converted to a measurable signal by the photomultiplier.

1.5   Tandem Mass Spectrometry

Several desorption and spray ionization methods can be used to convert synthetic polymers into intact molecular or quasimolecular ions (vide supra), whose exact m/z ratio identifies the composition of the polymer. For structural information about the polymer, the dissociation behavior or ion-molecule reactions of the polymer ions must be studied. Such reactions, which rarely take place during the *soft* ionization processes necessary to generate intact gas phase ions from synthetic macromolecules, are most conveniently assessed by tandem mass spectrometry (MS/MS).[3–5,7] With MS/MS, a specific precursor ion is mass-selected, so that its reactivity can be investigated without perturbation from the other ions formed upon ionization. The reaction products of this ion are then mass-analyzed and collected in the MS/MS spectrum. MS/MS studies on polymer ions have so far focused on their spontaneous ("metastable") or collision-induced fragmentation. The fragments arising in these reactions are displayed in metastable ion (MI) or collisionally activated dissociation (CAD) spectra, respectively.[4,5] Customarily, MI spectra acquired with a TOF mass analyzer have been named "post-source decay (PSD)" spectra;[27,45] similarly, CAD is often referred to as CID (collision-induced dissociation).[5]

The precursor ion selection, fragmentation, and product ion analysis can be separated in space or in time, as shown in Figure 1.29.[51] Separation in time requires trapped ions, as available in the quadrupole ion trap or the ion cyclotron resonance trap. Separation in space necessitates at least two physically distinct mass analyzing devices, one for precursor ion selection (MS-1) and one for product ion analysis (MS-2). The simplest in-space tandem instruments are the triple quadrupole mass spectrometer (*QqQ*), the double-focusing sector tandem mass spectrometer (*EB* or *BE*), and the reflectron time-of-flight mass spectrometer. In a triple quadrupole, the first and third quadrupoles (*Q*) are mass analyzers, while the center quadrupole (*q*) serves as the collision cell. In sector instruments, a collision cell is situated

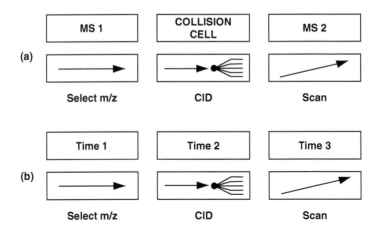

FIGURE 1.29
Tandem mass spectrometry with events separated (a) in space and (b) in time. (Reprinted (modified) from Ref. 51 with permission of John Wiley & Sons)

in the field-free regions following the ion source and/or in the interface region between the sectors which are the two mass analyzing devices (cf. Figure 1.21). A reflectron TOF mass spectrometer may contain a collision cell somewhere in the field-free region between the ion source and the reflecting lenses (cf. Figure 1.23); the ion gate (vide supra) and the reflectron serve as MS-1 and MS-2, respectively.

Any other in-space combination of the mass analyzers discussed in the previous section is possible. Common types are tri- and four-sector instruments (e.g., EBE, EBEB, etc.) and hybrid instruments (e.g., BEqQ, EBEqQ, Q-TOF, EBE-TOF, etc.). Such instruments can offer better precursor and product ion resolution as compared to double sector, triple quadrupole, or TOF mass spectrometers, and can accommodate multistage MS/MS experiments, such as MS^3 or MS^4. For the latter experiments, tandem-in-time is, however, more suitable and sensitive. Traps and quadrupole-based instruments involve ions with low kinetic energies (typically <<100 eV); in contrast, sector and TOF-based instruments employ ion beams with high kinetic energies (typically 3–25 keV). The former allow for low-energy dissociative (CAD) or reactive collisions, while the latter generally allow for high-energy CAD only; either type of mass spectrometer can be used for the study of the spontaneous decompositions of metastable ions.

Tandem mass spectrometry applications to FD-, FAB-, and MALDI-generated polymer ions from linear polyglycols, polyesters, and polystyrenes have demonstrated that valuable structural information can be gained from the fragmentation patterns observed.[22,23,52–61] Complementary fragment ion series are often produced, each containing one of the end groups; consequently, the *individual* end groups can be inferred from the m/z values of the fragments. In contrast, the m/z ratios of the molecular or quasimolecular ions reveal the composition of *both* end groups (which may not be unequivocally

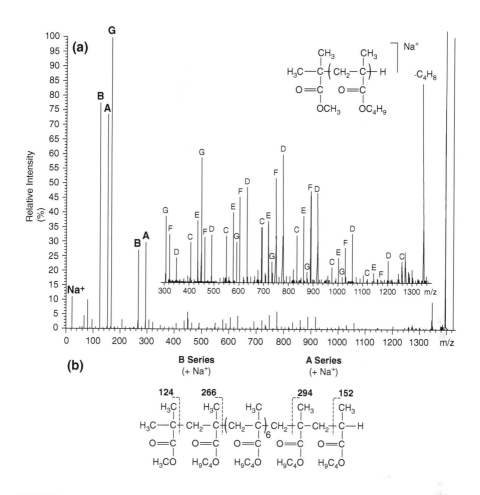

FIGURE 1.30
(a) MALDI-CAD spectrum of [M + Na]⁺, *m/z* 1403, of the 9-mer of PBMA A. Seven homologous series of fragment ions (**A, B, C, D, E, F,** and **G**) are observed, as discussed in the text. All fragment ions contain Na⁺. (b) Proposed fragmentation mechanism for the formation of series **A** and **B**. (Reprinted from Ref. 59 with permission of Elsevier Science)

divisible between the two ends). The potential of MS/MS studies on polymers is illustrated here with a few recent examples with precursor ions formed by MALDI.

Figure 1.30a depicts the MALDI-CAD spectrum of [M + Na]⁺ of the 9-mer of poly(butyl methacrylate), PBMA A, measured on an EBE-TOF instrument.[59] In this sector-orthogonal acceleration-TOF hybrid, the EBE and TOF sections are used for precursor ion selection (MS-1) and fragment ion analysis (MS-2), respectively, and CAD takes place in an intermediate collision cell at 800 eV collision energy. The CAD spectrum contains several series of fragments separated by 142 Da (repeat unit). Series **A** and **B** arise by charge-remote direct cleavages at either end, as shown in Figure 1.30b. Series **C/D** and **E/F**

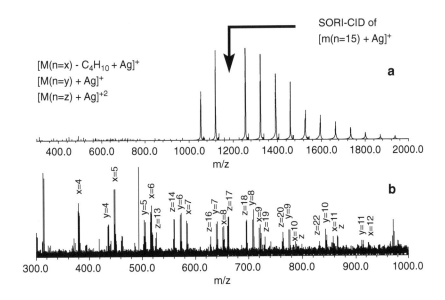

FIGURE 1.31
(a) MALDI-FTICR mass spectrum of poly(isoprene) 1000 after ejection of the ions below m/z 1,000 and SORI excitation of $[M + Ag]^+$ of the 15-mer. (b) SORI-CAD spectrum of $[M + Ag]^+$ of the 15-mer. (Reprinted from Ref. 61 with permission of Elsevier Science)

correspond to charge-remote H-rearrangements via six-membered rings with concomitant $HCO_2C_4H_9$ loss and charge (i.e., Na^+) retention at either side of the polymer chain. Finally, series **G** represents internal fragments containing neither end group; it presumably results from consecutive rearrangements of the **C** and **E** series. The masses of the end groups, viz. $(CH_3)_2C(CO_2CH_3)$– (101 Da) and $-CH(CH_3)(CO_2C_4H_9)$ (129 Da), are readily calculated from any member of the complementary series **A** and **B**. For example, m/z 266 (**B₂**, cf. Figure 1.27b) – 23 (Na^+) – 142 × 1 (one repeat unit) = 101 Da; similarly, m/z 294 (**A₂**) – 23 – 142 × 1 = 129 Da.

Figure 1.31 shows the CAD spectrum of $[M + Ag]^+$ from the 15-mer of poly(isoprene) 1000, acquired with an FTICR instrument.[61] First, all oligomers below m/z 1,000 are ejected from the cell. Then, Ar is introduced in the cell and the cationized 15-mer is selectively excited to higher kinetic energies using sustained off-resonance irradiation (SORI),[61] cf. Figure 1.31a. The nonejected oligomers (m/z > 1,000) are used for calibration of the mass scale. The fragment ions generated upon CAD are displayed in Figure 1.31b and consist of lower-mass oligomers or lower-mass oligomer fragments (as rationalized in the inset of Figure 1.31).

Figure 1.32 exemplifies an MS/MS experiment with a new generation TOF reflectron mass spectrometer.[62] Using the ion gate (cf. Figure 1.23), the $[M + Ag]^+$ ion of the 32-mer of poly(isoprene) 2500 (m/z 2345) is readily separated from all other ions produced upon MALDI (Figures 1.32a,b). The PSD spectrum of the selected ion contains two major series, **A** and **B**, arising by allylic cleavages

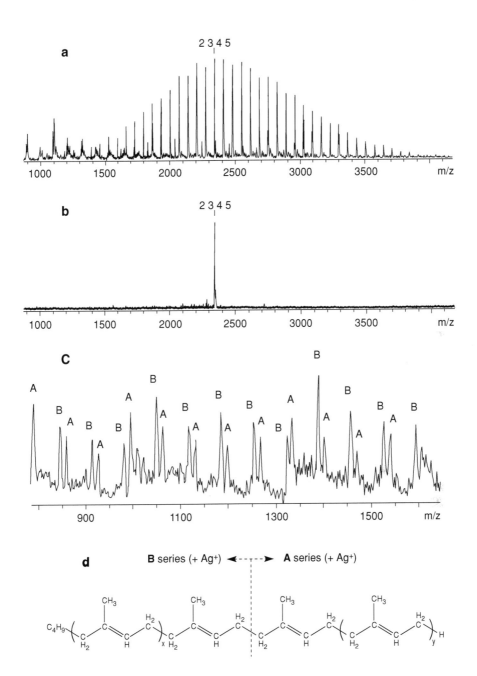

FIGURE 1.32
(a) MALDI-TOF mass spectrum of poly(isoprene) 2500. (b) Selection of [M + Ag]$^+$ of the 32-mer through the ion gate (cf. Figure 1.23) and (c) PSD spectrum of this ion. All fragment ions contain Ag$^+$. (d) Proposed fragmentation mechanism for the formation of series **A** and **B**.

along the polymer chain (Figures 1.32c,d). Again, each series contains just *one* end group, in contrast to the oligomers in the MS spectrum (Figure 1.32a) which contain *both* end groups.

The rather few polymer MS/MS studies reported so far have emphasized the information provided on the *individual end groups* of linear polymers. The fragmentation patterns and repeat units of the various fragment series may also enable the characterization of *block vs. random* copolymers, of *linear vs. branched* structures, and of *unsaturated vs. cyclic* oligomers.

Acknowledgments

We thank Paul Kowalski and Letitia Cornett (Bruker Daltonics) for helpful discussions and experimental assistance and the National Science Foundation (DMR-9703946) and the Ohio Board of Regents for generous financial support.

References

1. McLafferty, F. W. and Tureček, F., *Interpretation of Mass Spectra*, 4th ed., University Science Books, Mill Valley, CA, 1993.
2. Johnstone, R. A. W. and Malcolm, E. R., *Mass Spectrometry for Chemists and Biochemists*, 2nd ed., Cambridge University Press, Cambridge, 1996.
3. Hoffmann, E. de, Charette, J., and Stroobant, V., *Mass Spectrometry: Principles and Applications*, Wiley, Chichester, 1996.
4. McLafferty, F. W., Ed., *Tandem Mass Spectrometry*, Wiley, New York, 1983.
5. Busch, K. L., Glish, G. L., and McLuckey, S. A., *Mass Spectrometry/Mass Spectrometry*, VCH Publishers, New York, 1988.
6. (a) Schulten, H.-R. and Lattimer, R. P., Applications of mass spectrometry to polymers, *Mass Spectrom. Rev.*, 3, 231, 1984. (b) Montaudo, G., Mass spectrometry of synthetic polymers, *Trends Polym. Sci.*, 4, 81, 1996.
7. Chapman, J. R., *Practical Organic Mass Spectrometry*, 2nd ed., John Wiley & Sons, New York, 1993.
8. Kistemaker, P. G., Boerboom, J. H., and Meuzelaar, H. L. C., Laser pyrolysis mass spectrometry: some aspects and applications to technical polymers, *Dynamic Mass Spectrom.*, 4, 139, 1976.
9. Harrison, A. G., *Chemical Ionization Mass Spectrometry*, 2nd ed., CRC Press, Boca Raton, FL, 1992.
10. Daves, G. D., Jr., Mass spectrometry of involatile and thermally unstable molecules, *Acc. Chem. Res.*, 12, 359, 1979.
11. Udseth, H. R. and Friedman, L., The analysis of styrene polymers by mass spectrometry with filament-heated evaporation, *Anal. Chem.*, 53, 29, 1981.
12. Matsuo, T., Matsuda, H., and Katakuse, I., Use of field desorption mass spectra of polystyrene and polypropylene glycol as mass references up to mass 10 000, *Anal. Chem.*, 51, 1329, 1979.

13. Lattimer, R. P., Harmon, D. J., and Hansen, G. E., Determination of molecular weight distributions of polystyrene oligomers by field desorption mass spectrometry, *Anal. Chem.*, 52, 1808, 1980.
14. Rollins, K., Scrivens, J. H., Taylor, M. J., and Major, H., The characterization of polystyrene oligomers by field-desorption mass spectrometry, *Rapid Commun. Mass Spectrom.*, 4, 355, 1990.
15. Day, R. J., Unger, S. E., and Cooks, R. G., Molecular secondary ion mass spectrometry, *Anal. Chem.*, 52, 557A, 1980.
16. Benninghoven, A., Ed., *Ion Formation From Organic Solids*, Springer-Verlag, New York, 1983.
17. Hittle, L. R., Altland, D. E., Proctor, A., and Hercules, D. M. Investigation of molecular weight and terminal group effects on the time-of-flight secondary ion mass spectra of polyglycols, *Anal. Chem.*, 66, 2302, 1994.
18. Barber, M., Bordoli, R. S., Elliott, G. J., Sedgewick, R. D., and Tyler, A. N., Fast atom bombardment mass spectrometry, *Anal. Chem.*, 54, 645A, 1982.
19. Aberth, W., Straub, K. M., and Burlingame, A. L., Secondary ion mass spectrometry with cesium primary beam and liquid target matrix for analysis of bio-organic compounds, *Anal. Chem.*, 54, 2029, 1982.
20. Watson, J. T., Fast atom bombardment, in *Biological Mass Spectrometry—Present and Future*, Matsuo, T., Caprioli, R. M., Gross, M. L., and Seyama, Y., Eds., John Wiley & Sons, Chichester, 1994, p. 23.
21. Lattimer, R. P., Fast atom bombardment mass spectrometry of polyglycols, *Int. J. Mass Spectrom. Ion Processes*, 55, 221, 1983/1984.
22. Lattimer, R. P., Tandem mass spectrometry of poly(ethylene glycol) lithium-attachment ions, *J. Am. Soc. Mass Spectrom.*, 5, 1072, 1994.
23. Selby, T. L., Wesdemiotis, C., and Lattimer, R. P., Dissociation characteristics of $[M + X]^+$ ions (X = H, Li, Na, K) from linear and cyclic polyglycols, *J. Am. Soc. Mass Spectrom.*, 5, 1081, 1994.
24. Hillenkamp, F., Karas, M., Beavis, R. C., and Chait, B. T., Matrix-assisted laser desorption/ionization mass spectrometry of biopolymers, *Anal. Chem.*, 63, 1193A, 1991.
25. Räder, H. J. and Schrepp, W., MALDI-TOF mass spectrometry in the analysis of synthetic polymers, *Acta Polymer.*, 49, 272, 1998.
26. Siuzdak, G., *Mass Spectrometry for Biotechnology*, Academic Press, San Diego, CA, 1996.
27. Cotter, R. J., Time-of-flight mass spectrometry, *ACS Professional Reference Books*, Washington, DC, 1997.
28. Schriemer, D. C. and Li, L., Detection of high molecular weight narrow polydisperse polymers up to 1.5 million daltons by MALDI mass spectrometry, *Anal. Chem.*, 68, 2721, 1996.
29. Whittal, R. M., Schriemer, D. C., and Li, L., Time-lag focusing MALDI time-of-flight mass spectrometry for polymer characterization: oligomer resolution, mass accuracy, and average weight information, *Anal. Chem.*, 69, 2734, 1997.
30. Blakley, C. R. and Vestal, M. L., Thermospray interface for liquid chromatography/mass spectrometry, *Anal. Chem.*, 55, 750, 1983.
31. Fink, S. W. and Freas R. B., Enhanced analysis of poly(ethylene glycols) and peptides using thermospray mass spectrometry, *Anal. Chem.*, 61, 2050, 1989.
32. Whitehouse, C. M., Dreyer, R. N., Yamashita, M., and Fenn, J. B., Electrospray interface for liquid chromatographs and mass spectrometers, *Anal. Chem.*, 57, 675, 1985.

33. Smith, R. D., Loo, J. A., Edmonds, C. G., Barinaga, C. J., and Udseth, H. R., New developments in biochemical mass spectrometry: electrospray ionization, *Anal. Chem.*, 62, 882, 1990.
34. Kebarle, P. and Tang, L., From ions in solution to ions in the gas phase, *Anal. Chem.*, 65, 972A, 1993.
35. Gaskell, S. J., Electrospray: principles and practice, *J. Mass Spectrom.*, 32, 677, 1997.
36. Saf, R., Mirtl, C., and Hummel, K., Electrospray ionization mass spectrometry as an analytical tool for non-biological monomers, oligomers and polymers, *Acta Polymer.*, 48, 513, 1997.
37. McEwen, C. N., Simonsick, W. J., Larsen, B. S., Ute, K., and Hatada, K., The fundamentals of applying electrospray ionization mass spectrometry to low mass poly(methyl methacrylate) polymers, *J. Am. Soc. Mass Spectrom.*, 6, 906, 1995.
38. Burlingame, A. L., Boyd, R. K., and Gaskell, S. J., Mass spectrometry, *Anal. Chem.*, 70, 614, 1998.
39. O'Connor, P. B. and McLafferty, F. W., Oligomer characterization of 4-23 kDa polymers by electrospray Fourier transform mass spectrometry, *J. Am. Chem. Soc.*, 117, 12826, 1995.
40. Lambert, J. B., Shurvell, H. F., Lightner, D. A., and Cooks, R. G., *Organic Structural Spectroscopy*, Prentice Hall, Upper Saddle River, NJ, 1998, p. 375.
41. March, R. E. and Hughes, R. J., *Quadrupole Storage Mass Spectrometry*, John Wiley & Sons, New York, 1989.
42. McLuckey, S. A., Van Berkel, G. J., Goeringer, D. E., and Glish, G. L., Ion trap mass spectrometry of externally generated ions, *Anal. Chem.*, 66, 689A, 1994.
43. Louris, J. N., Brodbelt-Lustig, J. S., Cooks, R. G., Glish, G. L., Van Berkel, G. J., and McLuckey, S. A., Ion isolation and sequential stages of mass spectrometry in a quadrupole ion trap mass spectrometer, *Int. J. Mass Spectrom. Ion Processes*, 96, 117, 1990.
44. Vestal, M. L., Juhasz, P., and Martin, S. A., Delayed extraction matrix-assisted laser desorption time-of-flight mass spectrometry, *Rapid Commun. Mass Spectrom.*, 9, 1044, 1995.
45. Spengler, B., Post-source decay analysis in matrix-assisted laser desorption/ionization mass spectrometry of biomolecules, *J. Mass Spectrom.*, 32, 1019, 1997.
46. Boyle, J. G. and Whitehouse, C. M., Time-of-flight mass spectrometry with an electrospray ion beam, *Anal. Chem.*, 64, 2084, 1992.
47. Marshall, A. G. and Grosshans, P. B., Fourier transform ion cyclotron resonance mass spectrometry: the teenage years, *Anal. Chem.*, 63, 215A, 1991.
48. Holliman, C. L., Rempel, D. L. M., and Gross, M. L., Detection of high mass-to-charge ions by Fourier transform mass spectrometry, *Mass Spectrom. Rev.*, 13, 105, 1994.
49. Buchanan, M. V. and Hettig, R. L., Fourier transform mass spectrometry of high-mass biomolecules, *Anal. Chem.*, 65, 245A, 1993.
50. Watson, J. T., *Introduction to Mass Spectrometry*, Lippincott-Raven, Philadelphia, PA, 1997, p. 331.
51. Hoffmann, E. de, Tandem mass spectrometry: a primer, *J. Mass Spectrom.*, 31, 129, 1996.
52. Craig, A. G. and Derrick, P. J., Production and characterization of beams of polystyrene ions, *Aust. J. Chem.*, 39, 1421, 1986.
53. Lattimer, R. P., Münster, H., and Budzikiewicz, H., Tandem mass spectrometry of polyglycols, *Int. J. Mass Spectrom. Ion Processes*, 90, 119, 1989.

54. Lattimer, R. P., Tandem mass spectrometry of lithium-attachment ions from polyglycols, *J. Am. Soc. Mass Spectrom.*, 3, 225, 1992.
55. Lattimer, R. P., Tandem mass spectrometry of poly(ethylene glycol) proton- and deuteron-attachment ions, *Int. J. Mass Spectrom. Ion Processes*, 116, 23, 1992.
56. Jackson, A. T., Yates, H. T., Scrivens, J. H., Critchley, G., Brown, J., Green, M. R., and Bateman, R. H., The application of matrix-assisted laser desorption/ionization combined with collision-induced dissociation in the analysis of synthetic polymers, *Rapid Commun. Mass Spectrom.*, 10, 1668, 1996.
57. Jackson, A. T., Jennings, K. R., and Scrivens, J. H., Generation of average mass values and end group information of polymers by means of a combination of matrix-assisted laser desorption/ionization-mass spectrometry and liquid secondary ion-tandem mass spectrometry, *J. Am. Soc. Mass Spectrom.*, 8, 76, 1997.
58. Scrivens, J. H., Jackson, A. T., Yates, H. T., Green, M. R., Critchley, G., Brown, J., Bateman, R. H., Bowers, M. T., and Gidden, J., The effect of the variation of cation in the matrix-assisted laser desorption/ionization-collision induced dissociation (MALDI-CID) spectra of oligomeric systems, *Int. J. Mass Spectrom. Ion Processes*, 165/166, 363, 1997.
59. Jackson, A. T., Yates, H. T., Scrivens, J. H., Green, M. R., and Bateman, R. H., Utilizing matrix-assisted laser desorption/ionization-collision induced dissociation for the generation of structural information from poly(alkyl methacrylate)s, *J. Am. Soc. Mass Spectrom.*, 8, 1206, 1997.
60. Jackson, A. T., Yates, H. T., Scrivens, J. H., Green, M. R., and Bateman, R. H., Matrix-assisted laser desorption/ionization-collision induced dissociation of poly(styrene), *J. Am. Soc. Mass Spectrom.*, 9, 269, 1998.
61. Pastor, S. J. and Wilkins, C. L., Sustained off-resonance irradiation and collision-induced dissociation for structural analysis of polymers by MALDI-FTMS, *Int. J. Mass Spectrom. Ion Processes*, 175, 81, 1998.
62. Polce, M. J., Kowalski, P., and Wesdemiotis, C., unpublished results.

2

Polymer Characterization Methods

G. Montaudo and M.S. Montaudo

CONTENTS

0-8493-3127-7/02/$0.00+$1.50
© 2002 by CRC Press LLC

2.1   Introduction

Polymers can display a variety of structures, including linear, cyclic, and branched chains, copolymers in which different repeat units are aligned along the chain in different manners, and star polymers with different numbers of arms. The identification of molecular structure is the first step in the analysis of a polymeric material, which may actually be a homopolymer, a copolymer, or a blend of homopolymers and/or copolymers.

Usually, commercial polymers contain a number of additives (thermal and photo antioxidants, plasticizers, etc.) that can be suitably extracted from the polymer bulk and analyzed separately. Polymers may contain low-mass oligomers, both linear and cyclic, and the identification of these species can be achieved by analyzing the intact polymer, or by extracting the oligomers from the polymer bulk.

Contrary to the usual organic compounds, polymers are far from being homogeneuos materials (i.e., polymer chains do not possess the same molar mass and chemical structure). As matter of fact, many synthetic polymers are heterogeneous in several respects. Homopolymers may exhibit both molar-mass distribution (MMD) and end-groups (EG) distribution. Copolymers may also show chemical composition distribution (CCD) and functionality distribution (FTD) in addition to the MMD. Therefore, different kinds of heterogeneity need to be investigated in order to proceed to the structural and molecular characterization of polymeric materials.

The estimate of molar masses (MM) and of molar-mass distributions (MMD) is of primary interest in polymer characterization work, and much effort has been dedicated to develop suitable methods for their determination. End-group analysis provides important structural information for all the synthetic polymers. It allows also the estimation of molar masses in low polymers and gives clues about the procedure adopted in the synthesis. In fact, polymer samples having the same molecular structure may contain different end groups, due to synthetic routes or to end capping with different agents. If the end groups are chemically reactive, the polymer may be further modified to obtain different materials.

Determining the composition and sequence of comonomer units is essential in the case of copolymers, since both parameters influence the physical and chemical properties of these materials. Furthermore, comonomer sequence is related to the mechanism of copolymerization, and to the reactivity ratios of the comonomers. Among the techniques developed for polymer characterization, Mass Spectrometry (MS) is one of the most powerful. The mass spectrum of a polymer contains plenty of information on polymer properties such as the structure, the repeat units which constitute the macromolecular backbone, the length of macromolecular chains, the end groups which terminate the chains, the chemical heterogeneity, the sequence of copolymers and their composition heterogeneity.

MS is one of the newest methods developed for polymer analysis, and a variety of other methods are currently adopted as well,[1-7] here referred to as "conventional" methods. In everyday practice, MS data are routinely compared and integrated with the results of other conventional techniques for polymer analysis. Furthermore, MS is often coupled to polymer separation methods such as Size Exclusion Chromatography (SEC) and Liquid Chromatography (LC).

The use of MS to analyze polymeric materials requires therefore some knowledge of basic polymer science, and acquaintance with the conventional methods used in the characterization of polymeric materials is desirable. The field of polymer characterization is vast; nevertheless comprehensive books are available which cover the entire subject.[1-4] Other texts are slightly less general and deal with groups of methodologies for polymer analysis, such as the spectroscopic methods (IR, NMR),[5] or the group of chromatographic methods.[6] Textbooks in polymer science[8-12] provide the reader with an account of the most popular methods used in polymer analysis.

This chapter is intended to give background information on the major applications of MS analysis to the characterization of polymers, and to describe the most conventional techniques used to determine polymer size and structure.

Section 2.2 gives a brief account of the "conventional" techniques used to measure the average molar masses and their distribution, namely, Osmometry, Intrinsic Viscosity, Size Exclusion Chromatography, and Light Scattering. We then discuss how to determine the molar-mass distribution (MMD) of the polymer sample from its mass spectrum and how to calculate average molar masses. A comparison is also provided for the averages obtained by MS with the averages obtained by "conventional" methods.

Section 2.3 focuses on various aspects of the determination of polymer structure, namely assessing the composition of the macromolecular chains, discriminating cyclic from linear and branched chains, end-groups determination, and detection of stereoregularity in polymers.

In Section 2.4 we consider copolymers. The determination, by MS techniques, of various quantities (such as copolymer composition, sequence distribution, composition heterogeneity, reactivity ratios, and bivariate distribution of chain compositions and chain lengths) are of major interest for the characterization of copolymers.

2.2   Molar Mass and Molar-Mass Distribution

Polymers are mixtures of molecules of different sizes. Contrary to proteins, where the chain length is a constant, synthetic polymers (rubbers and plastics) and polysaccharides always possess a certain polydispersity, which is intrinsic to the polymerization process. An important step for polymer characterization is to measure the number and weight of macromolecular chains at size (length) s, at size ($s+1$), at size ($s+2$), at size ($s+3$), and so on. This is referred to as molar-mass distribution (MMD) determination[8-12] (the traditional term "molecular weight distribution" is also used, although obsolete).

The MMD of a polymer is of prime importance in its application. In most instances, there is a molar-mass (MM) range for which a given polymer property will be optimal for a particular application. The control of molar mass (expressed in g/mole) and of its distribution is essential for the practical application of a polymerization process, since its utility is greatly reduced unless the reaction can be carried out to yield polymer of a sufficiently high and specified molar mass.

Figure 2.1 illustrates the possible results of the MMD measurement. The resulting MMD may turn out to be narrow (Figure 2.1a), or broad (Figure 2.1b). The distribution may display a single maximum (Figures 2.1a, 2.1b) (referred to as unimodal), or two maxima (Figure 2.1c) (referred to as bimodal). Once the MMD is known, one can compute MMD averages.[8-12] Let's indicate by N_i the number of chains with mass m_i. In this case, the number-average molar

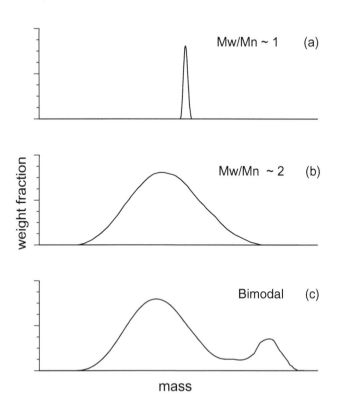

FIGURE 2.1
Types of molar mass distributions: (a) narrow, (b) broad, (c) bimodal.

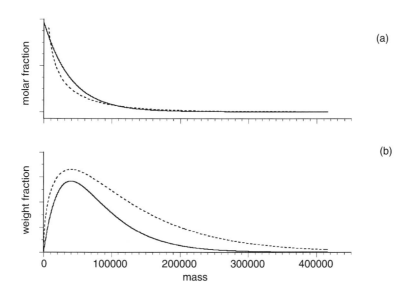

FIGURE 2.2
The molar mass distribution for two polymers. The first polymer (continuous trace) follows the Flory-Schulz MMD with $\overline{M}n = 40000$ ($\overline{M}w$ is the double, i.e., $\overline{M}w = 80000$). The second polymer (dotted trace) follows the Schulz-Zimm MMD with $\overline{M}n = 40000$ and $\overline{M}w = 120000$. The molar mass distribution is diplayed as the molar fraction (a) and weight fraction (b) vs. mass.

mass, $\overline{M}n$, is given by:

$$\overline{M}n = \left(\sum m_i N_i\right)\Big/\left(\sum N_i\right) \tag{2.1}$$

In a similar manner, the weight-average molar mass, $\overline{M}w$, is given by:

$$\overline{M}w = \left(\sum m_i^2 N_i\right)\Big/\left(\sum m_i N_i\right) \tag{2.2}$$

The z-average molar mass, $\overline{M}z$, is given by:

$$\overline{M}z = \left(\sum m_i^3 N_i\right)\Big/\left(\sum m_i^2 N_i\right) \tag{2.3}$$

The viscosity measurement yields the viscosity-average molar mass, $\overline{M}v$, which is given by

$$\overline{M}v = \left(\left(\sum m_i^{a+1} N_i\right)\Big/\left(\sum m_i N_i\right)\right)^{(1/a)} \tag{2.4}$$

where "a" is the exponent in the Mark-Houwink-Sakurada (MHS) equation (see below, Eq. 2.23). If the value of "a" is close to one, the $\overline{M}v$ comes close to $\overline{M}w$.

Once the $\overline{M}n$ and $\overline{M}w$ are known, one can compute the polydispersity index, D, defined as:

$$D = \overline{M}w/\overline{M}n \tag{2.5}$$

Since $\overline{M}n$ is always smaller than $\overline{M}w$, D cannot take values smaller than 1. Narrow MMD are characterized by D values close to 1, whereas broad MMD are characterized by D values about 2 or larger. A large value of D may indicate the presence of a low molar mass tail in the MMD, or it may be a symptom that the MMD is not unimodal.

2.2.1   Theory of Molar-Mass Distributions

If one knows the polymerization process used to produce a polymer, one can use the theory to predict the MMD of the resulting polymer. In the following we shall analyze the predictions in some detail.

2.2.1.1   The "Most Probable" Distribution

The MMD expected for condensation polymerization[8] (and also for free-radical polymerization with termination by disproportionation[8]) is the most probable (also referred to as Flory-Schulz) MM distribution:

$$I(n) = a_1(1-p)\,p^{n-1} \tag{2.6}$$

where p is the monomer conversion, $I(n)$ is the molar fraction of chains of size n, and a_1 is a suitable normalization factor.

The MM averages, computed using the definition, turn out to be:

$$\overline{M}n = M_0/(1-p) \tag{2.7}$$

$$\overline{M}w = 2\,M_0/(1-p) \tag{2.8}$$

where M_0 is the mass of the repeat unit. It can be seen that $\overline{M}w$ doubles $\overline{M}n$, so that the polydispersity index (Eq. 2.5) is two. Figure 2.2 reports the Flory-Schulz MMD for a polymeric sample where $\overline{M}n = 40000$ ($\overline{M}w$ is the double, i.e., $\overline{M}w = 80000$). Figure 2.2a reports the number fraction versus mass. Figure 2.2b reports the weight fraction versus mass. It can be seen that the two plots are quite different. Nevertheless, the two plots are connected, since the weight fraction is the product of the molar fraction times the mass.

2.2.1.2   Termination by Recombination

The MMD expected for the free-radical polymerization with termination by recombination is given by:[8–11]

$$I(n) = a_2 n \exp(-2\,n/DP_n) \tag{2.9}$$

where DP_n is the number-average degree of polymerization (i.e., the average number of units in the chains) and a_2 is a suitable normalization factor. The MM averages corresponding to this MM distribution are:

$$\overline{M}n = M_0/(1 - p) \tag{2.10}$$

$$\overline{M}w = 1.5\, M_0/(1 - p) \tag{2.11}$$

where the symbols have the usual meaning. The polydispersity index (obtained by taking the ratio between the two averages) is 1.5.

2.2.1.3   The Poisson Distribution

A large class of monomers exists which can be polymerized by anionic polymerization.[13] When the polymerization is carried out under suitable conditions, no termination occurs and the process is referred to as "living" polymerization. The MMD expected for "living" polymerization,[8] is the Poisson MMD:

$$I(n) = a_3\, (DP_n - 1)^n/n! \tag{2.12}$$

where DP_n is defined above and a_3 is a suitable normalization factor.

The MM averages can be computed easily using the definition. They turn out to be

$$\overline{M}n = DP_n^{*}\, M_0 \tag{2.13}$$

$$\overline{M}w = \overline{M}n + M_0 \tag{2.14}$$

where M_0 is the mass of the repeat unit. The polydispersity index of the Poisson MMD is therefore given by $D = 1 + M_0/\overline{M}n$. At higher masses, $\overline{M}w$ approaches $\overline{M}n$, and the Poisson distribution becomes almost monodisperse. In practice, however, various factors tend to increase the polydispersity index.[13]

2.2.1.4   Empirical Distribution Functions

Empirical MMD functions possess two or more adjustable parameters and the MM averages are given by closed expressions.[14,15] Empirical MMD functions are used to compare MMD data obtained by different methods and especially to compare data from viscosity measurements with data from Size Exclusion Chromatography or Osmometry.[14,15] They are also used in Dynamic Light Scattering to obtain $\overline{M}n$ and $\overline{M}w$ (see section on Light Scattering below).

The most important empirical MMD functions are:[14,15] the Schulz-Zimm, the Log-normal, and the Generalized Exponential. The Schulz-Zimm MMD function is:

$$I(n) = a_4(n)^\alpha \exp(-n/y) \tag{2.15}$$

where α and y are two adjustable parameters and a_4 is a suitable normalization factor.

The MM averages, computed using the definition, turn out to be:

$$\overline{M}n = M_0\,(\alpha+1)/y \tag{2.16}$$

$$\overline{M}w = M_0\,(\alpha+2)/y \tag{2.17}$$

In this case, the polydispersity index is $(\alpha + 2)/(\alpha + 1)$. Figure 2.2 reports the Schulz-Zimm MMD for a polymer sample with $\overline{M}n = 40000$ and $\overline{M}w = 120000$. Figure 2.2a reports the number fraction versus mass, whereas Figure 2.2b reports the weight fraction. The two plots are quite different due to the fact that short chains are very abundant but their weight is small. Nevertheless, the two plots are connected, since the weight fraction is the product of the molar fraction times the mass. For brevity, the descriptions of the Log-normal MMD and the Generalized Exponential MMD are omitted. (They can be found elsewhere.[14,15])

2.2.1.5   The MM Distribution of Cyclic Oligomers

Various reactions, both of polymerization and of polymer degradation, can produce cyclic polymer molecules. A well-known process is the ring-chain equilibration reaction, which may be used to produce cyclic siloxanes and other cyclic polymers.[16,17] The linear chain reacts intramolecularly and yields a cyclic and a linear chain. In the initial stages, the molar fraction of cyclics increases at the expense of the linear chains. After some time, equilibrium conditions are achieved and the molar fraction of cyclics remains constant. In some cases, all the sites in the macromolecular backbone are equivalent and no peculiar bond exists which is preferentially attacked. This case is referred to as thermodynamically controlled cyclization.

The theory of thermodynamically controlled cyclization[16] predicts that, at the equilibrium, the abundance of cyclic macromolecules, $I(n)$, follows the law:

$$I(n) = b(n)^{-2.5} \tag{2.18}$$

where n is the number of repeat units in the cycle, and b is a constant independent of "n" but dependent on the dimensions and on the structure of the repeat unit. The above equation predicts that the abundance of cyclics decreases rapidly as the chain size grows. For instance the ratio between the abundance of the cyclics with $n = 20$ and $n = 200$ is given by $I(20)/I(200) = 310/1$.

In other cases, some bonds are preferentially attacked. This case is referred to as kinetically controlled cyclization. The theory of kinetically controlled cyclization predicts[16,17] that the abundance of cyclic macromolecules, $I(n)$, follows the law:

$$I(n) = a_5 (n)^{-1.5} \qquad (2.19)$$

where a_5 is a suitable normalization factor.

The quantity $I(n)$ becomes vanishingly small at high masses. However, the drop of $I(n)$ is less pronounced than for the thermodynamically controlled cyclization. The relative abundance of cyclic oligomers formed in the two cases discussed above are shown in Table 2.1.

2.2.2 Conventional Methods of Molar Mass Measure

The determination of Molar Masses and Molar Mass Distributions in Polymers has always been a most difficult problem in polymer science. Initially, suitable methods were simply nonexistent (Victor Meyer's method is for gaseous substances), and all the methods subsequently developed proved

TABLE 2.1

Relative Abundance of Cyclic Oligomers Generated in Polymerization Reactions

Size	Abundance[a]	Abundance[b]
1	100	100
2	17.7	35.3
3	6.4	19.2
4	3.1	12.5
5	1.8	8.9
6	1.1	6.8
7	0.8	5.4
8	0.5	4.4
9	0.4	3.7
10	0.3	3.2
11	0.25	2.7
12	0.20	2.4
13	0.16	2.1
14	0.13	1.9

[a] Values calculated for the thermodynamic control in cyclization reactions (Eq. 2.18).
[b] Values calculated for the kinetic control in cyclization reactions (Eq. 2.19).

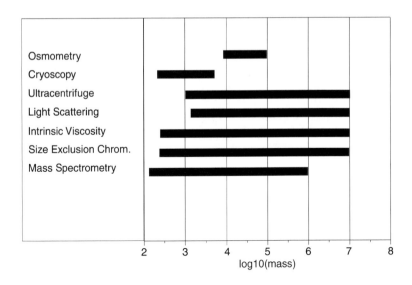

FIGURE 2.3
Some methods used for the measurement of MM averages and MMD along with the mass range of the samples that can be analyzed.

to be somehow complex and, to a certain extent, unsatisfactory either theoretically or experimentally. Osmometry, Light Scattering, Ultracentrifuge, all work once set up carefully and operated by expert hands, but in everyday practice polymer chemists experience problems in determining the molar mass of the polymers synthesized in their labs. Thus, researchers resorted to indirect methods to estimate the molar mass of polymers. Indirect methods, although requiring calibration standards, have encountered an enormous success.

Figure 2.3 reports some methods used currently, along with the mass range of the samples that can be analyzed. It can be seen that some techniques cannot be applied to high MM polymers, whereas other techniques, such as Light Scattering, do not work for low MM polymers. The method based on mass spectrometry (MS) for measuring $\overline{M}n$, $\overline{M}w$ and the MMD of a polymer sample came much later. It cannot be considered "conventional," since MS of polymers is a young discipline. In the following, we describe the most widely used conventional methods.

2.2.2.1   Direct Methods

Among conventional methods, some need calibration (a set of polymers of known $\overline{M}n$ and $\overline{M}w$), and are therefore "indirect" methods, whereas other techniques do not need calibration and we refer to them as "direct." Thus, Intrinsic Viscosity and Size Exclusion Chromatography (SEC) are indirect, whereas Osmometry, Ultracentrifuge, and Light Scattering (LS) are direct.

2.2.2.1.1 Osmometry

In osmometry one measures the osmotic pressure, Π, of a polymeric solution at various dilutions. The quantity Π/RTC is then computed and plotted versus C (C is the concentration of the polymer solution, T is the temperature, and R is the gas constant). The resulting plot is a straight line and its intercept is $1/\overline{M}n$, since Π is related to $\overline{M}n$ as follows:[8]

$$\Pi/RT = C/\overline{M}n + A_2 C^2 + A_3 C^3 \qquad (2.20)$$

where A_2 and A_3 are the second and the third coefficients of the virial expansion.

The field of application of Osmometry is limited to polymers with masses below 100,000, since the precision depends on the mass, and therefore it becomes inaccurate at high masses. Furthermore, being a colligative method, osmometry yields $\overline{M}n$ but not $\overline{M}w$, and therefore it cannot be used to discriminate a narrow MMD from a broad MMD nor a unimodal MMD from a bimodal MMD.

2.2.2.1.2 Light Scattering

In the Light Scattering (LS) method,[1,12,18] one shines a monochromatic light beam at the polymeric solution and measures the intensity of the light scattered by the macromolecules at various angles θ. The scattered intensity is then normalized for the sample-detector distance and for the incident beam intensity, and the Rayleigh Ratio, R_θ, is obtained.

The procedure for obtaining $\overline{M}_w$ from LS data follows.[1,12,18] First, one plots $K_s C/R_\theta$ vs. $\sin^2(\theta/2)$ for different concentrations (here C is the concentration of the polymer solution and K_s is a factor carrying experimentally known constants). The resulting plot is called a Zimm plot. Thereafter, simply extrapolating $K_s C/R_\theta$ at low θ values, $1/\overline{M}w$ can be obtained as the intercept with the ordinate axis. In fact, R_θ is connected to $\overline{M}w$ by the following relationship:

$$K_s C/R_\theta = 1/(P(\theta)\,\overline{M}w) + 2A_2 C \qquad (2.21)$$

where A_2 is the second virial coefficient and $P(\theta)$ is a shape factor.[1]

In Multi Angle Laser Light Scattering (MALLS), the light scattered at various different angles is simultaneously measured and the extrapolation at low θ values is done automatically.

One of the drawbacks of the LS method is that it requires extensive filtering of the solution to eliminate dust particles which may contribute to the scattering and may therefore cause errors. Another drawback is that the sensitivity for low MM chains is very low, since R_θ increases linearly as $\overline{M}w$ grows. If the sample is made of macromolecules which form random coils, and if the distribution of the end-to-end distances is Gaussian, or the shape

factor $P(\theta)$ in Eq. 2.21 is known, LS data can yield also $\overline{M}n$ of the sample.[1]
However, LS data do not allow the determination of the entire MMD, and
therefore LS cannot be used to distinguish a unimodal MMD from a bimodal
MMD.

Normally, this technique cannot be applied to copolymers. In fact, the LS
detector systematically underestimates the abundance of one of the repeat
units.[12,18] It has been shown[18] that for copolymers the LS detector measures
an apparent molar mass, M_{app}, which differs from $\overline{M}w$ by a term ΔM:

$$\Delta M = M_{app} - \overline{M}w \qquad (2.22)$$

In some cases[18] M_{app} turned out to be eleven times larger than $\overline{M}w$.

LS can be used to detect branched structures. In fact, one of the polymer
properties that can be extracted from the Zimm-Plot is the radius of gyration,
$\overline{R}g$. Comparing it with $\overline{M}w$ one can have a measure of the average degree
of branching.[1]

In the Dynamic Light Scattering method (also known as Photon Correlation
Spectroscopy),[19,20] one measures the diffusion of macromolecular chains in
solution. The diffusion coefficient is related to the molar mass. The method
suffers a number of difficulties. Raw experimental data do not lend themselves to immediate interpretation and must be fitted using a model
MMD.[19,20] Furthermore, Dynamic LS cannot be used to choose between a
unimodal MMD and bimodal MMD. In fact data are fitted using a model
MMD[19] and the number of parameters to be fitted increases quickly, going
from unimodal to bimodal (a bimodal MMD requires finding the best fit of
six parameters).

2.2.2.2 Indirect Methods

Among conventional methods for determining MM averages and MMD,
indirect methods are those that measure a quantity (such as the elution
volume or the solution viscosity) which is related to the MM. Indirect methods need calibration. A set of polymers of known $\overline{M}n$ and $\overline{M}w$ is used for
this purpose.

2.2.2.2.1 Intrinsic Viscosity

In the viscometry method[8–12,15] one measures the viscosity, η, of a polymer
solution at various degrees of dilution. The intrinsic viscosity, $[\eta]$, of the
polymer-solvent system is obtained by plotting the quantity $\{C^{-1}(\eta - \eta_0)/\eta_0\}$
vs. C (here η_0 is the viscosity of the solvent and C is the concentration of the
polymer solution) and extrapolating the resulting curve to zero concentration. Then one applies the Mark-Houwink-Sakurada (MHS) equation:

$$[\eta] = K (\overline{M}v)^a \qquad (2.23)$$

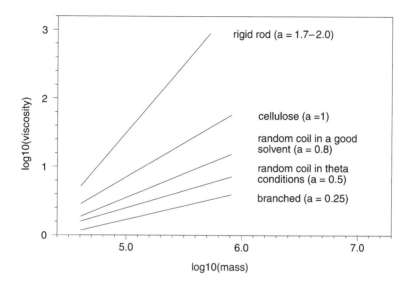

FIGURE 2.4

The exponent "*a*" in the Mark-Houwink-Sakurada (MHS) equation $[\eta] = K(\overline{M}v)^a$ for various polymer conformations.

where $\overline{M}v$ is the viscosity-average molar mass (defined above) and where K and "*a*" are two constants that are peculiar for a given polymer-solvent combination.[15]

Figure 2.4 reports the values of the constant "*a*" for different macromolecular shapes. The values of "*a*" are in the range 0.5–0.8 for random coils. For other shapes, such as rods, the "*a*" value is larger. The constants K and "*a*" are determined by obtaining (usually by SEC fractionation) a series of polymers (referred to as the calibrants) which possess different MM, a narrow MMD, and the same structure of the polymer to be analyzed. K and "*a*" have been tabulated for many polymer-solvent combinations.[15]

The method based on viscometry yields a single MM average, namely $\overline{M}v$ (see Eq. 2.4), and it does not give any hint of the polydispersity. Therefore it cannot be used to discriminate a narrow MMD from a broad MMD nor a unimodal MMD for a bimodal MMD. Another drawback is that the MHS constants are not known for a large number of polymer-solvent combinations and, in the case of condensation polymers the MHS constants are rarely reported.[15]

When viscometry is applied to copolymers, a number of additional problems arise. The MHS constants for copolymers are usually not known, and furthermore they change as the copolymer composition varies. When viscometry is used to analyze branched and star-shaped polymers, it yields $\overline{M}v$ values that are systematically underestimated. In fact, the MHS constants for branched and star-shaped polymers are usually not known,[15] and when using the constants K and "*a*" for linear chains one obtains erroneous values.

2.2.2.2.2   Size Exclusion Chromatography

Size Exclusion Chromatography (SEC) uses columns packed with a porous, inert material (with pore dimensions similar to the dimensions of the macromolecules to be analyzed) which does not interact with the polymer.[1-4,21-32] The SEC method is based on the fact that when a polymeric solution is in the presence of a porous particle, smaller polymer molecules penetrate the pores more deeply than the larger ones.

The polymer solution is injected into a continuous stream of solvent that is flowing through the column. Since the dimension of a macromolecule reflects its molar mass, polymer molecules which possess low mass are eluted by the flowing solvent at later times. Polymer solutions in the concentration range 0.1–20 mg/mL are adopted. These are injected into the column and the resulting chromatographic trace (i.e., the detector response as a function of the eluted volume, Ve) , is recorded.

The standard SEC apparatus is equipped with a solvent delivery system and a differential refractive index (RI) detector, which measures the difference between the refractive index n of the solution eluted at volume Ve and the refractive index n_0 of the solvent. The detector's response is proportional to the weight of the polymer dissolved.[1]

Expanding the refractive index (n) in a Taylor series as a function of the concentration C of the solution eluted at volume Ve, and neglecting the higher-order terms of the expansion, one has:

$$n - n_0 = \alpha_R \, C \tag{2.24}$$

where α_R is a shorthand notation[1] for $\left(\frac{dn}{dc}\right)$ the refractive index increment, which depends on the wavelength and on the polymer type.

The concentration C of the solution is proportional to the weight of the polymer dissolved. This implies that the refractive index detector measures the weight fraction of macromolecules. The molar fraction of macromolecules is not directly determined, and it must be derived by applying suitable transformation algorithms.[27] Examples of such transformation have been reported in Section 2.1, when dealing with theoretical distributions. More specifically Figure 2.2b reports the weight fraction for two model distributions, whereas Figure 2.2a reports the molar fraction.

SEC is an indirect method of measuring the MM of polymers, since the SEC trace provides only the elution volumes of the polymers, and these elution volumes need to be converted into MM. In other words, the SEC traces need mass calibration. It is generally observed that the mass (M) of the macromolecules eluted at a given elution volume (Ve) decreases very quickly as Ve grows. The relationship used is usually of the type:[21]

$$\log(M) = b_0 - b_1 \, Ve \tag{2.25}$$

where b_0 and b_1 are two constants. It is experimentally observed that b_0 and b_1 vary when the columns are changed and when the one switches from a solvent to a different one.

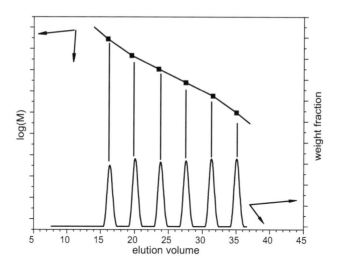

FIGURE 2.5
Construction of the calibration line for the SEC.

The method for measuring b_0 and b_1 is described schematically in Figure 2.5. A mixture of five or more polymer samples with the same repeat units, possessing a narrow MM distribution and known mass (the so-called SEC primary standards), is prepared. The mixture is injected in the SEC apparatus and the resulting chromatogram (Figure 2.5) is recorded. Measuring the elution volumes and plotting them against the logarithm of the mass (Figure 2.5), the calibration line (Eq. 2.25) is obtained. Actually, the slope b_1 is relatively insensitive to the type of polymer injected, whereas b_0 displays a more complex behavior, namely b_0 changes from a polymer type to another.[22–25]

Figure 2.6 shows the calibration curves[22] for poly(n-butyl methacrylate) (PnBMA), poly(decyl methacrylate) (PDMA), poly(methyl methacrylate) (PMMA), poly(*tert*butyl methacrylate) (PtBMA). It can be seen that the lines are almost parallel to each other, which implies that the slope b_1 is approximately constant, whereas b_0 varies. The changes in b_0 correspond to a change in the molar mass, and they may be quite large. For instance, when tetrahydrofuran (THF) is used as a solvent, polystyrene (at a given elution volume) has a mass that is about two times the mass of polycarbonate.[23] It can be concluded that every polymer has its own calibration line, corresponding to a specific set of values for b_0 and b_1 in Eq. 2.25.

When the polymer sample is monodispersed, straightforward application of Eq. 2.25 leads to the MM determination. However, the most common case is that of a polydispersed polymer, where the significant quantities are the MM averages ($\overline{M}n$, $\overline{M}w$) and the MMD. The SEC trace contains this information, but the process of extracting such information is more complex.

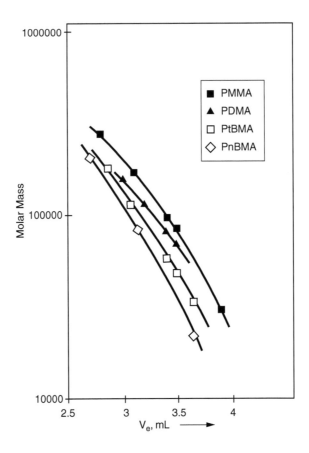

FIGURE 2.6
SEC calibration curves for PMMA, PDMA, PtBMA, PnBMA (reproduced with permission from
Ref. 22).

By inspection of the SEC trace, one finds the most probable molar mass,
Mp, i.e., the mass at which the trace takes its maximum (see Figure 2.1):

$$Mp = \text{MAX}(m\ N_i) \tag{2.26}$$

where the symbols have the same meaning as above. The usefulness of this
calculation relies on the fact that if the sample's polydispersity index is not
extremely large (say, it is smaller than 2.5), *Mp* falls very close to $\overline{Mw}$, and
hence a measure of this polymer property is obtained. In a second step, one
uses software (commercial packages are available), which receives the input
data (the elution volumes and the chromatographic trace) and gives as
output $\overline{Mn}$, $\overline{Mw}$, and the entire MMD.[27]
   In fact, the software digitizes the SEC trace and transforms it into a series
of points. Each point corresponds to an elution volume and to an intensity
value (RI response). The elution volumes are then converted into masses by

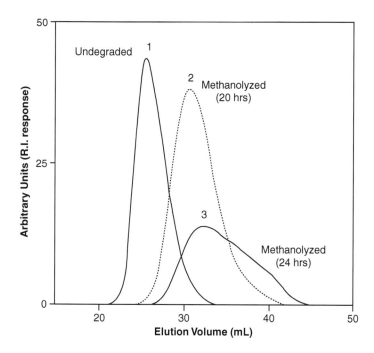

FIGURE 2.7
SEC traces of a polyester sample subjected to partial degradation in order to reduce its molar mass (reproduced with permission from Ref. 26).

making use of the calibration parameters (Eq. 2.25) previously determined for the specific polymer to be analyzed, and the SEC trace is calibrated with respect to the molar mass.[27] At this point, the calculation of $\overline{M}n$ and $\overline{M}w$ is performed by the software using Eq. 2.1 and Eq. 2.2.

SEC can be used to follow the polymerization processes. As the reaction proceeds, the MMD of the sample varies and the SEC trace changes accordingly.[8–10] In a similar manner SEC can be used to follow depolymerization processes.

Figure 2.7 reports the SEC traces of a polymeric sample subjected to partial degradation in order to reduce its molar mass.[26] The intact polymer (trace 1) is eluted between 18 and 33 mL, whereas the partially degraded polymer is eluted at higher volumes, namely in the range 25–42 mL and in the range 28–45 mL (20 and 24 hours methanolysis, respectively). In a similar manner, the SEC trace for the intact polymer takes its maximum at Ve = 25 mL, whereas in the SEC traces for the partially degraded polymer, the peak shifts toward higher volumes, namely Ve = 32 mL and Ve = 33 mL.

2.2.2.2.3 *The Universal Calibration Concept*

The reliability of SEC results strongly depends on the availability of a set of polymers as known mass and narrow MM distribution, which possess the same structure as the polymer of interest. However, a set of such calibration

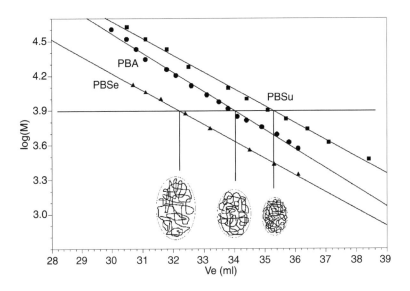

FIGURE 2.8
SEC calibration curves for PBSe, PBA, PBSu (reproduced with permission from Ref. 25).

standards is often unavailable for each specific polymer. As an approximation, some authors assume that b_0 and b_1 in Eq. 2.25 are constant, independent of the polymer structure, and a common procedure is to use a mixture of poly-styrene standards (which are easily available) to build an SEC calibration line valid for any type of polymer. Unfortunately this assumption is rarely valid.

Figure 2.8 shows the SEC calibration lines for three polyesters, namely poly(butylene succinate) (PBSu), poly(butylene adipate) (PBA), and poly (butylene sebacate) (PBSe), along with a schematic representation of the solvated coils formed by three polymers in solution.[25] Taking coils of the same MM ($\log M = 3.9$ in Figure 2.8), the solvated coil of PBSu is seen to have smaller dimensions with respect to the other two polyesters and there-fore it is eluted at a later time, whereas for PBSe the solvated coil is bigger and therefore it is eluted first. Furthermore, the solvated coil of PBAd has an intermediate dimension and therefore it is eluted between PBSe and PBSu.

In order to solve this difficulty, a variant of the SEC method based on the "universal calibration" concept is adopted.[1,28] The method is based on the discovery that flexible macromolecules with the same hydrodynamic volume are eluted at the same time.[28] The hydrodynamic volume of a chain is the dimension of a hard sphere that has the same migration velocity as the macromolecule in question, and it is in turn proportional to the product $[\eta]M$.

The hydrodynamic volume is of particular relevance since a peculiar rela-tionship between the mass, M, of the macromolecule and its elution volume, Ve, is used, namely:

$$\log([\eta]M) = Q_0 + Q_1 \, Ve \qquad (2.27)$$

where [η] is the intrinsic viscosity of the sample, and Q_0 and Q_1 are "universal" constants. It is experimentally observed that Q_0 and Q_1 vary when the columns are changed, but they remain unchanged when going from one polymer type to another.[1,28]

In this variant, one measures the intrinsic viscosity ([η]) of a series of narrow distribution calibration standards (usually, anionic PS, PMMA, and PEG). Thereafter, one injects them into the SEC apparatus, records the elution volume (Ve) and plots the product [η]M vs. elution volume. The plot is called "universal calibration" plot. Figure 2.9 shows a typical "universal calibration" plot, obtained[28] using dozens of different polymers.

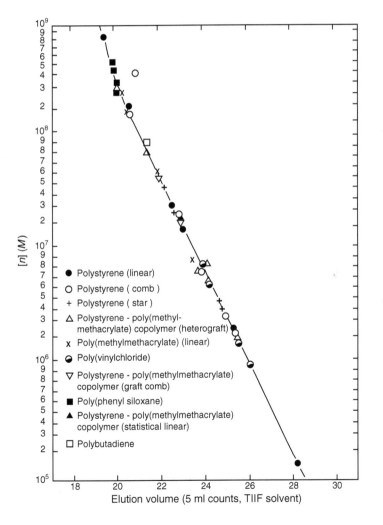

FIGURE 2.9
Universal calibration plot (reproduced with permission from Ref. 28).

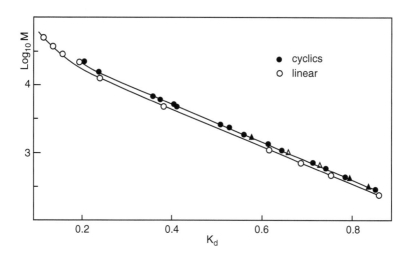

FIGURE 2.10
SEC calibration lines for cyclic PDMS and linear PDMS (reproduced with permission from Ref. 17).

2.2.2.2.4   SEC of Non-Linear Polymers and of Copolymers

When one injects a branched polymer into the SEC column, it will elute at early times (with respect to its unbranched analog that possesses the same molar mass). This is due to the fact that branched polymers are very compact. In particular, the slope b_1, which appears in the calibration equation (Eq. 2.25), is large. Also cyclic oligomers elute earlier than do linear oligomers of the identical type.

Figure 2.10 reports the calibration lines for cyclic and linear polydimethylsilsiloxane (PDMS), respectively.[17] It is quite evident that the intercept b_0 for rings is different from the intercept b_0 for linear chains. For cyclic polymers the theory is able to predict that the cyclic polymer (at a given elution volume) has a mass that is about 1.25 times larger than the linear one, and this has been verified experimentally for a number of polymers.[16,17]

In the experimental practice, SEC of copolymers is identical to that of homopolymers, but some complications arise in the data-handling process. First of all, the refractive index signal systematically underestimates the abundance of one of the two repeat units.[15] In fact, in this case (see Eq. 2.24) one must take into consideration the refractive index increment for units A (referred to as α_A) and the corresponding increment for units B (referred to as α_B) and in practice, the detector measures an apparent concentration, C_{APP}, which differs[15] from the true concentration by ΔC:

$$\Delta C = C - C_{APP} \tag{2.28}$$

If $\alpha_A > \alpha_B$, then the abundance of B units will be underestimated. The above relationship implies that the error in concentration estimation can be high, especially in those cases in which α_A is much larger than α_B or when α_A is a positive number and α_B is a negative number.[15]

A second complication arises from the fact that mass calibration requires some additional effort compared to homopolymers, since the elution volume of chains having the same mass changes if the copolymer composition varies. This is due to differences in the hydrodynamic volumes of two copolymer chains with the same mass but a different ratio of A/B units. Furthermore, calibration standards with narrow distribution, known composition, and known $\overline{M}n$ are often not available.

These difficulties can be solved using a MALDI mass spectrometer as a SEC detector; this topic will be discussed in Chapter 10, devoted to MALDI.

2.2.2.2.5 Hyphenated SEC Methods

Hyphenated methods use SEC apparatus equipped with two or three detectors, mounted in parallel or in series.[6,29–33] In the *SEC/Intrinsic Viscosity* method, one places a viscometer after the SEC apparatus and records two traces, namely, the RI response and the viscometer response.[6,28] This technique encounters some problems. For instance, the pistons of the solvent delivery system cause systematic pressure drops that disturb the measurement. Furthermore, it cannot be applied to samples with broad MMD (say with a polydispersity index $\overline{M}w/\overline{M}n$ larger than 1.8–2.0). In fact, it has been shown by computer simulation[29] that for broad MMD samples, the time-lag between the two detectors is overestimated, which implies that there is a discrepancy, ΔC, given by

$$\Delta C = C_{\text{visc}}(\text{Ve}) - C_{RI}(\text{Ve})$$ (2.29)

where $C_{\text{visc}}(\text{Ve})$ and $C_{RI}(\text{Ve})$ are the concentration profiles of the viscometry detector and of the RI detector. This means that $\overline{M}n$, $\overline{M}w$, and the whole MMD measurement are affected by an error. It has been shown that it is possible to eliminate this error, but the proposed modification[30] is quite elaborate.

In the *SEC/LS* method, one places an LS equipment after the *SEC* apparatus,[6] thus recording the RI response and the LS response. The data from the two detectors are processed and give as output $\overline{M}n$, $\overline{M}w$, and the entire MMD. The method is quite popular and yields nice results for narrow MMD samples. For broad MMD samples, the time-lag between the two detectors causes a discrepancy, ΔC, given by

$$\Delta C = C_{LS}(\text{Ve}) - C_{RI}(\text{Ve})$$ (2.30)

where $C_{LS}(\text{Ve})$ is the concentration profiles of the LS detector.[6]

Another drawback is that, for copolymers, the LS detector causes an error in the mass calibration and the software is not able to correct it. The reason for incorrect mass calibration is that the *LS* detector measures an apparent molar mass, M_{app}, as discussed above.

In the *SEC/Evaporative* method, one places an Evaporative Scattering Detector after the RI detector. This method is gaining wide success, since it quite reliable.[22]

2.2.3   Mass Spectrometric Methods of Molar Mass Measure

As discussed above, "conventional" methods for MM and MMD determination suffer from a series of limitations. It is therefore fruitful to look to other devices and techniques, such as MS, where such limitations might not be encountered. The MS method is based on the fact that ions strike the detector and produce a current that reflects the number of ions. The number of ions of mass M is proportional to the *number of molecules* at that mass, which implies that the detector measures *the molar fraction* of molecules of mass M. For polymers, one records the mass spectrum, tabulates MS peak intensities and computes $\overline{M}n$ and $\overline{M}w$ using Eqs. (2.1) and (2.2). Furthermore, one can directly check from the whole spectrum if the MM distribution is unimodal or bimodal.

In the following we illustrate the straightforward application of the MS method to various polymers.

Figure 2.11 reports the Field Desorption mass spectrum of an anionic polystyrene sample[34] with a narrow distribution. There are a series of mass spectral peaks in the mass range 1000–9000 Daltons. In addition to singly charged ions, doubly, triply and quadruply charged ions are also visible. The peaks in the mass range 3000–8000 Daltons are due to ions of the type H-$(St)_n$-$C_4H_9^+$. The authors obtained $\overline{M}n$ and $\overline{M}w$ using Eqs. (2.1) and (2.2). The result was in good agreement with the value obtained by VPO and with the value obtained by SEC. This represents a success, since it demonstrates that polymers with narrow distributions can be analyzed by FD-MS.

Figure 2.12 illustrates the mass spectrum of a narrow-dispersed polystyrene sample recorded using a Secondary Ion Mass Spectrometer equipped with a time-of-flight detector (TOF-SIMS).[35] There are a series of mass spectral peaks due to ions of the type H-$(St)_n$-C_4H_9, Ag^+ (silver was added as a cationization agent). Equation (2.1) yielded $\overline{M}n = 4550$, which is within 8% with the value $\overline{M}n = 4964$ obtained by VPO.

Figure 2.13 shows the mass spectrum of a polystyrene sample recorded using an instrument equipped with a Direct Chemical Ionization (DCI) Ion Source.[36] The peaks are due to protonated ions. The authors obtained $\overline{M}n$ and $\overline{M}w$ using Eqs. (2.1) and (2.2). The result was $\overline{M}n = 3906$ and $\overline{M}w = 4090$, which compares well with the values $\overline{M}n = 3744$ and $\overline{M}w = 4142$ obtained by SEC. DCI can be used for the analysis of narrow-distributed samples in the range below 5 kDa. A series of PEG samples obtained by anionic polymerization was analyzed using an instrument equipped with an Electrospray Ion Source and an FT-ICR cell (ESI-FT),[37] and $\overline{M}n$ and $\overline{M}w$

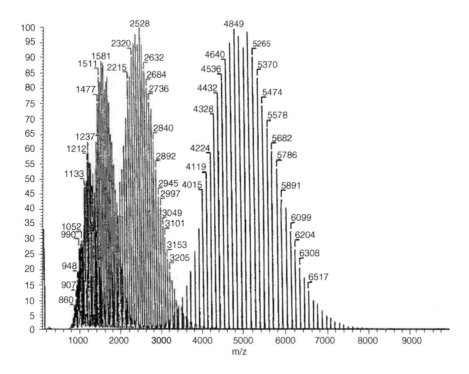

FIGURE 2.11
FD mass spectrum of a polystyrene sample (reproduced with permission from Ref. 34).

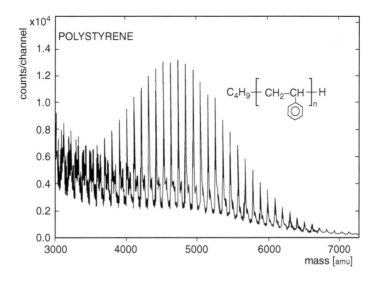

FIGURE 2.12
TOF-SIMS mass spectrum of a polystyrene sample (reproduced with permission from Ref. 35).

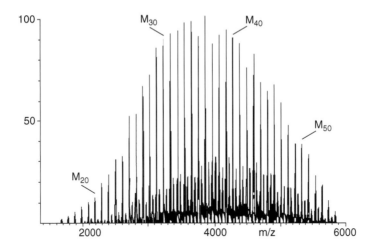

FIGURE 2.13
DCI mass spectrum of a polystyrene sample (reproduced with permission from Ref. 36).

data were derived from the mass spectra using Eqs. (2.1) and (2.2). The resulting values were in agreement with $\overline{M}n$ and $\overline{M}w$ obtained by conventional methods, and they concluded that electrospray-FT is well-suited for the analysis of anionic PEG samples.

Figure 2.14 reports the MALDI-TOF mass spectrum[38] of a polymethylmeth acrylate with a narrow distribution centered at about 15000 g/mol. The mass spectral intensities were used to compute $\overline{M}n$ and $\overline{M}w$ of the sample, and these values were compared with the values obtained by osmometry, SEC, and viscometry. The agreement is good and it further demonstrates the usefulness of MS in the case of anionic polymers.

Mass spectrometry is useful also in those cases in which the scope of the analysis is not simply to measure $\overline{M}n$ and $\overline{M}w$ of the sample, but the determination of the entire MMD.[39–43] For instance, in the case of polymers obtained by Pulsed Laser Polymerization, the determination of $\overline{M}n$ and $\overline{M}w$ of the sample is not the primary goal. In fact, one has to determine the point of inflection M_{inf} (see chapter on MALDI) and MS can be used for this purpose. One plots the oligomer's abundance against mass, then takes the second derivative.[38] The point at which the second derivative changes from negative to positive is the point of inflection.

Also in the case of polymers made principally of cyclic molecules the determination of $\overline{M}n$ and $\overline{M}w$ of the sample is not the primary goal. In fact, one has to determine the "trend" followed by the MMD and whether the abundance of cyclic molecules changes as the ring size grows.[16,17] MS can be used for this purpose, by fitting the logarithm of the abundance against log(n) and fitting the points with a straight line. The slope of the line is the exponent.[44] Although any ionization technique may be used, most of the investigations

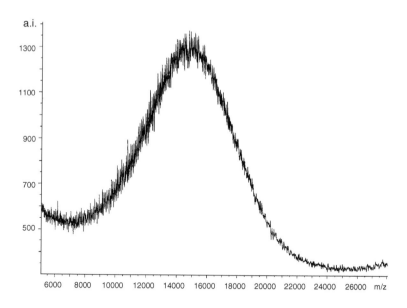

FIGURE 2.14

MALDI-TOF mass spectrum of a PMMA sample (reproduced with permission from Ref. 38).

on cyclic oligomer distributions have been performed by FAB-MS, and therefore the topic is discussed in Chapter 7.

The analysis of polydisperse polymers by MS methods poses some problems that have been only recently solved. Polydisperse polymers (including many industrial polymers) are made of a mixture of macromolecular chains that have quite different sizes, ranging from dimers and trimers, up to chains with thousands of units. Using MS methods, all the oligomer chains can be ionized, but the ratio between the number of ions of a given size and the number of molecules of that size (i.e., the ion yield as a function of chain size) is not constant. Actually, the ion yield decreases with chain length, implying that Eqs. (2.1) and (2.2) cannot be reliably used to compute $\overline{M}n$ and $\overline{M}w$.

Therefore, in the case of polydisperse polymers, the usual method cannot be applied and a modified method must be adopted. The variant will be discussed to some extent in Chapter 10 (MALDI-TOF MS), and therefore only a brief account is given here. It consists in fractionating polydisperse polymers by SEC and collecting the fractions. These fractions, which possess a narrow MMD, are then analyzed by MALDI-TOF MS and the MM averages of each fraction are computed. Since the fractions are monodispersed, ion yield problems do not exist, and the MM values thus obtained are correct. When the $\overline{M}n$ of each fraction is plotted vs. the corresponding SEC elution volume, an accurate calibration for the chromatogram is obtained. The trace and the calibration are fed into the SEC software that yields as an output $\overline{M}n$ and $\overline{M}w$ of the polymer.

2.3 Structure and End-Groups

Polymers can display a variety of structures, including linear and branched chains, copolymers with different sequences, and star polymers with different numbers of arms. Due to the variety of possible structures, the process of analyzing a polymer has to answer several questions and proceeds by steps.

The first step deals with the determination of the chemical structure of the backbone. The second step consists in finding out if the chains possess branching points and in the determination of the degree of branching. The third step involves finding out which end-groups lie at the chain ends. The last step entails detecting the cyclic oligomers that may be present in a linear (or branched) polymer sample. Using MS for elucidating the polymer structure, one can obtain the chemical structure of the backbone and detect the end-groups and the cyclic oligomers, all at one time. In fact, polymers possessing different structures yield mass spectra that are different for the position (mass numbers) at which MS peaks appear.

2.3.1 Structure Determination

The polymer structure determination consists in finding out the chemical units that constitute the backbone and the type of chemical bonds linking the monomers together. Elemental Analysis yields the empirical formula of the polymer. However, it does not allow guessing the structural formula (there are many possible compounds that possess approximately the same empirical formula).

MS can be applied to structure analysis and polymer identification, since the mass spectrum of a polymer possesses "characteristic" peaks that are peculiar to a polymer structure.[45,46] The determination of polymer structure by MS can be performed also on insoluble and intractable samples. This is a particularly attractive feature of MS, since it is more versatile than other techniques that strictly require sample dissolution. Often one records the mass spectrum of the polymer using an instrument equipped with a hard-ionization source such as electron impact (EIMS) or secondary ion (SIMS) source operated at high energy. Hard-ionization produces ion fragments and fragmentation patterns that are usually highly specific for a polymer type and therefore are useful for determining the polymer structure. Recently, it has been shown[47,48] that the post source decay (PSD) MALDI mass spectrum of the polymer may display fragmentation patterns very similar to those seen in EI. Thus PSD can be useful in finding out the structure of the repeat unit.

Furthermore, mass spectra of two polymers possessing different repeat units will produce widely different mass spectra, since the spacing between peaks will be different. Thus, in poly(ethylene glycol) the spacing is 44.05 g/mol., in poly(lactic acid) the spacing is 72.1 g/mol., in poly(dimethyl siloxane) the spacing is 74.1 g/mol., in poly(ethylene terephthalate) the spacing is 196.2 g/mol., in poly(butylene terephthalate) the spacing is 220.2 g/mol. This peculiar feature of polymer mass spectra can be used for polymer identification (see Chapters 5 and 8). In some cases, soft-ionization techniques can be useful for structure elucidation and polymer recognition.

The method, referred to as "high resolution MS" or "atomic composition MS," has been developed to identify small molecules. First, one records the MS of the sample and measures the mass of the ions with high accuracy down to the third (and even to the fourth) decimal digit. Thereafter one generates (preferably by an isotope calculation program) the masses of all compounds $C_xO_yN_wH_z$, where x, y, w, z are integers, until the generated mass coincides with the observed mass.[50]

Double-focussing magnetic sector mass spectrometers were originally used (and are still used) for "atomic composition" analysis. Spectrometers equipped with a high-field magnet FT-ICR ion selector are nowadays also used for such kind of measurements. The instrument has a very useful option (narrow band detection) that allows recording a small portion of the spectrum (a width of few Daltons) at high resolution, so that one can apply the "atomic composition" method.[50] Various examples (see Chapter 9) can be found of mass spectra of polymers recorded on a high-field magnet FT-ICR instrument.

"Atomic composition" MS becomes less definitive at higher masses. This is due to the fact that the number of formulas that yield masses close to the measured mass increases dramatically as the ion mass increases.

Discriminating branched and star polymers from linear ones is sometimes possible using MS. A typical limitation occurs in the case of polyolefins, where MS is of little help since branched and linear polyolefins have the same mass. However, when the polymers possess end-groups different from hydrogen atoms, and the spectra are mass resolved, the mass of branched polymers differs from linear ones. In fact, trifunctional (or multifunctional) units have different masses than the corresponding linear polymer (this circumstance often occurs in grafted copolymers). Figure 2.15 shows the MALDI-TOF mass spectrum of a commercial polyester[49] produced from a dicarboxylic acid (1–4 ciclohexane dicarboxylic acid), a diol (neopentylglycol), and a triol (trimethylolpropane). The peaks due to branched chains are indicated by a star.

Discriminating branched and star polymers from linear ones can always be achieved by measuring the properties in dilute solution. In fact, molecules having the same molar mass but different macromolecular architectures exhibit different transport and light scattering properties. More specifically, a branched macromolecule is more compact than a linear molecule having the same molar mass, and therefore it will display less friction and will diffuse more easily in the solvent. Viscometry can be used to detect branched structures, since the Mark-Houwink-Sakurada exponent (Eq. 2.23) for branched and star-shaped polymers is lower than that for the corresponding linear chain. Unfortunately, in order to measure the difference, one must have a sample made exclusively

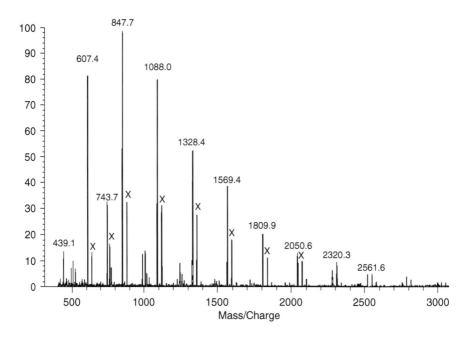

FIGURE 2.15

MALDI-TOF mass spectrum of a commercial polyester which is made of a mixture of linear and branched chains. The peaks due to branched chains are indicated by a star (x) (reproduced with permission from Ref. 49).

of linear chains, which may be of limited availability. Furthermore, the determination of the MHS exponent may be labor intensive, since it may imply using narrow polymer fractions obtained by fractionation.

LS can be used to detect branched structures. In fact, one of the polymer properties that can be extracted from the Zimm plot is the radius of gyration,[1] $\bar{R}g$. Comparing $\bar{R}g$ to $\bar{M}w$ can provide a measure of the average degree of branching.[13]

MS can be used to characterize hyperbranched polymers and dendrimers, since it is able to discriminate between acyclic and cyclic hyperbranched polymer chains, and it enables the investigator to follow the various stages of dendrimer growth.[51–56]

2.3.2 End-Group Determination

End-group identification, including species present to minor amounts in a polymer sample, is so crucial in polymer analysis that its importance cannot be overemphasized. Knowledge of the structure of chain end-groups may yield indications on the synthetic process, and it is essential for performing chemical modifications of the polymer structure by further reactions.

Determining the amount of end-groups in a polymer sample leads to the MM estimate. Wet chemistry methods can be used for end-group quantitation.[2-4] For instance, carboxyl end-groups content can be measured by conductometric titration, amino end-groups can be estimated by reacting the polymer with 2,4-dinitro-fluorobenzene and then recording the UV-visible spectrum at two different wavelengths.[2,3] However, wet chemistry methods are time-consuming, are specific for one particular kind of end-group, and are limited to low MM.

The NMR technique[7] is very often used for end-group determination. The method is quite general because it does not depend on a particular kind of end-group, although the detection of OH end-groups often presents a problem in ^{1}H-NMR.

However, NMR determination of end-groups is limited to low MM. As the size of the macromolecular chain increases, the signal due to the polymer backbone becomes larger and larger and the intensity of the signal due to end-groups in the NMR spectrum becomes buried in the noise.

Another disadvantage of NMR is that cyclic oligomers do not have end-groups, and therefore only the polymer backbone signal is seen in the NMR spectrum. This causes errors in the MM determination by NMR as well.

Mass spectrometry is able to look at the mass of individual molecules in a mixture of homologs, and therefore it is specifically suited for end-group analysis. In fact, the latter has been one of the most popular applications of MS to polymer analysis, and it can be carried out virtually with any MS ionization technique (see Chapters 4, 6, 7, 8, 9, and 10).

The general structure of the ions detected by MS is of the type:[57-60]

$$G1\text{-}AAAAAAA\text{-}G2 \dots C^+$$

where G1 and G2 are end-groups, C is a proton or a cation, and A is the repeat unit.

End-group determination by MS is done as follows. One considers the mass number of one of the MS peaks, subtracts the mass of C, then subtracts many times the mass of the repeat unit, until one obtains the sum of the masses of G1 + G2. Linear best-fit can also be used to find the sum of the masses of G1 + G2. As a simple example, Figure 2.11 is the FD spectrum of a polystyrene sample. The individual peaks are molecular ions (M^+), and the repeat unit is C_8H_8, with a mass of 104.15 Da. If one subtracts an integral number of monomer units from the observed molecular ion, a "residual" mass of 58 Da is obtained (e.g., $4849 - 46 \times 104.15 = 58$). This represents the sum of the end-group masses. A mass of 58 Da is consistent with a butyl group on one end of the chain and a hydrogen on the other: $H\text{-}(St)_n\text{-}C_4H_9$. In this instance, the polymer was prepared anionically using *n*-bytyl lithium.

One should note that the lowest residual mass (in this case 58 Da) is not necessarily the correct sum of the end-group masses. That is, in principle the sum of the end-group could be 58 Da plus one or more monomer units

(58, 162, 266 Da, etc.). One may need additional information to elucidate the correct end-groups; for example, other spectroscopic or chromatographic data and/or a knowledge of the synthetic procedure that was used.

Synthetic polymers are often made of a mixture of chains with different end-groups (polymers with end-group heterogeneity), and it is of interest to establish the relative abundance of each species.

In the case of high molar mass oligomers, the weight of the end-group is small with respect to the weight of the oligomer and therefore the number of ions is approximately proportional to the relative abundance of chains with different end-groups, therefore allowing a quantitative estimate.

For instance, Figure 2.16 reports the MALDI mass spectrum of a PEG sample terminated by fluorescent end-groups.[48] It displays a large number

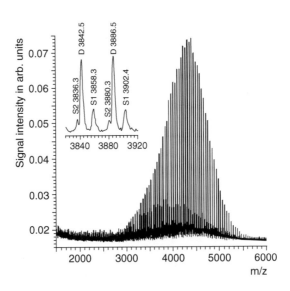

FIGURE 2.16
MALDI mass spectrum of a PEG sample with different types of end-groups (reproduced with permission from Ref. 48).

of peaks in the mass range 3000–5500 Da. The inset reports an expansion of the mass region 3800–3920 Da. Some peaks (S1 and S2) are due to PEG chains terminated with one fluorescent group, whereas other peaks (D) are due to PEG chains terminated with two fluorescent groups. This evidence demonstrates that 76% of the chains are doubly labelled.[48] The PEG sample was also analyzed by SEC equipped with a double detector (RI and UV), and the inspection of the trace[48] allowed the conclusion that 80% of the chains are doubly labeled.

In other cases, the number of ions does not reflect the relative abundance of the species, because a "selective suppression" phenomenon is occurring at the sample surface.[61,62] In such cases, different species are competing for space at the sample surface and some species are able to suppress the presence of others. This phenomenon has been observed, for instance, in the FAB mass spectra of exactly alternating copolyesters,[61] and in the MALDI mass spectra of a mixture containing polyether chains with and without fullerene in the main chain.[62]

Sometimes Tandem Mass Spectrometry[46–49] is used to analyze polymers that possess two different end-groups attached to the same chain. The MS/MS spectrum of the polymer then displays a series of peaks due to ion fragments that have experienced the loss of a segment containing one end-group, mixed to peaks due to fragments that have lost a segment containing the other end-group. By the analysis of the MS/MS spectrum, one may recover the sum of the masses of both end-groups and the individual mass of each end-group.

2.3.3 Stereoregularity

Stereoregularity is a feature of many synthetic and natural polymers. Among natural polymers, polysaccharides, proteins and nucleic acids are examples of stereoregular polymers. As far as synthetic polymers are concerned, stereoregularity is usually obtained in condensation polymers using optical active monomers that maintain their chirality along the macromolecular chain. In addition, polymer stereoregularity is achieved by controlling the reaction of double-bond opening (or of ring opening) in the monomers (stereoselection). Chains may possess different tacticity, being named isotactic or syndiotactic, or lack stereoregularity, being atactic. Stereoregularity is a different property from structural regularity. In fact, the repeat units in vinyl or olefin polymers may be connected in a head-to-tail mode (HT-HT) or in a head-head-tail-tail mode (HH-TT). When both type of connections are present along the same polymer chain, the polymer is said to be structurally irregular.

NMR is the method of choice to determine polymer stereoregularity,[7] whereas MS is notoriously not sensitive to this property. In the case of homopolymers, the application of NMR analysis to determining the stereosequences is straightforward.

For copolymers, instead, the analysis of the stereosequences is often complicated by the effect of copolymer sequence distribution. In these cases, a parallel MS investigation of the copolymer sequence distribution by MS (see Section 4) may be of help. MS is insensitive to the stereosequence, but it provides independent information on the sequence of the comonomer units, and this knowledge can be used to help interpret the NMR spectra.

Py-GC/MS is the only MS technique capable of yielding information on the stereoregularity of polymers. In the Py-GC/MS technique (Chapter 3), the polymer is partially pyrolyzed, and the products are then injected into a gas chromatography (GC) column using a carrier gas. The gas-liquid retention mechanism is such that short oligomers (dimers, trimers, tetramers) flow through the column in less time than long oligomers. The retention time also depends on the stereoregularity of the oligomers. At the end of the column, a detector measures the relative abundance of each stereo-oligomer.

2.4   Copolymer Composition and Sequence

Sequencing monomer distributions, determining the composition of copolymers, and relating them to material properties is one of the most common demands posed to polymer analysts. The concept of copolymer sequence may have a double meaning. In fact, in proteins, nucleic acids, and other biopolymers, the comonomer units are aligned into a well-determined sequence that is constant for each copolymer chain contained in a specific material. The sequence of amino acids in a protein is invariant, and it is usually called the "primary" structure of the protein.

In constrast, for synthetic copolymers (with 2, 3, or more components), the sequence of comonomer units present in each copolymer chain is no more constant. Furthermore, we can define only an average sequence length of alike monomers for these synthetic materials. Chain statistics quantities, such as the probability matrices (average number of identical repeating units) that are related to composition, to mechanism of copolymerization, and to reactivity ratios are useful in order to deal with this average copolymer sequence.

In the early 1940s when the polymerization theory was developed, the ideal, terminal, and penultimate models for the copolymerization were established; also the possible distribution laws for the monomer sequence along the copolymer chains were defined: Bernoullian, first- and second-order Markoffian.[63,64]

In a Bernoullian distribution the sequence of comonomer units is completely random, since all the comonomers enjoy the same probability of being added to the growing chain. In the Markoffian distribution the probability of being added to the growing chain is in favor of one of the comonomers, and therefore block or segmented copolymers are generated.

The kinetic approach[63,64] was the first to provide insights into the copolymer sequence, then modern NMR analysis provided a direct access to determine copolymer sequence by associating peak intensities to specific sequence probabilities in the Bernoullian or Markoffian statistics.[65-88] Both NMR[65-88] and MS[89-137] techniques allow determining the key chain statistics quantities as the molar fraction of A and B units in the copolymer, and the average number of alike adjacent repeating units. While NMR is an extremely powerful method, it cannot handle all the structural problems. For instance, the characterization of sequence arrangements in aliphatic condensation polymers having large comonomer subunits is not easily handled by current NMR methods, which otherwise proved of general utility in the case of vinyl, olefin, and diene copolymers.[70-80]

Determination of copolymer sequence by MS, meant to complement the information obtained from NMR, is a most desirable step toward mastering this problem, and this task has been recently accomplished.[114,115] It is now possible to calculate the copolymer sequence starting from the experimental mass spectra,[100-119] analogous to what is currently done in the case of NMR spectra.[65-88]

The analysis of copolymers is more complex compared to the analysis of homopolymers, since one has to determine the "usual" quantities (MM and MMD averages, structure of the repeat units, branch points, end-groups), plus additional quantities. These are: (i) the average molar fraction of *A* and B units in the copolymer; (ii) the number average length of *A* and *B* blocks; (iii) the variation of copolymer composition as the molar mass of the macromolecular chain grows (usually called "composition drift"); (iv) the weight of copolymer chains which possess a given mass and composition (referred to as the "bivariate distribution" with respect to molar mass and composition).

If one knows the polymerization process used to produce a copolymer, one can use the theory to predict the molar fraction of a given sequence in the resulting copolymer. When the copolymerization method varies, a different sequence distribution is to be expected.

The sequence distribution followed by copolymers obtained by condensation polymerization is that of Bernoulli statistics,[65-67] which produces random copolymers. In the Bernoulli distribution there is only one independent parameter, namely c_A (or c_B, since $c_A + c_B = 1$), which represents the molar fraction of A in the copolymer. The molar fraction of a specific sequence A_nB_m is given by $(c_A)^n(c_B)^m$, i.e., the product of c_A (taken "m" times) and c_B (taken "n" times), where m and n are the number of times A and B appear in the sequence. The number average lengths of like monomers are given[65-67] by:

$$\langle n_A \rangle = 1/c_B \qquad \langle n_B \rangle = 1/c_A \qquad (2.31)$$

Copolymers with three and four components contain oligomers of the type $A_mB_nC_p$, $A_mB_nC_pD_q$, respectively, where m, n, p, q are the numbers of each unit in the molecule. If the copolymer is obtained by condensation copolymerization, it follows Bernoulli statistics and the molar fraction of the sequence $A_3B_8C_4$ is $(c_A)^3(c_B)^8(c_C)^4$. Furthermore, the number average lengths of like monomers are given by:[66,67]

$$\langle n_A \rangle = 1/(1 - c_A) \qquad \langle n_B \rangle = 1/(1 - c_B) \qquad \langle n_C \rangle = 1/(1 - c_C) \qquad (2.32)$$

Block copolymers are produced when an anionic initiator is first reacted with monomer A and a homopolymer segment is formed. Then a second monomer B is added and the chain continues growing through the addition of the second monomer. This type of reaction is called "living copolymerization," since the chain propagation occurs without termination.[63] Varying the initiator to monomer ratio, one can regulate the block lengths.[63,64]

The conventional free radical copolymerization at low conversions yields copolymers that follow a first-order Markoff distribution.[65–67] According to this model, the sequence distribution of a two-component copolymer is completely defined when the four elements P_{AA}, P_{AB}, P_{BA}, P_{BB} of the probability matrix (P-matrix) are determined.[65–67] The P-matrix elements vary between 0 and 1, and are related by the following conditions:

$$P_{AA} + P_{AB} = 1 \qquad P_{BA} + P_{BB} = 1 \qquad (2.33)$$

The copolymer composition, c_A, associated with P-matrix elements is

$$c_A = P_{BA}/(P_{BA} + P_{AB}) \qquad (2.34)$$

The molar fraction of a sequence is given by a recursive formula,[65–67] which makes use of P-matrix elements. For instance the molar fraction of the sequence BAABA is: $c_B P_{BA} P_{AA} P_{AB} P_{BA}$. It can be shown that the number average lengths of like monomers are given[65–67] by:

$$\langle n_A \rangle = 1/P_{AB} \qquad \langle n_B \rangle = 1/P_{BA} \qquad (2.35)$$

In the copolymerization reactions, the affinity of the free radical growing chains for monomers A and B may be substantially different, and therefore the composition of the resulting copolymer may differ from the A/B ratio used in the feed. When one has copolymer samples for which the A/B ratio in the feed is known, one can determine the reactivity ratios (r_1, r_2).[63,64] Reactivity ratios regulate the copolymer composition as a function of the relative amounts of the comonomers in the reaction feed.[63,64]

The chain statistics method[65–87] to determine the reactivity ratios is based on the fact that the P-matrix elements are related to the molar fraction of

A units in the feed, f_A, as follows:

$$P_{AA} = f_B/(f_A + f_B/r_1) \tag{2.36}$$

$$P_{BB} = f_B/(f_B + f_A/r_2) \tag{2.37}$$

The above equation can be rewritten as:

$$r_1 = (f_B/P_{AB} - f_B)/f_A \tag{2.38}$$

and can be used to determine the reactivity ratios associated with the copolymerization of common monomers, for instance methyl methacrylate (MMA) and butyl acrylate (BA).[84]

The measurement of reactivity ratios is important from a practical point of view, since it allows knowing which feed must be used to synthesize copolymers with a given composition. If the copolymerization reaction is carried out up to high conversions, one must take into consideration that one monomer is consumed more quickly than the other, and the feed composition varies as the conversion increases.

The overall copolymer composition is obtained by integrating the instantaneous composition over conversion.[63–67,70–87] Figure 2.17 reports an illustrative calculation for acrylonitrile/butadiene copolymerization at high conversion. At low conversions, the molar fraction of acrylonitrile in the copolymer is about 0.5 and differs considerably from the molar fraction of acrylonitrile in the feed, which is about 0.8. As the conversion increases, the molar fraction

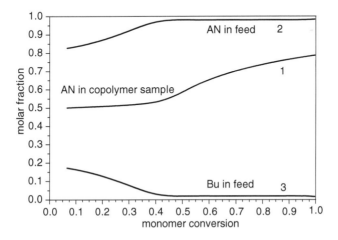

FIGURE 2.17
Acrylonitrile/butadiene copolymerization at high conversion. Expected molar fraction of acrylonitrile in the copolymer (trace1), molar fraction of acrylonitrile in the feed (trace2), and molar fraction of butadiene in the feed (trace3).

of butadiene in the feed falls and becomes vanishingly small. The feed becomes richer and richer in acrylonitrile, and the copolymer composition drifts slowly toward higher acrylonitrile contents. At 100% conversion, the molar fraction of acrylonitrile in the copolymer reaches the value of about 0.8, coincident with the initial composition of the feed.

Copolymers can also be obtained by allowing macromolecules to react.[86,87] Considerable interest has arisen in the reactive blending of polyamides, polycarbonates, polyesters (and also combinations of the above) and in the exchange reactions that may occur during the melt mixing processes.[86,87] The exchange reactions leading to the formation of copolymer molecules from two homopolymers can be seen as the result of two consecutive processes. First is the formation of the copolymer from homopolymers by intermolecular exchange, and then rearrangement of the copolymer (by intramolecular exchange) to a different sequence of A and B units along the copolymer chain:

$$-A-A-A-A + -B-B-B-B- \rightarrow A-A-A-A-B-B-B-B- \qquad (2.39)$$

$$-A-A-A-A-B-B-B-B- \rightarrow -A-B-B-A-A-B-B-A-B- \qquad (2.40)$$

A copolymer sample produced by reactive blending of two homopolymers is made of three components: the two homopolymers and the copolymer formed. As the reaction goes on, the two homopolymer sequences disappear and the copolymer sequences become abundant.

Kotliar[86] developed a model for this exchange process. He showed that $\langle n_A \rangle$ and $\langle n_A \rangle$ decrease as the process goes on. In the later stages, the sequence distribution exhibits a change, namely it goes from Markoffian to Bernoullian.[86] The extent of exchange, EE, is defined as :

$$EE = P_{BA} + P_{AB} \qquad (2.41)$$

One can obtain EE by measuring the value of the P-matrices P_{BA} and P_{AB}. In the initial stages the EE value is close to zero, then it grows and, at the final stages, it may reach the value one.

Another interesting case[85] is the ring-opening copolymerization of lactones (AA) and depsipeptides (AB), where the feed is made of AA and AB structures. At first glance, the copolymerization should lead to a random copolymer, but this is not the case. Schematically, the reaction is:

$$AA + AB \rightarrow (AA)_X(AB)_Y \qquad (2.42)$$

This copolymerization reaction is peculiar, since the synthetic route adopted to produce the copolymer is such that the abundance of the triad ABB and of the pentads BBBAB, BBABA, ABABB, BBABB is zero, and therefore the Bernoullian distribution is not followed.[122]

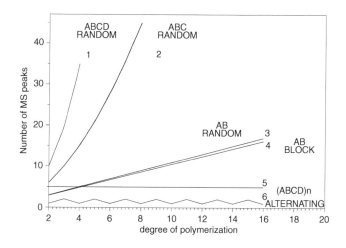

FIGURE 2.18

Number of peaks expected in the mass spectrum vs. the length of the chain for various types of copolymers.

2.4.1 Mass Spectrometric Methods

Mass spectrometry is able to look at the mass of individual molecules in a mixture of homologs, and this is of great advantage in copolymer analysis. Mass spectra of copolymers often display a large number of peaks due to the different oligomers. It is generally observed that the number of peaks in the mass spectrum increases as the number of different repeat units grows, and the number of peaks increases as the size of the copolymer chain becomes longer and longer. Figure 2.18 shows the number of MS peaks that are expected in the mass spectrum of different copolymers, namely for a random copolymer containing four different units (trace 1), for a random copolymer containing three different units (trace 2), for a random copolymer containing two different units (trace 3), for a block copolymer containing two different units (trace 4), for a sequential copolymer of the type $(ABCD)_n$ (trace 5) and for an alternating copolymer of the type $(AB)_n$ (trace 6). From Table 2.2 it can be seen that we expect 5 tetramers and 6 pentamers in a random AB copolymer. On the other hand, we expect 15 tetramers and 21 pentamers in an ABC copolymer and 35 tetramers and 56 pentamers in an ABCD copolymer.[115]

The intensities of the peaks appearing in the mass spectrum of a copolymer are directly bound to the relative abundance of oligomers present in the copolymer. Therefore, MS peak intensities can be used to determine the copolymer composition, provided that the ionization method used to desorb and ionize these oligomers does not produce significant ion fragmentation.[118] A certain extent of ion fragmentation actually occurs in nearly all the MS techniques (Chapter 1). However, if the extent of ion fragmentation is independent of the oligomers molar mass (at least within the short mass interval used for the

TABLE 2.2

Number of Peaks Corresponding to Specific Oligomers
in the Mass Spectra of Random Copolymers

Oligomer	AB	ABC	ABCD	ABCDE
Dimers	3	6	10	15
Trimers	4	10	20	35
Tetramers	5	15	35	70
Pentamers	6	21	56	126
Hexamers	7	28	84	210
Heptamers	8	36	120	330
Octamers	9	45	165	495
Nonamers	10	55	220	715
Decamers	11	66	286	1001
11-mers	12	78	364	1365
12-mers	13	91	455	1820
13-mers	14	105	560	2380
14-mers	15	120	680	3060
15-mers	16	136	816	3876

measurements; see below), the determination of the sequence in copolymers can be successfully carried out by MS.[115–133] Meeting this essential condition means that the intensities of the peaks appearing in the mass spectrum reflect the relative abundances of the oligomers present in the copolymer.

After recording the mass spectrum of the copolymer, one finds an assignment for each spectral peak. Usually this task is done by hand, although computer programs are available to automatically find MS peak assignment[89] and to determine the isotopic distribution.[121,122]

Once the table of assignments is generated, one may give a quick glance to the most intense peaks in the mass spectrum and derive a gross estimate for the copolymer composition. Figure 2.19 shows the mass spectrum of the cluster due to oligomers with 40 repeat units, related to a 70/30 AB random copolymer. The most intense peaks in the mass spectrum are $A_{30}B_{10}$, $A_{29}B_{11}$, $A_{28}B_{12}$, $A_{27}B_{13}$ and $A_{26}B_{14}$, which possess associated compositions of 75/25, 73/27, 70/30, 67/33, 65/35, respectively. Thus the copolymer composition falls between 65/35 and 75/25, the most probable value being 70/30. The method is extremely attractive for its simplicity.

If one knows that the sample under consideration is a block copolymer, then assignment of the MS peak is revealing, in the sense that it shows the sequences. This is due to the fact that in a block copolymer A_mB_n means a block of "m" consecutive units of A followed by a block of "n" consecutive units of B. In this case, the method also yields the number average lengths of like monomers $\langle n_A \rangle$ and $\langle n_B \rangle$:

$$\langle n_A \rangle = c_A \, \overline{M}n/m \tag{2.43}$$

$$\langle n_B \rangle = c_B \, \overline{M}n/m \tag{2.44}$$

where "m" is the average between the masses of the two repeat units.

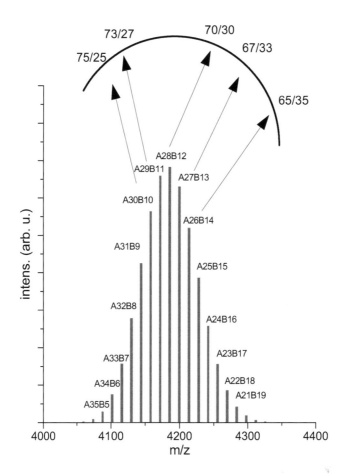

FIGURE 2.19
Expected mass spectrum of the cluster due to oligomers with 40 repeat units, related to a 70/30 AB random copolymer.

Two general methods have been developed for the determination of copolymer composition and sequence, namely a method which uses a combination of mass spectral intensities[106,115,116] and the chain statistics approach.[115,117,121,131]

In the approach based on a combination of MS intensities one computes the average copolymer composition (molar fraction of A units in the copolymer), c_A:

$$c_A = (1/Y_1) \sum \sum m I(A_m B_n) \tag{2.45}$$

where Y_1 is $\sum\sum (m + n)I(A_m B_n)$, the summations are over m and n, $I(A_m B_n)$ is the mass spectral intensity of the oligomer $A_m B_n$. The above equation is referred to as the "composition estimates" equation.[115,116]

TABLE 2.3

Composition Estimates for an AB Copolymer

Oligomers	Symbol	Formula
Monomers	c_1^A	$I(A)/\{I(A)+I(B)\}$
Dimers	c_2^A	$\{2I(A_2)+I(AB)\}/\{2I(A_2)+2I(AB)+2I(B_2)\}$
Trimers	c_3^A	$\{3I(A_3)+2I(A_2B)+I(AB_2)\}/$ $\{3I(A_3)+3I(AB_2)+3I(A_2B)+3I(B_3)\}$
Tetramers	c_4^A	$\{4I(A_4)+3I(A_3B)+2I(A_2B_2)+I(AB_3)\}/$ $\{4I(A_4)+4I(A_3B)+4I(A_2B_2)+4I(AB_3)+4I(B_4)\}$

Table 2.3 reports the composition estimates equations for monomers, dimers, up to tetramers and it can be seen that the numerator has "s" terms whereas the denominator has "s + 1" terms. The composition estimates for *ABC* and *ABCD* copolymers (copolymers with three and four components), are reported elsewhere.[115]

This approach has also been used to estimate composition heterogeneity in low MM copolymers.[117–119] The process is comprised of two steps. In a first step, one computes the "composition estimates," $c_A^{(1)}, c_A^{(2)}, c_A^{(3)}, c_A^{(4)}, c_A^{(5)}, c_A^{(6)},\ldots$, which represent the molar fraction of A units in all oligomers of size 1, 2, 3,….

$$c_A^{(s)} = (1/Y_2)\sum mI(A_mB_n)\delta_{m+n} \tag{2.46}$$

where $I(A_mB_n)$ is the mass spectral intensity of the oligomer A_mB_n, and where the summation is over m and where δ_{m+n} is equal to 1 when the size of the oligomer is $(m + n)$ and δ_{m+n} is equal to 0 otherwise and where Y_2 is $\Sigma\,(m + n) \times I(A_mB_n)\,\delta_{m+n}$. Thereafter, one plots $c_A^{(s)}$ vs. chain size (s). In this way, one can directly determine if the copolymer composition is homogeneous by measuring how the composition varies as the chain length grows.

The second method, based on chain statistics, has been widely applied to the analysis of copolymers (see Section 2.4.2). It consists in fitting the observed mass spectral intensities with the intensities derived from model sequence distributions. The relative abundance of all the oligomers of a defined chain length (dimers, trimers or higher oligomers) reflects the composition and monomer sequence present in the copolymer.[113–117] Thus, the estimate of sequence might be done restricting the analysis only to one group of oligomers. Of course, it is good practice to take the average of the single estimates (see below), and to keep in mind that higher oligomers are much more sensitive to subtle sequence differences.[115–122]

When using a spectroscopic technique to obtain the copolymer sequence and composition, the essential step is to generate a theoretical spectrum, to be compared with the experimental one. The chain statistics approach allows discriminating among different sequence distribution models.[113–117] The process is described in Table 2.4. For each copolymer composition, an arrangement

TABLE 2.4

Modeling Process Based on Chain Statistics

---

1. Choosing Distribution Models:

If one considers a number of different distribution models, the oligomers' abundances
can be generated according to the selected model (Bernoullian, Markoffian 1° and
2° Sequential).

2. Generating Theoretical Spectra:

Theoretical spectra corresponding to a specific copolymer composition and sequence
can be generated by assuming that peak intensities reflect relative oligomers'
abundances.

3. Iteration and Best Fit Minimization

The experimental spectrum is obtained. A series of theoretical spectra are originated
and, by comparing the experimental peak intensities with those calculated for a
specific model, the most likely copolymer sequence and composition can be
determined.

---

of comonomer units along the chain is generated, according to a predefined
model. Starting from any sequence, a theoretical spectrum can be generated,
based on the assignment of each mass peak to a set of sequential arrangements of monomers.

The quantity to be minimized is the agreement factor (AF):

$$\text{AF} = \left\{ \sum \{[I(A_m B_n) - I_{\text{expe}}]^2\} / \sum [I_{\text{expe}}]^2 \right\}^{1/2} \qquad (2.47)$$

where I_{expe} and $I(A_m B_n)$ are the experimental and the theoretical molar fraction
of copolymer chains of the type $A_m B_n$ and where the summation spans over
all mass spectral peaks considered. The best-fit minimization procedure
follows the scheme described (the parameters of the model are varied iteratively
until convergence occurs), and yields copolymer composition and sequence.
Following this iter, one records the spectrum, selects a model, compares the
experimental and theoretical intensities, performs a best fit minimization,[115–122] finds the minimum, and writes down the result. Then one selects
a different model, finds the minimum, and writes down the result. At the
end, one selects the model that gives the best result.

A problem frequently encountered in copolymer analysis is that the MS
peaks can be assigned to two or more isobaric structures. In this case, the
peak intensity experimentally observed comes from several contributions.
An automated procedure to find composition and sequence of the copolymers analyzed has been developed to cope with this problem of determining
the sequence of copolymers when a mass spectroscopic peak has a multiple
structural assignment.[113–117]

Monte Carlo simulation represents an alternative approach to generating a macromolecular chain of infinite length, having a sequence of comonomer units along the chain.[136,137] In this method, once a theoretical sequence (called "linear strip") has been generated, one may simply count the type and frequency of oligomers present in the linear strip and compare them with those deduced from the experimental mass spectrum of a copolymer sample.[136,137] Generating a series of linear strips by the Monte Carlo method, and matching them with the experimental spectra, is much more complex as compared to using the chain statistics models. In spite of this, the Monte Carlo method has distinct advantages in certain cases.[136,137]

The Bernoulli statistics predict[114,115] that the molar fraction, $I(A_mB_n)$, of the oligomer A_mB_n is given by:

$$I(A_mB_n) = \{(m + n)!/[(m)!\,(n)!]\}\,(c_A)^m(c_B)^n \tag{2.48}$$

where c_A and c_B are the molar fractions of A and B units. The above equation is the well-known Newton formula and it predicts that the most abundant oligomer is that with x units of the type A and y units of the type B, i.e., the oligomer A_xB_y, where $x = (m + n)c_A$ and $y = (m + n)c_B$.

Copolymers with three and four components contain oligomers of the type $A_mB_nC_p$, $A_mB_nC_pD_q$, respectively, where m,n,p,q are the number of units in the molecule.

The Bernuolli model predicts[114,115,121] that the molar fraction, $I(A_mB_nC_p)$, of oligomer $A_mB_nC_p$ is given by:

$$I(A_mB_nC_p) = g_{ABC}(c_A)^m(c_B)^n(c_C)^p(c_D)^q \tag{2.49}$$

whereas the molar fraction, $I(A_mB_nC_pD_q)$, of oligomer $A_mB_nC_pD_q$ is given by:

$$I(A_mB_nC_p) = g_{ABCD}(c_A)^m(c_B)^n(c_C)^p(c_D)^q \tag{2.50}$$

where c_A, c_B, c_C, c_D are the molar fractions of A, B, C, D units in the copolymers and $g_{ABC} = (m + n + p)!/[(m)!(n)!(p)!]$; $g_{ABCD} = (m + n + p + q)!/[(m)!(n)!(p)!(q)!]$. The above equations are referred to as the Liebniz formulas.

Equations 2.48 through 2.50 allow one to generate theoretical mass spectra, which show identical peak intensity patterns for any random copolymer of a given composition, a remarkable result. Figure 2.20 shows the theoretical mass spectrum for an equimolar *AB* copolymer, and Figure 2.21 reports those for three and four component copolymers that follow Bernoullian statistics.

The theoretical mass spectrum for the two component copolymer (Figure 2.20) consists of 3 dimers, 4 trimers, 5 tetramers, and 6 pentamers. The one for the three-components copolymer (Figure 2.21a) consists of 6 dimers, 10 trimers, and 15 tetramers. In the theoretical mass spectrum of the four component copolymer (Figure 2.21b) dimers and trimers are reported exclusively, since the number of tetramers is too high (Table 2.2).

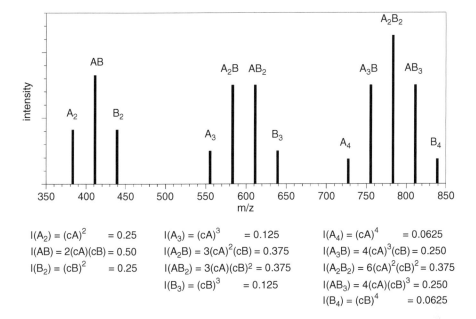

$I(A_2) = (cA)^2 \quad\quad = 0.25$
$I(AB) = 2(cA)(cB) = 0.50$
$I(B_2) = (cB)^2 \quad\quad = 0.25$

$I(A_3) = (cA)^3 \quad\quad\quad = 0.125$
$I(A_2B) = 3(cA)^2(cB) = 0.375$
$I(AB_2) = 3(cA)(cB)^2 = 0.375$
$I(B_3) = (cB)^3 \quad\quad\quad = 0.125$

$I(A_4) = (cA)^4 \quad\quad\quad\quad = 0.0625$
$I(A_3B) = 4(cA)^3(cB) = 0.250$
$I(A_2B_2) = 6(cA)^2(cB)^2 = 0.375$
$I(AB_3) = 4(cA)(cB)^3 = 0.250$
$I(B_4) = (cB)^4 \quad\quad\quad\quad = 0.0625$

FIGURE 2.20
Theoretical mass spectra for an AB random copolymer which possess a 50/50 composition, repeat units 172 and 200 g/mol. The peak intensities calculated according to Eq. 2.48 are also shown.

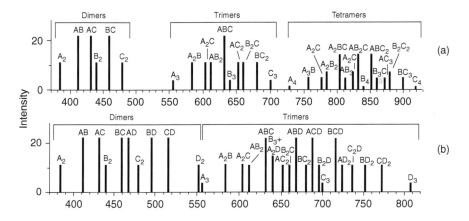

FIGURE 2.21
Theoretical mass spectra (a) for an ABC random copolymer which possess a 33/33/33 composition, repeat units 172, 200, and 220 g/mol (b) for an ABCD random copolymer which possess a 25/25/25/25 composition, repeat units 172, 200, 220, and 256 g/mol.

Figure 2.22a reports the DCI mass spectrum of a microbial copolyester, poly(β-hydroxybutyrate-co-15% β-hydroxyvalerate)(PHB/HV).[135] The relative intensities of several series of oligomers in the spectrum, from dimers to hexamers, are compared with theoretical intensities (Figure 2.22b). Each series

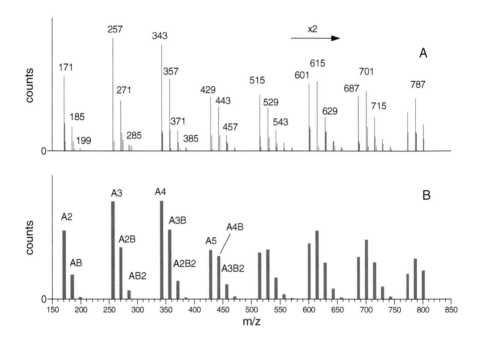

FIGURE 2.22

DCI mass spectrum of a copolymer containing units of HB and HV. Experimental mass spectrum (A) and theoretical mass spectrum (B) (reproduced with permission from Ref. 135).

of oligomers (dimers, trimers, etc.) allows an independent calculation of the copolymer composition and sequence distribution. The MS method provides an excellent way to evaluate the precision of these measurements.[135] The best matching of the experimental with the theoretical peak intensities was obtained for a Bernoullian distribution and a molar ratio of 85/15 between the two comonomers.[115] The random copolymer produced according to the Bernoulli model is composition homogeneous, i.e., the composition of the copolymer does not vary as the chain length increases.

The sequence distribution followed by copolymers produced by the conventional free radical processes at low conversions is the first-order Markoff distribution,[114–120] which possesses an associated P-matrix. This model predicts[114–116,118–120] that the molar fraction of dimers is given by:

$$I(A_2) = c_A \, P_{AA} \tag{2.51}$$

$$I(AB) = 2c_A \, P_{AB} \tag{2.52}$$

$$I(B_2) = c_B \, P_{BB} \tag{2.53}$$

The corresponding equations for trimers and tetramers can be found elsewhere, along with the prediction for the number average lengths of like monomers.[114–116,118–120] The Markoff model predicts that the resulting copolymer is composition homogeneous i.e., that the composition of the copolymer does not vary as the chain length increases.

Mass spectra of alternating and sequential copolymers[96,119] display a scarce number of peaks since some of the sequences are forbidden. For these polymers the theory predicts[115] that the molar fraction, $I(A_mB_n)$, of the oligomer A_mB_n is given by:

$$I(A_mB_n) = \sum \gamma/(m + n) \qquad (2.54)$$

where the summation is over all sequences with "m" A units and "n" B units and where γ is the number of times that the sequence appears in an oligomer of length $(m+n)$. The latter can be computed applying the "rate of occurrence" method.[115]

The chain statistics method is of particular value when it is necessary to discriminate between a pure copolymer sample and a sample made from a physical mixture of two copolymers. This is a frequent case, since commercial copolymers are often obtained by mixing two different copolymer batches.

Let us consider a mixture of two random copolymers of the same chemical structure. The quantities of interest are d_A and e_A, which represent the molar fraction of A units of the first and of the second copolymer, and X the molar fraction of the first copolymer in the mixture. The theory shows that the copolymer sequence distribution followed by mixtures of two copolymers is peculiar[113–115,118,119] (sequences due to the presence of both components of the mixture are present). The overall composition of the copolymer sample (c_A) will be intermediate between the compositions of the two components:

$$c_A = [X](e_A) + [1 - X]d_A \qquad (2.55)$$

The cited model predicts[113–115,118,119] that the molar fraction, $I(A_mB_n)$, of the oligomer A_mB_n is given by:

$$I(A_mB_n) = \{(m + n)!/[(m)!\,(n)!]\}\,G_3 \qquad (2.56)$$

where G_3 is given by $X(e_A)^m(1 - e_A)^n + [1 - X](d_A)^m(1 - d_A)^n$.

From the above compact formula, the explicit expressions for each oligomer can be derived.[113–115,118,119] The above equations have been used for the analysis of MS spectra of copolyesters produced by bacteria.[113–115]

2.4.2 Applications of the MS Method

Nuwaysir et al.[106] analyzed the LD mass spectra of a series of three copolymer samples containing units of Styrene (referred to as A) and MMA (referred to as B). They measured peak intensities, applied Eq. 2.20, and found $c_A = 0.063$, $c_A = 0.164$, $c_A = 0.215$, which are in reasonable agreement with the composition data available from conventional methods[106] ($c_A = 0.09$, $c_A = 0.19$, $c_A = 0.28$).

They also analyzed a series of three copolymer samples containing units of BMA (referred to as *A*) and MMA (referred to as *B*). They applied Eq. 2.20 and found $c_A = 0.117$, $c_A = 0.163$, $c_A = 0.184$ for the samples. The agreement with composition values available for the samples ($c_A = 0.09$, $c_A = 0.19$, $c_A = 0.25$) is acceptable.[106]

Vitalini et al.[104] considered the FAB spectrum of a copolymer containing units of acrylonitrile (referred to as *A*) and units of butadiene (referred to as *B*). They used a variant in which the mere presence or absence of an oligomer in the MS can be used to find the composition. The calculation yielded $c_A = 0.33$, which comes very close to the NMR value ($c_A = 0.31$).

Raeder et al.[105] analyzed the MALDI-TOF mass spectrum of a sulfonated styrene sample containing units of styrene (*A*) and styrene-sulfonic acid (*B*). They measured the intensity of the MS peaks, inserted the intensity values in Eq. 2.20 and found that $c_A = 0.06$, which implies that the average degree of sulfonation is 94%.

The Field Ionization mass spectrum of a copolymer containing units of acrylonitrile (*A*) and units of butadiene (*B*) was recorded.[115,132] The composition estimates for trimers, tetramers, up to 16-mers were plotted vs. chain length and the composition was found to be constant in the cited range.[115]

Wilczek-Vera et al.[102] applied the composition estimates approach to the MALDI-TOF mass spectrum of a block copolymer containing units of α-methyl styrene (referred to as *A*) and units of styrene (referred to as *B*). The mass spectral intensities were inserted into the corresponding equation, and the resulting composition was $c_A = 0.29$, which comes close to the value $c_A = 0.31$ derived from NMR. The MS intensities were inserted by these authors also into Eqs. (2.43) and (2.44) and the result was $\langle n_A \rangle = 20.4$ and $\langle n_B \rangle = 7.3$. This implies that styrene blocks are much longer than α-methyl styrene blocks.[102]

Ramjit et al.[91] used characteristic peaks in the EI spectrum to follow the formation of a copolyester by ester-ester exchange. In this case, MS intensities do not reflect abundances. Nevertheless, intensities can be corrected and the composition is derived.

Van Rooij et al.[109] analyzed the FT-ICR mass spectrum of a block copolymer containing units propyleneoxide (referred to as *A*) and ethyleneoxide (referred to as *B*), using the composition estimates approach. The result for the composition was $c_A = 0.85$ and the number average block lengths turned out to be $\langle n_A \rangle = 16.4$ and $\langle n_B \rangle = 0.2$. These values compared well with the number average block lengths measured by the manufacturer,[109] namely $\langle n_A \rangle = 15.57$ and $\langle n_B \rangle = 0.17$.

Lee et al.[111] considered the MALDI spectrum of a block copolymer containing units of ethyleneoxide (referred to as *A*) and units of lactic acid (referred to as *B*). They used Eq. 2.43, which connects $\overline{M}_n$ with the number average length and the result is $\langle n_A \rangle = 44.6$.

The equation, which gives the composition drift (Eq. 2.46), was applied to a series of five copolymers[128] containing units of acrylonitrile (referred to as *A*) and units of butadiene (referred to as *B*). The results[128] showed that the drift is very small. For instance, for trimers, tetramers, pentamers, and hexamers the following values were found: $c_A^{(3)} = 0.19$, $c_A^{(4)} = 0.20$, $c_A^{(5)} = 0.19$, $c_A^{(6)} = 0.20$.

Guttman et al.[103] analyzed the MALDI-TOF mass spectrum of a copolymer containing units α-methyl styrene (referred to as A) and units of methyl methacrylate (referred to as B). The mass spectrum spans over a wide mass range (2000–13000 Da) and it displays well-resolved peaks up to high masses. The sample is a reference polymer named SRM1487, obtained by anionic polymerization using a bifunctional initiator, the dimer of α-methyl styrene (sodium salt). The MS intensities were used by the authors to detect changes in the average number of A units in the chain, referred to as T. As the size grows, T first decreases, then it reaches a minimum, and then it grows again.[103] The authors compared the result of the calculation to the T values determined by an SEC apparatus with a double detector and the agreement found is excellent.

The sequence and composition MS analysis of copolymers with MM higher than 10,000 Da has not been reported. This shortcoming derives from the fact that mass spectrometers used in the analysis of copolymer often possess a limited resolution, much lower than that needed for high MM copolymers. For instance, in order to obtain a mass-resolved spectrum of a copolymer at 28,000 Da with units of MMA and of St (the two repeat units differ by 4 masses), the analyzer must have a resolution higher than 7000 up to 28,000 Da, which is a formidable problem for most MS machines.

This problem can be circumvented by subjecting the copolymer to partial degradation to reduce its molar mass (Figure 2.23 shows a scheme for the degradation process of a copolymer chain). As the degradation goes on, the length of the chain decreases and new end-groups are generated. However, the sequence of the partially degraded copolymer will be identical to the initial copolymer. The relative abundance of the oligomers formed is then measured directly by MS, allowing one to derive the sequence and composition of the original copolymer.[90–101]

Of course, one has to take into account whether the partial degradation process is occurring in a totally selective, partially selective, or nonselective

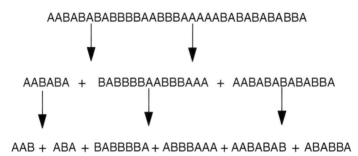

FIGURE 2.23
Scheme of the partial degradation of a copolymer.

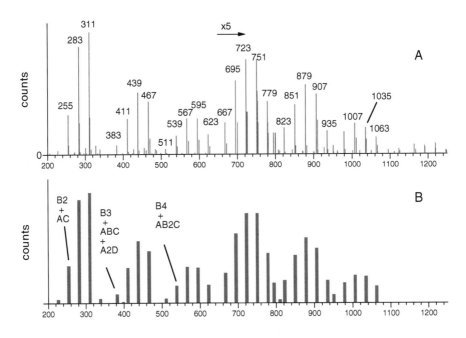

FIGURE 2.24
DCI mass spectrum of a copolymer containing four different repeat units. Experimental mass spectrum (A) and theoretical mass spectrum (B) (reproduced with permission from Ref. 135).

way, and adequate treatment has been developed for these cases (the non-selective degradation process being the simplest and most common case).[115]

A nonselective thermal cleavage occurred, for instance, in the pyrolytic process accompanying the recording of the DCI mass spectrum of a four-components random copolymer containing units of β-hydroxyvalerate (referred to as A), units of β-hydroxyheptanoate (referred to as B), β-hydroxynonaoate (referred to as C), and β-hydroxyundecanoate (referred to as D).[135] The authors applied chain statistics (Leibniz formula, Eq. 2.50) and found $c_A = 0.02$, $c_B = 0.67$, $c_C = 0.37$, $c_D = 0.02$, which matches well with values found by other methods. Figure 2.24a reports the experimental mass spectrum,[135] whereas Figure 2.24b reports the theoretical mass spectrum computed at the minimum.[135]

It is interesting to note that in the mass spectrum (Figure 2.24) there are many peaks that have masses corresponding to isobar structures, leading to ambiguous peak assignments. For instance B_2 and AC both contribute to the peak at $m/z = 255$, whereas B_3 and ABC both contribute to the peak at $m/z = 383$. The number of possible structures increases rapidly as the mass increases. For instance, the MS peak at mass 723 is the sum of three contributions, namely B_4C, AB_2C_2, A_2C_3.

Whereas the chain statistics approach[115] allows one to deconvolute the single contributions to peak intensity (by generating theoretical mass spectra corresponding to each composition and sequence), the composition estimates

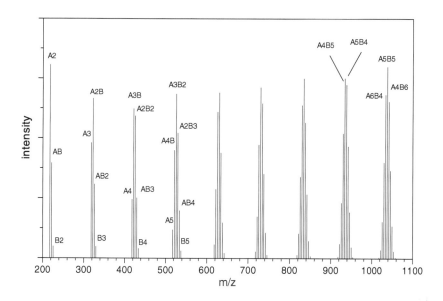

FIGURE 2.25
Theoretical mass spectrum of a random copolymer subjected to partially selective partial degradation.

approach gives incorrect results,[121] namely $c_A = 0.10$, $c_B = 0.50$, $c_C = 0.30$, $c_D = 0.10$. Most probably, this is due to the inability of the method (which is not based on a theoretical model) to account for the peak assignment uncertainties.

Figure 2.25 shows the theoretical mass spectrum of a random copolymer containing equimolar quantities of A and B units, which has been subjected to partially selective degradation. This means that adjacent A units (AA) are preferentially cleaved in the process, with respect to the AB or BB units contained in the copolymer chains. The molar fraction, $I(A_mB_n)$, of the oligomer A_mB_n is given by:[115]

$$I(A_mB_n) = \delta_{pref} \{(m + n)!/[(m)!\,(n)!]\}\,(c_A)^m(1 - c_A)^n \qquad (2.57)$$

where δ_{pref} takes into account that oligomers with end units A—A are preferentially produced in the cleavage.[115]

In Figure 2.25 the ratio among the peak intensities corresponding to dimers $(A_2\,AB,\,B_2)$ is strongly altered compared to the case of nonselective cleavage, where A_2 and B_2 would have the same intensity (see Fig. 2.20). The ratio among the peak intensities corresponding to $A_3\,A_2B,\,AB_2,\,B_3$ in Figure 2.25 is also altered: A_2B possesses a higher intensity compared to AB_2, whereas B_3 appears to have zero intensity. On the contrary, the peak clusters corresponding to nonamers and decamers are almost symmetrical, and appear therefore similar to the shape of the clusters generated by nonselective cleavage.[115] In detail, peaks A_4B_5, A_5B_4 have almost the same intensity, and

TABLE 2.5

MS Peak Intensity for a Totally Selective Cleavage
Process of an Equimolar AB Copolymer

Peak Intensity		
Number of Dimer Peaks	1	
A_2		$C_A{}^2$
AB		0
B_2		0
Number of Trimer Peaks	2	
A_3		$C_A{}^3$
A_2B		$C_A{}^2C_B$
AB_2		0
B_3		0
Number of Tetramer Peaks	3	
A_4		$C_A{}^4$
A_3B		$2C_A{}^3C_B$
A_2B_2		$C_A{}^2C_B{}^2$
AB_3		0
B_4		0
Number of Pentamer Peaks	4	
A_5		$C_A{}^5$
A_4B		$3C_A{}^4C_B$
A_3B_2		$3C_A{}^3C_B{}^2$
A_2B_3		$C_A{}^2C_B{}^3$
AB_4		0
B_5		0

the ratio among peaks A_6B_4, A_5B_{54}, A_4B_6 is that expected for a nonselective cleavage.[115]

The effect of partially selective cleavage is essentially that of altering the composition associated with dimers and trimers, whereas the composition associated with nonamers and longer chains does not differ from the nonselective cleavage values.[115] This result stresses the need to perform sequence studies on copolymers by analyzing the mass spectra of the higher oligomers.

In the case of totally selective cleavage, both the number and relative abundance of the oligomers generated are different with respect to the nonselective cleavage. Table 2.5 reports pertinent figures related to the case of totally selective process of an equimolar AB copolymer.[115] These numbers can be compared with those in Figure 2.20, relative to the case of nonselective cleavage.

The partial degradation method has given good results in the case of the MS analysis of aromatic copolyesters,[92,93] copolycarbonates,[94] aliphatic copolyesters of microbial origin,[95,99,100] copolyesters containing a photolabile unit in the main chain,[96,98] copolyamides,[97] and also of butadiene-styrene copolymers[101] and other copolymers.[90,91] Most of the investigations were performed by FAB and they are summarized in Chapter 7.

Plage and Schulten[132] applied the chain statistical method to the FD mass spectrum of a copolymer containing units of acrylonitrile (*A*) and units of butadiene (*B*). They assumed that the copolymer follows Bernoulli statistics and found that the copolymer composition was $c_A = 0.25$.

Spool et al.[133] recorded the TOF-SIMS spectrum of a copolymer with units of $-OCF_2-$ and $-OC_2F_4-$. The copolymer is commercially available under the trade name Fomblin Z and it is used in a variety of applications. They used Bernoulli statistics, generated a series of theoretical spectra and matched them with the experimental one. In order to have a more realistic comparison, they added a damping factor which depends on the chain size.

The Electrospray technique was employed to record[130] the mass spectrum of a copolymer containing units of β-hydroxybutyrate and β-hydroxyvalerate. The method based on chain statistics was employed and the minimization yielded an agreement factor AF = 6%.

Majumdar et. al.[124] recorded the EI mass spectrum of a copolymer with units of methyl isocyanate (referred to as *A*) and of butyl isocyanate (referred to as *B*). They considered MS peak intensities, applied chain statistics, and found that $c_A = .88$. They also computed the agreement factor, which turned out to be AF = 8%.

Servaty et al.[131] applied chain statistics to MALDI-TOF mass spectrum of a copolymer sample containing units of hydromethylsiloxane and dimethylsiloxane). The authors calculated the theoretical MS intensities using the Newton formula. They also determined the weight fraction of the chains that possess one functionalizable unit (hydromethylsiloxane), two functionalizable units, three functionalizable units, and so on. They found that the molar fraction of the chains, which do not possess any functionalizable unit at all, is by no means negligible[131] (it accounts for 25% of the total).

Zoller and Johnston[128] applied the above method to a copolymer containing units of acrylonitrile and butadiene. They used mass spectral data to discriminate between first-order Markoff and Bernoulli distributions. First they assumed Bernoulli, performed the best-fit, and found an agreement of AF = 0.11. Thereafter they assumed first-order Markoff and the minimization yielded AF = 0.09. As a consequence they concluded that first-order Markoff gives better results.

Abate et al.[141] analyzed the MALDI-TOF mass spectrum of a copolyester of microbial origin. They applied the Bernoullian statistics and performed a comparison between the experimental and the calculated MS intensities, as discussed in Chapter 10.

Suddaby et al.[125] recorded the MALDI-TOF mass spectrum of a series of five copolymers containing units of methyl methacrylate (MMA) and units of butyl methacrylate (BMA). Chain statistics was applied, yielding an estimate of the reactivity ratios (Eqs. 2.11 through 2.13), namely the $r_{MMA} = 1.09$,

$r_{BMA} = 0.77$, which agrees well with literature values and with the reactivity ratios determined by ^{1}H NMR ($r_{MMA} = 0.75$, $r_{BMA} = 0.98$).

Chen et al.[127] recorded the DCI mass spectrum of a copolymer with units of difluoromethyl isocyanate (referred to as *A*) and of trifluorobutyl isocyanate (referred to as *B*). They applied first-order Markoff statistics and found that the best P-matrix turned out to possess the following matrix elements: $P_{AA} = 0.99$, $P_{AB} = 0.01$, $P_{BA} = 0.56$, $P_{BB} = 0.43$.

Montaudo et al.[121] considered a group of three mass spectra of copolymers containing methyl methacrylate (MMA) and butyl acrylate (BA) units. The samples were polymerized using different MMA/BA molar ratios in the feed, namely 86.4/13.6; 73.8/26.2; 62.1/37.9. They used the mass spectral intensities and chain statistics to compute the reactivity ratios, which turned out to be $r_{MMA} = 2.1$, $r_{BA} = 0.7$.

The chain statistics method was applied to the FAB spectrum of a copolymer containing units of PC and units of resorcine, which was supposed to be an exactly alternating copolymer.[115] Various sequence models were tested in order to fit the mass spectrum. The exactly alternating model failed to give a good fit. The AF for a mixture of exactly alternating and Bernoulli was 8%, whereas the AF for Markoff was 6%. It was therefore concluded that a Markoffian distribution gives a more detailed description of the sequence.

Montaudo et al. examined the FAB mass spectrum[113] of a copolymer sample containing units of β-hydroxybutyrate and β-hydroxyvalerate (HB/HV), with composition 87/13. Actually, the sample had been prepared as an equimolar mixture of an HB homopolymer and an HB/HV copolymer. The composition of the second component (HB/HV) was known to be 72/28. They applied the chain statistics approach developed to fit the mass spectra of copolymer mixtures, and found the correct composition values, namely 50% of β-hydroxybutyrate and 50% of (HB/HV) copolymer.[113] Remarkably, the difference among peak intensities calculated for a pure copolymer and for a mixture of copolymers became more striking as the oligomer mass increased. Thus, limiting the analysis to dimers, trimers, and tetramers would have led to poor discrimination between the two possibilities. Fortunately, the FAB mass spectra allowed the authors to also use the peak intensities of the higher oligomers, which were found to be quite sensitive to minor changes in the oligomers' distribution, permitting an excellent discrimination.[113]

Figure 2.26 illustrates the results of this study. The agreement factor AF (i.e., the error between observed mass spectrum and that calculated according to the Bernoullian model), was monitored as a function of the HB content (mole%) for each set of MS peak intensities from the dimers to the hexamers. Figure 2.26a shows the AF values corresponding to a pure HB/HV copolymer, and it is seen that all the oligomers fall in the same minimum (85% mole HB), with an excellent AF value.[113] Figure 2.26b shows the AF values corresponding to the mixture of copolymers, and it can be immediately visualized that the AF values increase for the higher oligomers (for the hexamer the error is about 40%), and also the minimum shifts to higher composition values.

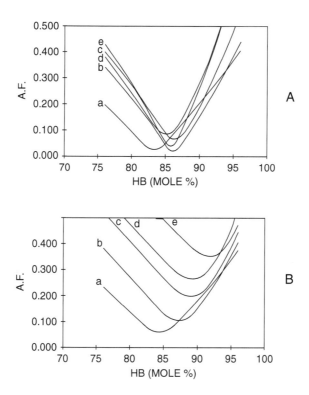

FIGURE 2.26
The agreement factor for a genuine copolyester (A) and a mixture of a homopolyester and a copolyester (B) (reproduced with permission from Ref. 113).

When the composition estimates approach was used to find the composition of the HB/HV copolymer,[121] the result was 63/37 instead of 72/28. Therefore, the composition estimates approach gave an erroneous result. The chain statistics fitting gave a more accurate value for the second component of the mixture, HB/HV = 74/26.

The intensity of the mass spectral peaks in the mixture originates from two different polymers. Therefore, the simple summation procedure on which the composition estimates method is based (Eq. 2.45) may be affected by error, because it is not able to deconvolute the contributions of isobaric oligomers to the intensity of the single peaks.[113]

The chain statistics method has been applied to the case[115] of a copolyamide with units of adipoyl piperazine (referred to as *A*) and of truxilloyl piperazine (referred to as *B*). Photolysis was used for the copolymer degradation. This is a case of totally selective cleavage, since the cyclobutane ring is split quantitatively in two parts by photolysis. Figure 2.27a reports the experimental mass spectrum, whereas Figure 2.27b reports the calculated spectrum. The copolymer composition turned out to be $c_A = 0.80$, with an agreement factor AF = 6%.

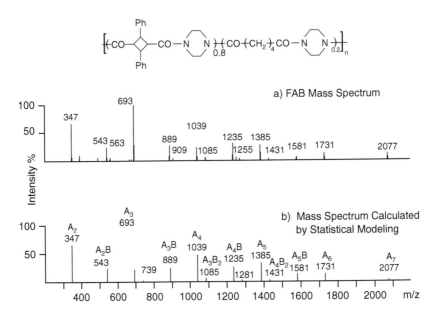

FIGURE 2.27
FAB mass spectrum of a random copolyamide with units of adipoyl piperazine and of truxilloyl piperazine. Experimental mass spectrum (A) and theoretical mass spectrum (B) (reproduced with permission from Ref. 115).

Remarkably, the spectrum in Figure 2.27a has very few peaks, which accurately correspond to the oligomers predicted by the theory of the totally selective process (compare with Table 2.5).

Shard et al.[126] analyzed by the chain statistical method the SIMS mass spectrum of a copolymer with units of glycolic acid residue (referred to as *A*) and of glycine (referred to as *B*). In this case, isobar structures were present, and many peak intensities were due to two contributions. They subtracted the unwanted contribution, applied chain statistics, and compared theoretical intensities with the experimental ones.[126]

Another study[122] concerned a sample obtained by ring-opening copolymerization of two cyclic monomers, namely dilactide and depsipeptide of lactic acid and glycine. The MALDI mass spectrum was recorded (Chapter 10), and chain statistics was applied to determine the sequence. The sequence turned out to follow first-order Markoff. The composition associated with the P-matrix elements is $c_A = 0.77$, which compares well with the composition determined by NMR,[122] $c_A = 0.76$.

Polyethylene terephthalate (PET) and polyethylene truxillate (PETx) were melt-mixed, yielding a copolyester.[117] The FAB mass spectrum was recorded, chain statistics was applied (Markoff distribution), and the extent of exchange (see above) was calculated.[117] This study showed the potential of the MS method that allowed calculating the sequence and composition of the copolymer formed in the melt-mixing process, even in the presence of sizeable amounts of unreacted homopolymers. This allows one to avoid the conventional time-consuming practice of solvent extraction and fractionation of the melt-mixing products before the sequence analysis.

In a similar study, the FAB mass spectrum[116] of a copolyester obtained by melt-mixing polyethylene terephthalate (PET) and of polyethylene adipate (PEA) was recorded. The authors computed the number average length of adipate and terephthalate blocks by applying chain statistics. The result was: $n_{EA} = 7.1$; $n_{ET} = 2$ implying that the initial copolymer, obtained by melt-mixing of the two polyesters, has a Markoffian (block) distribution of the monomer sequence.[116]

EI-MS was used[92] to analyze an aromatic copolyester with units of p-oxybenzoate (PO) and of ethylene terephthalate (ET). The EI mass spectra were recorded up to 480°C. At these temperatures, the chain cleavage is totally selective, since PO units are stable and ET units are not. Chain statistics was applied to the EI mass spectra and a good agreement was found between the computed and the known composition.[115]

An important quantity that may be determined for a copolymer sample is the weight of copolymer chains at each composition and at each length, i.e., the "Bivariate Distribution," with respect to molar mass and composition.[138,139] Information on the Bivariate Distribution is important in high conversion copolymers, since the rates of consumption of monomers *A* and *B* during the reaction do often change and one of the two monomers is preferentially incorporated into the copolymer chain. This implies that the spread in composition increases with conversion, and that high conversion copolymers display composition heterogeneity. Furthermore, the average length of the chains produced at high conversion may be higher than the average length of chains produced at low conversion.

Figure 2.28 reports the Bivariate Distribution calculated for a 70/30 mixture of an AB random copolymer with an *A* homopolymer produced by anionic living polymerization.[142] Figure 2.28 shows that the surface possesses two maxima, illustrating the usefulness of the bivariate plot.

The method traditionally used for measuring the Bivariate Distribution is the "Chromatographic Cross Fractionation" (also referred to as "Two-dimensional Chromatography" or "Orthogonal Chromatography"). Macromolecules having different composition are separated in a silica column, and an SEC column is then used to elute chains having different sizes.[139]

Being able to discriminate among different masses and possessing a remarkably high sensitivity, mass spectrometry may also be used for measuring the bivariate distribution. Suddaby et al.[125] computed the bivariate distribution from the MALDI spectrum of a low mass copolymer with units of MMA and units of *n*-butyl acrylate. Wilczek-Vera et al.[102] computed the bivariate distribution from the MALDI spectrum of a block copolymer containing units of α-methyl styrene and units of styrene. Van Rooji et al.[109] analyzed

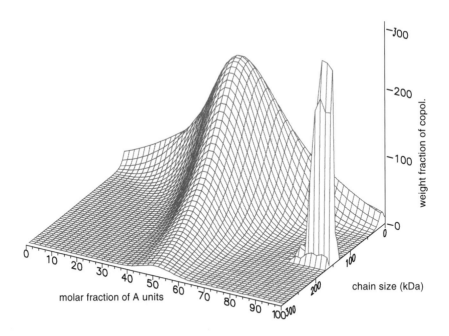

FIGURE 2.28

Bivariate distribution of chain sizes and composition expected for a mixture of an AB random copolymer and a homopolymer produced by anionic polymerization.

by FT-ICR a copolymer with units of EO and PO and computed the bivariate distribution.

All of these authors[102,109,125] plotted the molar fraction of chains for each oligomer present in the sample, since the mass spectra directly yield this information. Therefore, their bivariate distribution representation is not directly comparable with the traditional plots dealing with the weight fraction of chains with a given length and composition.[138,139]

Montaudo[142] computed the bivariate distribution from the laser desorption (LD) mass spectra[106] of several BMA/MMA copolymers with $\overline{M}n$, $\overline{M}w$, below 10000 Da, and plotted the weight fraction of chains with a given length and composition. Figure 2.29 reports the original LD mass spectra[106] of the MMA/ BA copolymers MB73, MB82, MB91, whereas in Figure 2.30 the resulting bivariate distribution of chain lengths and compositions are shown. Note how the bivariate plots provide an immediate picture of the different molecular properties of these copolymers.

In the case of copolymers of high molar mass, some problems arise due to the loss of resolution at mass above 10 kDa in the MALDI spectra. Nevertheless, a variant of the SEC-MALDI method (see above and also Chapter 10) can be used to overcome the problem. The variant consists[140] in fractionating the whole copolymer by SEC, collecting the fractions, and recording both the MALDI and NMR spectra. In fact, due to the high sensibility of the MS and

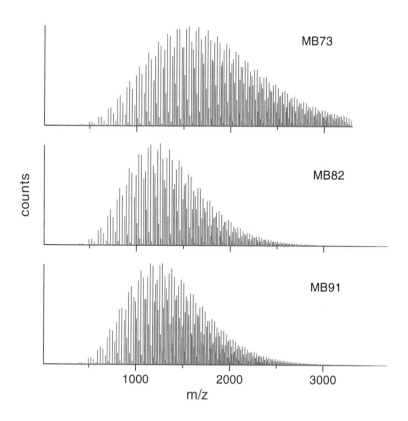

FIGURE 2.29
Laser desorption FT-ICR mass spectra[106] for three MMA/BA copolymers referred to as MB73, MB82, and MB91.

NMR methods, the sample amount contained in a narrow fraction provided by an analytical SEC device is sufficient to run both type of spectra.

Two BMA/MMA random copolymer samples with high $\overline{M}n$ and $\overline{M}w$ were analyzed by SEC-MALDI and SEC-NMR.[140] Figure 2.31 reports the bivariate distribution of chain lengths and compositions. Sample *A* is a low conversion copolymer, whereas sample *B* is a high conversion copolymer. The sample reacted at low conversion (Figure 2.31a) possesses a symmetrical bivariate distribution, whereas the sample reacted at high conversion (Figure 2.31b) is not symmetrical, thus showing the composition drift expected for samples reacted up to high conversion.[138,140]

2.4.3 NMR Methods

NMR has received much attention as a method for copolymer analysis. Both copolymer composition and sequence can be determined.[65–88] ^{1}H and ^{13}C-NMR are both used, but often ^{13}C-NMR is preferred, since the signals possess better resolution.

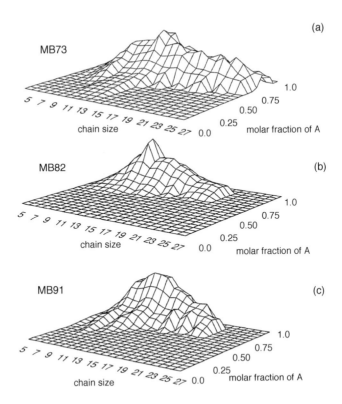

FIGURE 2.30
Bivariate distributions for three MMA/BA copolymers referred to as MB73, MB82, and MB91.

The NMR method has been extremely successful when applied to sequencing addition copolymers with carbon atom backbone, such as ethylene, propylene, butadiene, acrylonitrile, vinyl acetate, methyl methacrylate, styrene, methylstyrene, vinyl chloride, vinyl fluoride (in this case, ^{19}F-NMR can be used).[65–67]

Condensation copolymers such as polyurethanes, polyesters, and polyamides have been analyzed by ^{1}H and ^{13}C NMR.[65–67] Excellent reviews have appeared on this topic,[65–67] the literature on the subject is always growing,[65–88] and the instrumental progress is fast.[83]

The determination of composition and sequence by NMR, using the chain statistics, is often reported.[65–87] It has been applied to copolymers containing styrene,[65–67] to methyl acrylate/diphenylethylene copolymers,[74] to styrene-acrylonitrile copolymers,[75] to isobutylene/isoprene copolymers,[70] to methacrylate-acrylamide copolymers,[72] and to styrene-maleic anhydride copolymers.[76] A "perturbed Markoff" model distribution has been introduced by Cheng, and this model enables one to explain the NMR spectra of copolymer samples reacted at high conversions.[78]

When the scope of the investigation is merely to find the number average lengths and the molar fraction of *A* units, one may consider triads and take a

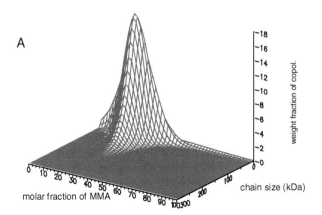

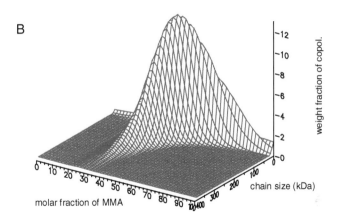

FIGURE 2.31
Bivariate distribution obtained by SEC-MALDI, SEC-NMR of MMA-BA reacted at low conversion (A) and at high conversion (B) (reproduced with permission from Ref. 140).

linear combination of peak intensities. For reasons of brevity, the pertinent equations are omitted. They can be found elsewhere.[65–67] However, the method of linear combination of peak intensities has limitations. For instance, it does not enable discrimination between pure copolymers and mixtures of homopolymers and copolymers.[65–68,70,72–75,78–84]

Contrary to MS, where the mass number associated with a given oligomer can be determined *a priori*, the NMR chemical shift of a dyad or triad sequence does not have a simple relationship with sequence properties. To determine the copolymer sequence distribution by NMR one first proceeds to the assigned dyad and triad peaks (or higher sequences). The task of peak assignment is error-prone and, in any case, it may require the analysis of model structures (such as the exactly alternating structure or structures with *A* and *B* blocks), which may be difficult to obtain.

In some cases, NMR can "see" only short sequences. For instance, in copolymers with large aliphatic units (often encountered in condensation copolymerization), the distance between a repeat unit and the successive can be of several Ångstroms, and therefore the two adjacent units interact very weakly. Thus, the chemical shift in the NMR spectrum remains unaffected when the first neighbor is replaced by another comonomer, and the NMR spectrum gives little information on copolymer sequence.

An example of this type of difficulty is the case of multicomponent copolyesters containing higher β-hydroxyalkanoates, such as β-hydroxyheptanoate, β-hydroxyoctanoate, β-hydroxynonaoate, β-hydroxydecanoate and β-hydroxyundecanoate, where the ^{13}C-NMR signals of the β-methyl groups fall at the same chemical shift. It has not been possible to determine the sequence and composition of these materials by ^{13}C-NMR.[73]

Due to the great amount of NMR data available on this subject,[65–88] we shall limit ourselves to illustrating here only an example where the composition and sequence distribution of poly(ether-sulfone)/poly(ether/ketone) (PES/PEK) copolymers was determined by performing comparative ^{13}C-NMR and MS measurements.[119] A series of four PES/PEK samples possessing different composition and/or sequence were studied.

The ^{13}C-NMR spectra of the PEK/PES copolymers are shown in Figure 2.32, and all the predicted peaks at the dyad and triad level appear present and well-resolved, owing to the long-range transmission effect of the ether-ketone units. Figure 2.32 also shows the structural formulas of the copolymers' repeat units and the respective peak assignments. Diphenylsulfone units are named *S*, and diphenylketone units are named K. Therefore, the symbol SS or SK specifies a dyad, and the symbol KKK or SKS specifies a triad.

The signals between 117 and 120 ppm in Figure 2.32 were assigned to the four dyads centered on the tertiary carbons (a). The signals between 131 and 138 ppm were assigned to the eight triads centered on the quaternary carbons (b). Peak assignments, their relative intensities, and the computed compositions are shown in Table 2.6. These data were used to determine the sequence distribution of the PEK/PES copolymer samples by the usual chain statistics methods. The results indicate that the sequential arrangement of ether-sulfone/ether-ketone units present in these materials corresponds to a random distribution. The copolymer composition values, estimated by the above-described minimization procedure, are reported in Table 2.6.[119]

A parallel MS analysis was then performed on these PES/PEK copolymers. Methoxidation was used to reduce the molar mass of the samples by partial degradation. The mass spectral analysis was performed only at the dimers and trimers level in order to make the results directly comparable with those from the ^{13}C-NMR analysis.

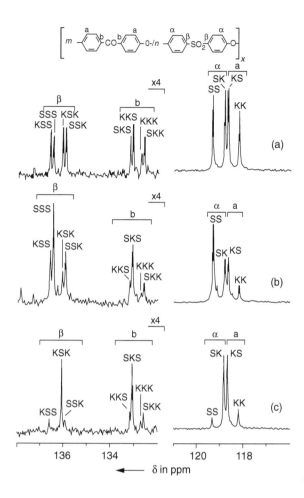

FIGURE 2.32
NMR spectrum of three PES/PEK copolymer samples referred to as C1, C2, and C3 (reproduced with permission from Ref. 119).

The negative ion FAB mass spectra of the PES/PEK oligomers obtained from the methoxidation of the copolymer samples are reported in Figure 2.33. Three and four peaks, respectively, are expected for dimers and trimers in the mass spectra (see Table 2.2). However, in the spectra in Figure 2.32 this number is doubled because two mass series are present (i.e., two types of oligomers with different end-groups are formed in the methoxidation process). Table 2.7 reports the MS peak assignments, their relative intensities, and the computed compositions.

By comparing the results in Table 2.6 and Table 2.7, it can be seen that the figures obtained by the ^{13}C-NMR analysis agree well with those obtained by MS.[119]

TABLE 2.6

Chemical Shifts, Experimental and Calculated Intensities, and Results of Best-Fit Minimization for ^{13}C NMR data[119]

Sequence	δ in ppm	Sample C1		Sample C2		Sample C3	
		Exp.	Calc.	Exp.	Calc.	Exp.	Calc.
KK	118.2	1480	1671	450	560	480	477
KS	118.7	2070	1860	1260	1138	1840	1772
SK	118.9	2020	1860	1240	1138	1870	1772
SS	119.3	1890	2069	2200	2314	310	477
KKK	133.0	570	871	290	266	90	293
KKS	133.3	1100	968	400	541	540	293
KSK	136.0	1170	968	550	541	1870	1884
KSS	132.6	1120	1078	1100	1099	360	293
KKK	133.6	970	968	410	541	500	293
SKS	133.2	1280	1078	1300	1099	1760	1884
SSK	135.9	1150	1078	1130	1099	350	293
SSS	136.5	950	1199	2240	2235	60	293
Agreement factor		0.13		0.08		0.13	
Mole fraction of ketone units		0.48		0.33		—	

TABLE 2.7

Assignments of the Oligomers Observed in the Negative FAB-MS Spectra of the Methoxidated PES/PEK Copolymers, and Best-Fit Minimization Results of the MS Data[119]

Structure	Mass	Sample C1		Sample C2		Sample C3	
		Exp.	Calc.	Exp.	Calc.	Exp.	Calc.
H-S$_2$-OH	482	446	432	1053	978	204	273
H-SK-OH	446	663	622	639	709	1307	1293
H-K$_2$-OH	410	168	224	124	128	329	273
H-S$_3$-OH	714	39	71	71	141	79	61
H-S$_2$K-OH	678	140	152	143	154	242	402
H-SK$_2$-OH	642	115	110	71	56	320	402
H-K$_3$-OH	606	64	26	72	7	186	61
H-S$_2$-OCH$_3$	496	301	332	643	725	242	273
H-SK-OCH$_3$	460	532	478	594	525	1737	1737
H-K$_2$-OCH$_3$	424	147	171	109	95	304	273
H-S$_3$-OCH$_3$	728	27	45	24	66	535	416
H-S$_2$K-OCH$_3$	692	91	98	56	72	340	339
H-SK$_2$-OCH$_3$	656	77	70	44	26	267	339
H-K$_3$-OCH$_3$	620	36	17	43	3	337	416
Agreement factor		0.11		0.12		0.10	
Mole fraction of ketone units		0.43		0.26		—	

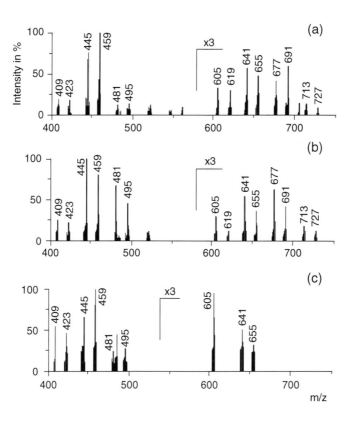

FIGURE 2.33
Mass spectrum of three PES/PEK copolymer samples referred to as samples C1, C2, and C3 (reproduced with permission from Ref. 119).

References

1. Allen, G. and Bevington J. (eds), *Comprehensive Polymer Science,* Pergamon, Oxford, 1989, Vol 1 (*Polymer Characterization*).
2. Pethrick, R.A. and Dawkins, J.V. (eds.), *Modern Techniques for Polymer Characterization,* Wiley, New York, 1999.
3. Mitchell, J. Jr (ed.), *Applied Polymer Analysis and Characterization,* Hanser Publ., Munich, 1987.
4. Bark, L.S. and Allen, N.S. (eds), *Analysis of Polymers Systems,* Applied Science, London, 1982.
5. Koenig, J.L., *Spectroscopy of Polymers,* 2nd edition, Elsevier, New York, 1999.
6. Provder, T., Barth, H.G., and Urban, M.W. (eds.), *Chromatographic Characterization of Polymers,* ACS, Washington DC, (1995).
7. Bovey, F. and Mirau, H., *NMR of Polymers,* Academic Press, New York, 1996.
8. Flory, P., *Principles of Polymer Chemistry,* Cornell University Press, Ithaca, New York, 1971.
9. Billmeyer, F.W. Jr, *Textbook of Polymer Science,* Wiley, New York, 1971.

10. Painter, P.C. and Coleman, M.M., *Fundamentals of Polymer Science*, Technomic Publ., Lancaster, 1997.

11. Rempp, P. and Merril, E.W., *Polymer Synthesis*, 2nd edition, Huthig-Wepf, Basel, 1991.

12. Boyd, R.H. and Phillips, P.J., *The Science of Polymer Molecules*, Cambridge University Press, New York, 1993.

13. Hsieh, K. and Quirk, Q., *Anionic Polymerization*, Marcel Dekker, New York, 1996.

14. Allen, G. and Bevington, J. (eds), *Comprehensive Polymer Science*, Pergamon, Oxford, 1989, Vol 3, chapter 4.

15. Brandrup, J. and Immergut, E.H. (eds.), *Polymer Handbook*, Wiley, New York, 1989.

16. Semlyen, J.A., *Adv. Polym. Science*, **21**, 41, (1976).

17. Semlyen, J.A. (ed.), *Cyclic Polymers*, Elsevier, London, 1986.

18. Bushuk, F. and Benoit, H., Light Scattering studies in Copolymers I. Effect of the heterogeneity of chain composition on molecular weight, *Canadian J. Chem.*, **36**, 1616 (1958).

19. Faraone, A., Magazu, S., Maisano, G., Villari, V., and Maschio, G., Possibilities and Limits of Photon Correlation Spectroscopy in Determining Polymer Molecular Weight Distributions, *Macromol. Chem. Phys.*, **200**, 1134 (1999).

20. Shalaby, S.W., McCormick, C.L., and Butler, G.B. (eds.), *Water-Soluble Polymers*, ACS, Washington, DC (1991), chap. 18.

21. Uglea, C.V., *Liquid Chromatography of Oligomers*, Marcel Dekker, New York, 1996.

22. Pasch, H., Hyphenated techniques in liquid chromatography of polymers, *Advan. Polym. Sci.*, **150**, 1 (2000).

23. Bruessau, R.J., Experiences with interlaboratory GPC experiments, *Macromol. Symp.*, **110**, 15 (1996).

24. Montaudo, G., Montaudo, M.S., Puglisi, C., and Samperi, F., Molecular Weight Determination and Structural Analysis in Polydisperse Polymers by Hyphenated Gel Permeation Chromatography/MALDI-TOF Mass Spectrometry Method, *Internat. J. of Polym. Anal. Charact.*, **3**, 177 (1997).

25. Montaudo, G., Montaudo, M.S., Puglisi, C., and Samperi, F., Molar Mass Distribution and Hydrodynamic Interactions in Random Copolyesters Investigated by Size Exclusion Chromatography Matrix Assisted Laser Desorption/Ionization, *Macromolecules*, **31**, 3839 (1998).

26. Montaudo, M.S., Puglisi, C., Samperi, F., and Montaudo, G., Partially Selective Methanolysis of Sebacic Units in Biodegradable Multicomponent Copolyesters, *Macromol. Rapid Commun.*, **203**, 455 (1998).

27. Dubin, P.L. (ed.), *Aqueous Size Exclusion Chromatography*, Elsevier, New York, 1988.

28. Grubisic, Z., Rempp, P., and Benoit, H., The Universal Calibration in Size Exclusion Chromatography of Polymers, *J. Polym. Sci.*, **5**, 753 (1967).

29. Jackson, C. and Yau, W. W., *Computer Simulation Study of Multidetector SEC*, chapter 6 in ref. 6.

30. Suddaby, K.G., Sanayei, R.A., O'Driscoll, K.F., and Rudin, A. *Eliminating Lag Time Estimation in Multidetector SEC*, chapter 7 in ref. 6.

31. Runyon, J.R., Barnes, D.E., Rudd, J.F., and Tung, L. H. Multiple detectors for Molecular weight and compositional analysis of Copolymers by Gel Permeation Chromatography, *J. Applied Polymer Sci.*, **13**, 2359 (1969).

32. Trathnigg, B. and Yan, X., Copolymer analysis by SEC with dual detection. Coupling of density detection with UV and RI detection, *Chromatographia*, **33**, 467 (1992).

33. Dumelow, T., Holding, S.R., Maisey, L.J., and Dawkins, J.V., Combining SEC and Light Scattering for the characterization of copolymers, *Polymer,* **27**, 1170 (1986).
34. Rollins, K., Scrivens, J.H., Taylor, M.J., and Mayor, H., The Characterization of Polystyrene Oligomers by Field Desorption Mass Spectrometry, *Rapid Commun. Mass Spectrom.,* **4**, 355 (1990).
35. Bletsos, I.V., Hercules, D.M., Van Leyen, D., and Benninghoven, A., Time-of-Flight Secondary Ion Mass Spectrometry of Polymers in the Mass Range 500–10000, *Macromolecules,* **20**, 407 (1987).
36. Vincenti, M., Pellizzetti, E., Guarini, A., and Costanzi, S., Determination of Molecular Weight Distributions of polymers by desorption chemical ionization Mass Spectrometry, *Anal. Chem.,* **64**, 1879 (1992).
37. O'Connor, P.B. and McLafferty, F.W., Oligomer Characterization of 4-23 kDa Polymer by Electrospray Fourier Transform Mass Spectrometry, *J. Am. Chem. Soc.,* **117**, 12826 (1998).
38. Montaudo, G., Montaudo, M.S., Puglisi, C., and Samperi, F. Characterization of polymers by matrix-assisted laser desorption/ionization time-of-flight mass spectrometry: molecular weight estimates in samples of varying polydispersity, *Rapid Commun. Mass Spectrom,* **9**, 453 (1995).
39. Kapfenstein, H.M. and Davies, T.P., Studies in the application of MALDI-TOF MS to the Determination of Chain Transfer Coefficients in Free Radical Polymerization, *Macromol. Chem. Phys.,* **199**, 2403 (1998).
40. Zammit, U.D., Davies, T.P., Haddleton, D.M., and Suddaby, K.G., MALDI determination of Rate of Free-Radical Rate Coefficients, *Macromolecules,* **30**, 1915 (1997).
41. Coote, M.L., Zammit, U.D., and Davies, T.P., Determination of Free-Radical Rate Coefficients using Pulsed-Laser Polymerization, *Trends in Polym Sci.,* **6**, 189 (1996).
42. Schweer, J., Sarnecki, J., Mayer Posner, F., Mullen, K., Rader, H.J., and Spickermann, J., Pulsed-Laser Polymerization/Matrix-Assisted Laser Desorption/Ionization Mass Spectrometry. An approach toward Free-Radical Propagation Rate Coefficients of Ultimate Accuracy, *Macromolecules* **29**, 4536 (1996).
43. Danis, P.O., Karr, D.E., Westmoreland, D.G., Piton, M.C., Christie, D.I., Clay, P.A., Kable, S.H., and Gilbert, R.G., Measurement of Propagation Rate Coefficients Using Pulsed-Laser Polymerization and Matrix-Assisted Laser Desorption/Ionization Mass Spectrometry, *Macromolecules,* **26**, 6684 (1993).
44. Montaudo, G., Mass Spectral Determination of Cyclic Oligomers Distributions in Polymerization and Degradation Reactions, *Macromolecules,* **24**, 5289 (1991).
45. Statheropoulos, M., Georgakopoulos, K., and Montaudo, G., The Interpretation of Mass Spectra of Polymers Using a Hybrid Software System Based on Library Searching with Euristic, *J. Anal. Appl. Pyrolysis,* **20**, 65 (1991).
46. Garozzo, D. and Montaudo, G., Identification of Polymers by Library Search of Pyrolysis Mass Spectra and Pattern Recognition Analysis, *J. Anal. Appl. Pyrolysis,* **9**, 1 (1985).
47. Przybilla, L., Rader, H.J., and Mullen, K., Post source decay Fragment Ion Analysis of polycarbonates by MALDI-TOF, *Europ. Mass Spectrom,* **5**, 133 (1999).
48. Whittal, R. M., Li, L., Lee, S., and Winnik, M. A., Characterization of pyrene end-labeled poly(ethylene glycol) by high resolution MALDI time-of-flight mass spectrometry, *Macromol. Rapid Commun.,* **17**, 59 (1996).
49. Hunt, S.M., Sheil, M., and Derrick, P.J., Comparison of Electrospray Ionization-MS and MALDI-MS and Size Exclusion Chromatography for the Characterization of Polyester Resins, *Eur. Mass Spectrom.,* **4**, 475 (1998).

50. Van Rooij, G.J., Duursma, M.C., Heeren, R.M.A., Boon, J.J., and de Koster, C.G., High Resolution End Group Determination of Low Molecular Weight Polymers by Matrix-Assisted Laser Desorption Ionization on an External Source Fourier Transform Ion Cyclotron Resonance Mass Spectrometry, *J. Am. Soc. Mass Spectrom.*, **7**, 449 (1996).

51. Goodeen, J.K., Gross, M.L., Mueller, A., Stefanescu, A.D., and Wooley K.L., Cyclization in hyperbranched Polymer Synthesis: Characterization by MALDI-TOF Mass Spectrometry, *J. Am. Chem. Soc.*, **120**, 10180 (1998).

52. Yu, D., Vladimirov, N., and Frechet, J.M.J., MALDI-TOF in the Characterization of dendridic linear Block Copolymers and Stars, *Macromolecules*, **32**, 5186 (1999).

53. Leon, J.W., Frechet, J.M.J., Analysis of Aromatic Polyether Dendrimer and Dendrimer-Linear Block Copolymers by MALDI-MS, *Polym. Bullett.*, **35**, 449 (1995).

54. Hult, A., Johansson, M., and Malmstrom, E., Hyperbranched Polymers, *Advan. Polym. Sci.*, **143**, 1 (1999).

55. Hawker, C., Dendridic and Hyperbranched Macromolecules. Precisely Controlled Macromolecular Architectures, *Advan. Polym. Sci.*, **147**, 133 (1999).

56. Blais, J.C., Turrin, C.O., Caminade, A., and Majorat, J.P., MALDI-TOF Mass Spectrometry for the characterization of Phosphorus-containing dendrimers. Scope and limitations, *Anal. Chem.*, **72**, 5097 (2000).

57. Jackson, A.T., Yates, H.T., Scrivens, J.H., Critchley, G., Brown, J., Green, M.R., and Bateman, R.H., The application of Matrix-Assisted Laser Desorption/Ionization combined with Collision-Induced Dissociation in the analysis of synthetic polymers, *Rapid Commun. Mass Spectrom.*, **10**, 1668 (1996).

58. Jackson, A.T., Jennings, K.R., and Scrivens, J.H., Generation of average Mass Values and End Group Information of Polymers by means of a Combination of Matrix-Assisted Laser Desorption/Ionization-Mass Spectrometry and Liquid Secondary ion-tandem mass spectrometry, *J. Am. Soc. Mass Spectrom*, **8**, 76 (1997).

59. Scrivens, J.H., Jackson, A.T., Yates, H.T., Green, M.R., Critchley, G., Brown, J., Bateman, R.H., Bowers, M.T., and Gidden, J., The effect of the variation of cation in the Matrix-Assisted Laser Desorption/Ionization-Collision Induced Dissociation (MALDI-CID) spectrum of oligomeric systems, *Int. J. Mass Spectrom. Ion Processes*, **165/166**, 363 (1997).

60. Jackson, A.T., Yates, H.T., Scrivens, J.H., Green, M.R., and Bateman, RH., Utilizing Matrix-Assisted Laser Desorption/Ionization-Collision Induced Dissociation for the generation of structural information from Polyalkylmethacrylates, *J. Am. Soc. Mass Spectrom*, **8**, 1206 (1997).

61. Montaudo, G., Scamporrino, E., and Vitalini, D., Synthesis and Structural Characterization of Exactly Alternating Copolyesters Containing Photolabile Units in the Main Chain, *Polymer*, **30**, 297 (1989).

62. Vitalini, D., Mineo, P., and Scamporrino, E., Synthesis and Characterization of Some Copolyformals Containing, in the main chain, Different Amounts of Fullerene Units, *Macromolecules*, **32**, 4247 (1999).

63. Odian, G., *Principles of Polymerization*, McGraw-Hill, New York, 1970.

64. Elias, H.G., *Macromolecules*, Plenum Press, New York, 1984.

65. Bovey, F., *High Resolution NMR of Macromolecules*, Academic Press, New York, 1972.

66. Randall, H., *Polymer Sequence Determination (the ^{13}C Method)*, Academic Press, New York, 1977.

67. Tonelli, A.E., *NMR Spectroscopy and Polymer Microstructure*, VCH Publishers, New York, 1989.
68. Price, F.P., Copolymerization mathematics and the description of stereoregular polymers, *J. Chem. Phys*, **36**, 209 (1962).
69. Ross, J.F., Copolymerization Kinetic Constants and Their Prediction from Dyad/ Triad Distributions, *J. Macromol. Sci.-CHEM*, **A21**, 453 (1984).
70. Cheng, H.N. and Bennett, M.A., General Analysis of the Carbon-13 Nuclear Magnetic Resonance Spectra of vinyl copolymers by the Spectral Simulation approach, *Anal Chem*, **56**, 2320 (1984).
71. San Roman, J. and Valero, M., Quantitative evaluation of sequence distribution and Stereoregularity in ethylacrylate-methyl methacrylate copolymers by ^{13}C-NMR spectroscopy, *Polymer*, **31**, 1216 (1990).
72. Kaim, A. and Oracz, P., Penultimate model in the study of the bootstrap effect in the methyl methacrylate-acrylamide copolymerization system, *Polymer*, **38**, 2221 (1997).
73. Kamiya, N., Yamamoto, Y., Inoue, Y., Cuhjo, R., and Doi, Y., Microstructure of bacterially synthetized Poly(3 hydroxybutyrate-co-3 hydroxy valerate), *Macromolecules*, **22**, 1676 (1989).
74. Litt, M. and Seiner, J.A., The role of monomer charge transfer complexes in free radical copolymerization III. Methyl acrylate diphenyl ethylene copolymers. Comparison with penultimate model, *Macromolecules*, **4**, 308 (1971).
75. Hill, D.J.T., O'Donnell, J.H., and O'Sullivan, P.W., Analysis of the mechanism of copolymerization of styrene and acrylonitrile, *Macromolecules*, **15**, 960 (1982).
76. Ha, N.T.H., The effect of the number of meaningful experimental on the fitness of the theoretical copolymerization models, *Compututat. Theoret. Polym Sci*, **9**, 11 (1999).
77. Lopez-Gonzalez, M.M.C., Fernandez-Garcia, M., Barrales-Rienda, J.M., Madruga, E.L., and Arias, C., Sequence distribution and stereoregularity in methyl methacrylate-methyl acrylate copolymers at high conversions, *Polymer*, **34**, 3123 (1993).
78. Cheng, H.N., Perturbed Markoffian probability Models, *Macromolecules*, **25**, 2351 (1992).
79. Galimberti, M., Piemontesi, F., Fusco, O., Camurati, I., and Destro, M., Ethene/ Propene Copolymerization with High Product of Reactivity Ratios from a Single Center, Metallocene-Based Catalytic System, *Macromolecules*, **31**, 3409 (1998).
80. Van der Burg, M.W., Chadwick, J.C., Sudmejer, O., and Tulleken, H.J.A.F., Probabilistic multisite modelling of propene polymerization using Markoff models, *Makromol. Chem. Theory Simul*, **2**, 399 (1993).
81. Segre, A.L., Delfini, M., Paci, M., Raspolli-Galletti, A.M., and Solaro, R., Optically active Hydrocarbon Polymers with Aromatic Side chains 13. Structural Analysis of (S)-4-methyl-1-hexene/Styrene Copolymers by ^{13}C NMR spectroscopy, *Macromolecules*, **18**, 44 (1985).
82. Matlengiewicz, M., Tetrad Distribution of an Aromatic Copolyterephthalate by ^{1}H NMR, *Macromolecules*, **17**, 44 (1984).
83. Cheng, N.H., Two-dimensional NMR Characterization of propylene copolymers, *J. Polym. Sci., Polym. Phys. Ed*, **25**, 2355 (1987).
84. Aerdts, A.M., German, A.L., and Van der Velden, G.P.M., Determination of the Reactivity Ratios, Sequence Distribution and Stereoregularity of butylacrylate-Methyl Methacrylate Copolymers by means of Proton and Carbon ^{13}C NMR, *Magnet. Reson. Chem*, **580**, 32 (1994).

85. Helder, J., Kohn, F.E., Sato, S., vanderBerg, J.W., and Feijen, J., Copolymers of glicine and lactic obtained by stannous octanoate, *Makromol. Chem. Rapid Comm,* **6**, 9 (1985).

86. Kotliar, A.M., Block Sequence distribution and homopolymer content for condensation polymers undergoing interchange reactions, *J. Polym. Sci., Polym. Chem. Ed,* **13**, 973 (1975).

87. Fakirov, S., (ed), *Transreactions in Condensation Polymers,* VCH Publ., New York, 1998.

88. Polic, A.L., Duever, T.A., and Penlidis, A., Case Studies and Literature review on the estimation of copolymerization reactivity ratios, *J. Polym. Sci., Polym. Chem. Ed,* **36**, 812 (1988).

89. Danis, P.O. and Huby, F.J., The Computer-Assisted Interpretation of Copolymers Mass Spectra, *J. Amer. Soc. Mass Spectrom.,* **6**, 1112 (1995).

90. Zoller, D.L., Sum, S.T., Johnston, M.V., Haffieldt, G.R., and Qian, K., Determination of Polymer Type and Comonomer Content in Polyethylenes by Pyrolysis-Photoionization Mass Spectrometry, *Anal. Chem,* **71**, 866 (1999).

91. Ramjit, H.G. and Sedgwig, R.D., Exchange reactions in polyesters studied by mass spectrometry, *J. Macromol. Sci. Chem.,* **A10**, 173 (1987).

92. Garozzo, D., Giuffrida, M., Montaudo, G., and Lenz, R.W., Mass Spectrometric Characterization of Random Ethylene terephthalate and p-Hydroxybenzoic Acid Copolyesters, *J. Polym. Sci. Polym. Chem. Ed.,* **25**, 271 (1987).

93. Montaudo, G., Scamporrino, E., and Vitalini, D., Structural Characterization of Copolyesters by FAB-MS of the Partial Amminolysis Products, *Makromol. Chem. Rapid Commun,* **10**, 411 (1989).

94. Montaudo, G., Puglisi, C., and Samperi, F., Sequencing Aromatic Copolycarbonate by Partial Ammonolysis and FAB-MS Analysis of the Products, *Polymer Bulletin,* **21**, 483 (1989).

95. Ballistreri, A., Garozzo, D., Giuffrida, M., Impallomeni, G., and Montaudo, G., Sequencing Bacterial Poly(β-Hydroxybutyrate-co-β-Hydroxyvalerate) by Partial Methanolysis, HPLC Fractionation and FAB-MS Analysis, *Macromolecules,* **22**, 2107 (1989).

96. Montaudo, G., Scamporrino, E., and Vitalini, D., Synthesis and Structural Characterization of Exactly Alternating Copolyesters Containing Photolabile Units in the Main Chain, *Polymer,* **30**, 297 (1989).

97. Montaudo, G., Scamporrino, E., and Vitalini, D., Characterization of Copolymer Sequences by FAB-MS. I. Identification of Oligomers Produced in the Hydrolysis and Photolysis of Random Copolyamides Photolabile Units in the Main Chain, *Macromolecules,* **22**, 623 (1989).

98. Montaudo, G., Scamporrino, E., and Vitalini, D., Characterization of Copolymer Sequences by FAB-MS. II. Identification of oligomers contained in the Alternating and Random Copolyesters with Photolabile Units in the main chain, *Macromolecules,* **22**, 627 (1989).

99. Ballistreri, A., Garozzo, D., Giuffrida, M., Montaudo, G., Microstructure of Bacterial Poly(β-hydroxybutyrate-co-β-hydroxyvalerate) by Fast Atom Bombardment Mass Spectrometry. Analysis of their Partial Degradation Products in: *Novel Biodegradable Microbial Polymers,* E.A. Dawes (ed), Kluwer, 1990, pp. 49–64.

100. Ballistreri, A., Impallomeni, G., Montaudo, G., Lenz, R.W., Kim, Y.B., and Fuller, R.C., Sequence Distribution of beta-Hydroxyalkanoate Units with Higher Alkyl Groups in Bacterial Polyesters, *Macromolecules,* **23**, 5059 (1990).

101. Montaudo, G., Scamporrino, E., and Vitalini, D., Structural Characterization of Butadiene/Styrene Copolymers by Fast Atom Bombardment Mass Spectrometry Analysis of the Partial Ozonolysis Products, *Macromolecules*, **24**, 376 (1991).

102. Wilczek-Vera, G., Danis, P.O., and Eisenberg, A., Individual Block Length distributions of Block Copolymers of poly(alfa methyl styrene)-block-poly(styrene) by MALDI-TOF Mass Spectrometry, *Macromolecules*, **29**, 4036 (1996).

103. Guttman, C.M., Blair, W.R., and Danis, P.O., Mass Spectroscopy and SEC of SRM 1487, a low molecular weight Poly(methylmethacrylate) Standard, *J. Polym. Sci. Part B-Polym Phys*, **35**, 2409 (1997).

104. Vitalini, D. and Scamporrino, E., Structural characterization of butadiene/acrylonitrile copolymers by FAB mass spectrometric analysis of the partial ozonolysis products, *Polymer*, **33**, 4597 (1992).

105. Rader, H J., Spickermann, J., and Mullen, K., MALDI-TOF Mass Spectrometry in polymer analytics. I. Monitoring the polymer analogous sulfonation reaction of polystyrene, *Macromol. Chem Phys*, **196**, 3967 (1995).

106. Nuwaysir, L.M., Wilkins, C.L., and Simonsick, W.J., Analysis of Copolymers by Laser Desorption Fourier Transform Mass Spectrometry, *J. Am. Soc. Mass Spectrom*, **1**, 66 (1990).

107. Schriemer, D.C., Whittal, R.M., and Li, L., Analysis of structurally complex polymers by time-lag focusing MALDI-TOF MS, *Macromolecules*, **30**, 1955 (1997).

108. van der Hage, E.R.E., Duursma, M.C., Heeren, R.M.A., Boon, J.J., Nielen, M.W.F., Weber, A.J.M., de Koster, C.G., and de Vries, N.K., Structural Analysis of Polyoxyalkyl-eneamines by Matrix-Assisted Laser Desorption/Ionization on an External Ion Source FT-ICR–MS and NMR, *Macromolecules*, **30**, 4302 (1997).

109. van Rooji, G.J., Duursma, M.C., de Koster, C.G., Heeren, R.M.A., Boon, J.J., Schuyl, P.J.W., and Van der Hage, E.R.E., Determination of Block Length Distributions of Poly(oxypropylene) and Poly(oxyethylene) Block Copolymers by MALDI-FT-ICR Mass Spectrometry, *Anal. Chem*, **70**, 843 (1998).

110. Wilczek-Vera, G., Yu, Y., Waddell, K., Danis, P.O., and Eisenberg, A., Analysis of Diblock Copolymers of poly(alfa methyl styrene)-block-poly(styrene) by Mass Spectrometry, *Macromolecules*, **32**, 2180 (1999).

111. Lee, H., Lee, W., Chang, T., Choi, S., Lee, D., Ji, H., Nonidez, W.K., and Mays, J.W., Characterization of poly(ethyleneoxide)-block-poly(lactide) by HPLC and MALDI-TOF mass spectrometry, *Macromolecules*, **32**, 4143 (1999).

112. Yoshida, S., Yamamoto, S., and Takamatsu, T., Detailed Structural Characterization of Modified Silicone Copolymers by Matrix-Assisted Laser Description/Ionization Time-of-flight Mass Spectrometry, *Rapid Comm Mass Spectrom.*, **12**, 535 (1998).

113. Ballistreri, A., Montaudo, G., Garozzo, D., Giuffrida, M., and Montaudo, M.S., Microstructure of Bacterial Poly(β-hydroxybutyrate-co-β-hydroxyvalerate) by Fast Atom Bombardment Mass Spectrometry Analysis of the Partial Pyrolysis Products, *Macromolecules*, **24**, 1231 (1991).

114. Montaudo, M.S., Ballistreri, A., and Montaudo, G., Determination of Microstructure in Copolymers. Statistical Modeling and Computer Simulation of Mass Spectra, *Macromolecules*, **24**, 5051 (1991).

115. Montaudo, M.S. and Montaudo, G., Further Studies on the Composition and Microstructure of Copolymers by Statistical Modeling of Their Mass Spectra, *Macromolecules*, **25**, 4264 (1992).

116. Montaudo, G., Montaudo, M.S., Scamporrino, E., and Vitalini, D., Mechanism of Exchange in Polyesters. Composition and Microstructure of Copolymers Formed in the Melt-Mixing Process of Poly(ethylene terephthalate) and Poly(ethylene adipate), *Macromolecules*, **25**, 5099 (1992).

117. Montaudo, G., Montaudo, M.S., Scamporrino, E., and Vitalini, D., Composition and Microstructure of a copolyester Formed in the Melt-Mixing of Poly(ethylene terephthalate) and Poly(ethylene truxillate), *Die Makromol. Chem*, **194**, 993 (1993).

118. Montaudo, M.S. and Montaudo, G., Microstructure of Copolymers by Statistical Modeling of Their Mass Spectra, *Die Makromol. Chem. Makromol. Symp*, **65**, 269 (1993).

119. Montaudo, G., Montaudo, M.S., Puglisi, C., and Samperi, F., Sequence Distribution of Poly(Ether-Suphone)/Poly(Ether-Ketone) Copolymers by Mass Spectrometry and ^{13}C-NMR, *Macromol. Chem. Phys*, **196**, 499 (1995).

120. Montaudo, G., Garozzo, D., Montaudo, M.S., Puglisi, C., and Samperi, F., Molecular and Structural Characterization of Polydisperse Polymers and Copolymers by Combining MALDI-TOF Mass Spectrometry with GPC Fractionation, *Macromolecules*, **28**, 7983 (1995).

121. Montaudo, M.S., Puglisi, C., Samperi, F., and Montaudo, G., Structural Characterization of Multicomponent Copolyesters by Mass Spectrometry, *Macromolecules*, **31**, 8666 (1998).

122. Montaudo, M.S., Sequence Constraints in Glycine-Lactic acid Copolymer Determined by MALDI-MS analysis, *Rapid Comm Mass Spectrom*, **13**, 639 (1999).

123. Montaudo, M.S. and Samperi F., Determination of Sequence and Composition in Poly(butylene adipate-co-butyleneterephthalate) by MALDI-TOF, *Europ. Mass Spectrom*, **4**, 459 (1999).

124. Majumdar, T.K., Eberlin, M.N., Cooks, R.G., Green, M.M., Munoz, B., and Reidy, M.P., Structural Studies on Alkyl isocyanate Polymers by Thermal Degradation Tandem Mass Spectrometry, *J. Am. Soc. Mass Spectrom*, **2**, 130 (1991).

125. Suddaby, K.G., Hunt, K.H., and Haddleton, D.M., MALDI-TOF Mass Spectrometry in the study of Statistical Copolymerization and its application in examining the free Radical Copolymerization of Methylmethacrylate and *n*-Butyl methacrylate, *Macromolecules*, **29**, 8642 (1996).

126. Shard, A.G., Volland, C., Davies, M.C., and Kissel, T., Information on Monomer Sequence of poly(lactic acid) and random copolymers of Lactic Acid and Glycolic Acid by examination of Static Secondary Ion Mass Spectrometry Ion Intensity, *Macromolecules*, **29**, 748 (1996).

127. Chen, G., Cooks, R.G., Jha, S.K., Oupicky, D., and Green, M.M., Block Microstructural Characterization of Copolymers formed from Fluorinated and non-Fluorinated Alkyl Polyisocyanates using Desorption Chemical Ionization Mass Spectrometry, *Int. J. Mass Spectrom. Ion Proc*, **165/166**, 391 (1999).

128. Zoller, D.L. and Johnston, M.V., Composition and Microstructure of Acrylonitrile–Butadiene Copolymers by Pyrolysis-Photoionization Mass Spectrometry, *Anal. Chem*, **69**, 3791 (1997).

129. Zhuang, H., Gardella, J.A. Jr., and Hercules, D.M., Determination of the Distribution of Poly(dimethylsiloxane) Segment Lengths at the Surface of Poly[(dimethylsiloxane)-urethane]-Segmented Copolymers by Time-of-Flight Secondary Ion Mass Spectrometry, *Macromolecules*, **30**, 1153 (1997).

130. Haddleton, D., Feeney, E., Buzy, A., Jasieczek, and Jennings, K.R., Electrospray ionization mass spectrometry (ESI MS) of poly(methyl methacrylate) and acrylic statistical copolymers, *Chem Comm.*, 1157, **May** (1996).

131. Servaty, S., Kohler, W., Meyer, W.H., Rosenhauer, C., Spickermann, J., Rader, H.J., Wegner, G., and Weier, A., MALDI-TOF MS Copolymer Analysis: Characterization of a Poly(dimethylsiloxane)-co-poly(hydromethylsiloxane) as a Precursor of a Functionalized Silicone Graft Copolymer, *Macromolecules*, **31**, 2468 (1998).
132. Plage, B. and Schulten, H.R., Sequences in Copolymers Studied by High-Mass Oligomers in Pyrolysis-Field Ionization Mass Spectrometry, *Angew Makromol Chem*, **184**, 133 (1991).
133. Spool, A.M. and Kasai, P.H., Perfluoropolyethers: Analysis by TOF-SIMS, *Macromolecules*, **29**, 1691 (1996).
134. Adamus, G., Sikorska, W., Montaudo, M.S., Scandola, M., and Kowalczuk, M., Sequence Distribution and Fragmentation Studies of Bacterial Copolyester—Characterization of PHBV Macroinitiator by Electrospray Ion-Trap Multistep Mass Spectrometry, *Macromolecules*, **33**, 5797 (2000).
135. Abate, R., Garozzo, D., Rapisardi, R., Ballistreri, A., and Montaudo, G., Sequence Distribution of beta-hydroxyalkanoate Unit in Bacterial Copolyesters Determined by Desorption Chemical Ionization Mass Spectrometry, *Rapid Comm. Mass Spectrom*, **6**, 702 (1992).
136. Montaudo, M.S., Montecarlo Simulation of Copolymer Mass Spectra, *Die Makromol. Chem, Theory and Simul*, **2**, 735 (1993).
137. Montaudo, M.S., Montecarlo Modeling of Exchange Reactions in Polyesters. Dependence of Copolymer Composition from the Exchange Mechanism, *Macromolecules*, **26**, 2451 (1993).
138. Stockmayer, W.H., Bivariate Distribution of chain lenghts and compositions, *J. Chem Phys*, **13**, 199 (1945).
139. Mori, S., Determination of Chemical Composition and Molecular weight distributions in high-conversion Styrene methyl methacrylate copolymers by liquid adsorption and Size exclusion chromatography, *Anal. Chem*, **60**, 1125 (1988).
140. Montaudo, M.S. and Montaudo, G., Bivariate Distribution in PMMA/PBA Copolymers by combined SEC/NMR and SEC/MALDI measurements, *Macromolecules*, **32**, 7015 (1999).
141. Abate, R., Ballistreri, A., Montaudo, G., Garozzo, D., Impallomeni, G., Critchley, G., and Tanaka, K., Quantitative Applications of Matrix-assisted Laser Desorption/Ionization with Time-of-flight Mass Spectrometry: Determination of Copolymer Composition in Bacterial Copolyesters, *Rapid Comm. Mass Spectrom.*, **7**, 1033 (1993).
142. Montaudo, M.S., unpublished results.

3

Pyrolysis Gas Chromatography/Mass Spectrometry (Py-GC/MS)

Shin Tsuge and Hajime Ohtani

CONTENTS

3.1   Introduction

Pyrolysis-GC/MS (Py-GC/MS) has increasingly been utilized in the field of structural characterization of versatile polymeric materials. This technique often provides a simple but rapid and extremely sensitive tool not only for ordinary solvent soluble polymers, but also for intractable cured polymers with three-dimensional networks.

Py-GC, in its early stage, however, had some serious limitations such as difficulty in attaining specific pyrolysis of samples, insufficient chromatographic separation of degradation products, and poor identification and interpretation of resultant chromatograms (pyrograms). Therefore, Py-GC has been regarded for a long time as a relatively crude technique for the characterization of polymers.

However, owing to recent developments in highly efficient separation columns for GC, and various powerful hyphenated identification techniques for GC such as GC/MS, GC/FTIR, and GC/atomic emission detector (AED), Py-GC has made great strides toward becoming a powerful tool for the structural characterization of polymeric materials. Therefore, Py-GC, in particular Py-GC/MS, now plays a very important role among various methods developed in the field of analytical pyrolysis.

In this chapter, the history and scope of analytical pyrolysis are presented first, and then the instrumental and methodological aspects of Py-GC/MS are briefly discussed. Then, some recent typical applications of Py-GC/MS to the structural characterization of various polymeric materials will be discussed in detail.

3.2   History and Scope of Analytical Pyrolysis

In 1948, the first reports on the off-line pyrolysis-MS (Py-MS) of polymers were published by Madorsky and Straus,[1] and Wall.[2] In 1953, Bradt et al. described on-line Py-MS for which pyrolysis of polymer samples was effected within the instrument in vacuo.[3] Thus valuable structural information about the samples became obtainable.

Two years after the introduction of GC by James and Martin in 1952,[4] Davison et al. reported the first work on off-line Py-GC of polymers.[5] These workers demonstrated that Py-GC was quite effective for the characterization of polymeric materials. In 1959, on-line Py-GC systems and their applications to polymer analysis were reported independently by three research groups: Lehrle and Robb,[6] Radell and Strutz,[7] and Martin.[8] These achievements triggered a boom in Py-GC.

The high-resolution capillary columns introduced by Golay in 1958 had a strong impact on Py-GC.[9] However, general application of their effectiveness was held back until the advent of the chemically inert fused silica capillary columns in 1979,[10] since the earlier metal capillary columns were not suitable for the separation of polar and/or higher boiling point compounds. The chemical inertness and thermal stability of capillary separation columns are among the most important factors in achieving high-resolution pyrograms of polymer samples by Py-GC/MS since the thermal degradation products of polymers usually consist of compounds with a wide range of boiling points which often contain fairly polar compounds such as carboxylic acids, amines, nitriles, epoxides, and so on. Modern fused silica capillary columns and recently developed metal capillary ones with deactivated inner walls have drastically improved the situation of Py-GC/MS.

In the early stage of Py-GC, significant interlaboratory discrepancies between pyrolysis data (pyrograms) were reported even for the same polymer types. This was mainly because of a diversity of pyrolysis devices operated under varied conditions. Owing to continued improvement of pyrolyzers and fundamental studies to control the operating conditions and obtain reproducible and characteristic degradation of the studied materials, most of the commercially available pyrolyzers now have made the interlaboratory discrepancies a minor problem. Now, various flash filament-, furnace-, and Curie-point type pyrolyzers are utilized for both Py-MS and Py-GC.

The dissemination of GC-MS after 1965 strongly accelerated the development of Py-GC. In 1966, Simon and co-workers reported the first directly coupled Py-GC/MS system using a Curie-point pyrolyzer, a metal capillary column GC, and a rapid scanning MS.[11] In 1968, Schuddemage and Hummel introduced field ionization MS(FI-MS) to simplify the mass spectra resulting from Py-MS.[12] This technique dramatically improved the effectiveness of direct Py-MS, because of the simpler fragmentation in FI-MS.

Standardization and reliable compilation of a standard database for various series of standard samples are among the important factors to promote interlaboratory data comparison in analytical pyrolysis. Recently, a trial standard database for 136 kinds of typical polymers compiled under the same conditions using Py-GC was published by the authors,[13] of which a Chinese translation is already available,[14] and an English version is now in preparation.

Now, analytical pyrolysis is most extensively practiced using Py-MS and Py-GC (or Py-GC/MS), where the characterization of the original samples is carried out through on-line separative analysis of the resulting complex pyrolyzates. Representative applications of modern analytical pyrolysis are found in various fields, including polymer chemistry, biochemistry, geochemistry, forensic science, food science, toxicology, environmental science and energy conservation, extraterrestrial studies, and other fields.

3.3 Py-GC/MS Measuring System for Polymer Characterization

Figure 3.1 illustrates a typical Py-GC/MS measuring system, where a microfurnace pyrolyzer, a gas chromatograph equipped with a chemically inert capillary column, and a quadrupole mass spectrometer with an electron impact ionization (EI) and/or chemical ionization (CI) source are directly connected in series.[15] Usually a given polymer sample weighing about 10–100 μg is instantaneously pyrolyzed at about 400–600°C with or without catalytic and/or reactive reagents under a flow of N_2 or He carrier gas. The resulting degradation products transferred into the separation column are separated to yield a pyrogram. The individual components on the pyrogram are continuously identified by use of the observed mass spectra. In addition, total ion monitoring

TABLE 3.1

Representative Subjects of Polymer Characterization by Py-GC/MS

I. *Identification of Polymers: Qualitative Analysis*

II. *Structural Characterization of Polymers:*

- (a) Composition
- (b) Average molecular weight
- (c) Monomer inversion in polymer chains
- (d) Chain-end structures
- (e) Branching structures
- (f) Ster eoregularity
- (g) Sequence distributions in copolymers
- (h) Degree of cure or cross-linking
- (i) Others

III. *Mechanisms and Kinetics of Polymer Degradation*

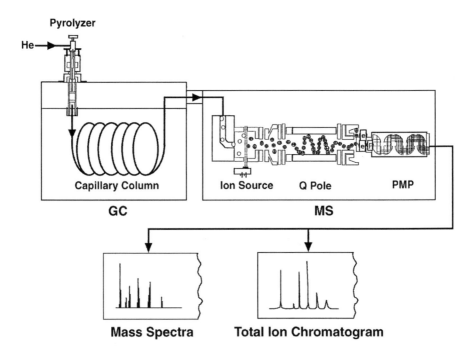

FIGURE 3.1
Py-GC/MS measuring system using a microfurnace pyrolyzer, GC equipped with a capillary separation column, and a quadrupole mass spectrometer, from Ref. 15.

(TIM) and selected ion monitoring (SIM) often provide supplemental and/or complement information of peak identification on the pyrogram.

The pyrolysis products of a given polymer sample are usually composed of very complex components each of which might reflect the original structures to some extent. Therefore, in the case of Py-MS, the resulting mass spectra

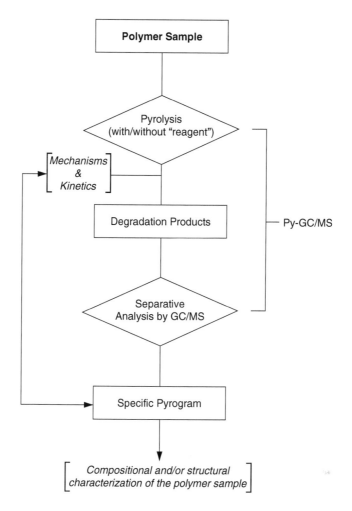

FIGURE 3.2
Flow diagram for polymer characterization by Py-GC/MS, from Ref. 15.

are often too complicated to obtain equivocal structural information of the polymer sample owing to the overlapping of fragment ion peaks from the ionization process of the complex degradation products. On the other hand, Py-GC/MS utilizing a chemically inert capillary column often provides highly efficient separation of those complex pyrolyzates to yield a specific pyrogram of which individual peak components can be identified based on their mass spectra.

Figure 3.2 illustrates the flow diagram in Py-GC/MS for polymer characterization.[15] The resulting specific pyrograms often provide valuable information regarding the composition and/or chemical structures of the original polymer sample as well as on the degradation mechanisms and related kinetics. Table 3.1 summarizes the representative subjects of polymer characterization to which the modern Py-GC/MS is applicable. Typical applications will be demonstrated in the following section.

3.4   Applications of Py-GC/MS to Polymer Characterization

3.4.1   Stereoregularity of Vinyl Polymers

3.4.1.1   Polystyrene

Polystyrene (PS) is one of the most widely used polymers in industry, because of its easy workability, transparency, high insulation resistance, and other properties. PS is classified as isotactic (iso), syndiotactic (syn), and atactic (at) PS, depending on its stereoregularity. It is known that tacticity greatly influences the material properties of PS. Up until recently, at-PS has been mostly utilized in industry. Very recently, syn-PS with high stereoregularity synthesized using new catalysts was expected to show enhanced properties such as higher heat resistance and tensile strength. The characterization of polymer tacticities has been most extensively carried out using NMR, but this usually requires a relatively large sample size and fairly long measuring time. Therefore, rapid and sensitive analytical methods for the determination of tacticity of PS would be of great interest.

Recently, the characteristic diastereoisomers, such as tetramers and pentamers, formed from various PS samples were separated and identified by Py-GC/MS by the use of capillary separation columns at high oven temperatures.[17] The relative peak intensities of the observed diastereotetramers were interpreted in terms of the stereoregularities of the PS samples by comparison with the results obtained by ^{13}C-NMR.

Figure 3.3 shows the pyrogram of at-PS at 600°C detected by a flame ionization detector (FID). The main pyrolyzates on the pyrogram (A) are the styrene monomer (about 80% of all the peak intensities), dimer (about 6%), and trimer (about 5%). Although the trimer has one asymmetric carbon atom, it does not have any diastereoisomers which reflect tacticity. The minimum requirement for a diastereoisomer is the inclusion of more than two asymmetric carbons in the molecule; this means that tetramers are the smallest possible candidates. Although styrene tetramers have fairly high boiling points, they are visible in the expanded pyrogram (B) even in very minute amounts (about 0.05%).

Figure 3.4 shows the expanded partial pyrograms of the tetramer region of the PS samples having different tacticities measured under the same condition. The four characteristic peaks (A, A'B and B') on the pyrograms apparently reflect the differences in tacticities among the samples.

The chemical constituents of the two main peaks (A and B) proved to be the same component by Py-GC/MS. Figure 3.5 shows the mass spectrum of peak B. Although the observed mass spectrum is rather noisy, the molecular

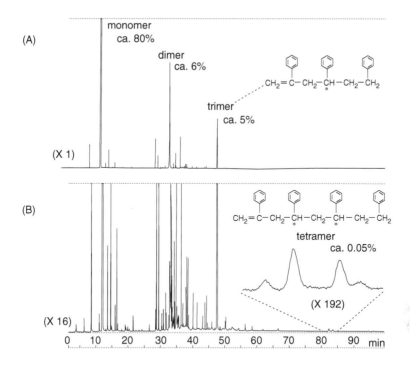

FIGURE 3.3
Pyrograms of atactic polystyrene at 600°C, from Ref. 17. (A) ordinary pyrogram (×1), (B) expanded pyrogram (×16). Column: fused silica capillary (HP Ultra 1; 0.20 mm I. D. × 50 m, 0.33 μm of immobilized PDMS) programmed from 50 to 280°C at 6°C/min. Sample size: ca. 0.2 mg. Detector: FID.

ion peak of the tetramer at $m/z = 416$ is still clearly observed together with the expected fragment peaks from the molecular ion. The mass spectrum of peak A is basically the same as that of peak B. The assignment of A and A' to meso, and B and B' to racemo diastereoisomers was inferred by the fact that the peak intensities of A and A' (having shorter retention times) become stronger for iso-PS sample. The retention differences between A and A' and between B and B' can be attributed to possible differences in the position of double bonds in these components.

Table 3.2 shows the average tacticity data for the various PS samples obtained from the relative intensities of the tetramers in the pyrograms and those obtained by ^{13}C-NMR. Although the results obtained by Py-GC/MS are not completely consistent with those by ^{13}C-NMR, the general trends in both results are in fairly good correlation. The discrepancy between the data obtained by the two methods can be mostly attributed to the stereoisomerization accompanying the formation of the tetramers through radical transfers.

In addition to the radical transfer mechanisms which involve stereoisomerization, repeated simple scissions followed by termination where the original tacticity is conserved, even for the degradation products, competitively contribute to the tetramer formation. The relative contributions of

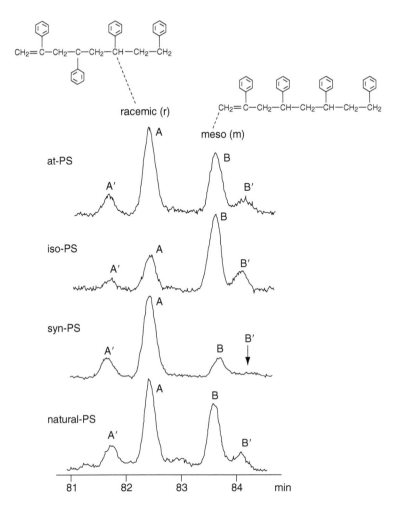

FIGURE 3.4
Partial pyrograms of various PS samples in the tetramer region, from Ref. 17. Py-GC conditions are the same as those in Figure 3.3.

these competitive mechanisms can be regarded as constant under a given pyrolysis temperature. Therefore, once a calibration curve is prepared by use of a series of standard samples whose tacticities are well-characterized, unknown PS samples could be quantitatively interpreted in terms of their stereoregularities.

Figure 3.6 shows the relationship between the relative peak intensity in the tetramer region and the average tacticity calculated from the blending composition for various mixtures of iso- and syn-PS. It is noted that this relationship shows a fairly good linearity. Furthermore, the plot for an at-PS sample also nearly falls on the line. Thus the tacticity of an unknown PS sample could be determined using only a small sample size in the order of submilligram.

TABLE 3.2

Comparison of the Observed Average Tacticity
of PSs as Determined by Py–GC and ^{13}C NMR

| | Py–GC | | ^{13}C NMR | |
Sample	r(%)	m(%)	r(%)	m(%)
At-PS	57.6[a]	42.4	67.5	32.5
Iso-PS	31.0	69.0	0.0	100
Syn-PS	84.9	15.1	98.0	2.0
Natural-PS	57.3	42.7	66.5	33.5

From Ref. 17.

[a] C.V. for five runs = 2.7%.

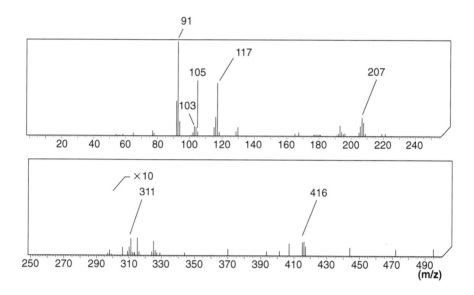

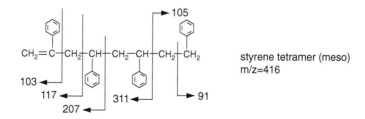

styrene tetramer (meso)
m/z=416

FIGURE 3.5
MS spectrum of peak B in Figure 3.4, from Ref. 17.

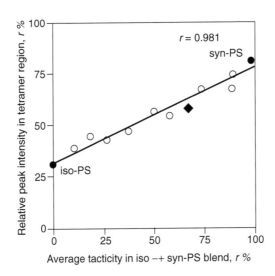

FIGURE 3.6

Relationship between relative peak intensities of racemo products in tetramer region and average r% in mixture of iso- and syn-PS. O: mixture of iso- and syn-PS; ♦: at-PS.

3.4.1.2   *Poly(methyl methacrylate)*

Basically the same Py-GC technique developed to study the tacticity of PS was applied to determine the tacticity of various stereoregular poly(methyl methacrylate) (PMMA) samples by separating the associated diastereomeric-tetramers of which identification was carried out by a directly coupled Py-GC/MS system.[18] Figure 3.7 shows a typical pyrogram of PMMA observed in the total-ion monitor (TIM) by a Py-GC/MS system in CI mode, where various minor peaks of MMA dimers, trimers, tetramers, and pentamers are recognized in addition to the main monomer one.

Figure 3.8 shows the EI and CI mass spectra corresponding to the two strong tetramers, A at ca. 45 min and B at about 46 min in the pyrogram of Figure 3.7. Although the expected quasi-molecular ions are not observed even in the CI spectra, the common ions at $m/z = 369$ can be attributed to $[M-OCH_3]^+$. Thus, both A and B should have the same molecular weight ($M_W = 400$). Furthermore, in the EI spectra, the tetramer A shows a fairly strong peak at $m/z = 301$, while B exhibits a prominent peak at $m/z = 315$. The possible bond cleavages are shown at the bottom of the figure together with the possible structures for the isomers. Additionally the small satellite peaks (A' and B') appearing at earlier retention times than those of the main tetramers (A and B) proved to have exactly the same chemical structures between A and A', and B and B', suggesting that they are the stereoisomers, respectively.

Figure 3.9 shows the expanded partial pyrograms of the tetramer region observed by FID for the PMMA samples (S-1 ~ S-4). By comparing the pyrogram

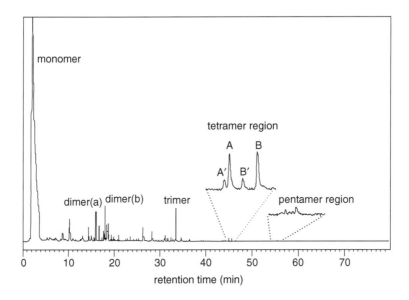

FIGURE 3.7
Pyrogram of syndiotacticity-rich poly(methyl methacrylate) observed in total-ion monitor by Py-GC/MS in CI mode at 500°C, from Ref. 18. Column: fused silica capillary (HP Ultra 1, 0.2 mm I. D. × 25 m, 0.33 μm of immobilized PDMS) programmed from 50 to 280°C at 4°C/min. Sample size: ca. 1 mg.

for the highly syndiotactic S-1 with that for the highly isotactic S-4, it is apparent that the diastereoisomers with meso (m) configuration always appear at earlier retention times than those with a racemo (r) configuration.

Provided that these diastereoisomers reflect the original stereoregularity of the PMMA samples, we can estimate the diad tacticity from the relative peak intensities between A' (m) and A (r), or B' (m) and B (r). The diad tacticity values thus determined are summarized in Table 3.3, together with the reference values obtained by ^{1}H-NMR. Here, fairly good reproducibility of the measurement by Py-GC as CV = 2.0% was obtained for seven repeated runs with S-3.

Thus, observed tacticity values using either tetramer pair, A and A', or B and B', are in fairly good agreement with those by ^{1}H-NMR. This fact suggests that any appreciable thermal isomerization does not contribute to the thermal degradation of PMMA to yield the tetramers since the associated radical transfers to yield the tetramers occur only at methyl or methylene carbons which are not asymmetric in the polymer chain. Moreover, it has been demonstrated that the diad tacticity of MMA sequences can be also precisely determined by basically the same method even in the copolymers of MMA and various acrylates and their crosslinked polymers,[19] which are difficult to characterize by NMR.

TABLE 3.3

Comparison of the Diad Tacticity (%) of PMMAs as Determined by Py-GC and ^{1}H NMR

| | Py-GC | | | | ^{1}H NMR | |
| | From A Peaks | | From B Peaks | | | |
Sample	m	r	m	r	m	r
S-1	7.3	92.7	7.4	92.6	5.6	94.4
S-2	21.3	78.7	24.1	75.9	24.0	76.0
S-3	79.7	20.3	81.8	18.2	82.8	17.1
S-4	98.5	1.5	97.3	2.7	97.2	2.8

From Ref. 18.

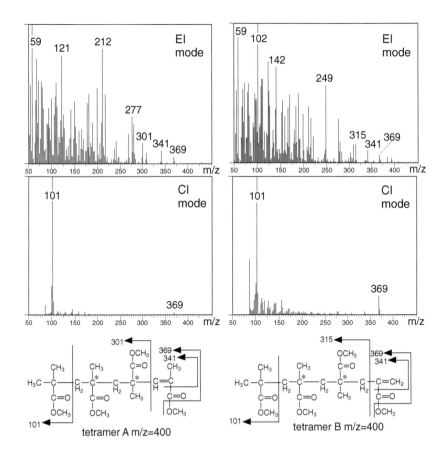

FIGURE 3.8
EI and CI mass spectra of methyl methacrylate tetramers observed in Figure 3.7, from Ref. 18. EI: 70 eV at 250°C; CI: 180°C using isobutane as the reagent gas.

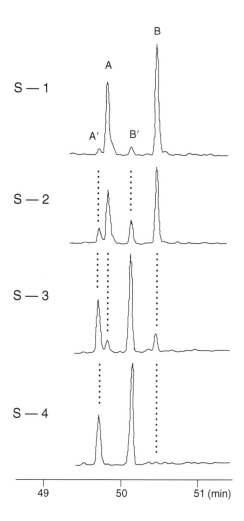

FIGURE 3.9
Partial pyrogram of the tetramer region for various PMMA samples, from Ref. 18. Pyrolysis temp: 500°C. Column: metal capillary column (Frontier Lab Ultra Alloy-1, 0.25 mm I. D. × 50 m, 0.15 μm of immobilized PDMS) and programmed from 50 to 340°C at 4°C/min. Sample size: ca. 0.4 mg. Detector: FID.

3.4.2   Terminal Groups of Polymer Chain

Since end-groups in polymers are generally attributed to an initiator and/or chain transfer and terminating agent incorporated into polymer chains, analysis of end-groups is one of the most substantial approaches for assessing the mechanism of polymerization. Furthermore, the presence or absence of specific end-groups often causes significant changes in the polymer properties, and thus precise characterization has been eagerly sought in recent multifunctionalization of polymeric materials. The characterization of end-groups in a high molecular weight (M_W) polymer sample, however, is not an easy task because of their very low relative concentration. Generally, NMR has been most extensively used for characterization of end-groups in polymers. MALDI-MS has been extensively used in recent years for end-group determination in lower molecular weight polymers. However, their sensitivity and resolution have not always been adequate for quantitative analysis of end-groups in high-M_W polymers. Recently, Py-GC/MS has also been recognized as an excellent and complement technique to approach the characterization of end-groups in polymers.

3.4.2.1 *Radically Polymerized PMMA*

The polymerization reagents incorporated into the polymer chain ends were able to be characterized by Py-GC/MS, where PMMA samples radically polymerized in toluene with benzoyl peroxide (BPO) as an initiator and dodecanethiol as a chain transfer reagent were studied.[20]

Figure 3.10 shows a pyrogram at 460°C detected by FID for a PMMA sample prepared in the presence of BPO and dodecanethiol in toluene. Since PMMA has a tendency to depolymerize mostly into the original monomer at elevated temperatures around 500°C, the main pyrolysis product on the pyrograms (more than 90%) is the MMA monomer. Among various minor peaks, however, peaks A-I were assigned from their mass spectra to the fragments of polymerization reagents incorporated into the polymer chain.

The dissociation of BPO during the polymerization reactions yields both benzoyloxy and phenyl radicals, both of which initiate radical polymerization. Furthermore, chain transfer from polymeric radicals to solvent (toluene) can

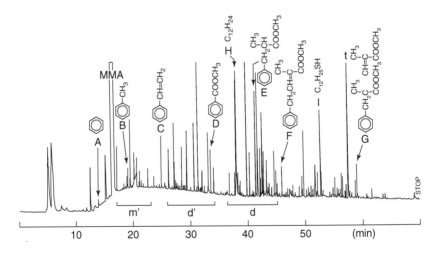

FIGURE 3.10
Pyrogram at 460°C of radically polymerized PMMA prepared in toluene with 0.3% of benzoyl peroxide and 1.5% of dodecanethiol, from Ref. 20. MMA: monomer, m': monomer-related products, d': dimer-related products, d: MMA dimers, t: MMA trimer. Column: fused silica capillary (HP Ultra 1, 0.2 mm I. D. × 50 m, 0.33 μm of immobilized PDMS) programmed from 0 to 250°C at 4°C/min. Sample size: ca. 0.5 mg, Detector: FID.

take place to form benzyl radicals that also trigger another polymerization reaction. On the other hand, dodecanethiol added as a chain transfer reagent also initiates the other polymerization reaction after it is converted into a dodecylthioradical. As a result, at least three types of aromatic chain ends and one dodecylthio chain end should exist in the PMMA sample.

As shown in Figure 3.10, the peaks A-G can be attributed to those aromatic chain ends. Methyl benzoate (peak D) should be responsible for the benzoyloxy-initiated chain ends, while peaks A (benzene), E, and G should be responsible for the phenyl-initiated ones. On the other hand, peaks B (toluene), C (styrene), and F can be attributed to the benzyl-initiated ones. The peaks of 1-dodecene (H) and dodecanethiol (I) on the pyrogram directly reflect the thiol-initiated chain ends. From the relative peak intensities of the above-mentioned characteristic peaks on the resulting pyrogram, it becomes possible not only to estimate the amounts of polymerization reagents in feed but also to discuss the associated polymerization mechanisms.

As an example, Figure 3.11 illustrates the observed relationships between the relative peak intensities of the characteristic aromatic products and the

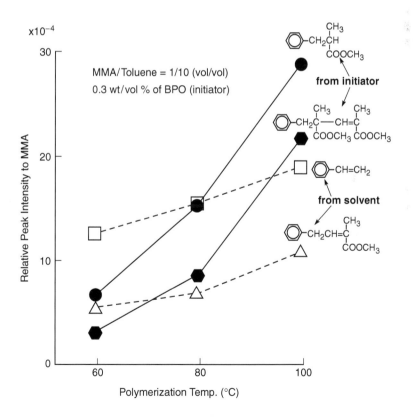

FIGURE 3.11
Relationship between the relative peak intensities characteristic of the aromatic end groups and the polymerization temperatures of PMMA samples polymerized using BPO in toluene, from Ref. 21.

polymerization temperature for samples polymerized at different temperatures between 60 and 100°C.[21] The fact that the peak intensities of all the
aromatic products increase with the rise in temperature indicates that the
temperature dependence of the initiation reaction is much greater than
that of the propagation reaction. Furthermore, the fact that the initiator-
incorporating chain ends (solid lines) increase more rapidly with the rise
in temperature than the solvent-incorporating ones (dotted lines) suggests
that the temperature dependence of the initiation reaction is also greater
than that of the chain transfer reaction to toluene.

Basically the same Py-GC/MS technique was also applied to elucidation of
M_W dependence of the end-groups of radically polymerized PMMA,[22] the end-
groups characterization of prepolymers of PMMA prepared by radical chain
transfer reaction using thio-carboxylic acid and macromonomers derived from
the prepolymers,[23,24] the quantification of end-groups in anionically polymerized PMMAs with narrow M_W distribution in the range of average M_W from
10^4 to 10^6,[25] and characterization of branched alkyl end-groups in PMMA[26] and
the end-groups and their adjacent structures in styrene-MMA copolymers.[27]

3.4.2.2   *Anionically Polymerized PS*

Absolute determination of end-groups in anionically polymerized monodisperse PS samples with M_W between 1000 and a few million was carried out
by Py-GC/MS.[28] Figure 3.12 shows typical pyrograms of PS samples at 600°C
for PS-2 with number average M_W (M_n) = 3090 (A), and PS-3 with M_n = 9000 (B).
At this temperature, polystyrenes were decomposed mainly to the styrene
monomer (ca. 80%), with some dimer (ca. 6.5%) and trimer (ca. 4.0%). Among
these many peaks, some characteristic ones might exist that reflect the structures of the *n*-butyl end-groups. When comparing the pyrograms of the two
PS samples with different M_W, the peaks 1–5 eluting before the monomer
and the peaks 6–9 eluting between the monomer and the dimer cluster could
be specifically attributed to the end-group moieties since those relative intensities became smaller in the pyrogram for the PS-3 with larger M_W.

The positive identification of these peaks carried out by Py-GC/MS is
summarized in Figure 3.13, together with the corresponding end-group moieties in the polymer chain. Thus, it proved that peaks 1–5 are the products
derived from the *n*-butyl end-group moiety, and peaks 6–9 are those from
the *n*-butyl end-group moiety plus a styrene unit. Based on these results, M_n
values could be estimated using the relative molar intensities of these nine
characteristic peaks against those of the major peaks from the main chain
(mostly monomer + dimers + trimer).

Table 3.4 summarizes the M_n values for nine PS samples, together with
the reference values provided by the supplier. The observed values of M_n
by Py-GC/MS are generally in fairly good agreement with the reference
values up to about a few tens of thousands of M_n (PS-1-4). For PS samples
with larger MW (PS-5-8), however, the estimated values are much too low.
This deviation should be attributed mostly to the contribution of the thermal
degradation of the main chain to form, to some extent, the components of

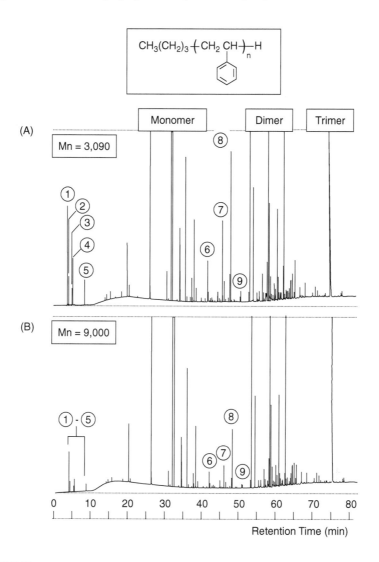

FIGURE 3.12
Pyrograms of polystyrenes having an *n*-butyl end group at 600°C, from Ref. 28. (A) PS-2, M_n = 3090, (B) PS-3, M_n = 9000. Column: fused silica capillary (HP Ultra 1, 0.2 mm I. D. × 50 m, 0.33 μm of immobilized PDMS) programmed from 0 (10 min) to 280°C at 5°C/min. Sample size: ca. 0.1 mg. Detector: FID.

the side reaction which was empirically estimated by Py-GC/MS measurement of different PS samples having a benzyl end-group rather than samples with *n*-butyl end-groups. As shown in Table 3.4, thus corrected M_n values are in fairly good agreement with PS-4 before and after correction at about 3.1 and 5.7%. These results suggest that the Py-GC/MS method can be used to estimate absolute M_n values of PS samples very rapidly using only about 0.1 mg of the samples.

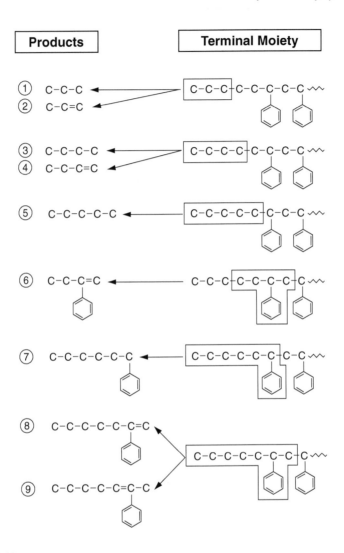

FIGURE 3.13
Assignment of the characteristic peak in 1–9 in Figure 3.12 together with the corresponding *n*-butyl end group moieties in the PS chain, from Ref. 28.

3.4.3 Proof of Ring Structure in Cyclic Polymers

Cyclic polymers having no chain end are expected to have quite different natures from those of the corresponding linear ones. Therefore, the study of physical properties of cyclic polymers has been of interest for many years. Cyclic polymers are usually prepared by the reaction of the living polymers having functional groups on both chain ends with bifunctional linking agents. However, the direct proof of the ring structures of the cyclic polymers has not been an easy task. Recently, the existence of the ring structure in cyclic poly(2-vinylpyridine) (P2VP) was confirmed by using Py-GC/MS.[29] In this

TABLE 3.4

Number-Average Molecular Weight of Polystyrenes Estimated by
SEC and Calculated from End Groups Measured by Py-GC

		M_n	
Sample	By SEC	By Py–GC ($D_p \times 104$) Before Correction	By Py–GC ($D'_p \times 104$) After Correction
PS-1	1190	1180	1240
PS-2	3090	3170	3250
PS-3	9000	8520	9100
PS-4	27600	23200[a]	28200[b]
PS-5	64600	41700	61400
PS-6	152000	65100	130000
PS-7	419000	107000	587000
PS-8	979000	116000	1090000

From Ref. 28.

[a] CV = 3.1% (for five measurements).
[b] CV = 5.7% (for five measurements).

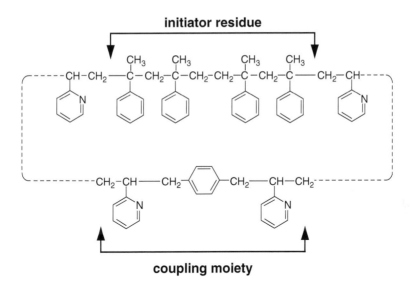

FIGURE 3.14
Possible structure of cyclic poly(2-vinylpyridine), from Ref. 29.

work, the P2VP sample containing ring structure was tentatively prepared
by the reaction of linear P2VP having two living ends, which was polymerized by anionic polymerization of 2-vinylpyridine (2VP) using dipotassium
salt of α-methylstyrene tetramer as a bifunctional initiator, with α,α'-
dibromo-*p*-xylene as a bifunctional coupling agent. Thus prepared, crude
product was fractionated by preparative size exclusion chromatography in
order to enrich the cyclic polymer portion by removing multiply coupled
products. Figure 3.14 shows the possible molecular structure of the cyclic

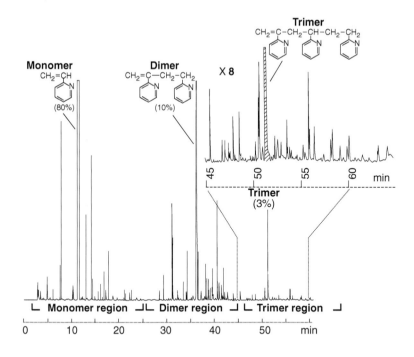

FIGURE 3.15
Pyrogram of poly(2-vinylpyridine) containing cyclic polymer fraction (sample R-1) at 650°C, from Ref. 29. Column: fused silica capillary (HP Ultra 1, 0.2 mm I. D. × 50 m, 0.33 μm of immobilized PDMS) programmed from 50 to 290°C at 5°C/min. Sample size: ca. 0.3 mg, Detector: FID.

P2VP which might be contained to some extent in the fractionated P2VP sample, R-1. Then the proof of the ring structure in the sample R-1 was carried out by detecting the characteristic fragments reflecting the coupling moiety of the polymers at which the coupling agent combined the living 2VP units in both functional points.

Figure 3.15 shows the pyrogram of R-1 at 650°C detected by FID. The characteristic degradation products are 2VP monomer, dimer, and trimer, of which relative peak intensity among total one is about 80%, 10%, and 3%, respectively. These main products, which are mostly associated with the main chain of the molecule, do not provide any useful information about the ring structure at all. According to the anticipated ring structure, however, the possible smallest characteristic fragments that reflect the coupling moiety in the cyclic P2VP, if any, should be observed in the trimer region because the coupling agent combined with the two 2VP units in both sides might yield a hybrid trimer containing three aromatic rings of which molecular weight is to be very close to 2VP trimer with three aromatic rings. Therefore, the trimer region was focused in the following discussion. As shown in Figure 3.15, the expanded pyrogram in the trimer region consists of a number of peaks in addition to the strongest 2VP trimer (VVV).

Here three kinds of model linear P2VP (L-1, L-2, and L-3) were also used in order to comparatively discriminate the characteristic fragments that reflect the coupling moiety in the cyclic P2VP from the other many degradation products observed in the pyrogram. The model polymer named L-1 was synthesized by initiating with α-methylstyrene tetramer dipotassium followed by terminating with MeOH, L-2 by initiating with *n*-BuLi, followed by coupling with α,α'-dibromo-*p*-xylene, and L-3 by initiating with *n*-BuLi, followed by terminating with α-bromo-*p*-xylene.

Figure 3.16 illustrates the possible trimer structures formed from the cyclic P2VP and the model linear P2VPs (L-1, L-2, and L-3), where 2VP, α-methylstyrene, and *p*-xylene units are abbreviated as V, S, and X, respectively. The possible degradation products from R-1 in the trimer region should comprise the fragments containing the residue of the initiator (SSS, VSS, and VVS) and those containing the moiety of the coupling agent (VVX) and (VXV) as well as the strongest 2VP trimer (VVV). On the other hand, as was illustrated in Figure 3.16, L-1, L-2, and L-3 should yield the respective characteristic trimers corresponding to the original molecular structures.

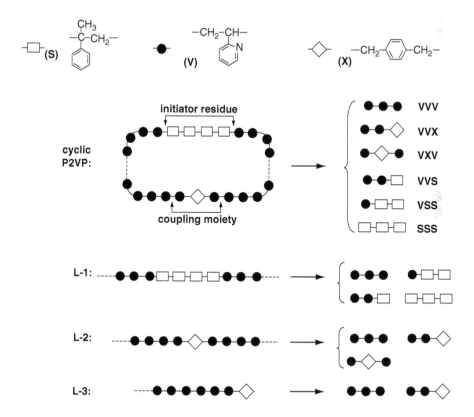

FIGURE 3.16

Possible trimer structure formed from cyclic poly(2-vinylpyridine) [P2VP] and model linear P2VP (L-1–L-3), from Ref. 29.

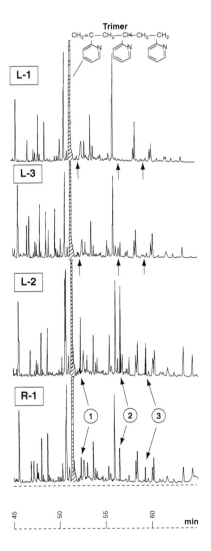

FIGURE 3.17
Detailed pyrograms in the trimer region of poly(2-vinylpyridine) [P2VP] containing cyclic polymer fraction (sample R-1) and model linear P2VP (L-1–L-3), from Ref. 29. Py-GC conditions are the same as those in Figure 3.15.

Therefore, when the retention data of the peaks in the trimer region for L-1, L-2, and L-3 are carefully compared with each other, the possible candidates for the key trimers of VXV that might reflect the coupling moiety in the ring polymer can logically be discriminated as follows.

Figure 3.17 shows the observed partial pyrograms in the trimer region for L-1, L-2, L-3, and R-1. The trimers which are observed on the pyrogram of L-2 but are missing on that of L-3 can be the candidates of VXV. These, however, might have the same retention times by chance as those of the other trimers,

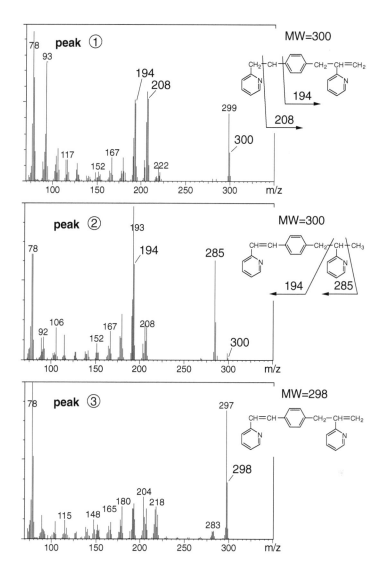

FIGURE 3.18

EI mass spectra of the products reflecting the coupling moiety observed in the trimer region (peaks 1, 2, and 3 in Figure 3.17), from Ref. 29.

including S unit such as VVS, VSS, and SSS, of which the contribution, if any, can easily be eliminated by comparing with the pyrogram of L-1. By this logical comparison of the pyrograms of the three model polymers (L-1, L-2, and L-3), it was deduced that the peaks 1, 2, and 3 appearing on the pyrogram of L-2 should be attributed to the peaks of the key trimers of VXV.

Then, by considering the deduction process mentioned above, the mass spectra of those three peaks (1, 2, and 3) were carefully interpreted. Figure 3.18

shows the observed EI mass spectra for peaks 1, 2, and 3. The spectrum of peak 1 indicates the M^+ ion at m/z 300 and fragment peaks at m/z 208 and 194. That of peak 2 indicates the same M^+ ion at m/z 300 and the base peak at m/z 285 and a fragment peak at m/z 194. That of peak 3 indicates the M^+ ion at m/z 298. As was expected, the possible structural formulae of the three peaks thus estimated, shown in Figure 3.18, have the common basic structure as VXV.

As shown in Figure 3.17, the specific peaks, 1, 2, and 3 are also clearly observed in the pyrogram of R-1. Therefore, it can be unequivocally concluded that R-1 should include a certain amount of the polymers with the anticipated ring structure illustrated in Figure 3.15. Here, it can be empirically estimated that about 30% of the fractionated P2VP sample (R-1) should have the ring structure by comparing the relative intensities of the specific peaks 1, 2, and 3 between R-1 and L-2, taking the following conditions into consideration: i) each molecule of L-2 comprises one coupling moiety, and ii) R-1 and L-2 have almost comparable M_n, 1.5×10^4 and 1.3×10^4, respectively.

3.4.4   Thermal Degradation of Mechanisms of PS

Py-GC/MS is one of the most powerful techniques to study the thermal degradation mechanism of polymers. In particular, the use of isotope-labeled polymer samples is often effective for such investigation. Here the detailed study of the thermal degradation mechanism of PS is described using a deuterium-labeled PS sample.

As shown in Figure 3.3, PS thermally degrades mainly to monomer (ca. 70–80%) with some dimer (2.4-diphenyl-1-butene) and trimer (2.4.6-triphenyl-1-hexene). The thermal degradation of PS at elevated temperatures is initiated by a random scission of the polymer main chain to give primary and secondary macroradicals (Eq. 3.1).

$$
\sim C-C-C-C \!\!\mid\!\! C-C-C-C\sim \quad \rightarrow \quad
\begin{array}{l}
\sim C-C-C-C^{\cdot} \quad \text{primary macroradical} \\
\;\;|\quad\;\;| \\
\text{Ph}\;\;\;\text{Ph} \\
\\
+ \\
\\
\sim C-C-C^{\cdot} \quad \text{secondary macroradical} \\
\;\;|\quad\;\;| \\
\text{Ph}\;\;\;\text{Ph}
\end{array}
\tag{3.1}
$$

Both macroradicals chiefly depolymerize to the monomer. Macroradicals can also engage in hydrogen abstraction. The formation of the dimer and the trimer has been mostly explained through intramolecular 1,3- and 1,5-transfers, respectively, from the terminal groups of the secondary macroradical to the reactive tertiary carbon followed by β-scission (Eq. 3.2).

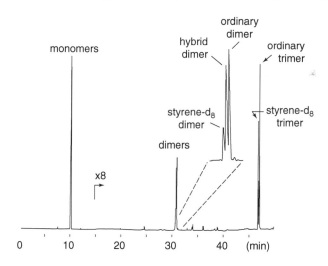

$$(3.2)$$

Among these, the latter 1,5-transfer is reasonably accepted as the main path to form the trimer by a back-biting mechanism through a six-membered ring intermediate. However, the former 1,3-transfer has not been clearly accepted as the main path of the dimer formation by a reasonable mechanism and/or intermediates.

Recently, the thermal degradation of a block copolymer of ordinary and perdeuterated styrene, poly(styrene-b-styrene-d_8) (styrene/styrene-d_8 = ~1/1) was investigated by Py-GC/MS and by pyrolysis-field ionization MS (Py-FIMS) to clarify the detailed mechanism of the dimer formation.[30] Figure 3.19 shows the pyrograms of the poly(styrene-b-styrene-d_8) at 500°C.

FIGURE 3.19

Pyrogram of poly(styrene-b-styrene-d_8) at 500°C, from Ref. 30. Column: fused silica capillary (HP Ultra 1, 0.2 mm I. D. × 50 m, 0.33 μm of immobilized PDMS) programmed from 50 to 180°C at 8°C/min, then 2°C/min to 210°C, and finally 8°C/min to 300°C. Sample size: ca. 0.1 mg, Detector: FID.

The main degradation products are styrene-d_8 and the ordinary styrene monomers, which are not completely separated under the given gas chromatographic conditions, although the retention times of the deuterated products are always slightly shorter than those of ordinary hydrogenated ones. The trimer peaks in Figure 3.19 are composed of a doublet that corresponds to a styrene-d_8 trimer and an ordinary styrene trimer. Hybrid trimers consisting of both ordinary styrene and styrene-d_8 units are negligible.

The dimer peaks are composed of a triplet with relative intensities of 14%:45%:41%. Almost the same triplet of the dimer peaks are also observed in the pyrogram of the blend of styrene and styrene-d_8 homopolymers. The first and the last peaks of the triplet are homo-dimers of styrene-d_8 and ordinary styrene, respectively; the central peak proved to be the hybrid dimer of both styrene units. The fairly strong peak intensity of the hybrid dimer (45%) cannot be fully explained through simple chain scissions around the minute amount of the junctions of the two types of styrene units in the block copolymer chains, but must also result from the contribution of some additional intermolecular reactions to form the dimers.

Figure 3.20 shows two EI-mass spectra measured at (a) the relatively early and (b) the later retention part of the hybrid dimer peak, which apparently consists of a single peak. Interestingly, the spectrum of the first part of the peak differs from that of the latter part. This fact suggests that the hybrid dimer peak would contain at least two components. The spectrum of the first part indicates the base peak at m/z 91 and the most intense M^+ ion at m/z 217, whereas that of the latter indicates the base peak at m/z 98 and the most intense M^+ ion at m/z 215. These observations suggest that the possible hybrid dimers should be $C_{16}H_7D_9$ ($M_W = 217$) and $C_{16}H_9D_7$ ($M_W = 215$) whose structures are shown in Figure 3.20.

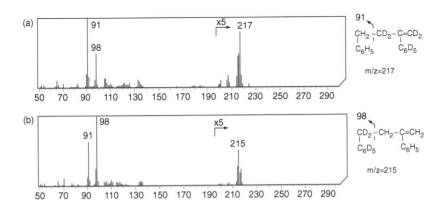

FIGURE 3.20
EI mass spectra of the hybrid dimer peak in Figure 3.19, from Ref. 30. (a) early part of the peak, (b) latter part of the peak.

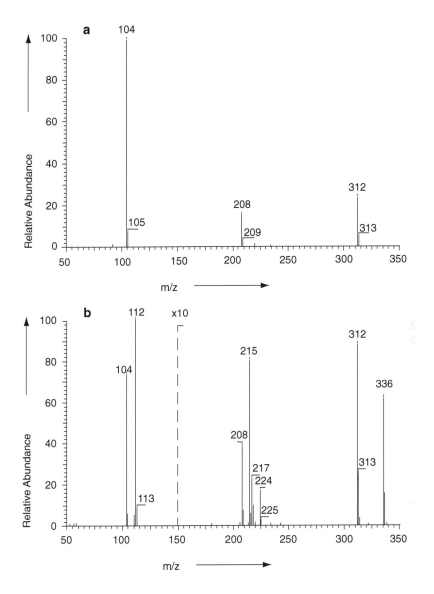

FIGURE 3.21
Py-FIMS spectra of polystyrenes, from Ref. 30. (a) ordinary polystyrene, (b) poly(styrene-b-styrene-d$_8$). Pyrolysis heating rate: 50 to 700°C at 1.2°C/sec. Sample size: ca. 0.05 mg.

These results are also supported by Py-FIMS for the same sample. Figure 3.21 shows Py-FIMS spectra of (a) ordinary PS and (b) the poly(styrene-b-styrene-d$_8$). In the spectrum of the block copolymer (b), signals resulting from molecular ions of ordinary styrene ($m/z = 104$) and styrene-d$_8$ ($m/z = 112$) monomers and the respective homo-trimers ($m/z = 312$ and 336) are observed, whereas hybrid trimers are completely missing. The observed hybrid dimer peaks are mostly composed of the signals at m/z 215 and 217, whereas that at m/z 216($C_{16}H_8D_8$) is fairly small. This result suggests that the hybrid dimers are not mainly formed through the simple dimerization of monomers generated during pyrolysis.

On the basis of the results observed by Py-GC/MS and Py-FIMS, the most probable mechanisms for formation of the dimers and the trimers during the thermal degradation of PS are proposed as follows. Among the transfer reactions of the macroradicals formed initially, the intramolecular 1,5-transfer of the secondary radical (R_2) followed by β-scission occurs most frequently, whereas the primary radical mostly undergoes depolymerization. Thus, a trimer and R_2 radical are proposed by the scission at β_1 (Eq. 3.3).

$$\underset{\substack{| \\ Ph}}{\sim C-C}\overset{\beta_1}{\underset{|}{\biggl|}}\underset{\substack{| \\ Ph}}{C-C}-\underset{\substack{| \\ Ph}}{C-C}-\underset{\substack{| \\ Ph}}{C-\dot{C}} \quad \longrightarrow \quad \underset{\substack{| \\ Ph \\ R_2}}{\sim C-C^{\cdot}} \; + \; \underset{\substack{| \\ Ph}}{C=C}-\underset{\substack{| \\ Ph}}{C-C}-\underset{\substack{| \\ Ph}}{C-C} \qquad (3.3)$$

In this case, only homo trimers are formed even from the poly (styrene-b-styrene-d_8). On the other hand, a 1,3-diphenylpropyl radical (R_3) and an unsaturated polymer chain end are formed by the scission at β_2 (Eq. 3.4).

$$\underset{\substack{| \\ Ph}}{\sim C-C}-\underset{\substack{| \\ Ph}}{C-C}\overset{\beta_2}{\underset{|}{\biggl|}}\underset{\substack{| \\ Ph}}{C-C}-\dot{C} \quad \longrightarrow \quad \underset{\substack{| \\ Ph}}{\sim C-C}-\underset{\substack{| \\ Ph}}{C-C=C} \; + \; \underset{\substack{| \\ Ph}}{^{\cdot}C-C}-\underset{\substack{| \\ Ph}}{C} \qquad (3.4)$$

$$\qquad\qquad\qquad\qquad\quad \text{unsaturated chain end} \qquad R_3$$

The R_3 radical should give rise to 1,3-diphenylpropane by hydrogen abstraction. However, the fact that the observed peak of 1,3-diphenylpropane is fairly small suggests that the R_3 radical might undergo further decomposition to form a monomer and a benzyl radical (Eq. 3.5).

$$\underset{\substack{| \\ Ph}}{^{\cdot}C-C}-\underset{\substack{| \\ Ph}}{C} \quad \longrightarrow \quad \underset{\substack{| \\ Ph}}{\dot{C}} \; + \; \underset{\substack{| \\ Ph}}{C=C} \qquad (3.5)$$

$$\qquad R_3 \qquad \text{benzyl radical} \qquad \text{monomer}$$

Although the benzyl radical could give rise to toluene by hydrogen abstraction, the toluene peak observed in the pyrogram is also very small. This result suggests that the benzyl radical might attack the other unsaturated chain ends formed by the β-scission described previously; the dimer is then formed by the β-scission to the double bond (Eq. 3.6).

$$
\begin{array}{ccc}
\sim\!\!\overset{|}{C}\!\!\!+\!\!C\!-\!C\!=\!\!\overset{|}{C} & \overset{.}{C} & \\
\underset{Ph}{|} \quad \underset{Ph}{|} + \underset{Ph}{|} & & \longrightarrow \\
\text{another chain-end} & &
\end{array}
\qquad
\begin{array}{ccc}
\sim\!\!\overset{.}{C} & C\!=\!C\!-\!\overset{|}{C}\!-\!\overset{|}{C} & \\
+ & \underset{Ph}{|} \quad \underset{Ph}{|} & \\
R_2 & \text{dimer} &
\end{array}
\qquad (3.6)
$$

Thus, the hybrid dimers can be formed from the block copolymer if the ordinary or deuterated benzyl radical attacks the other unsaturated chain ends formed during pyrolysis as follows (Eq. 3.7).

$$
\begin{array}{l}
\sim\!CD_2\!-\!\underset{C_6D_5}{\overset{|}{CD}}\!-\!CD_2\!-\!\underset{C_6D_5}{\overset{|}{C}}\!=\!CD_2 \;+\; \underset{C_6D_5}{\overset{|}{\overset{.}{C}D_2}} \longrightarrow CD_2\!=\!\underset{C_6D_5}{\overset{|}{C}}\!-\!CD_2\!-\!\underset{C_6H_5}{\overset{|}{CD_2}} \quad \text{styrene-}d_8 \text{ dimer}
\end{array}
$$

$$
\begin{array}{l}
\sim\!CD_2\!-\!\underset{C_6D_5}{\overset{|}{CD}}\!-\!CD_2\!-\!\underset{C_6D_5}{\overset{|}{C}}\!=\!CD_2 \;+\; \underset{C_6H_5}{\overset{|}{\overset{.}{C}H_2}} \longrightarrow CD_2\!=\!\underset{C_6D_5}{\overset{|}{C}}\!-\!CD_2\!-\!\underset{C_6H_5}{\overset{|}{CH_2}} \quad \text{hybrid dimer } (M_w = 217)
\end{array}
$$

$$
\begin{array}{l}
\sim\!CH_2\!-\!\underset{C_6H_5}{\overset{|}{CH}}\!-\!CH_2\!-\!\underset{C_6H_5}{\overset{|}{C}}\!=\!CH_2 \;+\; \underset{C_6D_5}{\overset{|}{\overset{.}{C}D_2}} \longrightarrow CH_2\!=\!\underset{C_6H_5}{\overset{|}{C}}\!-\!CH_2\!-\!\underset{C_6D_5}{\overset{|}{CD_2}} \quad \text{hybrid dimer } (M_w = 215)
\end{array}
$$

$$
\begin{array}{l}
\sim\!CH_2\!-\!\underset{C_6H_5}{\overset{|}{CH}}\!-\!CH_2\!-\!\underset{C_6H_5}{\overset{|}{C}}\!=\!CH_2 \;+\; \underset{C_6H_5}{\overset{|}{\overset{.}{C}H_2}} \longrightarrow CH_2\!=\!\underset{C_6H_5}{\overset{|}{C}}\!-\!CH_2\!-\!\underset{C_6H_5}{\overset{|}{CH_2}} \quad \text{ordinary dimer}
\end{array}
$$

$$(3.7)$$

The most reliable mechanism for the dimer formation from PS proved to be associated with the intermolecular reaction of the benzyl radicals, which are formed from the secondary macroradicals, rather than the intramolecular 1,3-transfer of the secondary macroradical.

In a similar manner, thermal degradation mechanisms of fully aromatic polyesters were investigated in detail by Py-GC/MS using a partially deuterated polymer sample.[31] Moreover, the mechanisms of reactive pyrolysis of the same aromatic polyester in the presence of tetramethylammonium hydroxide (TMAH) was also studied by Py-GC/MS.[32] In this case, the contribution of solvent (methanol) to the reaction was confirmed by use of a deuterated methanol solution of TMAH.

3.4.5 Thermal Degradation Mechanisms of Flame-Retarded Polyesters

Halogenated organic compounds in the presence of antimony oxide (Sb_2O_3) have been widely used as effective flame-retardants in many polymeric systems. It is generally thought that the flame-retarding effect of these additive systems is due mainly to the synergism between the halogenated compounds and antimony oxide to form volatile antimony halides such as $SbCl_3$ and $SbBr_3$, which act as free radical traps around the flame. Therefore, the formation mechanisms of the antimony halides are of great interest.

On the other hand, evolved gas analysis (EGA) can observe the evolution behavior of the thermal degradation products as a function of programmed temperature. This temperature-programmed pyrolysis (TPPy) technique would provide detailed information to understand the synergistic flame-retardancy of halogenated organic compounds/Sb_2O_3 system. Recently, the thermal degradation process of poly(butylene terephthalate) (PBT) containing a synergistic flame-retardant system based on brominated polycarbonate (Br-PC), consisting of tetrabromobisphenol-A and Sb_2O_3, was investigated by means of the TPPy technique.[33]

A discussion of detailed degradation processes on the basis of the chemical information about the volatile products follows. First TPPy-GC/MS was applied to identify the thermal degradation products formed from flame-retarded (FR)-PBT during the programmed heating. In a TPPy-GC/MS system, a temperature-programmable microfurnace pyrolyzer was directly coupled with a GC/MS equipped with a metal capillary separation column. During the heating of the furnace from 60 to 700°C at a rate of 10°C/min, the degradation products were trapped in a part of the separation column near the injection port by immersing in liquid nitrogen in a Dewar vessel, and then the trapped components were analyzed by GC/MS.

Figure 3.22 shows (a) the TIC chromatogram of cold trapped products evolved from FR-PBT, and (b) the mass chromatogram for *m/z* 362 corresponding to the parent ion of $SbBr_3$ observed for FR-PBT. The TIC chromatogram for FR-PBT mostly consists of the common pyrolysis products observed both for PBT and for Br-PC alone. Among these, various flame-retarding species such as HBr and various bromophenols originated from Br-PC were observed. Moreover, as might be expected for the synergistic reaction between Br-PC and Sb_2O_3, a prominent peak of $SbBr_3$ is found out on the mass chromatogram at around 11 min shown in Figure 3.22(b), although the corresponding peak was hardly observed in the TIC chromatogram. The identification of $SbBr_3$ was confirmed by the facts that the retention time and the mass fragmentation pattern for this peak shown in Figure 3.22 basically coincided with those observed for the $SbBr_3$ reagent sample.

The molecular evolution profiles of the thermal degradation products were then studied by a TPPy-MS system in which the capillary separation column was replaced by a deactivated stainless steel capillary transfer line maintained at 300°C. Figure 3.23 shows the thermal degradation profiles using

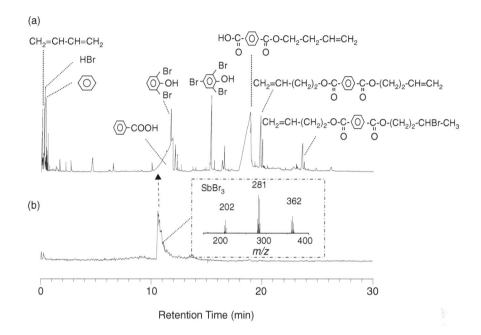

FIGURE 3.22
Typical gas chromatogram of the degradation products evolved from (a) flame-retarded (FR)-PBT, and (b) the mass chromatogram for *m/z* 362 corresponding to the parent ion of SbBr₃ observed for FR-PBT along with the mass spectrum of SbBr₃, from Ref. 33. Column: a metal capillary (Frontier Lab Ultra Alloy-1, 0.25 mm I. D. × 15 m, 0.15 μm of immobilized PDMS) programmed from 35 to 300°C at 10°C/min.

single-ion monitoring (SIM) curves traced mostly at the molecular ion peaks of the main degradation products, together with the corresponding TIC curves obtained for (a) PBT, (b) Br-PC and (c) FR-PBT.

As shown in Figure 3.23(a), the SIM curves at *m/z* 54 (1,3-butadiene) and at *m/z* 203 (butylene terephthalate) and TIC curve for PBT show apparent one-stage degradation at ca. 380°C. On the other hand, as shown in Figure 3.23(b), the SIM curves at *m/z* 82 (HBr) and at *m/z* 252 (2,6-dibromophenol) for Br-PC show that the degradation takes place stepwise in at least two stages. At the early degradation stage, brominated phenols such as 2,6-dibromophenol is evolved in a temperature range from 400 to 460°C with a peak at ca. 450°C. At a slightly higher temperature above the TIC peak, a steep evolution peak for HBr is observed with a maximum at ca. 460°C. These observations suggest that in the early degradation stage the degradation of Br-PC takes place through scission of ester linkages, while at the higher temperatures above ca. 460°C dehydrobromination, forming char residue, occurs predominantly.

Here, it is interesting to note that the observed TIC curve with a peak-top at ca. 380°C for FR-PBT shown in Figure 3.23(c) is very close to that for pure PBT, although the former exhibits significant leading evolution starting at

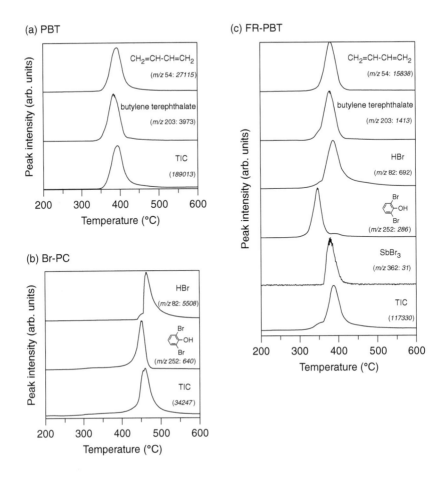

FIGURE 3.23
SIM curves for representative degradation products and TIC curved observed for (a) PBT, (b) brominated polycarbonate (Br-PC), and (c) FR-PBT measured by TPPy-MS, from Ref. 33. The italic numbers in parentheses represent the peak heights.

ca. 320°C. The observed substantial temperature lowering shift for the evolution of both HBr and 2,6-dibromophenol from Br-PC moiety clearly suggests a contribution of the synergistic effects of the flame-retardant additives. These synergistic effects are mostly attributed to the catalytic effect of Sb_2O_3 to promote the thermal degradation of carbonate linkages of Br-PC.

These observations also clearly indicate that in this flame-retardant system, brominated phenols are first evolved at slightly lower temperatures than those of the flammable product such as 1,3-butadiene and butylene terephthalate evolved from the substrate polymer, PBT, to cause the initial flame-retarding effect. In addition, HBr would then be evolved over the whole degradation temperature range for PBT to cause free radical trap in volatile phase and char formation to decrease flammable products.

Furthermore, it is worthwhile to note that the observed initiation temperature of the flame-poisoning SbBr$_3$ (*m/z* 362) evolution nearly coincides with that of HBr evolution, suggesting that the formation of SbBr$_3$ might be triggered by the evolution of HBr. These observations support the reported mechanisms that the SbBr$_3$ formation is closely related to the evolution of HBr in the flame-retardant systems containing brominated flame retardants in the presence of Sb$_2$O$_3$.

3.5 Conclusions

Py-GC/MS has long been appreciated as a simple, but rapid and extremely sensitive technique for the characterization of almost any type of polymers, including insoluble thermosetting and crosslinked polymers and complex biopolymers, as well as ordinary synthetic polymers. Owing to the recent progress in the measuring system such as pyrolyzers, separation columns and detection systems comprising MS, the modern Py-GC/MS has made a great stride toward becoming a powerful tool not only in polymer identification and compositional determination of polymer blends and copolymers but also in various microstructural characterization of high polymers. The structural information such as branching, stereoregularity, end-group, network, and copolymer sequence is often exclusive and/or complementary to that obtained by the conventional spectroscopic methods such as FT-IR and NMR. Moreover, pyrolysis in the presence of some specific reagents and/or catalysts has recently acquired a stable position among various analytical pyrolysis methodologies. For example, so-called "thermally assisted hydrolysis and methylation" technique using organic alkaline reagents is widely utilized for reliable and informative characterization of various condensation-type polymers that are often intractable for the conventional pyrolysis techniques.[34] Thus, advanced Py-GC/MS will increasingly play a very important role in the fields of practical material characterization as well as basic polymer specification.

References

1. Madorsky, S. L. and Straus, S., *Ind. Eng. Chem.*, **5**, 848, 1948.
2. Wall, L. A., Mass Spectrometric Investigation of the Thermal Decomposition of Polymers, *J. Res. Natl. Bur. Std.*, **41**, 315, 1948.
3. Bradt, P., Dibeler, V. H., and Mohler, F. L., A New Technique for the Mass Spectrometric Study of the Pyrolysis Products of Polystyrene, *J. Res. Natl. Bur. Std.*, **50**, 201, 1953.
4. James, A. T. and Martin, A. J. P., Gas-Liquid Partition Chromatography, *Analyst.*, **77**, 915, 1952.

5. Davison, W. H. T., Slaney, S., and Wragg, A. L., A Novel Method of Identification of Polymers, *Chem. Ind.*, 1356, 1954.

6. Lehrle, R. S. and Robb, J. C., Direct Examination of the Degradation of High Polymers by Gas Chromatography, *Nature*, **183**, 1671, 1959.

7. Radell, E. A. and Strutz, H. C., Identification of Acrylate and Methacrylate Polymers by Gas Chromatography, *Anal. Chem.*, **31**, 1890, 1959.

8. Martin, S. B., Gas Chromatography Application to the Study of Rapid Degradative Reactions in Solids, *J. Chromatogr.*, **2**, 272, 1959.

9. Golay, M. J. E., in *Gas Chromatography*, Coates., V. J., Ed., Academic Press, New York, 1, 1958.

10. Dandneau, R. and Zerner, H. E., An Investigation of Glasses for Capillary Chromatography, *J. High Res. Chromotagr. Chromatogr. Commun.*, **2**, 351, 1979.

11. Vallmin, J., Kriemler, P., Omura, I., Seible, J., and Simon, W., *Microchem. J.*, **11**, 73, 1966.

12. Schmuddege, D. R. and Hummel, D. O., *Adv. Mass Spectrom.*, **4**, 73, 1968.

13. Tsuge, S. and Ohtani, H., *Pyrolysis-Gas Chromatography of Polymers—Fundamentals and Data Compilations*, Techno System Co., Tokyo, 1989.

14. Translation of Reference 13 by Jin, X. and Luo,Y. F., Chinese Science and Technology Publ. Co., Beijing, 1992.

15. Tsuge, S. and Ohtani, H., Structural Characterization of Polymeric Materials by Pyrolysis-GC/MS, *Polym. Degrad. Stab.*, **58**, 109, 1997.

16. Tsuge, S., Analytical Pyrolysis—Past, Present and Future, *J. Anal. Appl. Pyrolysis*, **32**, 1, 1995.

17. Nonobe, T., Ohtani, H., Usami, T., Mori, T., Fukumori, H., Hirata, Y., and Tsuge, S., Characterization of Stereoregular Polystyrene by Pyrolysis-Gas Chromatography, *J. Anal. Appl. Pyrolysis*, **33**, 121, 1995.

18. Nonobe, T., Tsuge, S., Ohtani, H., Kitayama, T., and Hatada, K., Stereoregularity of Poly(methyl methacrylate)s Studied by Pyrolysis-Gas Chromatography/Mass Spectrometry, *Macromolecules*, **30**, 4891, 1997.

19. Kiura, M., Atarashi, J., Ichimura, K., Ito, H., Ohtani, H., and Tsuge, S., Tacticity of Methacrylic Copolymers and Their Crosslinked Polymers Studied by Pyrolysis-Gas Chromatography, *J. Appl. Polym. Sci.*, **78**, 2410, 2000.

20. Ohtani, H., Ishiguro, S., Tanaka, M., and Tsuge, S., Characterization of Polymerization Reagents Incorporated into Poly(methyl methacrylate) Chains by Pyrolysis-Gas Chromatography, *Polym. J.*, **21**, 41, 1989.

21. Ohtani, H., Tanaka, M., and Tsuge, S., Pyrolysis-Gas Chromatographic Study of End Groups in Poly(methyl methacrylate) Radically Polymerized in Toluene Solution with Benzoyl Peroxide as Initiator, *J. Anal. Appl. Pyrolysis*, **15**, 167, 1989.

22. Ohtani, H., Tanaka, M., and Tsuge, S., End Group of Poly(methyl methacrylate) as a Function of Molecular Weight Determined by Pyrolysis-Gas Chromatography, *Bull. Chem. Soc. Japan*, **63**, 1196, 1990.

23. Tsukahara, Y., Nakanishi, Y., Yamashita, Y., Ohtani, H., Nakashima, Y., Luo, Y. F., Ando, T., and Tsuge, S., Investigation of the End Groups of Prepolymers/ Macromonomers Prepared by Radical Chain Transfer Reaction, *Macromolecules*, **24**, 2493, 1991.

24. Ohtani, H., Luo, Y. F., Nakashima, Y., Tsukahara, Y., and Tsuge, S., Determination of End Group Functionality in Poly(methyl methacrylate) Macromonomers by Pyrolysis Simultaneous Multidetection Gas Chromatography, *Anal. Chem.*, **69**, 1438, 1994.

25. Ohtani, H., Takehana, Y., and Tsuge, S., Quantification of End Groups in Anionically Polymerized Poly(methyl methacrylate)s by Pyrolysis-Gas Chromatography, *Macromolecules*, **30**, 2542, 1997.
26. Ito, Y., Tsuge, S., Ohtani, H., Wakabayashi, S., Atarashi, J., and Kawamura, T., Characterization of Branched Alkyl End Group of Poly(methyl methacrylate) by Pyrolysis-Gas Chromatography, *Macromolecules*, **29**, 4516, 1996.
27. Ohtani, H., Suzuki, A., and Tsuge, S., A Novel Approach to the Characterization of End Groups in Styrene-Methyl Methacrylate Copolymers by Pyrolysis-Gas Chromatography, *J. Polym. Sci.: Part A*, **38**, 1880, 2000.
28. Ito, Y., Ohtani, H., Ueda, S., Nakashima, Y., and Tsuge, S., Quantification of End Groups in Polystyrene by Pyrolysis-Gas Chromatography, *J. Polym. Sci.: Part A*, **32**, 383, 1994.
29. Ohtani, H., Kotsuji, H., Momose, H., Matsushita, Y., Noda, I., and Tsuge, S., Ring Structure of Cyclic Poly(2-vinylpyridine) Proved by Pyrolysis-GC/MS, *Macromolecules*, **32**, 6541, 1999.
30. Ohtani, H., Yuyama, T., Tsuge, S., Plage, B., and Schulten, H.-R., Study on Thermal Degradation of Polystyrenes by Pyrolysis-Gas Chromatography and Pyrolysis-Field Ionization Mass Spectrometry, *Eur. Polym. J.*, **26**, 893, 1990.
31. Sueoka, K., Nagata, M., Ohtani, H., Nagai, N., and Tsuge, S., Thermal Degradation Mechanisms of Liquid Crystalline Aromatic Polyester Studied by Pyrolysis-Gas Chromatography/Mass Spectrometry, *J. Polym. Sci.: Part A*, **29**, 1903, 1991.
32. Ishida, Y., Ohtani, H., and Tsuge, S., Effects of Solvents and Salts on the Reactive Pyrolysis of Aromatic Polyester in the Presence of Tetramethylammonium Hydroxide Studied by Pyrolysis-Gas Chromatography/Mass Spectrometry, *J. Anal. Appl. Pyrolysis*, **33**, 167, 1995.
33. Sato, H., Kondo, K., Tsuge, S., Ohtani, H., and Sato, N., Thermal Degradation of Flame-Retarded Polyester with Antimony Oxide/Brominated Polycarbonate Studied by Temperature Programmed Analytical Pyrolysis Techniques, *Polym. Degrad. Stab.*, **64**, 41, 1998.
34. Special issue THM, *J. Anal. Appl. Pyrolsyis*, in press.

a

4

Electrospray Ionization (ESI-MS) and On-Line Liquid Chromatography/ Mass Spectrometry (LC/MS)

Laszlo Prokai

CONTENTS

4.1 Introduction

The possibility of obtaining gaseous ions via the use of charged droplets has captured the attention of mass spectrometrists since the late 1960s. Projects by Dole and co-workers, involving spraying dilute solutions of macromolecules such as polystyrene through a syringe held at high potential into a gas-filled chamber for mass spectrometric analysis, have been the origin of today's electrospray ionization (ESI).[1,2] These early efforts ended up with the developers concluding that problems apparently due to multiply-charged species and cluster formation prevented the determination of accurate molecular weights of polymers, and work had been abandoned on ESI mass spectrometry until the 1980s.[3,4] Rediscovered by Fenn and co-workers, it has now been developed into a powerful technique used extensively to desorb at atmospheric pressure biomolecules dissolved in polar solvents. Ironically, the unique ability of the technique to produce multiply charged gaseous ions has been found a benefit for these analytes, which allowed for viewing of macromolecular ions on a quadrupole mass spectrometer having a mass range extending only to m/z 1500 in the early experiments. Synthetic polymers were among the early focus of research on ESI of high-mass species, and they have represented a significant application area of the technique.

A catalyst to the development of techniques involving sprays as methods of sample introduction and/or ionization has also been the pursuit of on-line coupling of liquid chromatography (LC) with mass spectrometry (LC/MS).[4] ESI is inherently suited for this hyphenation. Interfaces that initially accommodated liquid effluents at only about 10 μl/min have now been developed to be compatible with a wide range of flow rates—from nanoliters per minute (nanospray) to 1–2 ml per min (pneumatically assisted electrospray or ionspray). ESI mass spectrometry has become the method of choice in general, when on-line LC/MS of various synthetic polymers is considered. Atmospheric-pressure chemical ionization (APCI) is another technique that employs spray vaporization, but relies on external ionization, and has been applied to the on-line LC/MS analysis of selected oligomeric mixtures ("prepolymers"). Thermospray was an earlier method developed as an LC/MS interface that employed heat instead of an electric field for generation of charged droplets. This spray ionization technique (together with other methods to couple LC with mass spectrometry such as particle beam and flow-FAB interfaces) will not be discussed in detail, because it has now been superseded by ESI and APCI that are more sensitive and amenable to a much wider variety of analytes, including synthetic polymers, than thermospray.

In this chapter, an overview is presented about the principles and instrumentation of atmospheric pressure ionization mass spectrometry, followed by the application of ESI mass spectrometry and LC/MS to oligomers and polymers. Representative applications are discussed to highlight benefits and limitations of the techniques.

4.2 Electrospray, Ionspray, and APCI

4.2.1 Instrumentation

Today's mass spectrometers are available with electrospray interfaces of various designs. However, they all rely on the principle of nebulization of a liquid into an aerosol of charged microdroplets due to strong electric field.[5] The schematic diagram of a typical ESI interface is shown in Figure 4.1. The solution containing the analyte is injected at a constant flow rate into the spray chamber through a small-diameter (metal, fused silica, or glass) capillary or needle (isolated by fused silica tubing from the device maintaining

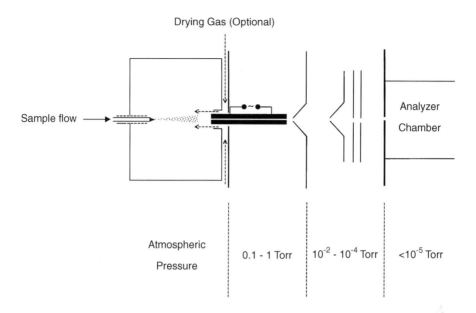

FIGURE 4.1
Schematic illustration of a typical ESI source.

the sample flow). The solution leaving the capillary is sprayed into the plume of charged microdroplets because of a ΔV difference (usually 1–5 kV) between the potential at the tip of the sprayer (V_n) and the potential of an inlet of a small internal-diameter capillary transfer tube (Figure 4.1) located a few millimeters away. A cone or flat plate with an orifice (V_c) may also serve as a counter electrode. The solution fed to the needle is normally drawn into a liquid cone (the so-called "Taylor cone") by the applied potential. If ΔV is positive, the microdroplets contain excess positive charge and positive ions are generated for mass analysis, while negatively charged microdroplets are formed upon a negative ΔV value and, ultimately, gas-phase negative ions are obtained.[6] The absolute potential values applied depend on the source design and on the mass analyzer, and they are normally adjusted ("tuned") to achieve a stable spray from the sample solution supplied. The electrospray process, in which only a high electric field at the surface of the liquid creates the electric stress generating the droplets, is limited to liquid flow rates of microliters per minute.

Efforts to develop routine electrospray interfaces have also concentrated on providing additional ways of stabilizing the production of the charged microdroplets and/or to increase liquid flow rates that the instrument can tolerate, compared to the basic design where only electrical forces are used for nebulization (Figure 4.2a). A high-velocity coaxial gas (usually nitrogen) flow can be used to assist the process of aerosol formation, as shown in Figure 4.2b. This technique has been referred to as ionspray[7] or pneumatically assisted electrospray. Arrangements that incorporate the infusion of an

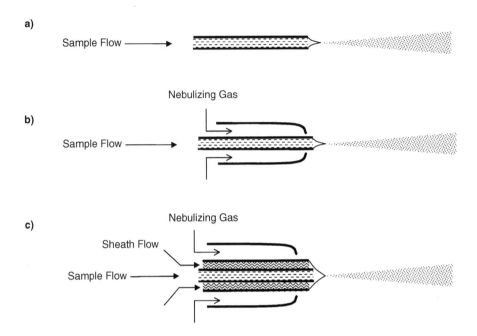

FIGURE 4.2
Spraying capillaries for ESI: (a) Nebulization and droplet charging by the electric field only,
(b) Pneumatically assisted electrospray or ionspray, and (c) Tri-coaxial probe with sheath flow.

additional solution or sheath flow (Figure 4.2c) during ESI have also been
constructed. In addition to enhancing spray stability under certain conditions, a liquid sheath flow may be used to add an agent necessary for ESI
of selected analytes in LC/MS applications without affecting the chromatographic separation.[8] Off-axis electrospray configurations and ESI sources that
employ capillaries pulled to a narrow (5 to 20 μm i.d.) tip and allow for spraying
at nL/min flow rates (nano-ESI) have also been developed. However, the
principle of ionization is the same for all spraying techniques.

Spraying in ESI is conducted at atmospheric pressure and ambient temperature. Therefore, no thermal effects causing the decomposition of the
analyte are observed. The process of ion formation is extremely soft; usually
no fragmentation occurs. Fragmentation can be induced by increasing the
kinetic energy of the ions leaving the droplet in specific parts of the ion
source and/or by employing collision-induced dissociation (CID). This may
be done, depending on the source design, by increasing the capillary—
skimmer or repeller—collimator potential difference, or increasing the "cone"
voltage. Ions entering through the skimmer or orifice are transferred to the
mass analyzer by using electrostatic lenses, further skimmer systems, collima-
tors, octapoles, and so on, where they are separated according to their mass-
to-charge ratio (m/z).

APCI, in contrast with ESI, employs a heated vaporizer that dispenses a
flowing liquid stream (up to 2 ml/min) in the form of small droplets in a carrier

gas, and the ionization of the vaporized sample molecules is carried out down-stream in the gas phase through an atmospheric-pressure ion–molecule reaction (i.e., by chemical ionization).[9,10] At a sufficiently high vaporizer and source temperature, the droplets are vaporized very rapidly, which allows intact molecules to evaporate or desorb with minimal thermal decomposition. The primary ionization, usually by a corona discharge maintained in the source via a sharp needle at kV potential, creates reagent ions from the solvent vapor that flows through the discharge region. Reagent ions consist of the protonated solvent ions in the positive-ion mode,[11] and solvated oxygen ions in the negative-ion mode.[12] Addition of modifiers such as a buffer may change reagent-ion composition; e.g., the addition of ammonium acetate buffer can make protonated ammonia and acetate ions the primary reagent ions in the positive and negative mode, respectively. Chemical ionization is very efficient at atmospheric pressure because of the high collision frequency, and the high gas pressure, as well as the moderating influence of solvent clusters on the reagent ions minimizing the internal energy transferred to the analyte ions formed, which reduces fragmentation. Nevertheless, multiply charged ions are not produced, and the upper molecular-weight limit of samples that can be addressed by APCI is much smaller than that of ESI, and APCI has been used mostly for LC/MS. APCI is usually available for instruments configured for ESI; changeover involves swapping of or switching to the respective spraying/nebulizing components, and both techniques may employ a common atmospheric pressure ionization (API) to analyzer interface ("API stack").[13] However, applications of APCI to LC/MS of synthetic polymers have been scarce.

4.2.2 Mechanism of Electrospray Ionization

The appearance of any particular compound in an ESI mass spectrum depends, first of all, on whether the compound is ionized in the solution being electrosprayed.[14] For biopolymers (peptides, proteins, oligonucleotides, and the like), ionization in the solution depends on the ionization constant (pK_a) of the analyte and on the pH of the solution. ESI produces multiply charged (protonated or deprotonated) ions from molecules that have multiple charge sites. Unlike biopolymers, most synthetic polymers have no acidic or basic functional groups that can be utilized for ion formation through acid–base equilibria. Cationization has been the preferred technique for producing gaseous ions from synthetic oligomers and polymers by electrospray ionization. A small amount (10^{-5} to 10^{-5} M) of an appropriate inorganic salt dissolved in the spraying solvent usually affords meaningful ESI mass spectra. Without the dissolved salt, the ESI process often may be erratic, and only relies on an uncertain amount of inorganic contaminants present in the sample for the formation of ionized molecules. On the other hand, ESI can be achieved from various spraying solvents.

Solvent evaporation upon heat transfer from the heated part of the ESI source via the ambient gas leads to the shrinking of the droplets and to the accumulation of excess surface charge that lead to the generation of gas-phase ions. According to the description in the previous section, ions of one polarity (depending on the sign of ΔV) are preferentially drawn into the droplets by the electric field as they are separated from the bulk liquid. (The separation is, however, incomplete, as each droplet contains many ions of both polarities.) The function of an ESI source, beyond the generation of the charged microdroplet, includes the removal of solvent vapor via a differentially pumped vacuum system and the transfer of analyte ions into the mass analyzer operating at high vacuum.

Models have been developed on ion formation in ESI, but there is still no consensus on the mechanism by which sample ions are obtained for mass spectrometric analysis. These models rely on the existence of preformed ions in solution; i.e., the ions observed in the mass spectra were presumed to be present originally as ionized molecules in solution. According to the charged residue model of Dole et al., the evaporation of solvent from a charged droplet increases the surface field until the Raleigh limit is reached:

$$d_{RL} \leq 8\gamma / \varepsilon_0 E^2 \qquad (4.1)$$

where d_{RL} is the diameter of the droplet of the Raleigh limit, γ is the surface tension, and ε_0 is the permittivity of the droplet's ambient medium, and E is the electric field at the surface of the droplet. At the Raleigh limit, Coulomb repulsion becomes of the same order as surface tension, and the resulting instability disperses the droplet into several smaller droplets in a process sometimes called "Coulomb explosion." These smaller droplets continue to evaporate until they also reach their Raleigh limit and, thus, disintegrate. A series of such solvent evaporation–Coulomb explosion sequences ultimately produces droplets small enough to contain a single molecule that holds some of its droplet's charge. This charged molecule becomes a gaseous ion when the last of the solvent molecules evaporate.

Another and more widely accepted model proposed by Iribarne and Thomson[15] assumes that solute ions from charged droplets are formed by field-assisted desorption (also referred to as "ion evaporation"). According to this model, the surface electric field becomes high enough (up to several V/nm)[16] to desorb analyte ions from the shrinking droplets before they reach their Raleigh limit. The strong electric field evolved on the surface of the droplet assists the solute ion in overcoming the energy barrier that blocks its escape.

Other theoretical aspects of ion formation during ESI have also been studied.[17] Most of them are beyond the scope of this chapter that concentrates on analytical application of the technique to synthetic oligomers and polymers. One aspect worthy of consideration is the extent of gas-phase charging in ESI. A basic model to explain multiple charging considers an ESI experiment involving PEGs of varying average molecular weight.[18] The charge capacity of a molecule is expected to reach its limit when, because of Coulomb repulsion by other charges, the electrostatic potential energy of the centermost charge equals the energy that binds the cation (Na^+) to the oxygen atoms on the $HO(CH_2CH_2O)_nH$ oligomers (which theoretically have $n + 1$ binding sites). The binding energy between Na^+ and the oxygen atoms is about 2.05 eV.[19] When the electrostatic potential energy of the central charge is taken to be the pairwise sum over all other charges of terms comprising A/x, where x is the distance between the charges (in Å) and A is the Coulomb constant (14.38 eV/Å), and the PEG molecule is considered of a linear "zigzag" configuration, the maximum number of charges according to this model should reach 10, 18, and 30 at PEG molecular masses of 3.6 kDa ($n \sim 80$), 8 kDa ($n \sim 180$), and 18 kDa ($n \sim 400$), respectively. However, the actual charge-holding capacities of these oligomers are 6, 10, and 22, respectively, which may be due to the folding of the molecules, making the distance between the charges less than that of the linear conformers assumed by the model. In another study, the extent of gas-phase protonation of an entire series of polyamidoamine (PAMAM) starburst dendrimers was found to exhibit a linear relationship to $M_r^{2/3}$, consistent with theoretical models predicting a spherical ion structure with maximum charging controlled by Coulombic effects.[20] In general, the extent of charging is expected to increase with the increase in the binding energy of the cation to its binding site on the macromolecule and, therefore, in the order (H^+), Li^+, Na^+, K^+, Rb^+, and Cs^+.

4.2.3 Interpretation of ESI Mass Spectra

ESI mass spectrometry has found applications for molecular-weight determination and structural analysis of biopolymers, especially proteins.[21] Fewer reports have been published on its application to synthetic polymers, despite the fact that the pioneering work of Dole and co-workers and Fenn et al. showed that macromolecules such as poly(ethylene glycol)s (PEGs) up to 5,000,000 Da could be ionized by this technique.[22] It was noted that ions in the ESI mass spectra of PEG fell, due to multiple charging, consistently within a certain mass-to-charge (m/z) window, regardless of the actual molecular weight distribution of the samples. Multiply charged molecules are produced by ionic species (H^+, Na^+, K^+, and the like) being attached to a neutral analyte in the positive-ion mode. In negative-ion ESI, removal of protons or attachment of anions yields the ions of the sample molecule. It is very crucial to recognize the ability of ESI to form multiply charged ions; therefore, ESI mass spectra may require an interpretation procedure.

Multiple charging and the interpretation method are illustrated by the ESI mass spectrum of insulin, recorded in the positive-ion mode at mass resolution of ≤1000 upon spraying from an aqueous-methanolic (1:1) solution that contained acetic acid (1%, v/v), in Figure 4.3. Peaks marked in the spectrum vary only by the number of attached protons or cations. When interpreting an ESI mass spectrum of an unknown compound, any pair of ions that differ in charge state by one can be used to calculate the charge state and to determine the relative molecular mass (M_r; usually referred to as the "molecular weight"). In general, the m_n mass-to-charge ratio of an ion

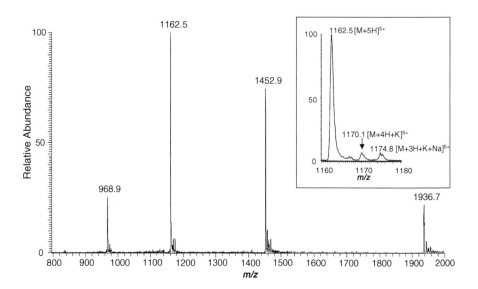

FIGURE 4.3
ESI mass spectrum of recombinant human insulin recorded at low mass resolution (M/ΔM < 1,000) by a quadrupole ion-trap instrument (spraying solution: 50/50 methanol/water containing 1% acetic acid, 3µL/min). Inset: Profile of the multiple-charged ions in the *m/z* 1160 to 1180 range.

with *n* positive charges can be calculated as

$$m_n = \frac{M_r + n \cdot M_a}{n} \tag{4.2}$$

where M_a is the relative mass of the attached ionic species (1 for H^+, 23 for Na^+, 39 for K^+, and so on). By expressing *m/z* for the ion with $n - 1$ positive charges,

$$m_{n-1} = \frac{M_r + (n-1) \cdot M_a}{n-1}, \tag{4.3}$$

and solving the above equations to *n*:

$$n = \frac{m_n - M_a}{m_{n-1} - m_n}. \tag{4.4}$$

After determination of *n*, M_r for the neutral analyte can be calculated:

$$M_r = n \cdot m_n - n \cdot M_a. \tag{4.5}$$

Taking *m/z* 1162.5 and 1452.9 as m_n and m_{n-1} and assuming the attachment of protons ($M_a = 1$) for the ESI mass spectrum of insulin (Figure 4.3), the calculated n and M_r are 5 and 5807.7, respectively. Repeating the calculation for the other m/z pairs as m_n and m_{n-1} (1452.9 and 1936.7; 968.9 and 1162.5), additional estimates of M_r may be obtained, and the calculated values may be averaged. (The standard deviation also can be used to judge the accuracy of the experimental M_r.)

Direct determination of the charge states can also be done by recording the ESI mass spectrum on an instrument that allows for the resolution of the isotope peaks of the analyte. (Chapter 1 in this book discusses mass resolution in greater detail.) Figure 4.4 shows the resolution of isotope peaks, by using ESI and Fourier-transform ion-cyclotron resonance mass spectrometry (FT-ICR) for the multiply charged ion of recombinant human insulin with m/z 1162.53 as a centroid. (The resolving power of the analyzer used to record

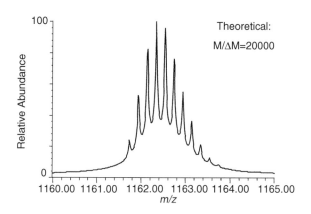

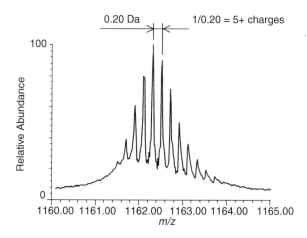

FIGURE 4.4

ESI mass spectrum showing the charge state of a multiply charged ion of recombinant human insulin (m/z 1162) directly by the resolution (at M/ΔM ≤ 20,000) of the isotope peaks on an FT-ICR instrument. Bottom trace: measured spectrum; top trace: predicted isotope pattern. (Courtesy of W. J. Simonsick, Jr., Du Pont Marshall Laboratory, Philadelphia.)

the ESI mass spectrum shown in Figure 4.3 was inadequate for mass separation of the isotope clusters.) The m/z difference between the peaks resolved by the analyzer is 0.2 u; by taking its reciprocal value, a charge-state of $n = 5$ can be obtained directly.

4.3   Electrospray Ionization Mass Spectrometry of Polymers

Few studies have addressed finding ESI mass spectrometric conditions optimal to polymer analysis. Factors to be considered include mainly the composition (solvent and the agent promoting the formation of sample ions) of the spraying solution and the sampling or focusing of the gaseous ions within the ESI source.

Compounds with high fluorine content are insoluble or only sparingly soluble in conventional solvent systems, making them difficult candidates for mass spectrometry studies.[23] In ESI, when precipitation is not an overriding factor, a minimal amount of water in the employed spraying solvent can vastly improve fluorinated polymer signal intensities compared to purely organic solvent systems. Dilution with small amounts of higher polarity solvents promotes the desorption of longer chain fluorocarbons, presumably due to augmented solvophobicity. However, a very high aqueous content (high polarity) may disfavor the desorption of longer chain fluorocarbons, especially at higher polymer concentration. This latter observation was attributed to preferential intermolecular aggregation of longer fluorocarbon chains. The presence of fluorinated groups offers the advantage of inductive stabilization of anionic charge sites for improved signals in negative-ion MS, while low molecular-weight halogenated solvents used for dissolution of fluorinated polymers can suppress the tendency toward discharge in negative-ion ESI. The solvent system was also found to influence the ESI mass spectrometric efficiency of polyesters.[24] Acetone-water, methanol-chloroform, and methanol-tetrahydrofuran mixtures (1:1, v/v) with 10^{-5} M polyester and sodium acetate concentrations gave good signal-to-noise ratios and reproducibility of the ESI mass spectra. However, the sodium acetate concentration was a particularly critical parameter in these solvent mixtures.

As discussed in the previous section of this chapter, cationization has been the preferred technique for producing gaseous ions from synthetic oligomers and polymers by electrospray ionization. Synthetic polymer samples usually contain alkali metal cations as impurities from the chemicals, solvents, or the glassware, and the attachment of these cations (with Na^+ adducts being the most common) may be observed. However, a small amount (10^{-5} to 10^{-4} M) of an appropriate inorganic salt dissolved in the spraying solvent is preferred to obtain ESI mass spectra. A lower salt concentration usually decreases the abundance of the analyte ions, while higher concentration of the inorganic salt may impair the ESI process.

A range of metal cations was shown to form adducts with polystyrene.[25] Both MALDI and ESI mass spectra contain adduct ion peaks corresponding to $[M + X]^+$, where PS represents a polystyrene molecule and X a cationic species. In addition, salt cluster complexes assigned as $[PS + X(XA)_n]^+$, where XA represents the metal salt, and $n = 1$ or 2, are observed in ESI spectra. The addition of K salts led to the most intense and reproducible ESI spectra for PS. Polysulfide oligomers, $H(SC_2H_4OCH_2OC_2H_4S)_nH$ with $n = 1$–24 also furnished the best signal-to-noise conditions, low-to-medium cone voltages, and a spraying solvent of acetone containing 0.5% KI.[26] Various cations can be used to generate the cationized species of poly(3-nitratomethyl-3-methyloxetane) also known as polynimmo, which contains up to 18 cyclic oligomers in addition to the dominant tetramer, the highest species detected containing 22 repeat units.[27] Highly specific cation-cyclic oligomer interactions were apparent in the adducts with Na^+, K^+, NH_4^+, and H^+, some of which have been rationalized through molecular modeling calculations.

Analytical ESI mass spectrometry of synthetic polymers concentrated mostly on the direct application of the technique. Mass spectra of polymers can be measured in minutes and provide far more detailed mass information than conventional methods. Polymers are complex mixtures with heterogeneity not only in size (molecular weight distribution), but in chemical composition and end-groups. Furthermore, distributions in architecture add another level of complexity. The characterization of polymer structure is important because it provides us with the basics for chemical and physical properties as well as the mechanism of polymerization. ESI mass spectrometry has been used for the qualitative analysis of various oligomeric mixtures to study their heterogeneity. Analysis of the mass spectra revealed, e.g., the presence of individual $H(SC_2H_4OCH_2OC_2H_4S)_nH$ oligomers, of certain oligomers with repeat units containing additional oxyalkylene groups (and in some cases a monosulfide link rather than disulfide), and the presence of end-groups such as epoxy in a complex oligomeric linear polysulfide.[28] In another application of the technique, linear polynimmo oligomers from the tetramer up to species of mass 3200 Da were detected, affording the characterization of several new combinations of end-groups. Application of tandem mass spectrometry has also been useful for the assignments of individual peaks in the ESI mass spectra and structural or end-group characterization of low molecular weight polymers.[29,30] Figure 4.5 shows the ESI mass spectrum of a polyester resin obtained by the condensation of 1,4-cyclohexanedicarboxylic acid (A), 2,2,4-trimethylpentane-1,3-diol (B), and a small amount of trimethylolpropane (TMP), along with CID product-ion MS/MS of the sodiated A_2B_3 oligomer. The molecular structure of the sodiated homopolyesters poly(dipropoxylated bisphenol-A/adipic acid) and poly(dipropoxylated bisphenol-A/isophthalic acid), as well as their copolyesters, was studied by electrospray ionization and sustained off-resonance irradiation CID on an FT-ICR instrument and six different dissociation mechanisms similar to those observed during the pyrolysis of these compounds were described.[31] The formation of the fragment ions observed in the CID spectra after the cleavage of the ester or ether

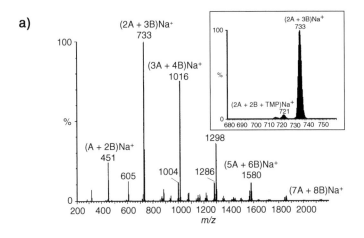

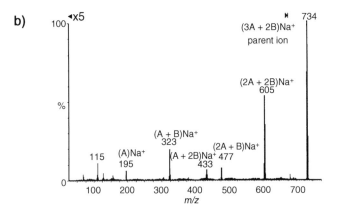

FIGURE 4.5

a) ESI mass spectrum (in 90% aqueous acetone doped with sodium acetate) of a polyester resin, a copolymer of 1,4-cyclohexanedicarboxylic acid (A), 2,2,4-trimethylpentane-1,3-diol (B) and a small amount of trimethylolpropane (TMP); b) MS/MS (product ion) spectrum of *m/z* 733. (Reprinted with permission from Ref. 29. Copyright ©1995 John Wiley & Sons, Inc., New York.)

bonds was found to be governed by the sodium affinity of the products. Although sequence-specific fragment ions among the CID products were present for some oligomers, such fragments were not present for all possible copolyester sequences. Higher-order fragment analysis or multistage tandem mass spectrometry (MS^n, $n \geq 3$) in combination with the ESI, which is routinely available on quadrupole ion-trap instruments, has also been employed recently for the structural characterization of mixed polyesters,[32,33] polyester-based functional polymers,[34] and poly(methyl methacrylate).[35] With MS^n, the sequence distribution and microstructure of mass-selected poly(3-hydroxybutyrate-co-3-hydroxyvalerate) (PHBV) biopolyester macroinitiator, obtained by partial alkaline depolymerization of natural PHBV containing 5 mol % of 3-hydroxyvalerate units, could be assessed from the dimer up to the oligomer with 22 repeat units, but the application of the technique was not able to furnish similar information for other mixed polyesters.

When applied to the determination of molecular weight distribution data, direct ESI mass spectrometric analyses face various obstacles. Similarly to biopolymers, synthetic polymers may also produce multiply charged ions when they are electrosprayed; therefore, their ESI mass spectra can be extremely complex, because the multiply charged ions of the oligomer distribution may overlap. In addition to "human data reduction," deconvolution techniques may be employed to provide molecular-weight information on many components from an unseparated mixture.[36–38] Mass spectra of multiply charged ions of a polyamidoamine starburst polymer were deconvoluted to provide molecular-weight information on many components in a nonseparated mixture, as shown in Figure 4.6.[39] These data were also used for determining the polydispersity value of this synthetic polymer system.

An increase in the polymer's average molecular weight requires an appropriate increase in the resolving power of the mass spectrometer to afford meaningful compositional information. While poly(ethylene glycol)s of M_n <5–10 kDa give ESI mass spectra containing unresolved oligomers due to overlapping charge-state envelopes of the polydisperse mixture, ESI-FT-ICR afforded resolved isotopic peaks representing the individual oligomers in samples up to an average molecular weight of 23 kDa. Approximately 5000 isotopic peaks of 47 oligomers in 10 charge states are identified in the 23 kDa spectrum, as well as <0.02% -$CH_2CH(CH_3)O$- monomer units in the 13 kDa spectrum (Figure 4.7).[40] As an unexpected advantage of ESI, the degree of mass discrimination was much less than that of mass/charge discrimination due to averaging of the values from different charge states. For the determination of molecular-weight distributions, geometric and entropy deconvolution methods yielded unacceptable artifact peaks and abundance discrimination, respectively. Combining their deconvolution attributes with isotopic peak restrictions for the 4.3 kDa polymer yielded a distribution similar to that from human data reduction, which was consistent with that from size-exclusion chromatography (SEC). While ESI-FT-ICR studies of PEGs certainly illustrate the power of the technique in the characterization of synthetic polymers, information on the chemical composition of copolymers can be obtained only by employing mass analysis at ultrahigh resolution available on FT-ICR instruments. For example, isobaric monomers glycidyl methacrylate (GMA) and butyl methacrylate (BMA) have the same nominal mass (142 Da) but differ in exact mass by 0.036 (the difference between O and CH_4). In addition to resolving the isotope peaks, isobaric resolution is required for detailed structural characterization. Although isobaric peaks could not be resolved at all by MALDI-FT-ICR at 3.0 Tesla (T),[41] isobaric resolution was obtained up to the hexamers (852 Da nominal mass) at 3.0 T with $m/\Delta m_{50\%} = 80,000$[42] and up to the 49-mers (molecular mass around 6965 Da) at 9.4 T with $m/\Delta m_{50\%} = 500,000$[43] in narrow molecular-weight fractions (obtained by off-line SEC fractionation). As shown in Figure 4.8,

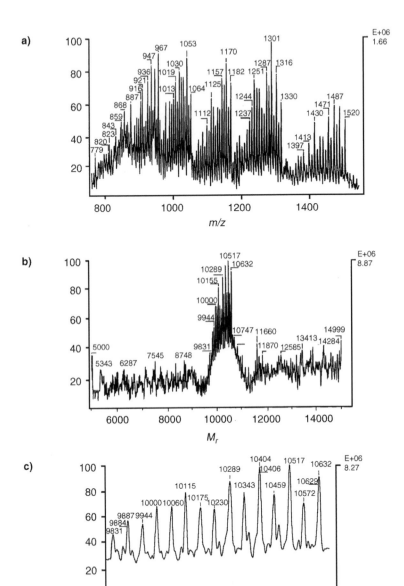

FIGURE 4.6
a) ESI mass spectrum of a polyamidoamine Starburst polymer (Generation 4); b) deconvoluted mass spectrum; c) expanded view of the deconvoluted mass spectrum. (Reprinted with permission from Ref. 39. Copyright ©1995 John Wiley & Sons, Ltd., Chichester, UK.)

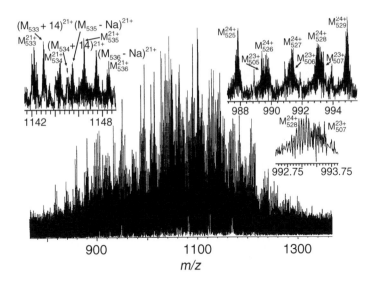

FIGURE 4.7
ESI-FT-ICR mass spectrum of PEG 20000, $M/\Delta M = 50,000_{FWHM}$. (Reprinted with permission from Ref. 40. Copyright ©1995 American Chemical Society.)

the copolymer products differing according to the end-groups (azo initiator, saturated/unsaturated) were clearly distinguished by ESI-FT-ICR, and the asymmetrical isobaric distribution for the n-mers $(GMA)_m(BMA)_{n-m}$, $0 \le m \le n$, also indicated that GMA was less reactive than BMA in the polymerization process. The determination of accurate end-group masses by FT-ICR has been another example, where the use of ESI may be preferred over MALDI.[44] The improvement is actually due to the combined use of multiple charge states observed with ESI, which results in an increased number of data points used for the regression procedure in the end-group determination. For poly(oxyalkylene)s, the end-group masses were determined with an error better than 5 and 75 millimass units in the molecular weight range of 400 to 4200 and 6200 to 8000 Da, respectively. End-group and monomer masses have been determined by ESI-FT-ICR analysis for PEG with average mass higher than 8000 Da that represented a two-fold increase over MALDI on the same instrument.

To reflect an accurate molecular weight distribution, the mass spectrometric method must not significantly discriminate over a wide mass and concentration range. Possible areas of discrimination are the ionization process itself, propensity to fragmentation of the ions formed, ion transfer/separation, and detection. To assess the reliability of obtaining accurate molecular-weight distribution data from ESI mass spectra, isolated monodisperse methyl methacrylate (MMA) oligomers (25 and 50 repeat units) were used to determine molar signal response and propensity for fragmentation.[45] The sum of the peak areas for the multiply charged MMA 50-mer was found to be only about 66% of the summed peak areas for the 25-mer for the same molar

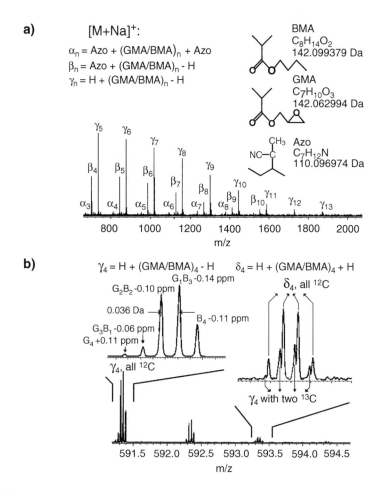

FIGURE 4.8
ESI-FT-ICR mass spectrum of a GMA/BMA copolymer: (a) 625 < *m/z* < 2100; (b) Ultrahigh resolution and accurate mass measurement for the GMA/BMA tetramer. (Reprinted with permission from Ref. 43. Copyright ©1995 American Chemical Society.)

concentration, but conversion of the multiply charged peak areas to the singly charged representation gave equal signal responses for the 25- and 50-mers with peak area compression taken into account. Therefore, it was proposed that transformation of the polymer spectra to the singly charged molecular-ion distribution should allow accurate calculation of average molecular weights, polydispersity, end-group mass, and repeat unit mass. However, both ESI and MALDI mass spectrometric methods appeared to emphasize the presence of low-mass material compared to gel-permeation chromatographic analysis, as revealed by studies on the thermal degradation of poly(propylene oxide) (PPO), Mn = 2000.[46] This was attributed to the different sensitivity of the techniques as a function of the molecular mass, and to some *in situ* fragmentation during mass spectrometric analysis.

Discrimination for lower mass against higher mass oligomers in ESI mass spectrometry has also been described by other publications. The discrimination was apparent when low molecular weight fractions were removed to allow for the observation of macromolecules with a high degree of polymerization.[47] While off-line fractionation can overcome these problems,[48] it is labor-intensive and time-consuming. A significant bias in transfer efficiencies of synthetic oligomer ions in monocomponent systems of polymeric mixtures from solution to gas phase has been found under routine ESI conditions (1 μL/min flow rate) even for two very similar polymers differing only in end-group functionality such as methyl-terminated vs. 2-methylpropyl-terminated poly(dimethylsiloxane) (PDMS).[49] For oligomers of similar mass but very different polarity such as PDMS and PEG, this bias was even more notable. This ionization dependence would pose a problem in quantitative analyses of polymer blends by ESI. The solution to gas phase bias correlates with the surface tension of the solutions containing the polymer blend; thus, it strongly depends on the molecular weight of the polymer and on the potential surface activity of the end-groups. Surface tension obviously influences ESI droplet dynamics, which may be manifested by the profound influence on the appearance of the ESI mass spectra of polymer blends by the desolvation conditions such as the application of a drying gas (see Figure 4.1 for the illustration of the drying gas in the ESI source). The ion signal from the higher surface-tension and more favorably solvated component such as PEG ($\gamma \approx 42$ dyne/cm) is expected to scale with the drying gas flow rate, and a component with the lower surface tension such as PDMS ($\gamma \approx 20$ dyne/cm) may be almost completely suppressed. Less bias in solution to gas phase transfer for diverse analytes should occur when nano-ESI conditions (in sources optimized for nL/min sample flow rates) are used.[42] In a nano-ESI source, the small-diameter (nanometer range) droplets with a lower volume solvent and a larger number of analyte molecules existing at the surface of the droplet introduce far less bias stemming from selective solution to gas phase ion transfer than ESI carried out routinely at μL/min flow rates. Therefore, nano-ESI has been suggested as the method of choice for the analysis of complex polymer blends,[42] although this technique is experimentally more demanding than conventional ESI.

Shifts in the distribution of multiply charged ions with changes in ion-optical conditions such as the cone potential have also been recognized in the multiple-charge distributions of proteins and other biomolecules. The effect of cone voltage on the molecular-weight distribution of poly(styrene) samples was studied, and relatively high cone voltages (120 V) were required to ionize this nonpolar polymer.[50] At lower cone voltages (30–60 V), cluster ions (e.g., $[M + 2X]^{2+}$ and $[2M + X]^+$) were observed. Singly charged species predominated only at high cone voltages, but no fragmentation of the oligomer ions was seen. The molecular weight distribution obtained by ESI was compared with that obtained by size-exclusion chromatography, and the agreement (generally within 10%) was cone-voltage dependent. The cone voltage applied in the ESI experiments also had significant effect on the relative abundances of high and low molecular-weight ions of polynimmo and on the relative amounts of H^+ and NH_4^+ adducts in cationization experiments with NH_4Cl. Despite giving highly accurate relative molar masses of individual species, the ESI technique failed to give accurate molecular-weight distributions, especially because no doubly or triply charged ions were formed. The effects of cone potential (related to ion transfer) have been modeled mathematically,[51] and it has been concluded that it exerts a focusing effect dependent on the m/z of ions. This focusing effect may determine the dependence of oligomer ion intensities upon the cone potential in the ESI mass spectra of polymers. For polymers with low polydispersities (e.g., narrow molecular weight PEG standards with $D < 1.5$), the variation in determinations of the molecular-weight averages tends to be small (typically <5%), whereas with synthetic polymers with polydispersities greater than 2, variations in cone potential can influence molecular-weight determinations significantly (by 100% or even more). Under nonideal, empirically determined operating conditions, mass discrimination effects were also shown to occur in the ESI-FT-ICR analysis of synthetic polymers as a function of external ion accumulation time, which required a multidimensional tuning process to eliminate the bias in obtaining accurate molecular weight distribution data by the technique.[52]

Many other factors that may influence the accuracy of the calculation of molecular-weight distribution data cannot be investigated rigorously. In general, there may be no major sources of inaccuracies, when calculation of molecular weight averages and polydispersity is based on multiple-charge distributions of a small number of species. However, obtaining realistic mass distributions and reliable molecular-weight averages from ESI mass spectra can be of major concern with most polydisperse synthetic polymers. Table 4.1

TABLE 4.1

Analytical Application of ESI Mass Spectrometry (Without On-Line Chromatographic Separation) to Synthetic Oligomers and Polymers

Analyte	Refs.
PEG (PEO)	53–55
PEG derivatives	56
PPG (PPO) and EO/PO copolymer	46, 57
Polyglycerols	58
Polyethers	27
Polysulfides	26, 59
Polyesters and acrylates	24, 29, 31–35, 43, 45, 60, 61
Polystyrene and stryrene copolymers	25, 62–64
Amidoamine and diaminobutane dendrimers	39, 65
Polysiloxanes	66
Fluorinated polyphosphazanes	23
Telechelics	67
Conducting polymers	68
Oligomeric surfactants	69

summarizes selected examples for studies on synthetic oligomers and polymers by ESI mass spectrometry that did not employ on-line chromatographic separation.

4.4 LC/MS of Polymers

LC techniques are widely used for the characterization of synthetic polymers.[70] With the need to control polymer composition and thus the resulting properties, the demand for more extensive polymer analysis and characterization has escalated.[71] Differential refractive index (RI) and evaporative light-scattering (ELS) detectors commonly used in LC, while good concentration detectors, provide very little information about the chemical composition. Other (such as ultraviolet and fluorescence) detectors can provide additional, albeit limited, chemical composition information. Mass spectrometry, on the other hand, is one of the most powerful detectors for polymers as it can provide chemical composition information at every chain length. The coupling of LC techniques with mass spectrometry addresses more than just the characterization of the polymer. Other fundamental analytical questions can be answered. By observing all the species eluting at a given time, the characteristics of the chromatographic system can be monitored and column behavior evaluated. In direct mass spectrometry, fragmentation can be a concern, especially for "fragile" polymers. It is sometimes unclear if peaks at low m/z are real components or fragment ions. The use of mass spectrometry alone often yields incorrect or unreliable quantitative data such as molecular weight averages and polydispersities, particularly for polymers with broad distributions. On the other hand, LC techniques are quantitatively reliable in polymer analysis.

There have been numerous successful approaches to LC/MS, ranging from mechanical transport of solute to the mass spectrometer after external solvent removal (belt/wire systems and particle-beam interfaces) to bulk solution introduction (with or without splitting) involving nebulization and ionization direcly from the solvent stream.[72,73] However, LC/MS has been surprisingly underutilized for the characterization of synthetic polymers despite its apparent advantages over the direct application of mass spectrometry.

ESI and APCI have become widely available interfaces on today's mass spectrometers, and on-line coupling to separation is straightforward, based on the design of the instruments as discussed in Section 2.1. Together, they cover a broad range of analyte polarities and molecular weights, are extremely versatile in accommodating various LC techniques, supercritical fluid chromatography (SFC), and even capillary electrophoresis (CE);[74] thus, they have essentially made obsolete other methods of on-line coupling of LC to mass spectrometry. Table 4.2 summarizes selected applications of LC/MS to the characterization of synthetic oligomers and polymers.

TABLE 4.2

Selected Publications on the Characterization of Oligomeric/
Polymeric Samples by ESI or APCI Mass Spectrometry
Combined with On-Line Separation

Analyte	Separation	Interface	Refs.
PEG	SEC	ESI	75
Poly(tetrahydrofuran)	SEC	ESI	8
Polystyrene	SEC	ESI	8
Polyesters	SEC	ESI	8, 42, 89, 96
	GPEC	ESI	88
	RPLC	APCI	76–78
Amino resins	SEC	ESI	73
	RPLC	ESI	79, 80
	RPLC	APCI	74, 80
	CE	ESI	74
Phenolic resins	SFC	APCI	81
	(SEC)	(PB-EI)	(75)
Poly(propylene imine) dendrimers	RPLC	ESI	82
Oligomeric surfactants	SEC	ESI	93–95

SEC: size-exclusion chromatography; RPLC: reversed-phase liquid
chromatography; GPEC: gradient polymer elution chromatography;
CE: capillary electrophoresis; SFC: supercritical-fluid chromatography; EI: electron ionization; PB: particle beam.

LC techniques for synthetic polymers can be categorized according to their
mode of operation. Figures 4.9 to 4.11 illustrate different modes that may
afford separation of oligomers and their mixtures according to specific molecular properties for an oligomeric surfactant Triton X-100 [octylphenoxy-
poly(ethoxy)ethanol] as an example. Figure 4.9 shows the total ion current
(TIC) chromatogram, along with the contour plot (*m/z* on the axis vs. chromatographic elution time on the horizontal axis, and shaded areas in the x-y
plane indicate ESI ions with intensity exceeding the threshold), for a normal-
phase separation by gradient elution. In this mode of LC, the oligomers are
separated, but the elution of different oligomeric series (**I-IV**) that reflect
chemical heterogeneities overlap.

A reversed-phase LC-ESI-MS analysis of Triton X-100 is shown in Figure 4.10,
which displays separation according to chemical heterogeneity practically
independent of molecular size. The chromatographic resolution of oligomeric
mixtures may also rely on liquid adsorption chromatography (LAC, performed usually on silica gel as a stationary phase) and gradient polymer
elution chromatography (GPEC). In LAC, all sample components initially
prefer to adsorb on the surface of the stationary phase, and the increase in
the percentage of a strong solvent (displacer) results in the sequential elution
according to the change in the adsorption equilibria involving the analyte
molecules, the stationary phase, and the mobile phase.[83] To date, no application of LAC coupled with ESI/APCI mass spectrometry has been reported.

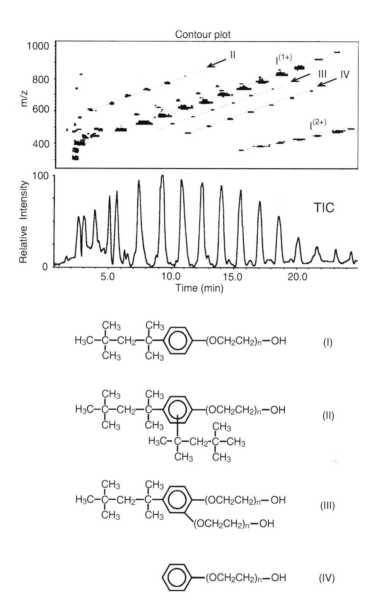

FIGURE 4.9

TIC chromatogram (bottom chart) and contour plot for the LC-ESI-MS analysis of Triton X-100, an oligomeric surfactant, by using gradient normal-phase chromatography (2 mm i.d. cyano-propylsilica column, from 95/5 hexane/dichloromethane to 50/40/10 hexane/dicholomethane/methanol in 20 min, 200 µL/min flow rate, no effluent split, nebulizer-assisted electrospray), and the oligomer series identified (**I-IV**). The m/z values of the peaks that belong to the major oligomer series (**I**) follow the formula $[M + NH_4^+] = 224 + 44n$, where n is the number of ethoxy units, and doubly charged $[M + 2NH_4]^{2+}$ ions are also present. (Courtesy of PE Sciex, Foster City, CA)

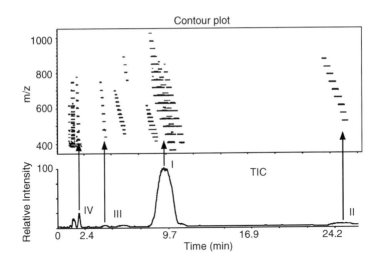

FIGURE 4.10

TIC chromatogram (bottom chart) and contour plot for the LC-ESI-MS analysis of Triton X-100, by using gradient reversed-phase chromatography [2 mm i.d. octadecylsilica column, 20/80 to 50/50 acetonitrile/10 mM ammonium acetate in 20 min, 200 μL/min flow rate, 3:1 effluent split, nebulizer-assisted electrospray]. See Figure 4.9 for the oligomer series (**I-IV**) separated. (Courtesy of PE Sciex, Foster City, CA)

In GPEC (also known as high-performance precipitation liquid chromatography),[84–87] the oligomeric mixture is dissolved in a good solvent of the analyte and injected onto the column equilibrated with a poor solvent ("nonsolvent") as an initial mobile phase that results in the precipitation of the polymer on the top of the column. An increasing percentage of the good solvent during gradient elution will redissolve the oligomer molecules according to both their molecular weight and chemical composition. Ideally, the stationary phase does not have an effect on the separation process. Therefore, GPEC may be performed on nonpolar stationary phases such as octadecylsilica (ODS) bonded phase.[88] GPEC/ESI-MS has been used to characterize dipropoxylated bisphenol A/adipic acid polyesters.[89] When liquid chromatography at the critical point of adsorption (LCCC) is used, the chromatographic elution becomes independent of the molecular weight and only depends on chemical heterogeneity. LCCC requires specific solvent composition and temperature; thus, method development is critical and may be tedious. Experimentally less demanding gradient separations in which the method is tuned for a mass-independent elution have been developed ("pseudo-LCCC"; the analysis of Triton X-100 presented in Figure 4.10 may be considered as an example) and used on-line with ESI mass spectrometry for the analysis of alkylated poly(ethylene glycol) and terephthalic acid/neopentyl glycol polyester resin. Because the separation does not significantly decrease the polydispersity of the analyte in this hyphenated technique, ESI mass spectrometry is useful mainly for identification of the oligomers, and as a support to LC method development and LC-based quantification.

SEC, which is also known as gel permeation chromatography (GPC), is the most commonly used method for determining polymer molecular weight distributions (MWD).[90] This LC method separates compounds based on their hydrodynamic volume in solution; larger molecular size materials, higher molecular weight, eluting first followed by the smaller molecules of lower molecular weight. The commonly used differential refractive index (RI) detector provides, again, very little information about the chemical composition, and molecular weight information obtained by the technique is highly dependent on the accuracy of the calibration procedure. Although a well-defined relationship exists only between the hydrodynamic volume (not molecular weight) of the solute and its retention volume (V_R), the common logarithm of relative molecular weight (log M_r) is correlated to V_R in practice. Well-characterized, narrow molecular weight distribution oligomer and polymer calibrants of similar chemical composition provide the most accurate results. Such calibrants are usually unavailable, and narrow molecular weight polystyrene standards are often used.[91] Besides, the mechanism of separation in SEC may involve solute-solvent-packing interactions that are not strictly dependent on molecular size,[92] and such interactions may lead to systematic errors in estimation of the molecular weight relying on calibration curves obtained by polystyrene standards when measuring polymers other than polystyrene. The SEC analyses of oligomeric mixtures may suffer the most from structure-dependent interactions. Oligomers of dissimilar chemical composition can also assume significantly different hydrodynamic volumes depending on their conformation in solution, even though their M_r is identical.

ESI mass spectrometry is compatible with the SEC conditions applied to the routine analysis of synthetic oligomers and polymers, and the coupling offers specific benefits in terms of obtaining chemical composition information and accurate molecular weight calibration.[93,94] Figure 4.11 shows the GPC/ ESI-MS analysis of Triton X-100. [No cationizing agent is added to the tetrahydrofuran mobile phase; therefore, the major ions represented in the contour plot are the protonated octylphenoxypoly (ethoxy)-ethanol oligomers with $m/z = 207 + 44n$.] Most GPC/ESI-MS applications have relied on the pre-column or post-column addition of a cationizing agent, most commonly NaI which has good solubility in the mobile phase. ESI mass spectrometry can directly handle effluents from analytical (7.8-mm i.d.) SEC columns with very little (<1%) of that effluent required (spectra are recorded from ~10 ng of sample during elution). This approach results in a significant decrease in the polydispersity of the analyte entering the ion source of the mass spectrometer, as demonstrated in Figure 4.11; therefore, problems associated with the ESI-MS analysis of polymeric samples with broad molecular weight distribution (discrimination according to cationization efficiency as a function of molecular weight, bias based on detection efficiencies, choice of experimental conditions, and so on) are eliminated or greatly reduced. With the mass spectrometer continually acquiring spectra as the molecules elute from the SEC, an on-line absolute molecular weight detector is employed for polymers

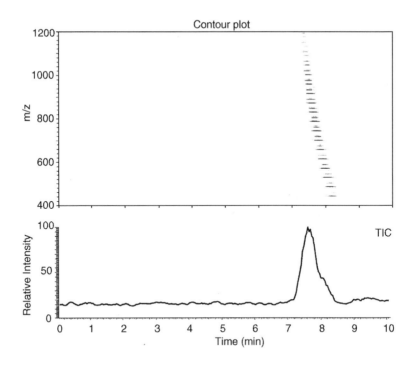

FIGURE 4.11

TIC chromatogram (bottom chart) and contour plot for the SEC-ESI-MS analysis of Triton X-100 (30 cm × 7.8 mm i.d. PLGel 3-μm Mixed-E column, tetrahydrofuran mobile phase at 1.0 mL/min, effluent split 1:100). Major ions are the protonated oligomers.

that have been size-separated by the SEC. From the mass spectra, the elution profiles of individual oligomers are determined from the reconstructed selected ion current as a function of elution time (t_R) or elution volume, V_R. Because ESI uses only a very small fraction of the effluent, conventional SEC detectors (RI, UV, and the like) can be operated parallel with mass spectrometry. The peak apex for each selected ion chromatogram or selected oligomer profile (the sum of ion intensities over different charge states) is used for an accurate elution volume for the given oligomer mass, which is then used to generate a calibration curve as shown in Figure 4.12. To better demonstrate the effect of using an SEC calibration obtained by coupling with ESI mass spectrometry vs. calibration with narrow-dispersity polystyrene standards, an octylphenoxypoly(ethoxy)ethanol sample with higher average molecular weights and broader molecular distribution (Igepal), compared to Triton X-100, was used as an analyte. In Table 4.3, a comparison of quantitative molecular weight distribution data obtained by direct ESI, analytical SEC with polystyrene calibration, and SEC after accurate ESI mass spectrometric calibration is presented. A recent development in GPC/ESI-MS includes miniaturization of the column (μSEC) that offers various advantages to the technique, such as low eluent consumption, low cost per column, reduced maintenance requirement, ability to interface to other chromatographic

a)

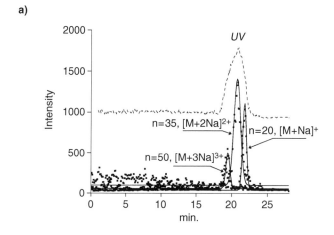

b)

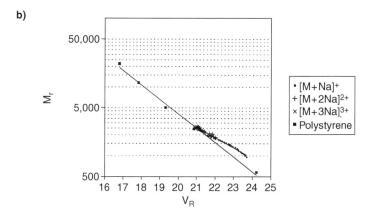

FIGURE 4.12

a) UV chromatogram (λ = 254 nm) and $[M + nNa]^{n+}$ selected-ion traces for octylphenoxypoly (ethoxy)ethanol oligomers separated on three SEC columns (30 cm × 7.8 mm i.d., 1000 Å, 500 Å, and 100 Å UltraStyragel) in series. The selected-ion trace for the triply charged n = 50 oligomer was obtained by summing m/z 824 to 826; the selected-ion trace for the doubly charged n = 35 oligomer was obtained by summing m/z 895 to 826; and the trace for the singly charged n = 20 oligomer was obtained by summing m/z 1108 to 1110 through the duration of the chromatogram. b) Calibration curves for octylphenoxypoly(ethoxy)ethanol: polystyrene vs. on-line ESI mass spectrometry. (Reprinted with permission from Ref. 94. Copyright ©2000 American Chemical Society.)

techniques (multidimensional LC) and the possibility of coupling to ESI mass spectrometry without the need for flow splitting.[95] In addition, better chromatographic performance can be achieved with microcolumns when compared to conventional-bore systems, which enables a better separation of sample constituents or significantly reduced time of analysis with separation power identical to conventional SEC columns. Newer mass analyzers such as orthogonal acceleration time-of-flight (oa-TOF)[89] and FT-ICR instruments[96] have been introduced into GPC/ESI-MS of polymers. Larger oligomers, such

TABLE 4.3

Molecular Weight Averages and Polydispersity of an Octylphenoxypoly(ethoxy)ethanol Oligomeric Surfactant by Direct ESI Mass Spectrometry, SEC with Polystyrene Calibration, and SEC with Calibration Via On-Line ESI Mass Spectrometry

	M_n	M_w	PD
ESI mass spectrometry, no separation	1736	1771	1.02
SEC, polystyrene calibration	2001	2162	1.15
SEC, calibration by on-line ESI-MS	1971	2016	1.08

as poly(methyl methacrylate) (PMMA) up to 9000 Da as $[M + 5Na]^{5+}$ ions, and minor impurities can be easily detected due to the extended mass range and high duty cycle of the oa-TOF analyzer, compared to GPC/ESI-MS on a quadrupole instrument or a quadrupole ion trap. Selected oligomer profiles for the sodiated (1+ through 5+ charge states) ions were also generated for a commercial, narrow molecular-weight-distribution PMMA sample, and they were used for obtaining a calibration curve and calculating accurate molecular-weight distribution data. In addition, GPC/ESI-FT-ICR mass spectrometry of PMMA allowed for an unequivocal end-group determination and characterization of a secondary distribution due to the formation of cyclic reaction products. A GMA/BMA copolymer with a broad molecular-weight distribution, where fractionation and high resolving power were required for adequate characterization, has also been analyzed by this hyphenated technique, and sodiated GMA/BMA oligomers in excess of 9 kDa were detected. End-groups resulting from the polymerization process were positively identified, and GPC/ESI-FT-ICR also allowed the accurate determination of the molecular weight distribution data.

4.5   Conclusion

In this chapter, the principles, instrumentation, and application of atmospheric-pressure ionization (principally ESI) mass spectrometry to synthetic oligomers and polymers are discussed through selected representative examples. The technique has proven potential in this application area from the structural and compositional characterization discussed here to, perhaps, preparative mass spectrometry to generate monodisperse synthetic polymers.[97] Direct use of the technique is appropriate for the qualitative analysis of samples with moderate complexity and molecular weight due to the phenomenon of multiple charging characteristic of ESI, and it is also well-suited under these conditions for sophisticated structural studies involving ultra-high resolution or tandem mass spectrometry. ESI mass spectrometry of mixtures with broad molecular weight distribution should benefit a prior separation, reducing the polydispersity of the analyte. The advantage of hyphenated LC/MS for obtaining information about chemical composition, resolution of overlapping charge envelopes in the ESI mass spectra of polymers, SEC calibration, and complex mixture analysis have been highlighted.

References

1. Dole, M., Mack, L. L., Hines, R. L., Mobley, R. C., Ferguson, L. D., and Alice, M. B. Molecular beams of macroions. *J. Chem. Phys*, 49, 2240, 1968.
2. Dole, M., Cox, H. L., Jr., and Giemic, *J. Adv. Chem. Ser.*, 125, 73, 1973.
3. Yamashita, M. and Fenn, J. B. Electrospray ion-source—Another variation on the free-jet theme. *J. Phys. Chem.*, 88, 4451, 1984.
4. Whitehouse, C. M., Dreyer, R. N., Yamashita, M., and Fenn, J. B. Electrospray interface for liquid chromatographs and mass spectrometers. *Anal. Chem.*, 57, 675, 1985.
5. Zeleny, J. On the conditions of instability of liquid drops, with applications to the electrical discharge from liquid point. *Proc. Camb. Phil. Soc.*, 18, 71, 1915.
6. Yamashita, M. and Fenn, J. B. Negative-ion production with the electrospray ion-source. *J. Phys. Chem.*, 88, 4671, 1984.
7. Bruins, A. P., Covey, T. R., and Henion, J. D. Ionspray interface for combined liquid chromatography/atmospheric pressure ionization mass spectrometry. *Anal. Chem.*, 59, 2642, 1987.
8. Nielen, M. W. F. Characterization of synthetic polymers by size-exclusion chromatography/electrospray ionization mass spectrometry. *Rapid Commun. Mass Spectrom.*, 22, 57, 1996.
9. Horning, E. C., Carroll, D. I., Dzidic, I., Haegele, K. D., Horning, M. G., and Stillwell, R. N. Atmospheric pressure ionization (API) mass spectrometry. Solvent-mediated ionization of samples introduced in solution and in a liquid chromatograph effluent stream. *J. Chromatog. Sci.*, 12, 725, 1974.
10. Carroll, D. I., Dzidic, I., Stillwell, R. N., Haegele, K. D., and Horning, E. C. Atmospheric pressure ionization mass spectrometry: Corona discharge ion source for use in liquid chromatograph-mass spectrometer-compute analytical system. *Anal. Chem.*, 47, 2369, 1975.
11. Huertas, M. L. and Fontan, J. Evolution times of tropospheric positive ions. *Atmospheric Environ.*, 9, 1018, 1975.
12. Horning, E. C., Carroll, D. I., Dzidic, I., Lin, S. N., Stillwell, R. N., and Thenot, J.-P. Atmospheric pressure ionization mass spectrometry. Studies of negative ion formation for detection and quantification purposes. *J. Chromatogr.*, 142, 481, 1977.
13. Bruins, A. P. Atmospheric-pressure ionization mass spectrometry. 1. Instrumentation and ionization techniques. *Trends Anal. Chem.*, 13, 81, 1994.
14. Kebarle, P. and Tang, L. From ions in solution to ions in the gas phase—The mechanism of electrospray mass spectrometry. *Anal. Chem.*, 65, 972A, 1993.
15. Iribarn, J. V. and Thomson, B. A. On the evaporation of small ions from charged droplets. *J. Chem. Phys.*, 64, 2287, 1976.
16. Fenn, J. B., Rosell, J., and Meng, C. K. In electrospray ionization, how much pull does an ion need to escape its droplet prison? *J. Am. Soc. Mass Spectrom.*, 8, 1147, 1997.
17. Fenn, J. B. Ion formation from charged droplets: Roles of geometry, energy, and time. *J. Am. Soc. Mass Spectrom.*, 4, 524, 1993.

18. Wong, S. F., Meng, C. K., and Fenn, J. B. Multiple charging in electrospray ionization of poly(ethylene glycols). *J. Phys. Chem.*, 92, 546, 1988.
19. Keese, R. G. and Castleman, A. W., Jr. *J. Phys. Chem. Ref. Data*, 15, 1011, 1986.
20. Schwartz, B. L., Rockwood, A. L., Smith, R. D., Tomalia, D. A., and Spindler, R. Detection of high molecular weight starburst dendrimers by electrospray ionization mass spectrometry. *Rapid Commun. Mass Spectrom.*, 9, 1552, 1995.
21. Smith, R. D., Loo, J. A., Ogorzalek-Loo, R. R., Busman, M., and Udseth, H. R. Principles and practice of electrospray ionization—Mass-spectrometry for large polypeptides and proteins. *Mass Spectrom. Rev.*, 10, 359, 1991.
22. Nohmi, T. and Fenn, J. B. Electrospray mass spectrometry of poly(ethylene glycols) with molecular weights up to five million. *J. Am. Chem. Soc.*, 114, 3241, 1992.
23. Latourte, L., Blais, J.-C., Tabet, J.-C., and Cole, R.B. Desorption behavior and distribution of fluorinated polymers in MALDI and electrospray ionization mass spectrometry. *Anal. Chem.*, 69, 2742, 1997.
24. Guittard, J., Tessier, M., Blais, J. C., Bolbach, G., Rozes, L., Maréchal, E., and Tabet, J. C. Electrospray and matrix-assisted laser desorption/ionization mass spectrometry for the characterization of polyesters. *J. Mass Spectrom.*, 31, 1409, 1996.
25. Deery, M. J., Jennings, K. R., Jasieczek, C. B., Haddleton, D. M., Jackson, A. T., Yates, H. T., and Scrivens, J. H. A study of cation attachment to polystyrene by means of matrix-assisted laser desorption/ionization and electrospray ionization-mass spectrometry. *Rapid Commun. Mass Spectrom.*, 11, 57, 1997.
26. Mahon, A., Kemp, T. J., Buzy, A., and Jennings, K. R. Mass-spectral characterization of oligomeric polysulfides by electrospray ionization combined with collision-induced decomposition. *Polymer*, 37, 531, 1996.
27. Barton, Z., Kemp, T. J., Buzy, A., Jennings, K. R., and Cunliffe, A. V. Mass-spectral characterization of linear and cyclic forms of oligomeric nitrated polyethers by electrospray ionization: Specific cationization effects in cyclic polyethers. *Polymer*, 38, 1957, 1997.
28. Mahon, A., Kemp, T. J., Buzy, A., and Jennings, K. R. Electrospray ionization mass spectrometric study of oligomeric linear polysulfides: Characterization of repeat units, end groups and fragmentation pathways. *Polymer*, 38, 2337, 1997.
29. Hunt, S. M., Binns, M. R., and Sheil, M. M. Structural characterization of polyester resins by electrospray mass spectrometry. *J. Appl. Polym. Sci.*, 56, 1589, 1995.
30. Yalcin, T., Gabryelski, W., and Li, L. Structural analysis of polymer end groups by electrospray ionization high-energy collision-induced dissociation tandem mass spectrometry. *Anal. Chem.*, 72, 3847, 2000.
31. Koster, S., Duursma, M. C., Boon, J. J., Nielen, M. W. F., De Koster, C. G., and Heeren, R. M. A. Structural analysis of synthetic homo- and copolyesters by electrospray ionization on a Fourier transform ion cyclotron resonance mass spectrometer. *J. Mass Spectrom.*, 35, 739, 2000.
32. Adamus, G., Sikorska, W., Kowalczuk, M., Montaudo, M., and Scandola, M. Sequence distribution and fragmentation studies of bacterial copolyester macromolecules: Characterization of PHBV macroinitiator by electrospray ion-trap multistage mass spectrometry. *Macromolecules*, 33, 5797, 2000.
33. Simonsick, W. J., Jr. and Prokai, L. ESI and tandem mass spectrometry of mixed polyesters. *Proceedings of the 48th ASMS Conference of Mass Spectrometry and Allied Topics*, Long Beach, CA, June 11–15, 2000, CD-ROM: MODpm.pdf.
34. Adamus, G. and Kowalczuk, M. Electrospray multistep ion trap mass spectrometry for the structural characterisation of poly[(R,S)-3-hydroxybutanoic acid] containing a β-lactam end group. *Rapid Commun. Mass Spectrom.*, 14, 195, 2000.

35. Stolarzewicz, A., Morejko-Buz, B., and Neugebauer, D. Study of the structure of poly(methyl methacrylate) obtained in the presence of potassium hydride. *Rapid Commun. Mass Spectrom.*, 14, 2170, 2000.

36. Mann, M., Meng, C. K., and Fenn, J. B. Interpreting mass spectra of multiply charged ions. *Anal. Chem.*, 61, 1702, 1989.

37. Reinhold, B. B. and Reinhold, V. N. Electrospray ionization mass spectrometry: Deconvolution by an entropy-based algorithm. *J. Am. Soc. Mass Spectrom.*, 3, 207, 1992.

38. Zhang, Z. and Marshall, A. G. A universal algorithm for fast and automated charge state deconvolution of electrospray mass-to-charge ratio spectra. *J. Am. Soc. Mass Spectrom.*, 9, 225, 1998.

39. Kallos, G. J., Tomalia, D. A., Hedstrand, D. M., Lewis, S., and Zhou, J. Molecular weight determination of a polyamidoamine starburst polymer by electrospray-ionization mass spectrometry. *Rapid Commun. Mass Spectrom.*, 5, 383, 1991.

40. O'Connor, P. B. and McLafferty, F. W. Oligomer Characterization of 4-23 kDa polymers by electrospray Fourier transform mass spectrometry. *J. Am. Chem. Soc.*, 117, 12826, 1995.

41. Ross, C. W., III. and Simonsick, W. J., Jr. *Proceedings of the 44th ASMS Conference on Mass Spectrometry and Allied Topics*, Portland, OR, May 12–16, 1996, p. 1270.

42. Simonsick, W. J., Jr., Aaserud, D. J., Grady, M. C., and Prokai, L. Gel permeation chromatography coupled to Fourier transform mass spectrometry for the characterization of the products of methacrylate polymerizations. *Polym. Prepr. (Am. Chem. Soc., Div. Polym. Chem.)*, 38, 483, 1997.

43. Shi, S. D.-H., Hendrickson, C. L., Marshall, A. G., Simonsick, W. J., Jr., and Aaserud, D. J. Identification, composition, and asymmetric formation mechanism of glycidyl methacrylate/butyl methacrylate copolymers up to 7000 Da from electrospray ionization ultrahigh resolution Fourier transform ion cyclotron resonance mass spectrometry. *Anal. Chem.*, 70, 3220, 1998.

44. Koster, S., Duursma, M. C., Boon, J. J., and Heeren, R. M. A. Endgroup determination of synthetic polymers by electrospray ionization Fourier transform ion cyclotron resonance mass spectrometry. *J. Am. Soc. Mass Spectrom.*, 11, 536, 2000.

45. McEwen, C. N., Simonsick, W. J., Jr., Larsen, B. S., Ute, K., and Hatada, K. The fundamentals of applying electrospray ionization mass spectrometry to low mass poly(methyl methacrylate) polymers. *J. Am. Soc. Mass Spectrom.*, 6, 906, 1995.

46. Barton, Z., Kemp, T. J., Buzy, A., and Jennings, K. R. Mass spectral characterization of the thermal degradation of poly(propylene oxide) by electrospray and matrix-assisted laser desorption ionization. *Polymer*, 36, 4927, 1995.

47. Saf, R., Mirtl, C., and Hummel, K. Electrospray ionization mass spectrometry as an analytical tool for non-biological monomers, oligomers and polymers. *Acta Polym.*, 48, 513, 1997.

48. Montaudo, G., Montaudo, M. S., Puglisi, C., and Samperi, F. Molecular weight determination and structural analysis in polydisperse polymers by hyphenated gel permeation chromatography matrix-assisted laser desorption ionization— Time of flight mass spectrometry. *Int. J. Polym. Anal. Charact.* 3, 177, 1997.

49. Maziarz, P. E. III, Baker, G. A., Mure, J. V., and Wood, T. D. A comparison of electrospray versus nanoelectrospray ionization Fourier transform mass spectrometry for the analysis of synthetic poly(dimethylsiloxane)/poly(ethylene glycol) oligomer blends. *Int. J. Mass Spectrom.*, 202, 241, 2000.

50. Jasieczek, C. B., Buzy, A., Haddleton, D. M., and Jennings, K. R. Electrospray ionization mass spectrometry of poly(styrene). *Rapid Commun. Mass Spectrom.*, 10, 509, 1996.

51. Hunt, S. M., Sheil, M. M., Belov, M., Derrick, P. J. Probing the effects of cone potential in the electrospray ion source: Consequences for the determination of molecular weight distributions of synthetic polymers. *Anal. Chem.*, 70, 1812, 1998.
52. Maziarz, E. P., III, Baker, G. A., Lorenz, S. A., and Wood, T. D. External ion accumulation of low molecular weight poly(ethylene glycol) by electrospray ionization Fourier transform mass spectrometry. *J. Am. Soc. Mass Spectrom.*, 10, 1298, 1999.
53. Saf, R., Mirtl C., and Hummel, K. Electrospray mass-spectrometry using potassium-iodide in aprotic organic solvents for the ion formation by cation attachment. *Tetrahedron Lett.*, 35, 6653, 1994.
54. Varray, S., Aubagnac, J.-L., Lamaty, F., Lazaro, R., Martinez, J., and Enjalbal, C. Poly(ethylene glycol) in electrospray ionization (ESI) mass spectrometry. *Analusis*, 28, 263, 2000.
55. Maekawa, M., Nohmi, T., Zhan, D., Kiselev, P., Fenn, J. B. Reflections on electrospray mass spectrometry of synthetic polymers. *J. Mass Spectrom. Soc. Jpn.*, 47, 76, 1999.
56. Nokwequ, G. M. and Bariyanga, J. Synthesis, characterization and biodegradability of a water-soluble poly(ethylene oxide) derivative polymer bearing carboxylic acid side chain function. *J. Bioact. Compat. Polym.*, 15, 503, 2000.
57. Chen, R., Tseng, A. M., Uhing, M., and Li, L. Application of an integrated matrix-assisted laser desorption/ionization time-of-flight, electrospray ionization mass spectrometry and tandem mass spectrometry approach to characterizing complex polyol mixtures. *J. Am. Soc. Mass Spectrom.*, 12, 55, 2001.
58. Crowther, M. W., O'Connell, T. R., and Carter, S. P. Electrospray mass spectrometry for characterizing polyglycerols and the effects of adduct ion and cone voltage. *J. Am. Oil Chem. Soc.*, 75, 1867, 1998.
59. Arakawa, R., Watanabe, T., Fukuo, T., and Endo, K. Determination of cyclic structure for polydithiane using electrospray ionization mass spectrometry. *J. Polym. Sci., Part A: Polym. Chem.*, 38, 4403, 2000.
60. Focarete, M. L., Scandola, M., Jendrossek, D., Adamus, G., Sikorska, W., and Kowalczuk, M. Bioassimilation of atactic poly[(R,S)-3-hydroxybutyrate] oligomers by selected bacterial strains. *Macromolecules*, 32, 4814, 1999.
61. Schwach-Abdellaoui, K., Heller, J., Gurny, R. Hydrolysis and erosion studies of autocatalyzed poly(ortho esters) containing lactoyl-lactyl acid dimers. *Macromolecules*, 32, 301, 1999.
62. Jasieczek, C.B., Buzy, A., Haddleton, D. M., and Jennings, K. R. Electrospray ionization mass spectrometry of poly(styrene). *Rapid Commun. Mass Spectrom.*, 10, 509, 1996.
63. vanHest, J. C. M., Delnoye, D. A. P., Baars, M. W. P. L., Elissen-Roman, C., vanGenderen, M. H. P., and Meijer, E. W. Polystyrene-poly(propylene imine) dendrimers: Synthesis, characterization, and association behavior of a new class of amphiphiles. *Chem. Eur. J.*, 2, 1616, 1996.
64. Scrivens, J. H. and Jackson, A. T. Characterization of synthetic polymer systems. *Int. J. Mass Spectrom.*, 200, 261, 2000.
65. Schwartz, B. L., Rockwood, A. L., Smith, R. D., Tomalia, D. A., and Spindler, R. Detection of high molecular weight starburst dendrimers by electrospray ionization mass spectrometry *Rapid Commun. Mass Spectrom.*, 9, 1552, 1995.
66. Maziarz, E.P., Baker, G.A., and Wood, T.D. Capitalizing on the high mass accuracy of electrospray ionization Fourier transform mass spectrometry for synthetic polymer characterization: A detailed investigation of poly(dimethylsiloxane). *Macromolecules*, 32, 4411, 1999.

67. Arslan, H., Adamus, G., Hazer, B., and Kowalczuk, M. Electrospray ionization tandem mass spectrometry of poly[(R,S)-3-hydroxybutanoic acid] telechelics containing primary hydroxy end groups. *Rapid Commun. Mass Spectrom.*, 13, 2433, 1999.
68. Dolan, A. R., Maziarz, E. P., III, and Wood, T. D. The analysis of conducting polymers by electrospray Fourier transform mass spectrometry. Part I: Ionene polymers. *Eur. J. Mass Spectrom.*, 6, 241, 2000.
69. Parees, D. M., Hanton, S. D., Clark, P. A. C., and Willcox, D. A. Comparison of mass spectrometric techniques for generating molecular weight information on a class of ethoxylated oligomers. *J. Am. Soc. Mass Spectrom.*, 9, 282, 1998.
70. Pasch, H. and Trathnigg, B. *HPLC of Polymers*, Springer, Berlin, 1998, pp. 224.
71. Provder, T., Barth, H. G., and Urban, M. *Chromatographic Characterization of Polymers: Hyphenated and Multidimensional Techniques;* ACS Advances in Chemistry Series 247; American Chemical Society: Washington, DC, 1995.
72. Jergey, A. L., Edmonds, C. G., Lewis, I. A. S., and Vestal, M. L. *Liquid Chromatography/Mass Spectrometry: Techniques and Applications*, Plenum Press, New York, 1990, pp. 316.
73. Niessen, W. M. A. *Liquid Chromatography—Mass Spectrometry*, Chromatographic Science Series, Volume 79, Marcel Dekker, New York, 1999, pp. 634.
74. Nielen, M. W. F. and van de Ven, H. J. F. M. Characterization of (methoxymethyl)melamine resins by combined chromatographic/mass spectrometric techniques. *Rapid Commun. Mass Spectrom.*, 10, 74, 1996.
75. Prokai, L., Myung, S.-W., and Simonsick, W. J., Jr. Coupling size exclusion chromatography with mass spectrometry: Strategies and applications to oligomers and polymers. *Polym. Prepr. (Am. Chem. Soc., Div. Polym. Chem.)*, 37, 288, 1996.
76. Bryant, J. J. L. and Semlyen, J. A. Cyclic polyesters: 6. Preparation and characterization of two series of cyclic oligomers from solution ring-chain reactions of poly(ethylene terephtalate). *Polymer*, 38, 2475, 1997.
77. Harrison, A. G., Taylor, M. J., Scrivens, J. H., and Yates, H. Analysis of cyclic oligomers of poly(ethylene terephtalate) by liquid chromatography/mass spectrometry. *Polymer*, 38, 2549, 1997.
78. Bryant, J. J. L. and Semlyen, J. A. Cyclic polyesters: 7. Preparation and characterization of cyclic oligomers from solution ring-chain reactions of poly(butylene terephtalate). *Polymer*, 38, 4531, 1997.
79. Marcelli, A., Favretto, D., and Traldi, P. High performance liquid chromatography combined with electrospray ionization ion trap mass spectrometry in the characterization of (methoxymethyl)melamine resins. *Rapid Commun. Mass Spectrom.*, 11, 1321, 1997.
80. Favretto, D., Traldi, P., and Marcelli, A. High performance liquid chromatography electrospray ionisation and atmospheric pressure chemical ionisation mass spectrometry for the analysis of butylated amino resins. *Eur. Mass Spectrom.*, 4, 371, 1998.
81. Carrott, M. J. and Davidson, G. Separation and characterisation of phenol-formaldehyde (resol) prepolymers using packed-column supercritical fluid chromatography with APCI mass spectrometric detection. *Analyst*, 124, 993, 1999.
82. van der Wal, S., Mengerink, Y., Brackman, J. C., de Brabander, E. M. M., Jeronimus-Stratingh, C. M., and Bruins, A. P. Compositional analysis of nitrile terminated poly(propylene imine) dendrimers by high-performance liquid chromatography combined with electrospray mass spectrometry. *J. Chromatogr. A*, 825, 135, 1998.

83. Mori, S. Size-exclusion and nonexclusion liquid chromatography for characterization of styrene copolymers. In *Chromatographic Characterization of Polymers: Hyphenated and Multidimensional Techniques* (Provder, T., Barth, H. G., and Urban, M., Eds.), ACS Advances in Chemistry Series 247; American Chemical Society: Washington, DC, 1995, p. 211.

84. Glöckner, G., Kroschwitz, H., and Meissner, C. H. HP precipitation chromatography of styrene-acrylonitrile co-polymers. *Acta Polym.*, 33, 614, 1982.

85. Glöckner, G. Control of adsorption and solubility in gradient high-performance liquid chromatography. 1. Principles of sudden transition gradients and elution characteristics of copolymers from styrene and methacrylates. *Chromatographia*, 37, 7, 1993.

86. Cools, P. J. C. H., Maesen, F., Klumperman, B., van Herk, A. M., and German, A. L. Determination of the chemical composition distribution of copolymers of styrene and butadiene by gradient polymer elution chromatography. *J. Chromatogr.*, 736, 125, 1996.

87. Philipsen, H. J. A., Wubbe, F. P. C., Klumperman, B., and German, A. L. Microstructural characterization of aromatic polyesters made by step reactions, by gradient polymer elution chromatography. *J. Appl. Polym. Sci.*, 72, 183, 1999.

88. Glöckner, G., Vandenberg, J. H. M., Meijerink, N. L. J., Scholte, T. G., and Koningsveld, R. Size exclusion and high-performance precipitation liquid chromatography of styrene-acrylonitrile copolymers. *Macromolecules*, 17, 962, 1984.

89. Nielen, M. W. F. and Buijtenhuijs, F. A. (Ab). Polymer analysis by liquid chromatography/electrospray ionization time-of-flight mass spectrometry. *Anal. Chem.*, 71, 1809, 1999.

90. Yau, W. W., Kirkland, J. J., and Bly, D. D. *Modern Size-Exclusion Liquid Chromatography.* Wiley-Interscience, New York, 1979.

91. Holding, S. R. In *Size Exclusion Chromatography* (Hunt, B. J. and Holding, S. R., Eds.), Blackie, Glasgow and London, 1989, p. 42.

92. Garcia Rubio, L. H., McGregor, J. F., and Hamielec, A. E. In *Polymer Characterization,* (Craver, C. D., Ed.), American Chemical Society, Washington, DC, 1983, p. 311.

93. Prokai, L. and Simonsick, W. J., Jr. Electrospray ionization mass spectrometry coupled with size-exclusion chromatography. *Rapid Commun. Mass Spectrom.* 7, 853, 1993.

94. Simonsick, W. J., Jr. and Prokai, L. Size-exclusion chromatography with electrospray ionization mass spectrometry. In *Hyphenated Techniques in Polymer Characterization,* T. Provder, H. Barth, and M. W. Urban, (Eds.) American Chemical Society, Washington, DC, 1995, p. 41.

95. Prokai, L., Aaserud, D. J., and Simonsick, W. J., Jr. Microcolumn size-exclusion chromatography coupled with electrospray ionization mass spectrometry. *J. Chromatogr. A*, 835, 121, 1999.

96. Aaserud, D. J., Prokai, L., and Simonsick, W. J., Jr. Gel permeation chromatography coupled to Fourier transform mass spectrometry for polymer characterization. *Anal. Chem.*, 71, 4793, 1999.

97. Siuzdak, G., Hollenbeck, T., and Bothner, B. Preparative mass spectrometry with electrospray ionization. *J. Mass Spectrom.*, 34, 1087, 1999.

5

Direct Pyrolysis of Polymers into the Ion
Source of a Mass Spectrometer (DP-MS)

Giorgio Montaudo and Concetto Puglisi

CONTENTS

5.1 Thermal Stability of Polymers

The thermal stability of polymers is dependent upon the strengths of their
constituent chemical bonds and on the ease of the hydrogen transfer reactions within them. For instance, polymers containing aliphatic units undergo
thermal degradation at lower temperatures compared to aromatic polymers
because (i) their C-C bonds are weaker and (ii) aliphatic hydrogen atoms are
transferred more easily. In the case of "high temperature" polymers (e.g.,
polyimides or other condensed polyaromatics), the thermal stability is the
result of the lack of labile hydrogen atoms within the polymer structure,

achieving a "quasi-char" structure already in the synthetic stage. As a consequence, these polymers start decomposing at relatively high temperatures yielding large amounts of charred residue.[1,2]

A great difference exists between the mechanisms of thermal decomposition of addition polymers (like polyolefins, polydienes, and polyvinyls) and those of condensation polymers. Addition polymers contain aliphatic hydrogen atoms, along the backbone, which are easily transferred following homolytic bonds cleavage in the temperature range of 200–500°C. The majority of addition polymers undergo thermal degradation through the formation of macro-radicals, which are very reactive species. Their decay may happen through several parallel routes, involving bond cleavages (e.g., beta scission, recombination, disproportionation, hydrogen elimination or abstraction.[3,4] Elimination of small stable molecules from side groups (e.g., HCl, H_2O) may play also a role.

Condensation polymers can be regarded as a sequence of monomer units containing functional groups immobilized into the polymer structure. Their decomposition pathways will often be dominated by the polarity and by the reactivity of the functional groups within their structure, and their thermal decomposition reactions will be ionic and selective, rather than radical and unselective. Such are the thermal degradation processes occurring in polyesters, polyamides, polycarbonates, polyurethanes, polyureas, and in several other cases.[1,2]

The specific thermal decomposition pathways occurring in these polymers largely depend upon the nature of the functional groups present and also upon the chemical structure of the monomer units, which causes drastic changes in the pyrolysis mechanisms even within the same class of condensation polymers.[1,2]

These ionic pathways typically occur at temperatures (150–300°C) that are below those of typical free-radical degradation reactions. However, if polymers belonging to the above classes are subjected to higher temperatures, radical reactions are likely to prevail. Furthermore, polymers containing less polar functional groups, like ethers and azomethynes, tend to undergo radical cleavage even when they decompose at relatively low temperatures.[1,2]

The moler mass of the pyrolysis products also varies with the thermal stability of the starting materials, and polymers decomposing at relatively low temperatures (200–400°C) usually yield large thermal fragments, whereas polymers thermally stable (~500°C) tend to generate fragments of smaller size.

The thermal degradation of polymers was an active field of investigation long before modern instrumental techniques were available. Earlier studies generally involved the pyrolysis of large samples and off-line degradation. The analytical methods often required sizeable quantities of compounds, and the subsequent identification of the pyrolysis products was difficult.

Contemporary work emphasizes the detection of primary thermal decomposition products, carrying significant structural information, i.e., polymer segments large enough to contain the repeat units or the sequence arrangement of sub-units in copolymers.[5–8]

This chapter is devoted to the discussion of the direct pyrolysis of polymers in the ion source of a mass spectrometer (DPMS).

5.2 Principles and Methodology of DPMS

Mass Spectrometry involves the formation of ions in the gas phase, and early ionization methods, such as electron impact (EI), chemical ionization (CI), and field ionization (FI), also require the gas-phase evaporation of the analyte before ionization.[9,10]

The low volatility of macromolecules initially precluded the widespread application of MS analysis to polymer systems. Nevertheless, meaningful structural information can be extracted from the mass spectra of unvolatile macromolecules by EI, CI, and FI techniques, and this is due to the considerable attention that has been focused on the methods of characterization of polymers by Direct Pyrolysis Mass Spectrometry (DPMS).[1,2,5–20]

In DPMS,[5,6,14] polymers are introduced into the ion source via the direct insertion probe (Figure 5.1a) and the temperature is increased gradually until thermal degradation reactions occur. The volatile products formed are ionized (both positive and negative ions are produced) and extracted from the ion source (Figure 5.1b) almost as soon as they are formed, by applying a positive or negative potential to the repeller.

Besides probe pyrolysis, a filament pyrolysis technique, usually called Desorption Chemical Ionization (DCI) is also used. The polymeric material is deposited on a filament, which is rapidly heated to the degradation temperature. The pyrolysis occurs very close to the electron beam, and therefore DCI is often referred to as an "in-beam" pyrolysis.[21–25]

A specific advantage of DPMS with respect to other pyrolysis techniques,[1,6] is that polymer pyrolysis is accomplished *"on-line"* under high vacuum at the μg level, so that pyrolysis products are quickly volatilized, providing low residence times in the hot zone. This reduces the probability of molecular collision and the occurrence of secondary reactions, allowing the detection of almost primary products (however, very reactive primary pyrolysis products may undergo further reaction in the condensed phase).[26]

The pyrolysis is achieved adjacent to the ion source, so that the primary pyrolysis products, once formed, are immediately ionized and accelerated toward the detector thus preventing further thermal rearrangements. In fact, the rapid evaporation/ionization process tends to "freeze" the structure of the pyrolysis products, which flee from the ion source to the detector in the order of a few μ-seconds.

In DPMS the overall residence in the hot zone and transport times are estimated in the order of milliseconds for probe pyrolysis (EI, CI, FI), and even

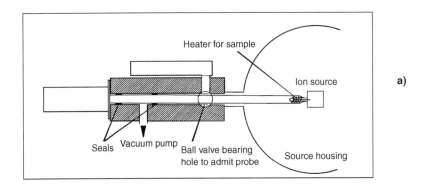

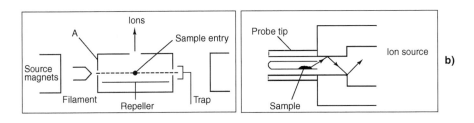

FIGURE 5.1
a) Direct insertion probe and vacuum lock; b) Schematic diagram of electron impact ion source and sample holder positioned outside the source. (Reproduced from Ref. 9, copyright 1995, by permission of John Wiley & Sons.)

less for filament pyrolysis (DCI). Therefore, it provides low residence times of the pyrolysis products in the hot zone, allowing the detection of thermally labile pyrolysis products with very short lifetimes at high temperatures.

Furthermore, ions of high mass, which are often essential for the structural characterization of the polymer, can be detected on-line whereas they might be lost using other techniques.

Relatively long residence times of the pyrolysis products in the hot zone may cause a further degradation of the compounds initially formed, especially in techniques such as Flash-Pyrolysis-GC/MS (Py-GC/MS, Chapter 3), where the pyrolysis products are not immediately converted to ions. In this case the thermal degradation processes may also occur during the transport time to the detecting device.

Py-GC/MS is an alternative technique differing from DPMS in (i) the fast heating of the probe, (ii) the inert-gas atmosphere of the pyrolysis chamber, and (iii) the separation of the pyrolysis products prior to MS analysis. In Py-GC/MS individual MS analysis of the isomeric products separated by GC is possible; nevertheless, only thermally stable products survive the high transport times and are volatile enough for the GC analysis.[27,28]

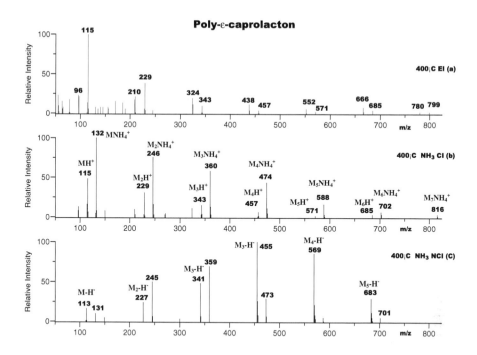

FIGURE 5.2

Mass spectra of pyrolysis compounds of poly-ε-caprolacton evolved at 400°C in (a) EI; (b) NH₃ CI; (c) NH₃ NCI ionization methods. (Reprinted with permission from Ref. 32. Copyright 1986 American Chemical Society.)

The time-scale of DPMS and Py-GC/MS is quite different and, in the case of thermally labile pyrolysis products the two techniques may detect different compounds.[27,28]

Ionization methods that minimize the EI fragmentation processes of the molecular ions, such as FI,[5,18,29] CI,[30–36] and DCI[21–25] are very valuable in DPMS since the resulting mass spectra contain more intense molecular ions and therefore the identification of pyrolysis compounds is easier.

In Figure 5.2a-c are reported the EI, ammonia CI, and NCI mass spectra of the pyrolysis products deriving from poly(ε-caprolacton) (PCL). The EI mass spectrum (Figure 5.2a) shows a small amount of monomeric caprolactone (m/z 114) and intense peaks due to EI fragmentation of pyrolysis compounds.[32] Instead, the ammonia CI mass spectrum (Figure 5.2b) is dominated by two series of intense peaks at 115 + n114 and m/z 132 + n114 which are assigned to protonated molecular ions $(M + H)^+$ and $(M + NH_4)^+$ adducts of cyclic oligomers of poly(ε-caprolacton), respectively. The pyrolysis mass spectrum of PCL obtained in ammonia negative CI (NCI) (Figure 5.2c) shows more intense peaks than the CI spectrum. The peaks at m/z 113 + n114 are assigned to $(M-H)^-$ molecular ions, generated by the loss of a proton from cycles.

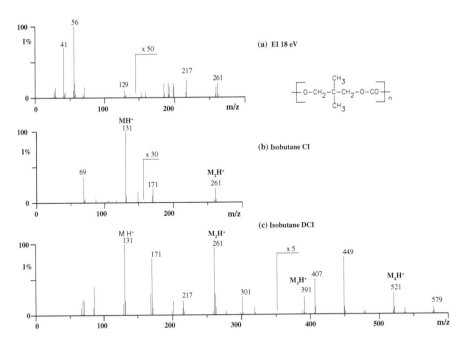

FIGURE 5.3
Mass spectra of pyrolysis compounds of poly(neopentylene carbonate) obtained by: (a) DPMS EI; (b) DPMS CI; (c) Isobutane DCI. (Reprinted from Ref. 37, copyright 1991, with permission from Elsevier Science.)

The DCI technique allows the detection of molecular ions of higher masses as compared to EI and CI and therefore enables the identification of polymer structure.

Figure 5.3a-c compares the EI, CI, and DCI pyrolysis mass spectra of neopentylpolycarbonate,[37] which decomposes producing cyclic oligomers (Eq. 5.1)

$$\text{wwww}-O\text{-}CH_2\text{-}\underset{\underset{CH_3}{|}}{\overset{\overset{CH_3}{|}}{C}}\text{-}CH_2\text{-}O\text{-}CO\text{----www} \quad \xrightarrow{\Delta} \quad \left(\!\!\left[\begin{array}{c} O\text{-}CH_2\text{-}\underset{\underset{CH_3}{|}}{\overset{\overset{CH_3}{|}}{C}}\text{-}CH_2\text{-}O\text{-}CO \end{array}\right]\!\!\right)_{\!n}$$

$$(5.1)$$

The EI mass spectrum (Figure 5.3a) shows only peaks due to ion fragmentation in the ion source, whereas molecular ions are absent. The isobutane CI spectrum shows instead the presence of protonated molecular ions at m/z 131 and m/z 261, due to the cyclic monomer and dimer, respectively. The isobutane DCI spectrum shows furthermore the presence of protonated molecular ions corresponding to cyclic trimer and tetramer at m/z 391 and m/z 521, respectively.[37]

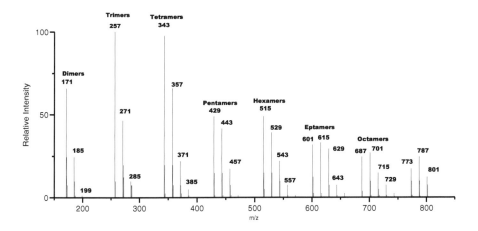

FIGURE 5.4
Ammonia negative DCI mass spectrum of poly(hydroxybutyrate-co-hydroxyvalerate). (Reprinted from Ref. 23.)

Excellent negative ion mass spectra have also been obtained in the DCI of poly(hydroxybutyrate-co-hydroxyvalerate) (HB/HV) copolymers.[23] The mass spectrum (Figure 5.4) shows negligible ion fragmentation and intense molecular ions corresponding to a series of oligomer clusters from dimers to octamers. The relative peak intensities corresponding to each oligomer are proportional to the relative abundance of the oligomers in the copolymer, as proved by comparison with independent NMR measurements. Therefore, these DCI data provide an accurate estimate of the copolymer composition.

An advantage of DPMS experiments is that they are carried out by gradual heating (typically at 10°C/min) of the probe and continuously scanning the MS. The total ion current (TIC) of each scan recorded as a function of the temperature produces a trace which compares with the derivative thermogravimetric (DTG) traces obtained at the same heating rate. Figure 5.5a,b shows the DTG and the TIC curves of a polyetherimide (PEI) sample. The DTG curve (Figure 5.5a), shows the onset of degradation at about 450°C and a maximum rate of decomposition at about 510°C, followed by a less marked decomposition step at higher temperatures (600–650°C).[38] The TIC curve of PEI (Figure 5.5b), obtained by DPMS, at the same heating rate used for the thermogravimetric experiment (10°C/min) nearly reproduces the DTG curve.[38]

The DPMS experiment also allows the collection of the single ion current (SIC) curve of specific pyrolysis compounds evolved as a function of the temperature. A typical application consists in the detection of residual low molecular mass compounds (oligomers, additives, solvents) eventually contained in the polymer sample. Due to the high vacuum, the existing preformed compounds may distill undecomposed, as the temperature increases.[15,20,39–42]

This experiment is easy to perform with nearly all kinds of polymers to look for the presence of traces of solvents and/or volatile additives. The evolution

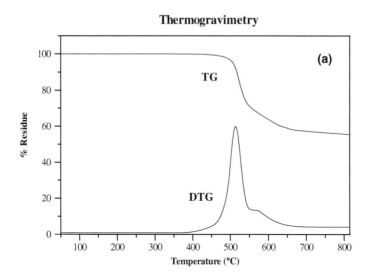

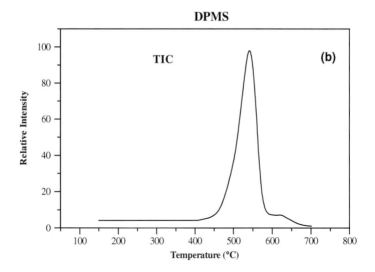

FIGURE 5.5
a) Thermogravimetric and Derivative Thermogravimetric Curves and b) Total Ion Current Curve of a PEI sample. (Reproduced from Ref. 38, copyright 1999, by permission of John Wiley & Sons.)

of volatiles is more rapid when the polymers are heated above their T_g, which increases the segmental mobility allowing the facile evolution of volatile compounds from the bulk.

Figure 5.6 reports the SIC curves corresponding to the distillation of a series of cyclic oligomers evolving from a crude sample of 1,3-polyphenylenesulfide (*m*-PPS). These preformed oligomers distill in the high vacuum of the MS source, as the temperature is raised.[39]

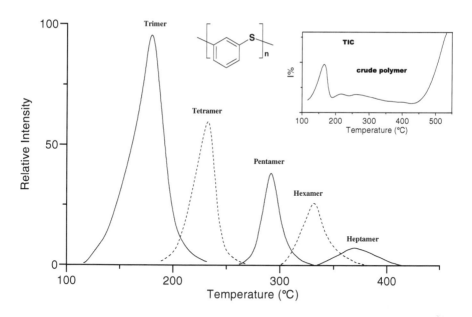

FIGURE 5.6

Single-ion current curves of cyclic oligomers evolved in the DPMS of a crude m-PPS sample. Inset: Total ion current curve. (Reprinted with permission from Ref. 39. Copyright 1986 American Chemical Society.)

The volatilization of these oligomers depends upon their molar mass and occurs at a relatively high temperature, which is nevertheless lower than the temperature at which the onset of the *m*-PPS pyrolysis is observed.[39]

When polymer pyrolysis ensues, the SIC curves of the evolving oligomers have instead a quite different appearance.

Figure 5.7a shows the TIC curve of a crude sample of an aliphatic-aromatic polyether, together with the evolution profiles of two preformed cyclic oligomers. The latter are distilling undecomposed before the onset of thermal degradation.[41] After oligomers extraction (by a solvent), the TIC curve (Figure 5.7b) shows negligible ion current below 400°C and the evolution of the cyclic oligomers (dimer and trimer) formed by pyrolysis occurs simultaneously. The evolution does not depend upon the mass of pyrolysis compounds produced. This fact is quite understandable, since the polymer decomposition starts above 400°C and reaches a maximum rate at about 500°C. Therefore, all the cyclic oligomers are formed at temperatures so high that they evolve simultaneously.[41,42]

One of the main problems connected with DPMS is that in the case of polymers producing pyrolysis compounds, which may have isobaric or isomeric structures (such as the case of cyclic and open chain oligomers), it is not possible to correct the assignments of the corresponding mass peaks. The advent of high resolution MS[5,9] and hyphenated methods such as tandem mass spectrometry (MS/MS)[43] has added much precision to DPMS for the

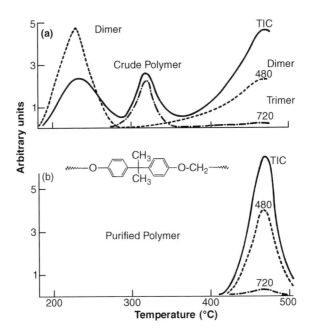

FIGURE 5.7
Total ion curve and evolution profiles of dimer (m/z 480) and trimer (m/z 720) of a polyether sample, (a) crude; (b) purified. (Reprinted with permission from Ref. 41. Copyright 1986 American Chemical Society.)

structural identification of the mass peaks. The MS/MS experiment is carried out selecting ions with the first mass analyzer, which are fragmented by a collision with a target gas located between the two analyzers. The collision-induced dissociation (CID) will generate product ions which will be detected by the second mass analyzer. CID product ions spectra can be also obtained in a double focusing MS with a forward geometry. The collision cell is located in the first free-field region between the source and the electrostatic analyzer. Selecting a specific ion and then reducing the electrostatic and magnetic fields together, maintaining the ratio B/E constant, a product ion mass spectrum is obtained.[43]

The structural identification of ion peaks arising from complex mixtures, such as those obtained in pyrolysis experiments, is achieved by the comparison of CID spectra of selected ions, with the CID product ion spectra of authentic model compounds.[44–53]

As an example, the application of the CID (B/E) method to the structure elucidation of mass peaks generated in the DPMS of poly-β-propiolactam (Nylon 3) is illustrated.[46] The pyrolysis mass spectra of Nylon 3 are characterized by a series of peaks at m/z 71 + n71 which can be assigned to cyclic lactams, formed by an intramolecular exchange process, or to open chain oligomers formed through a β-hydrogen transfer reaction (Scheme 5.1).

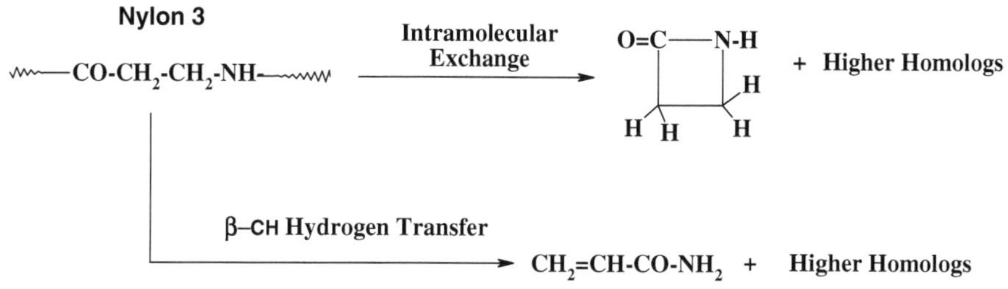

Nylon 3

$$\text{www} - CO-CH_2-CH_2-NH-\text{www}$$

Intramolecular
Exchange
→

$O=C$——$N-H$
| |
H
H H   H

+ **Higher Homologs**

β–CH **Hydrogen Transfer**
→ $CH_2=CH-CO-NH_2$ + **Higher Homologs**

SCHEME 5.1
Thermal degradation processes of Nylon 3.

Therefore, the ion with mass 71 (monomer) might correspond to two isomeric structures, namely acrylamide or cyclic propriolactam.[46]
Three CID MS/MS spectra were compared in this case (Figure 5.8a-c):

i) ion at *m/z* 71 from the EI DPMS mass spectrum of Nylon 3 (Figure 5.8a);

ii) ion at *m/z* 71 from the EI spectrum of an authentic sample of acrylamide (Figure 5.8b);

iii) ion at *m/z* 71 from an authentic sample of cyclic propio-β-lactam (Figure 5.8c).

The first two CID spectra appear very similar, while the third one shows a different fragmentation pattern and allows the assignments of the ion at *m/z* 71 (originated in the DPMS experiment) to acrylamide. As a consequence, this assignment helps to establish the mechanism of thermal decomposition of Nylon 3.[46]
Because of the importance of the information obtainable by DPMS, it has been widely used not only for the analysis of the compounds produced in the polymer pyrolysis, but also for polymer identification purposes.[54–58]
In general, the DPMS analysis of a polymeric material provides the following information:

i. the structure of the volatile compounds present in the sample (monomers, oligomers, additives, solvents, impurities), which usually distill undecomposed in the high vacuum of the MS ion source, before the polymer starts decomposing;

ii. the structure of the polymer, through the detection of a series of oligomer ions containing one or more structural units of the polymer. Although the molar masses of the oligomers produced by DPMS seldom exceed 1000–2000 Da, the information provided is generally sufficient for the structural identification of polymers;

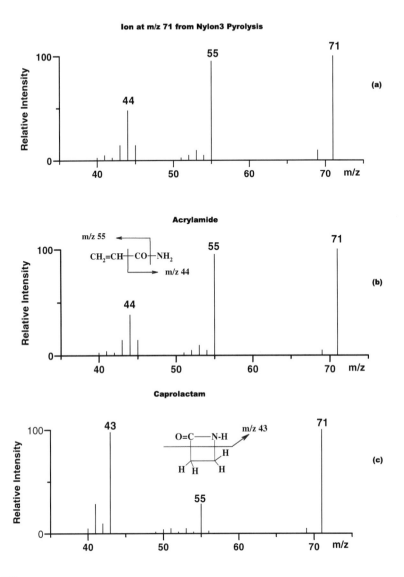

FIGURE 5.8
Collision induced dissociation (B/E) mass spectra of: (a) ion at *m/z* 71 from Nylon 3 pyrolysis; (b) acrylamide; (c) caprolactam. (Reprinted from Ref. 46, Copyright 1986, with permission from Elsevier Science.)

iii. the structure of the terminal groups produced in the pyrolysis. The structure of these end-groups generally provides insight on the mechanism of the thermal decomposition processes occurring in the pyrolysis of polymers.

The development of desorption/ionization techniques (FAB, SIMS, LD, MALDI), which allow the detection of intact polymer molecules, up high masses, has recently provided alternative and powerful methods for the structural analysis of polymeric materials. The advent of MALDI-TOF (Chapter 10) has recently opened the possibility of coupling DPMS and MALDI data to study the thermal degradation processes. In fact, one may proceed to partially degrade a polymeric sample, keeping it under inert atmosphere at a certain temperature, and then take the MALDI spectrum to observe the thermally induced changes. Intact polymer molecules can be desorbed in the MALDI mode, and therefore the spectrum will consist of a mixture of undegraded and degraded chains.

This off-line type of analysis may suffer from the fact that only the degradation products thermally most stable may survive to the heating at atmospheric pressure. DPMS data, being taken on-line and in a continuously evacuated system that provides very short transport times from the hot zone, may then complement the MALDI data by supplying information on less thermal stable pyrolysis products.[52,53,59]

Therefore, the uniqueness of DPMS today lies in the study of all kinds of catalyzed and uncatalyzed thermal decomposition processes of polymers, of polymer blends, and in the fire-retardant area, sharing this capability with Pyrolysis-GC/MS.

5.3 Selected Applications of DPMS to Polymers

In the following, we will discuss some examples, which better illustrate the applications of the DPMS and the power of this technique in the assessment of the thermal degradation mechanisms of polymers.

5.3.1 Polystyrenesulfide

The thermal degradation of poly(styrenetetrasulfide) (PST) occurs in the temperature range of 150°–250°C, and it was investigated in parallel by DPMS and by Py-GC/MS.[28] The results provided a nice illustration of capabilities and differences of the two techniques.

PST

The time-scale of the two pyrolysis techniques is quite different, and owing to the thermal lability of the sulfur-containing primary products, significantly different compounds were detected by the two methods. However, the results obtained were not contradictory and provided complementary information about the primary and secondary products formed in the thermal degradation of PST.[28]

The DPMS analysis of PST was performed both in EI and CI ionization modes, since they yield complementary information.

The EI and CI mass spectra of the degradation compounds originating from PST at the top of TIC maximum (200°C) are shown in Figure 5.9a-b, and the structural assignments for the most intense molecular ions are given in Table 5.1.

TABLE 5.1

Structural Assignments of Molecular Ions Detected in the DPMS of PST

Structure	M(x,y)
S_n	64(2), 96(3), 128(4), 160(5), 192(6), 224(7), 256(8)
$CH_2=S$	46
	104
	136(1), 168(2), 200(3), 232(4), 264(5), 296(6), 328(7)
	180
	212
	272(1,1), 304(2,1), 336(2,2), 368(3,2), 400(3,3), 432(4,3), 464(4,4)
408 440 472	

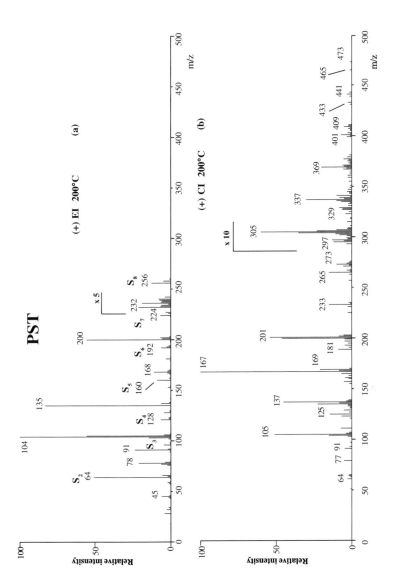

FIGURE 5.9

Mass spectra of pyrolysis compounds detected in the DPMS of polystyrenetetrasulphide: (a) EI; (b) CI. (Reprinted from Ref. 28, Copyright 1994, with permission from Elsevier Science.)

The EI mass spectrum (Figure 5.9a) shows the presence of sulfur peaks at m/z 64(S_2), 96(S_3), 128(S_4), 160(S_5), 192(S_6), 224(S_7), 256(S_8); the presence of styrene at m/z 104 (base peak) and a series of peaks at m/z 168, 200, and 232 that were assigned to styrene disulfide, trisulfide, and tetrasulfide, respectively (Table 5.1).

In the CI spectrum of PST (Figure 5.9b) peaks due to pure sulfur species are absent, because they are not ionized by the isobuthane plasma. Cyclic styrene sulfides appear as protonated molecular $(M + H)^+$ ions at m/z 137(S_1), 169(S_2), 201(S_3), 233(S_4), 265(S_5), 297(S_6), 329(S_7), containing up to seven sulfur atoms per ring (Table 5.1).

At higher masses are also present the $(M + H)^+$ ions corresponding to diphenyldithianes (i.e., cyclic styrene sulfide dimers) at m/z 273(S_2), 305(S_3), 337(S_4), 369(S_5), 401(S_6), 433(S_7), 465(S_8), together with ions at 409(S_3), 441(S_4), 473(S_5), corresponding to cyclic styrene sulfide trimers.[28] The presence of these pyrolysis compounds suggests that PST thermally decomposes through the formation of cyclic styrene tetrasulfide oligomers (Scheme 5.2).

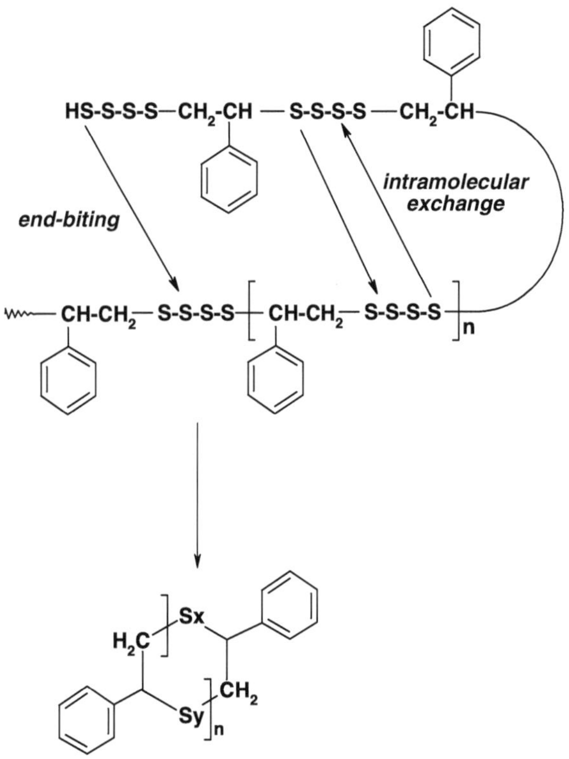

SCHEME 5.2
Thermal degradation processes of polystyrenetetrasulphide.

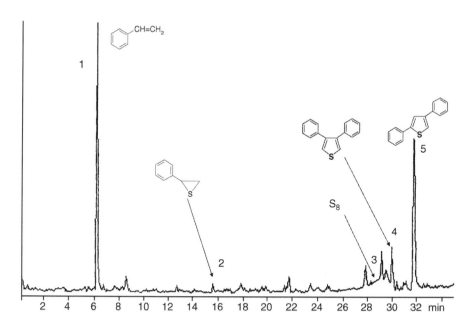

FIGURE 5.10
Flash pyrolysis-GC/MS pyrogram of polystyrenetetrasulphide obtained at 400°C. (Reprinted from Ref. 28, Copyright 1994, with permission from Elsevier Science.)

The pyrolysis products identified by DPMS leave little doubt that PST does indeed contain alternating styrene and sulfur units, although head-to-head and tail-to-tail structures cannot be distinguished.

The GC/MS pyrogram of PST (Figure 5.10) obtained by pyrolysis at 400°C shows a limited number of pyrolysis compounds with respect to DPMS. The most abundant pyrolysis compound (peak 1) is due to styrene, while peak 2 corresponds to styrene sulfide and peak 3 is due to S_8 (octasulfur ring). The broad and flat shape of the latter peak is most likely due to the condensation of elemental sulfur on cold spots and successive slow evaporation as S_8. The peaks 4 and 5 are due to two isomeric thiophenes: 2,4-diphenylthiophene and 2,5-diphenylthiophene.[28]

The striking difference among the pyrolysis products detected by DPMS and by Py-GC/MS may be explained by the specific difference of the two techniques. The DPMS method allows the gradual heating of the polymer, and the primary pyrolysis products are evolved and detected as soon as they are formed, at about 200–250°C in this case. In the Py-GC/MS method, PST is heated at 400°C and pyrolysis compounds may undergo further decomposition during the transport time to the detector, i.e., before the pyrolysis products get into the GC column or just within the GC column.

As a consequence, monomeric styrene polysulfides (from mass 168 up to m/z 328, Table 5.1) detected by DPMS, lose sulfur and yield the most thermally stable styrene sulfide (*m/z* 136), which is observed in the Py-GC/MS (Eq. 5.2).

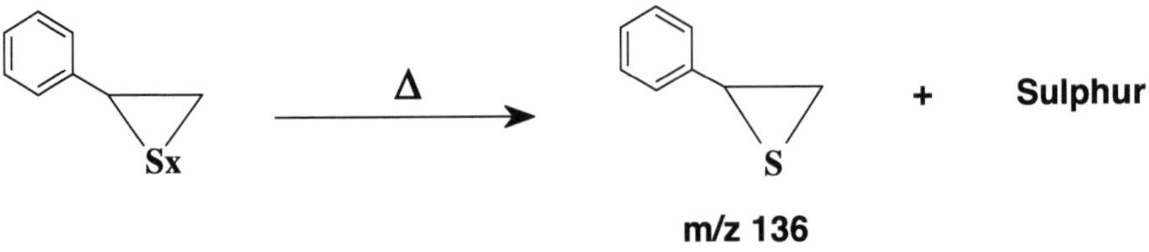

$$(5.2)$$

A similar mechanism applies to the diphenyldithianes, i.e., the cyclic styrene polysulfide dimers. However, in this case the most thermally stable compound is not the simplest cyclic dimer (*m/z* 272), but a thiophene derivative produced by thermal induced loss of H₂S and sulfur (Scheme 5.3). As a consequence, 3,4-diphenylthiophene can be generated from head-to-tail poly(styrene tetrasulfide) units and 2,4-diphenylthiophene from tail-to-tail units. The formation of thiophene derivatives from 1,4-dithiadiene rings, by thermal loss of sulfur, is well-known in the literature.[28]

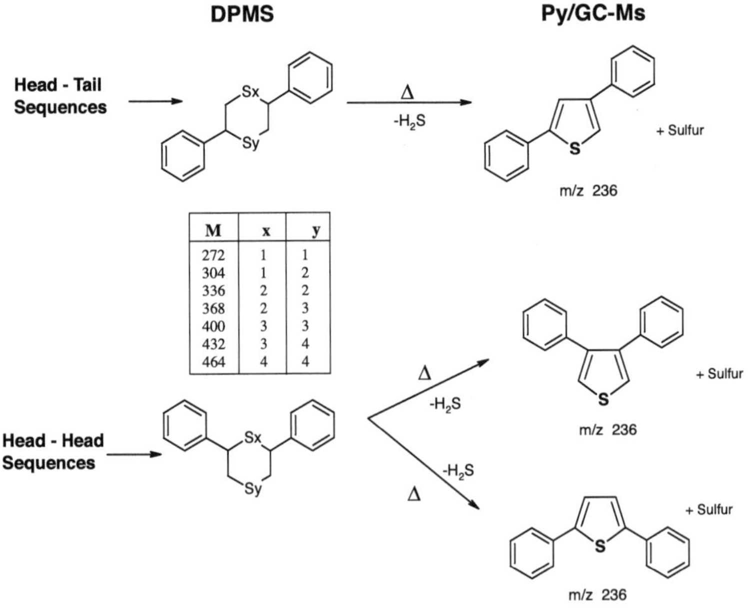

M	x	y
272	1	1
304	1	2
336	2	2
368	2	3
400	3	3
432	3	4
464	4	4

SCHEME 5.3
Further degradation reactions of cyclic compounds formed in the pyrolysis of PST. (Reprinted from Ref. 28, Copyright 1994, with permission from Elsevier Science.)

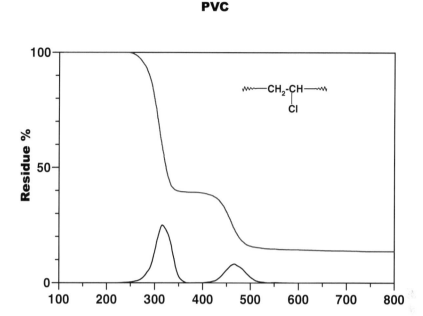

FIGURE 5.11
Thermogravimetric and derivative thermogravimetric curve of PVC. (Reprinted from Ref. 65, Copyright 1991, with permission from Elsevier Science.)

5.3.2 Evolution of Aromatics in the Thermal Degradation of PVC

The thermal decomposition of PVC involves the complete elimination of HCl, leading to the formation of macromolecular residues with polyene sequences. The latter then rearrange and decompose to yield sizeable amounts of aromatic hydrocarbons.[60-65]

Since the thermogravimetric curve of PVC (Figure 5.11) shows two well-defined degradation steps (maxima at 320°C and 450°C), it can be argued that the first weight loss step corresponds to the HCl evolution (theoretical content of HCl in PVC 58%, by weight), whereas the second degradation step would be due to the evolution of aromatic hydrocarbons.

Instead, the secondary thermal degradation of polyene sequences starts as soon as they appear along the PVC chains, and it does not wait for the complete dehydrochlorination. Aromatic compounds are therefore produced simultaneously with HCl, already in the first decomposition stage.[63-65]

In fact, the EI mass spectra in Figure 5.12 show that at 340°C unsubstituted aromatic compounds (benzene, naphthalene, anthracene) are present together with HCl, whereas at 450°C are mainly found alkyl aromatics (toluene, xylene, methylnaphthalene, and the like).[63-65]

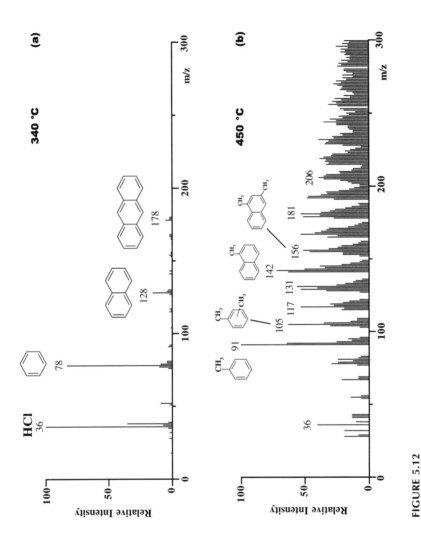

FIGURE 5.12
EI mass spectra of pyrolysis compounds evolved in the DPMS of PVC at: (a) 340°C and (b) 450°C.
(Reprinted from Ref. 65, Copyright 1991, with permission from Elsevier Science.)

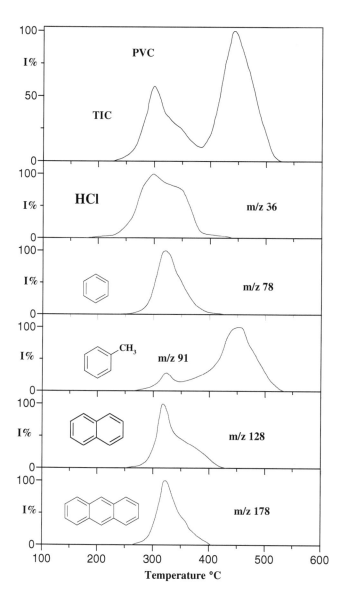

FIGURE 5.13
TIC curve and evolution profiles of some pyrolysis compounds evolved in the DPMS of PVC.
(Reprinted from Ref. 65, Copyright 1991, with permission from Elsevier Science.)

Similar to the DTG curve, the total ion current (TIC) curve (Figure 5.13) reveals two well-separated degradation steps. The evolution of HCl and unsubstituted aromatics occurs in the first stage, whereas toluene, together with other alkyl aromatics, are responsible for the second decomposition stage (Scheme 5.4).

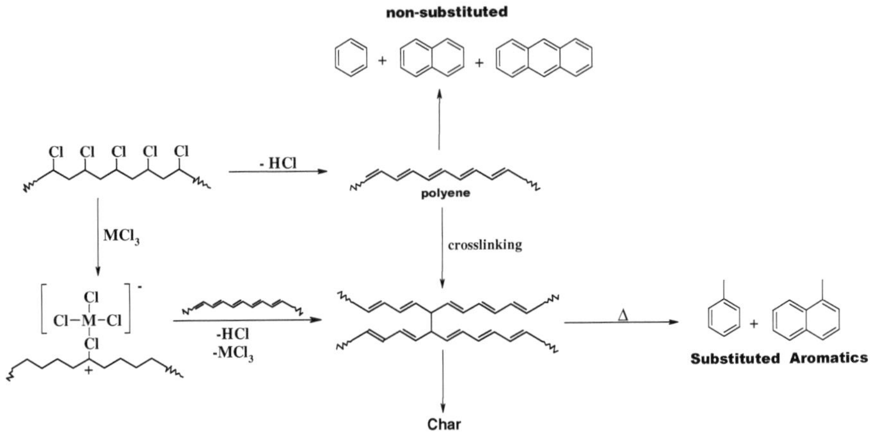

SCHEME 5.4
Thermal degradation processes of PVC.

Polyacetylene (PA) is a good model for studying the thermal decomposition of the polyene sequences formed from PVC by HCl loss.[65]

The TIC curve of PA (Figure 5.14) shows a maximum rate of polymer decomposition at about 470°C and a less pronounced broader decomposition step at lower temperature, centered at about 350°C. The maximum of evolution of benzene, naphthalene, and anthracene appears at about 350°C whereas, toluene (and the other alkyl-aromatics as well) are evolved at higher temperatures in a further decomposition process. Therefore, in the pyrolysis of PA are reproduced the main features characterizing the thermal behavior of PVC.

Further support to the hypothesis that the two families of aromatics originate from different chemical reactions comes from the effect of metal oxides. It is well-known that the addition of metal oxides to PVC produces a noticeable suppression of the amount of aromatic hydrocarbons evolved during the pyrolysis with a consequential increase of the char residue, allowing the explanation of the smoke suppressant action of these additives in the combustion of PVC. In the case of PA pyrolysis, metal oxides were not found able to suppress the evolution of aromatics, whereas metal chloride produced the selective suppression of benzene, naphthalene, and anthracene.[63–65]

These results show that the aromatic suppression agents are the metal chlorides and not the metal oxides. In fact the metal oxides were trasformed into the corresponding chlorides by the HCl evolved during the early stages of the PVC pyrolysis.

The chemical processes occurring in the thermal degradation of PVC are summarized in Scheme 5.4, which is designed to accommodate the two distinct processes that may cause the evolution of aromatic compounds and the mechanism of metal halide catalysis.

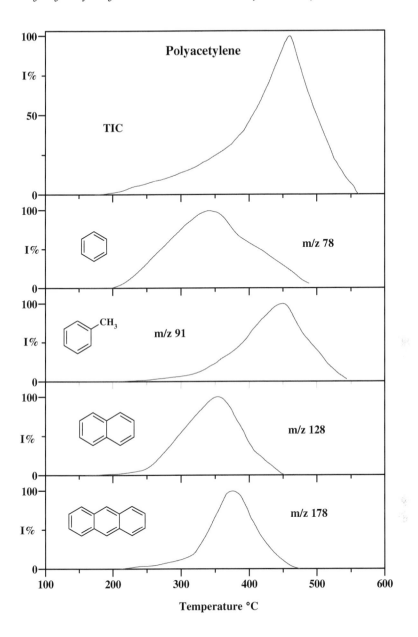

FIGURE 5.14
TIC curve and evolution profiles of some pyrolysis compounds evolved in the DPMS of poly-acetylene. (Reprinted from Ref. 65, Copyright 1991, with permission from Elsevier Science.)

In fact, linear polyenes are predicted to disappear through two competing processes: (i) intramolecular cyclization reactions producing only unsubstituted aromatics; (ii) intermolecular crosslinking leading to the formation of crosslinked polyene structures.

The amount of benzene, naphthalene, and anthracene evolved from PVC is subject to the relative rates of the two (intramolecular and intermolecular) processes.

Once formed, crosslinked polyene chains are predicted to undergo further reaction through two competing processes (Scheme 5.4): (i) further thermal decomposition to produce char residue; (ii) intramolecular cyclization reactions to produce alkyl-aromatics.

Therefore, alkyl-aromatics are generated from crosslinked polyacetylene sequences.

Since their formation requires not only the intramolecular cyclization of the polyenes but also the cleavage of some crosslinked structures, their evolution occurs at higher temperatures compared to the unsubstituted aromatics.

This reaction scheme also accounts for the *selective* suppression of the two families of aromatic compounds formed in the pyrolysis of PVC induced by metal oxides.[63–65] In fact, metal oxides (or better, the metal chlorides) are able to accelerate the rate of crosslinking of the linear polyene sequences and this process would compete with the parallel intramolecular cyclization of the polyenes. In the meantime, the amount of crosslinked polyene chains produced would be appreciably higher with respect to that formed in the pyrolysis of pure PVC. In fact, a reasonable increase of the residual char formed in the pyrolysis of PVC has also been observed.

The mechanisms concerning the chemical processes occurring in the thermal decomposition of PVC have been somewhat controversial in the past, but at present Scheme 5.4 is widely accepted in the literature.[66–68]

5.3.3 Nylon 6,6

Although DPMS has been valued as a method that allows the fast detection of primary thermal degradation products, in some cases this is not true. For instance, in some thermal degradation processes occurring through free radical mechanisms, macro-radicals undergo fast rearrangements and only secondary and tertiary products are detected.[2–4]

Another case may occur when primary pyrolysis products quickly react among themselves, in the condensed phase, so that they are detected together with the products of further reaction. The latter situation occurs in the pyrolysis of Nylon 6,6 (Ny66), where a specific structural effect due to the adipic acid unit is responsible for the formation of very reactive pyrolysis compounds.[69]

The presence of cyclopentanone, hexamethylenediamine, and cyclic monomer among the pyrolysis products of Ny66 was early recognized.[1,2] Recently MS studies suggested the occurrence of a β-CH hydrogen transfer process (Eq. 5.3), leaving unexplained the presence of cyclopentanone,

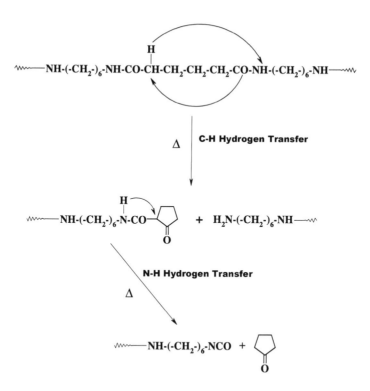

SCHEME 5.5
Thermal degradation processes of Nylon 66. (Reprinted with permission from Ref. 69. Copyright 1987 American Chemical Society.)

hexamethylenediamine, and other pyrolysis products.[1,2]

$$\text{NH-(CH}_2)_4\text{-CH-CH}_2\text{-NH-C-(CH}_2)_4\text{-CO} \longrightarrow$$

$$\text{CH}_2=\text{CH-(CH}_2)_4\text{-NH-CO-(CH}_2)_4\text{-CO} \longrightarrow$$
$$+$$
$$\text{H}_2\text{N-CO-(CH}_2)_4\text{-CO-NH-(CH}_2)_6\text{-NH} \longrightarrow$$

$$(5.3)$$

A DPMS investigation[69] finally established that the thermal degradation processes occurring in Ny66 are governed by the presence of the adipic acid units along the chains (Scheme 5.5). This process involves a hydrogen transfer to nitrogen, which causes the formation of oligomers with cyclopentadienone and amino end-groups. Further thermal scissions of these chain ends, then, account for the formation of cyclopentanone, hexamethylenediamine, and other compounds.[69]

The DPMS analysis was performed on a Ny66 sample free of cyclic oligomers to avoid the presence of residual cycles among the pyrolysis products.

The mass spectra were recorded in CI mode, because the EI mode does not produce sizeable molecular ions (similar to many other aliphatic polymers).

The CI mass spectrum of the pyrolysis products from Nylon 66, taken at 400°C (Figure 5.15), shows the presence of intense protonated molecular ions, which are identified in Table 5.2.

The identification of the structure of these pyrolysis products was a complex task, since there were several isobar structures corresponding to each peak in the CI spectrum.

The structural assignments were achieved by the comparison of the product ions (CID, B/E) spectra of authentic compounds, synthesized for this purpose, with those of products arising from the pyrolysis mixture.[69]

The ion at *m/z* 227 might be assigned to the cyclic monomer of Ny66; however, the CID (B/E) product ions mass spectrum of this ion arising from the pyrolysis mixture does not match with that of the cyclic monomer. Therefore, this ion is most likely due to a pyrolysis compound with amino and cyclopentanone as end-groups (Table 5.2), indicating that cyclic oligomers are not formed in the thermal degradation processes of Ny66.

Figure 5.16 reports the CID product ions mass spectrum of the ion at *m/z* 209 evolving from the pyrolysis of Nylon 6,6, which appear identical to the spectrum of the authentic cyclic compound with the same mass.[69]

The structure of this compound suggests that the ion at *m/z* 209 may originate from further thermal degradation of the primary pyrolysis compounds (molar mass 226), possessing cyclopentanone and amino chain ends, which are very reactive species and immediately react to produce Shiff-bases by water elimination (Scheme 5.6b). The monomeric oligomer also reacts

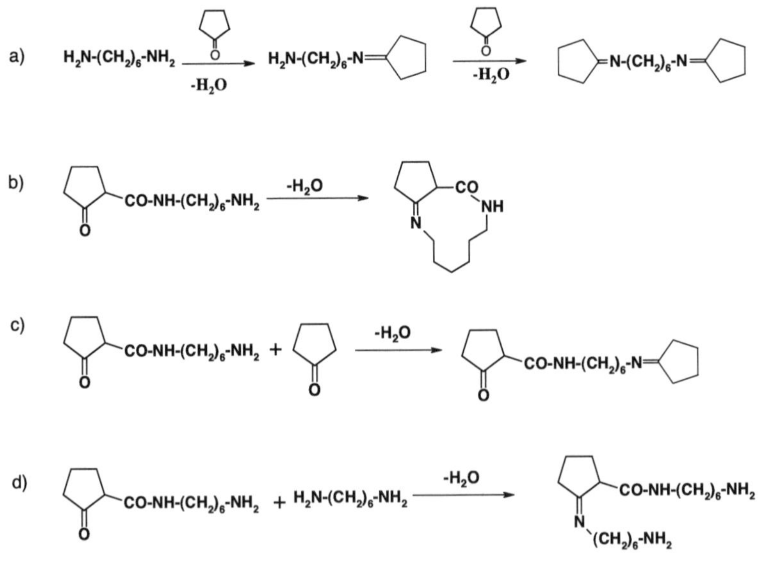

SCHEME 5.6
Further reactions of compounds formed in the pyrolysis of Nylon 66.

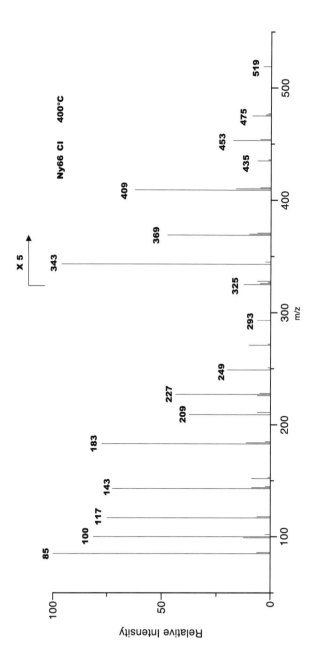

FIGURE 5.15
Isobuthane CI mass spetcrum of pyrolysis compounds evolved at 400°C in the DPMS of Nylon 66. (Reprinted with permission from Ref. 69. Copyright 1987 American Chemical Society.)

TABLE 5.2

Structural Assignments of Protonated Molecular Ions Appearing in the CI DPMS Mass Spectra of Ny66

Pyrolysis Products	n	MH$^+$
		85
		209
		325
H--[-HN-(CH$_2$)$_6$-NH-CO-(CH$_2$)$_4$-CO-]$_n$-NH-(CH$_2$)$_6$-NH$_2$	0 1	117 343
H--[-HN-(CH$_2$)$_6$-NH-CO-(CH$_2$)$_4$-CO-]$_n$-NH-(CH$_2$)$_6$-NCO	0 1	143 369
H--[-HN-(CH$_2$)$_6$-NH-CO-(CH$_2$)$_4$-CO-]$_n$-NH-(CH$_2$)$_6$-N=	0 1	183 409
H--[-HN-(CH$_2$)$_6$-NH-CO-(CH$_2$)$_4$-CO-]$_n$-NH-(CH$_2$)$_6$-NH-CO-	0 1	227 453
=N--[-(CH$_2$)$_6$-NH-CO-(CH$_2$)$_4$-CO-NH-]$_n$-(CH$_2$)$_6$-N=	0 1	249 475
=N-(CH$_2$)$_6$-NH--[-CO-(CH$_2$)$_4$-CO-NH-(CH$_2$)$_6$-NH-]$_n$-CO-	0 1	293 519

Ion at m/z 209 from Nylon 66 pyrolysis

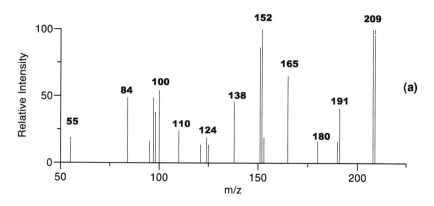

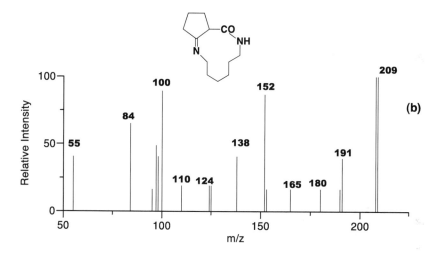

FIGURE 5.16

CID B/E mass spectra of the ions at *m/z* 209 obtained in the isobuthane CI of: (a) ion at *m/z* 209 from Ny66 pyrolysis; (b) authentic model compound. (Reprinted with permission from Ref. 69. Copyright 1987 American Chemical Society.)

with cyclopentanone and hexamethylenediamine producing the ions at *m/z* 293 and 343, respectively (Scheme 5.6c,d).

Remarkably, several of the compounds identified in Table 5.2 possess Shiff-bases as end-groups originating from the fast reaction of amino end-groups and cyclopentanone. This reaction was reproduced, and authentic samples of compounds with masses 182 and 248 (Scheme 5.6a) were prepared and used for the identification of the corresponding pyrolysis products by CID (B/E) product ions experiments.[69]

Further support for the thermal degradation mechanisms suggested for Ny66 (Schemes 5.5 and 5.6) comes from the structure of thermal degradation products of Nylon 11,6 and other polyamides containing the adipic acid unit. In fact, the CI mass spectra of these polymers show intense peaks due to pyrolysis products having cyclopentanone, amine, and azomethyne end-groups, as in the case of Nylon 6,6.[70–72]

5.3.4   Polyethyleneterephthalate

The mechanism of thermal degradation of polyetheleneterephthalate (PET) has been studied over many years by off-line analysis of the pyrolysis products[73–76] and by Py-GC/MS studies.[77] The results suggested a β-CH hydrogen transfer process leading to the formation of oligomers with olefin and carboxylic end-groups (Eq. 5.4).

$$(5.4)$$

However, the formation of cyclic oligomers in the pyrolysis of several aliphatic and aromatic polyesters is well-known,[1,2,15,20,78–80] and a distribution of cyclic compounds is expected if the thermal degradation proceeds through macrocyclization processes involving ester-exchange reactions.[15,20]

The presence of cyclic oligomers in melt PET was first reported in 1954.[81] Since then, the mechanism of rings formation from melt PET has been studied in detail,[82–87] and the molar cyclization equilibrium constants for rings, both in melt and in solution, have been measured.[86–88] Numerous studies[81,83,84] have established that the cyclization occurs in PET through an intramolecular alcoholysis reaction (end-biting), which is activated already in the temperature range of 250–300°C, implying the attack of hydroxyl ends on the inner ester groups of the polyester chain (Scheme 5.7).

Therefore it appears, that the intramolecular exchange reaction (an ionic process) leading to the formation of cyclic oligomers is the primary thermal process occurring in PET at temperatures below 300°C, and that the β-CH hydrogen transfer reaction occurs only as a secondary process. However, the temperature range explored in pyrolysis studies goes well beyond 300°C, and the question might rise if the cyclic oligomers are still the primary pyrolysis products at higher temperatures.

Thermally labile compounds, such as cyclic oligomers, which may have very short lifetimes at high temperatures, may escape detection. As a matter of fact, another important pathway of thermal decomposition of these polyesters was neglected in these studies: the intramolecular exchange process that leads to the formation of cyclic oligomers (Scheme 5.7).

Direct pyrolysis-MS (DPMS, an on-line technique) was used to investigate the thermal degradation of PET.[51] The EI mass spectra do not exhibit significant molecular ions of the cyclic oligomers.[41,51,78] On the contrary, the NCI mass

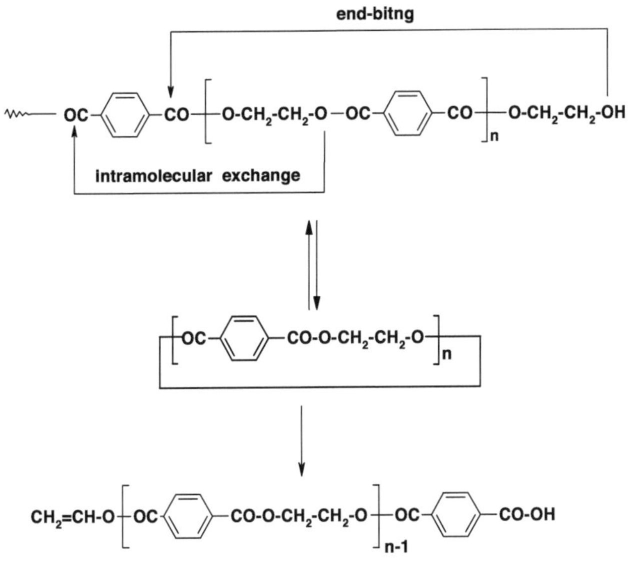

SCHEME 5.7
Thermal degradation processes of poly(ethyleneterephthalate).

spectrum of PET, reported in Figure 5.17, is dominated by a series of intense molecular ions at *m/z* 192, 384, 576, and 768 corresponding to oligomers of PET.[51]

However, the cyclic structure cannot be directly deduced from the mass of the molecular ion. In fact, the open-chain oligomers with an olefin and a carboxylic end-group, which are originated by a β-CH hydrogen transfer reaction (Eq. 5.4), are isomeric with the cyclic oligomers that might originate from an exchange process (Scheme 5.7).

The structure of the pyrolysis products has been identified by the comparison of their CID (B/E) product ion spectra with that of authentic samples of cyclic and open-chain oligomers with olefin and carboxyl end-groups.

Figure 5.18 reports the CID (B/E) product ion spectrum of the ion at *m/z* 576 arising from the pyrolysis mixture and those of the cyclic and open-chain trimers. Although the three spectra appear quite similar, they differ in some details, i.e., the ion at *m/z* 505 is absent in the MS/MS spectrum of cyclic trimer, whereas the ions at *m/z* 460, 546, and 558 are absent in the spectrum of the open-chain trimer. Therefore, these peaks can be considered as diagnostic for the structural identification of the two compounds. The CID (B/E) product ion spectrum of the ion at *m/z* 576 arising from PET pyrolysis shows the simultaneous presence of the peaks at *m/z* 505, 546, 488, and 460, indicating that both cyclic and open-chain oligomers are contributing to the formation of the fragmentation spectrum.

This evidence appears to support the hypothesis that the pyrolysis of PET proceeds through the primary formation of cyclic oligomers, which are

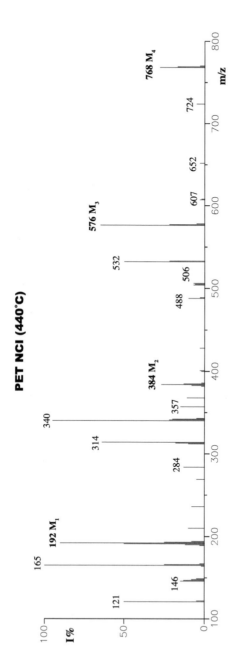

FIGURE 5.17

NCI spectrum of pyrolysis compounds detected in the DPMS of PET, at 440°C. (Reprinted from Ref. 51, Copyright 1993, with permission from Elsevier Science.)

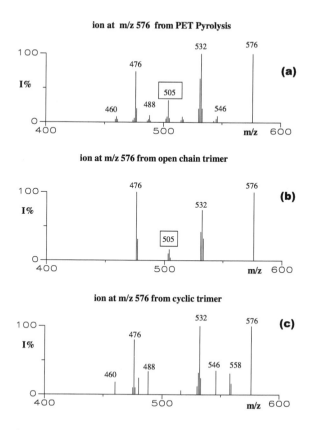

FIGURE 5.18

CID B/E product ion mass spectra of (a): ion at *m/z* 576 from PET pyrolysis; (b): ion at *m/z* 576 from authentic sample of open-chain PET trimer; (c): ion at *m/z* 576 from authentic sample of cyclic PET trimer. (Reprinted from Ref. 51, Copyright 1993, with permission from Elsevier Science.)

unstable at the temperatures necessary to further pyrolyze the polymer, and therefore, these decompose to generate open-chain oligomers (Scheme 5.7). The two series of oligomers are both present in the pyrolysis spectra, and they appear to be the predominant pyrolysis products in the time-scale of the DPMS experiments. If other pyrolysis techniques are used, where the residence time of the pyrolysis products in the hot zone (and also the transport time) is higher, then compounds originating from the further thermal degradation of the primary products may be exclusively detected.[76–78]

5.3.5   Poly(Bisphenol-A Carbonate)

The thermal decomposition processes occurring in poly(bisphenol-A carbonate) (PC) have received continued attention in the literature,[89–97] since this polymer is an important engineering thermoplastic that is subjected to injection molding operations at temperatures above 300°C. Degradation reactions are likely to occur at this temperature, and therefore the understanding of its thermal behavior is of crucial importance in the end-use application.

The thermal degradation of PC starts at about 350°C, with a maximum degradation rate at about 500°C. There is about 20–30% residue at 800°C, and there are no indications in the literature about its structure.

Davis and Golden[90,91] studied the isothermal pyrolysis of PC in the temperature range of 300–400°C and noticed a decrease in the molar mass of PC when it is thermally degraded in a sealed system, whereas in a continuously evacuated system PC rapidly crosslinks to form an insoluble gel. Rearrangement of the carbonate group to form a pendant carboxyl group, *ortho* to an ether link, which then leads to xanthone units and to a crosslinked structure with ester linkage between chains, have been hypothesized on the basis of the structure of pyrolysis compounds obtained from diphenylcarbonate, a model compound.

The thermal degradation of PC was also performed at higher temperatures (500°–850°C) by flash pyrolysis GC/MS (Py-GC/MS),[89,92,95,97] but only low molar mass compounds (H_2O, CO_2, bisphenol A, phenol, isopropenylphenol, diphenylcarbonate) were detected and therefore, they cannot allow the prediction of a complete pattern of the thermal decomposition processes of PC.

Since the mechanism of thermal cleavage of a polymer can be better inferred if larger primary pyrolysis products are detected, thermal degradation studies of PC were focused on DPMS,[96] which allows the detection of large primary products of pyrolysis and therefore the assessment of the thermal degradation processes. The structure of the pyrolysis compounds detected by DPMS suggests that PC undergoes thermal decomposition by a number of different pyrolysis processes (Scheme 5.8a-e), as can be recognized through the inspection of the time-temperature resolved evolution profiles of some pertinent pyrolysis compounds (Figure 5.19a-h).

The TIC curve (Figure 5.19a) shows the presence of three decomposition steps, with the maxima centered at about 380°C, 500°C, and 550°C, respectively.

In the initial stage of the thermal degradation of PC (about 300–400°C), cyclic oligomers are generated[94,96] through an intramolecular exchange reaction (Scheme 5.8a); this is a typical degradation pathway of polymers containing reactive functional groups. As a consequence, the evolution profile of the cyclic dimer (ion at m/z 508, Figure 5.19b) is included within the first TIC maximum.

At higher temperatures (500–700°C) other decomposition reactions are observed, with formation of open-chain compounds whose structures originate from molecular rearrangements or decomposition of the polycarbonate structural unit (Scheme 5.8).

The formation of CO_2, observed in the first decomposition stage (Figure 5.19c), is in large measure due to electron impact fragmentation processes, and to the hydrolytic cleavage of the carbonate group, which leads to the formation of hydroxyl terminated oligomers and CO_2 (Scheme 5.8b). CO_2 is also evolved in the second degradation step, most likely due to an intramolecular (1-3 shift) CO_2 elimination (Scheme 5.8c), yielding compounds with diphenylether linkages

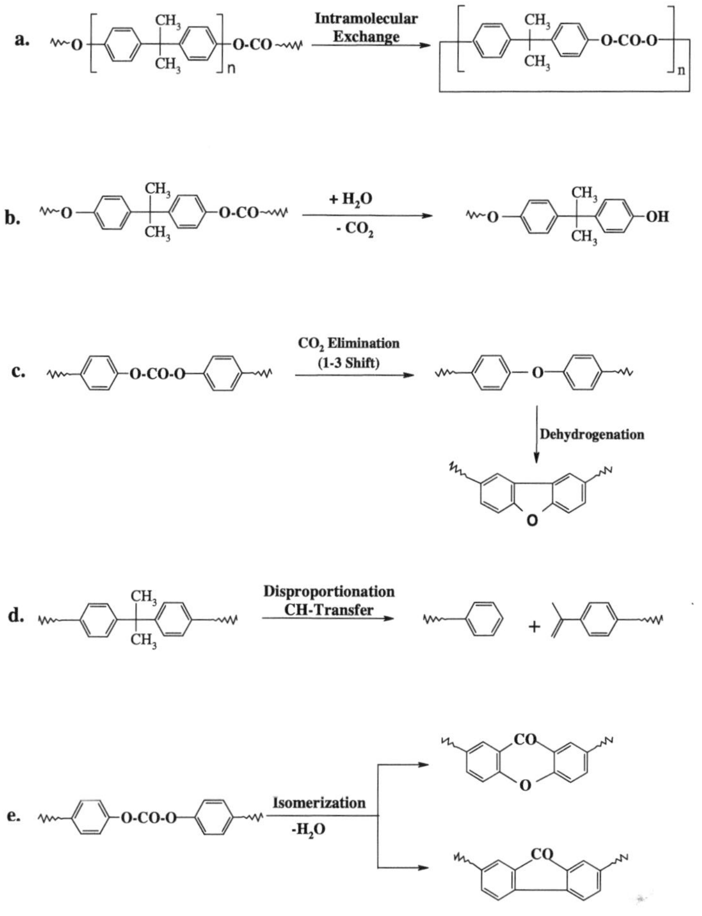

SCHEME 5.8
Thermal degradation processes occurring in poly(bisphenol-A-carbonate). (Reprinted with permission from Ref. 69. Copyright 1999 American Chemical Society.)

(ion at m/z 318, Figure 5.19e). At higher temperatures (third degradation step), the ether moieties undergo dehydrogenation (Scheme 5.8c) producing dibenzofuran units (ion at m/z 418, Figure 5.19g).

A disproportionation reaction of the bisphenol A unit (Scheme 5.8d), with consequent polymer chain cleavage, accounts for the formation of pyrolysis compounds with phenyl and isopropenyl end-groups, such as phenol (ion m/z 94, Figure 5.19d) and isopropenyl phenol.

Compounds containing xanthone units are detected in the second decomposition step (ion at m/z 370, Figure 5.19f), most likely formed through an isomerization of the carbonate group (Scheme 5,8e). This rearrangement may occur through a Fries rearrangement, which would also explain the formation of compounds containing fluorenone units (Scheme 5.8e), evolved in the third degradation step (ion at m/z 340, Figure 5.19h).[96]

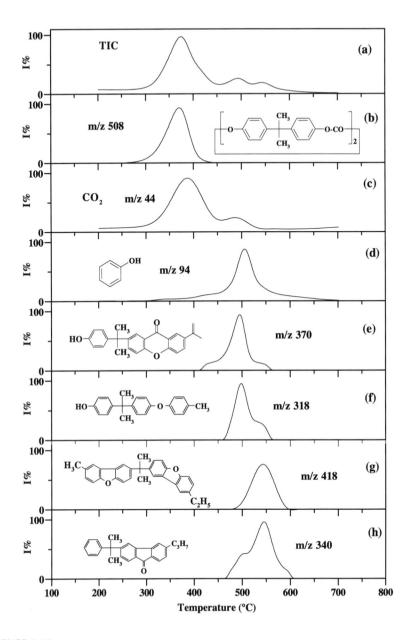

FIGURE 5.19
TIC curve and time-temperature resolved profiles of some selected pyrolysis compounds obtained in the DPMS of PC. (Reprinted with permission from Ref. 69. Copyright 1999 American Chemical Society.)

Furthermore, the analysis of the pyrolysis residue of PC obtained at 400°C, after 1 hour of isothermal heating, showed the presence of several consecutive xanthone units indicating that at this temperature the isomerization and condensation processes are quite extensive.[59,96]

Therefore, the pyrolysis residue can be considered as constituted by long ether/xanthone sequences that undergo aromatization and cross-linking processes, leading to a graphite-like charred residue as the temperature increases. This is analogous to several other bridged polyaromatics (PPO, PPS, PES, PEK),[98–100] which show a marked tendency to produce graphite-like pyrolysis residue.

5.3.6 Polyetherimide

Polyetherimide poly(2,2′-bis(3,4-dicarboxyphenoxy)phenylpropane)-2-phenylenediimide) (PEI) is a commercial engineering thermoplastic with excellent mechanical properties and high thermal stability.

PEI

This polymer is amorphous and thermoxidatively stable, with a Tg of 215°C, and it is subjected to injection molding operations at temperatures above 300°C. At these temperatures degradation reactions are likely to occur, and therefore the understanding of the thermal behavior of PEI is of crucial importance in the end-use applications.

A key problem in studying the thermal degradation processes of PEI is that this polymer has a repeat unit of 592 daltons, and therefore the detection of structurally significant pyrolysis products is a difficult task, since many will have masses beyond 600 Daltons.

In this respect, DPMS has the advantage of allowing the detection of high mass pyrolysis products that are crucial in defining structures and thermal degradation mechanisms of aromatic polymers bearing large repeat units.

Due to the phthalimide rings present in the backbone, PEI is thermally stable and starts decomposing at about 450°C (Figure 5.5a). It shows a temperature of maximum rate of decomposition (PDT) at about 510°C, followed by a less marked decomposition step in the temperature range of 600–650°C. At 800°C about 6% of charred residue is left.[38]

The total ion current (TIC) curve of PEI (Figure 5.5b), obtained by DPMS at the same heating rate used for the TG experiments (10°C/min), closely reproduces the two maxima appearing in the DTG curve.

The EI mass spectrum obtained at 520°C from PEI is reported in Figure 5.20 and shows the presence of mass peaks up to 830 Daltons. The structures of the pyrolysis compounds, given in Table 5.3, suggest the occurrence of several thermal degradation processes (Scheme 5.9a-g).

The most abundant pyrolysis compounds detected in the major degradation step (Figure 5.20) possess intact phthalimide rings containing the aromatic

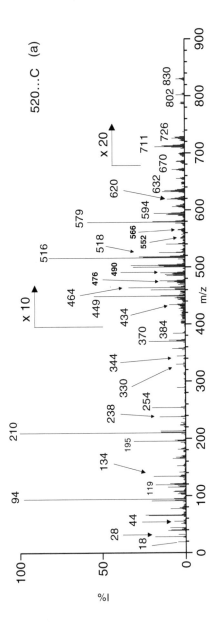

FIGURE 5.20
EI mass spectrum of pyrolysis compounds evolved in the DPMS of a PEI sample, at 520°C. (Reproduced from Ref. 38, Copyright 1999, by permission of Wiley-VCH Publications.)

TABLE 5.3

Structural Assignments of the Most Intense Molecular Ions Observed in the DPMS of Ultem Sample

Structures	X,Y	M⁺
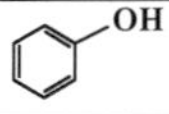 benzene-OH		94
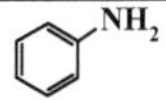 benzene-NH₂		93
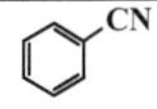 benzene-CN		103
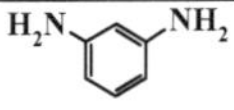 H₂N-benzene-NH₂		108
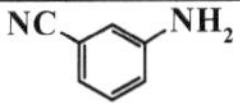 NC-benzene-NH₂		118
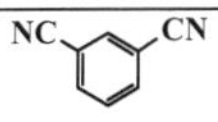 NC-benzene-CN		128
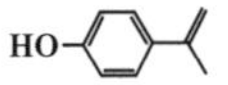 HO-benzene-isopropyl		134
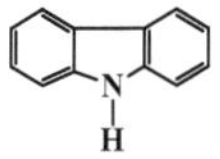 carbazole (N–H)		167
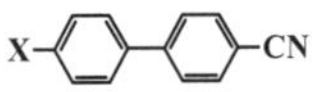 X-biphenyl-CN	H CH₃ C₂H₅	179 193 207
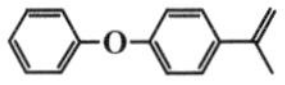 benzene-O-benzene-isopropyl		210
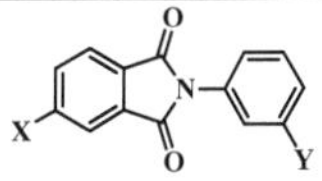 phthalimide structure X...Y	H,H H,NH₂ OH, NH₂	223 238 254
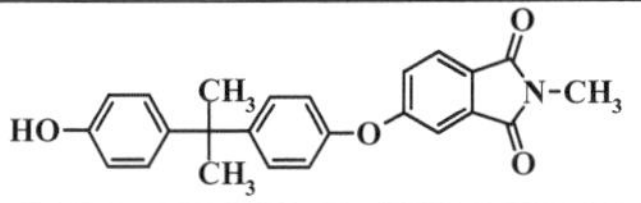 HO-benzene-C(CH₃)₂-benzene-O-phthalimide-N–CH₃		387
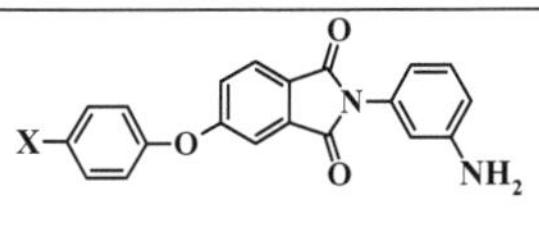 X-benzene-O-phthalimide-N-benzene-NH₂	H CH₃ C₂H₅ C₃H₅ C₃H₇	330 344 358 370 372
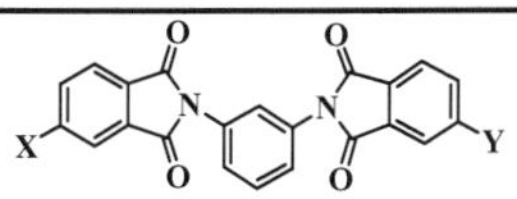 X-bis-phthalimide-benzene-Y	H,H OH,H	368 384
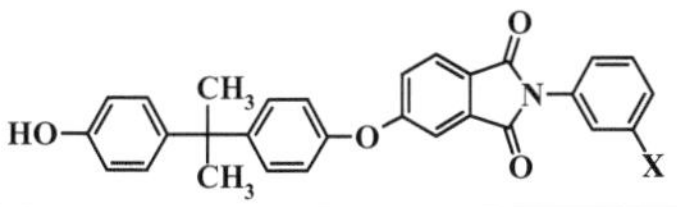 HO-benzene-C(CH₃)₂-benzene-O-phthalimide-N-benzene-X	H NH₂	449 464

(Continued)

TABLE 5.3

Structural Assignments of the Most Intense Molecular Ions Observed in the DPMS of Ultem Sample (Continued)

Structures	X,Y	M$^+$
(structure)	H,H	460
	H,CH$_3$	474
	H,C$_2$H$_5$	488
	H,C$_3$H$_5$	500
	H,C$_3$H$_7$	502
	OH, H	476
	OH, CH$_3$	490
	OH,C$_2$H$_5$	504
	OH,C$_3$H$_5$	516
	OH,C$_3$H$_7$	518
(structure)	H,H	552
	H,CH$_3$	566
	H,C$_2$H$_5$	580
	H,C$_3$H$_5$	592
	H,C$_3$H$_7$	594
	CH$_3$, C$_2$H$_5$	594
	CH$_3$, C$_3$H$_5$	606
	C$_2$H$_5$, C$_2$H$_5$	608
	C$_2$H$_3$, C$_3$H$_5$	620
	C$_3$H$_5$, C$_3$H$_5$	632
	C$_3$H$_5$, C$_3$H$_7$	634
(structure)	H	518
	(phenyl)	594
(structure)		594
(structure)	H,H	670
	H,OH	686
	CH$_3$, OH	700
	C$_2$H$_5$, OH	714
	C$_3$H$_5$, OH	726
	C$_3$H$_7$, OH	728
(structure)	H	802
	OH	818
(structure)		830

Reproduced from Ref. 38, copyright 1999, by permission of Wiley-VCH Publishers.

ether moiety, and bearing hydrogen, methyl, ethyl, ispropenyl, and isopropyl end-groups, generated by disproportionation of the isopropylidene bridge of the BPA units of PEI, followed by hydrogen transfer reactions. (Scheme 5.9a).

Furthermore, pyrolysis compounds containing phthalimide units, with the phenyl rings substituted with H/OH and/or bisphenol A, are formed by the scission of ether bridges (Scheme 5.9b–c), whereas compounds with N-H and/or N-phenyl as end-groups may be formed by the scission of phenyl-phthalimide bonds (Scheme 5.9d).

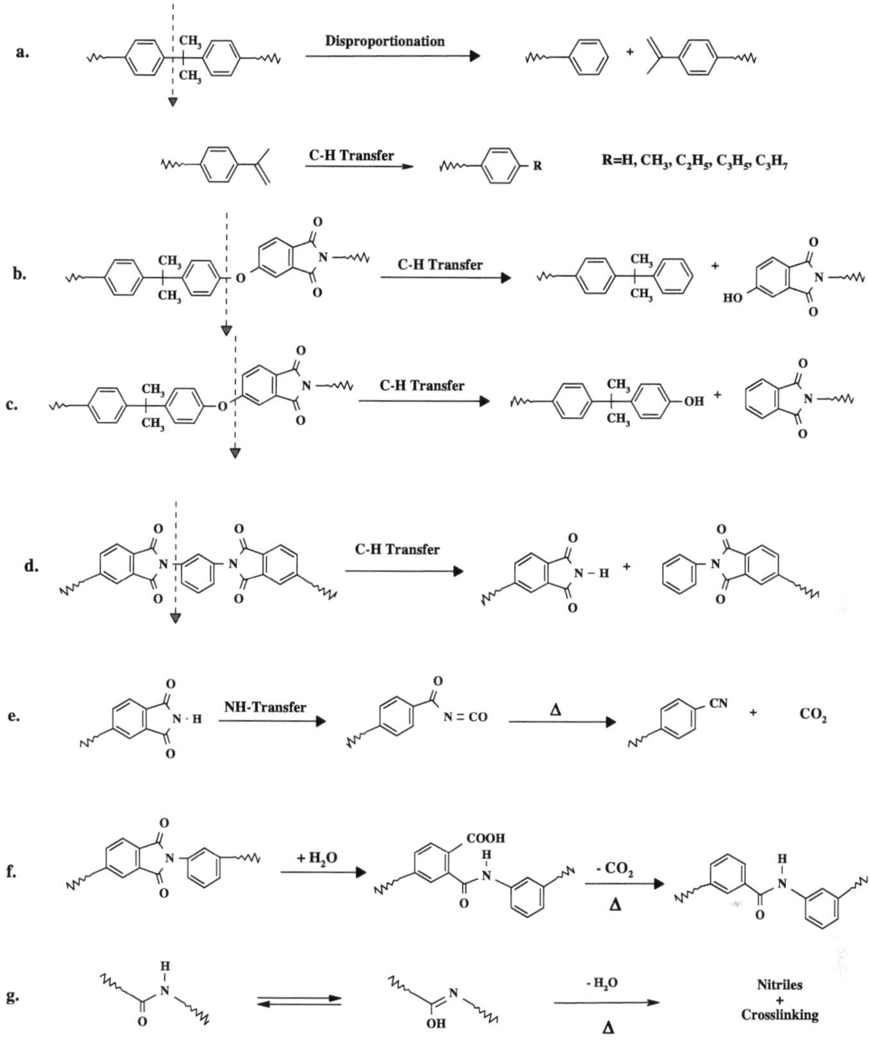

SCHEME 5.9
Thermal degradation processes of polyetherimide. (Reproduced from Ref. 38, Copyright 1999, by permission of Wiley-VCH Publishers.)

Concomitant extensive hydrogen transfer reactions may account for the high amount of char residue generated in the pyrolysis of PEI.

The temperature time-resolved evolution profiles of some relevant pyrolysis compounds of PEI (Figure 5.21) indicate that compounds generated from the scission of isopropenyl and ether bridges (e.g., ions at m/z 518 and 632) are evolved exclusively within the first TIC maximum, indicating that these bonds are the weakest units.

The low mass pyrolysis compounds such as aromatic nitriles (e.g., benzonitrile, m/z 103) and CO_2 (m/z 44) are evolved in both thermal degradation

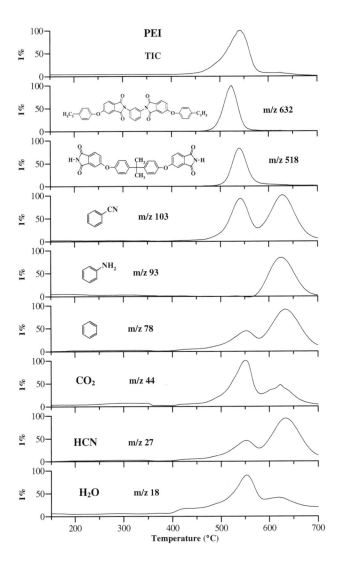

FIGURE 5.21
TIC curve and temperature-resolved evolution profiles of the ions at m/z 518, 632, 103, 93, 78, 44, 27, and 18 as obtained by DPMS of a PEI sample. (Reproduced from Ref. 38, Copyright 1999, by permission of Wiley-VCH Publishers.)

steps, and their formation most likely takes place from a high temperature secondary pyrolysis process due to the N-H phthalimide end-groups. This involves the formation of a thermally labile isocyanate intermediate that decomposes to generate nitriles and carbon dioxide (Scheme 5.9e).[38]

Another reaction that may account for the formation of char and CO_2 is the heat-induced hydrolysis of the phthalimide rings. This forms polyamic acid, which undergoes decarboxylation at above 600°C producing aromatic amide bonds (Scheme 5.9f). The tautomerization equilibrium of amide bonds is shifted toward iminolization at high temperatures, and the imino products undergo water elimination and crosslinking, providing an additional source of nitrile compounds and char (Scheme 5.9g).[38]

5.3.7 Polymer Blends

Blends of condensation polymers such as polyesters, polyamides, polycarbonates, and in general polymers bearing reactive backbone or pendant functional groups, may yield chemical exchange reactions in the molten state. These exchange processes may lead to the formation of copolymers that may drastically change the properties of the blends. At temperatures below 300°C, these reactions may be induced by the presence of catalysts or reactive terminal groups (OH, COOH, NH_2) originally present in the polymers or generated *in situ* by thermal and/or hydrolysis degradation reactions. At temperatures above 300°C, a thermally activated exchange (i.e., in the absence of catalyst) may occur, generating copolymers that may further decompose.

The DPMS method has been successfully used for monitoring the exchange reactions that may occur during the thermal treatment of polymer blends[101–107] because, being performed under vacuum and *on-line*, it allows (i) the detection of primary pyrolysis products of high mass (containing at least one repeat unit), and (ii) the prediction of the structure of the decomposing systems.[101–107]

5.3.7.1 PC/Ny6

The melt mixing of a PC/Ny6 blend at 240°C gives rise to the formation of copolymers through an exchange reaction involving the attack of amino terminals groups of Ny6 on the inner carbonate groups of PC (outer-inner, Scheme 5.10).[103] The yield of copolymer is small at the beginning of the

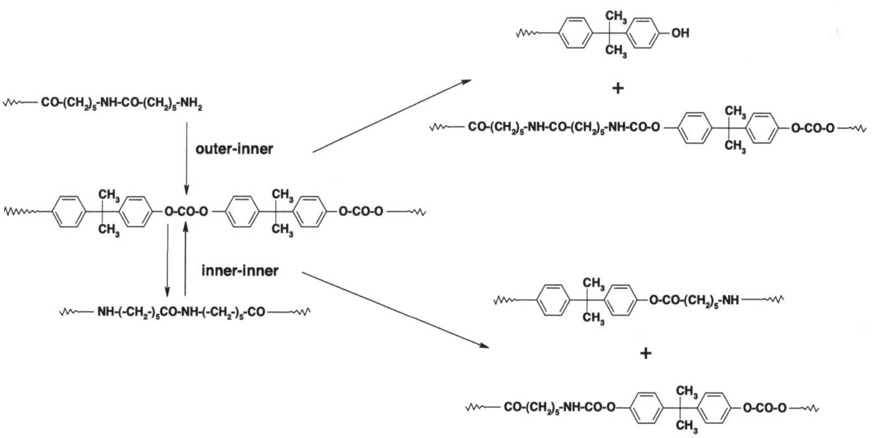

SCHEME 5.10
Exchange reactions occurring in the melt mixing of PC/Ny6 blends.

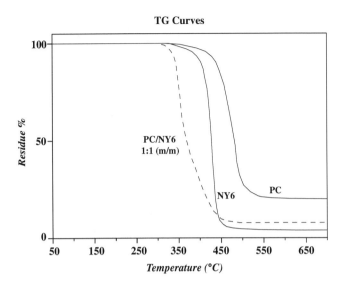

FIGURE 5.22
Thermogravimetric curves of Nylon 6, PC, and PC/Nylon 6 (1/1, mol/mol) blend obtained at
10°C/min under nitrogen flow. (Reprinted from Ref. 103.)

reaction, owing to the low amount of reactive amino end-groups initially
contained in the Ny6 sample. After twenty minutes of heating at 240°C,
amino terminal groups are generated by hydrolysis reactions of the Ny6
chains, and therefore the amount of copolymer increases with the reaction
time. The amount of urethane units increases as the reaction time increases,
with a consequent reduction of the size of PC and Ny6 blocks in the
copolymer.[103]

The direct amide/carbonate exchange reaction (inner-inner, Scheme 5.10)
cannot take place at 240°C because it would necessitate an appropriate
catalyst. To obtain information on the chemical reaction that may occur at
higher temperatures, TG and DPMS techniques were used.

The thermogravimetric curve of a PC/Ny6 blend (Figure 5.22) indicates a
marked lowering of the thermal stability as compared to that of the unmixed
polymers, with the onset of thermal degradation at about 300°C.

The structures of the thermal degradation products obtained in the
DPMS of PC/Ny6 blend permitted the identification of the chemical reactions occurring at higher temperatures.[103] The CI spectrum of PC/Ny6
blend (Figure 5.23) contains the ions at *m/z* 114 and 509 corresponding to
protonated molecular ions of caprolactam and cyclic dimer of PC, respectively. The peak at *m/z* 112 can be assigned to an unsaturated aliphatic
isocyanate whereas the ion at *m/z* 324 corresponds to an unsaturated ester
(Table 5.4). The peak at *m/z* 367 may possess an open-chain isocyanate-
ester or a cyclic urethane-ester structure. The pyrolysis compounds containing

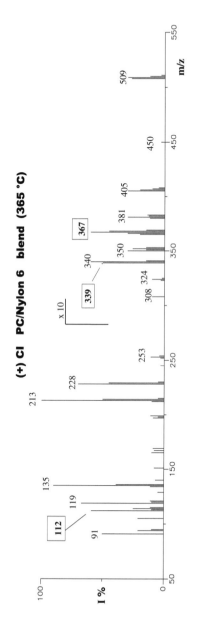

FIGURE 5.23

Isobutane CI mass spectrum of pyrolysis compounds evolved in the DPMS of a PC/Nylon 6 (1/1, mol/mol) blend sample, at 385°C. (Reprinted from Ref. 103.)

TABLE 5.4

Structural Assignments of Mass Peaks Observed in the CI DPMS Spectra of PC/Ny6 Blend

Pyrolysis Products	M^+	MH^+
or	367	
	339	340
	324	
O=C=N-(CH$_2$)$_3$-CH=CH$_2$	112	
		509
		114

the isocyanate and ester moieties are diagnostic for the identification of exchange reactions occurring in PC/Ny6 blend. In fact, the direct amide-carbonate exchange (Scheme 5.10, inner-inner) generate copolymer chains containing aliphatic-aromatic ester and urethane units, which undergo decomposition at these higher temperatures, producing isocyanate and OH

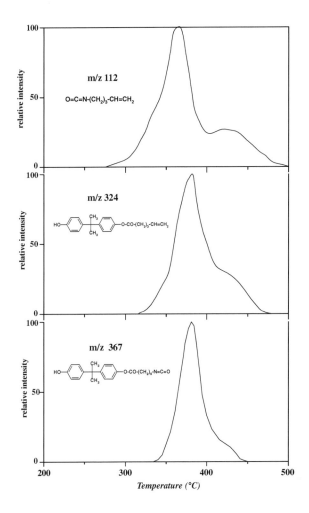

FIGURE 5.24
Temperature-time resolved evolution profiles of the ions at *m/z* 112, 324, and 367 as obtained by the DPMS of a PC/Nylon 6 (1/1, mol/mol) blend sample. (Reprinted from Ref. 103.)

phenolic groups (Eq. 5.5).[1,2]

$$(5.5)$$

The evolution profiles of these ions (Figure 5.24) indicate that the unsaturated aliphatic isocyanate (ion at *m/z* 112) is evolved at a lower temperature than the ester-containing pyrolysis compounds (ions at *m/z* 324 and 367). This suggests that the direct amide/carbonate exchange reaction takes place

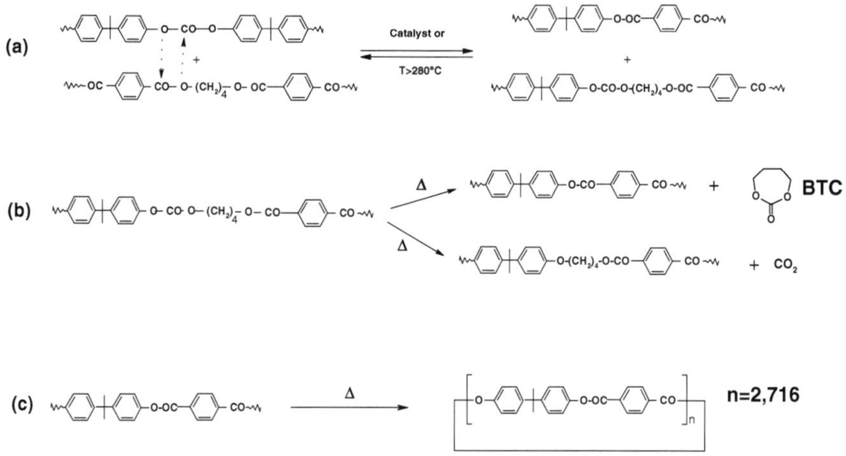

Mass Spectrometry of Polymers

(a) ... Catalyst or / T>280°C ...

(b) ... Δ ... + BTC ... Δ ... + CO₂

(c) ... Δ ... n=2,716

SCHEME 5.11
Chemical reactions occurring in the thermal treatment of PC/PBT blend.

at temperatures above 300°C and that the early evolution (below 300°C) of the isocyanate ion (m/z 112) is due to the decomposition of the urethane units that are generated by the attack of the amino end-groups of Ny6 on the carbonate groups of PC.[103]

5.3.7.2 PC/PBT

It has been observed that in the melt mixing of high molar mass PBT and PC (i.e., containing nondetectable amounts of reactive end-groups) at 240°C a direct ester-carbonate exchange occurs in the presence of a catalyst, producing a four-component copolymer at the equilibrium. Instead, in the absence of a catalyst, the exchange reaction between the two polymers takes place above 280°C (Scheme 5.11a).[107]

The DTG traces in Figure 5.25 (obtained at a heating rate of 10°C/min) show that PBT and PC undergo thermal decomposition in a single step, whereas their blend decomposes in three steps showing the maxima at 370°C, 410°C, and 500°C, respectively. The onset of thermal degradation of PC/PBT blend appears at a lower temperature (about 300°C), and this is an indication that the exchange reaction produces less thermally stable compounds compared to the unmixed polymers.[104] The structure of pyrolysis compounds detected by DPMS indicates that they arise by the thermal decomposition of copolymer chains formed through a thermal-activated exchange reaction (Scheme 5.11a), and also indicates that the equilibrium is affected by their evolution.[104]

The TIC curve of PC/PBT blend obtained by CI DPMS is shown in Figure 5.26, together with the temperature-time resolved evolution profiles of some of the most important pyrolysis compounds. The evolution of

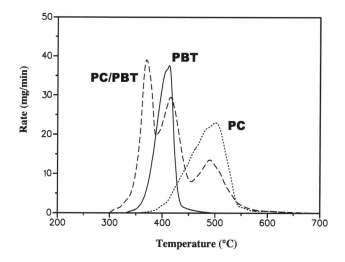

FIGURE 5.25
DTG curves of PC, PBT, and PC/PBT (1/1, mol/mol) blend sample. (Reprinted from Ref. 104.)

butylenecarbonate (BTC) (ion at m/z 117) takes place within the first maximum of the TIC curve. This is most likely due to the formation of copolymer sequences containing BTC units, which are the less thermally stable and decompose as soon as they are formed, by a rearrangement reaction involving the unimolecolar elimination of BTC molecules (Scheme 5.11b). This leaves the more stable aromatic moiety in the residue. This behavior is not unusual in condensation copolymers containing units with a marked difference of thermal stability.[1,2]

BTC units may also undergo CO_2 elimination (Scheme 5.11b), generating copolymer sequences with ether linkage along the chains, which are then decomposed at higher temperatures. In fact, the evolution of compounds with ether linkages (i.e., ion at m/z 547, Figure 5.26) occurs essentially in the second thermal decomposition step, whereas CO_2 (ion at m/z 44) is evolved in the first step.

Furthermore, the evolution of BTC-containing pyrolysis compounds (i.e., ion at m/z 475) and that of cyclic dimer of PC (i.e., ion at m/z 509) take place in the first degradation step. This is most likely due to the high rate of carbonate-carbonate exchange, which allows the faster reaction of PC with copolymer sequences containing butylenecarbonate.[104]

As the temperature increases, sequences containing totally aromatic polyester accumulate in the residue, indicating that the overall thermal reaction has resulted in the formation of the most thermally stable compound, i.e., a totally aromatic polyester. As a consequence, the last decomposition stage is solely due to the evolution of the cyclic dimer (ion at m/z 717) formed by the thermal degradation of the aromatic polyester sequences (Scheme 5.11c), as confirmed by the DPMS analysis of pure poly(bisphenol A terephthalate).[104]

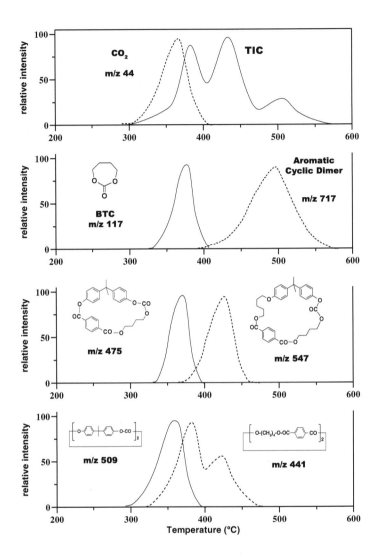

FIGURE 5.26
TIC curve and temperature-time resolved evolution profile of some selected pyrolysis compounds as obtained by CI DPMS of a PC/PBT blend sample. (Reprinted from Ref. 104.)

References

1. Montaudo, G. and Puglisi, C., Thermal Degradation Mechanisms in Condensation Polymers, in *Developments in Polymer Degradation*, N. Grassie, Ed. Vol. 7. Applied Science, London, 1987.
2. Montaudo, G. and Puglisi, C., Thermal Degradation of Condensation Polymers, in *Comprehensive Polymer Science*, First Supplement. Pergamon Press, Oxford, 1992. p. 227.

3. Rice, F. O. and Rice, K. K., *The Aliphatic Free Radicals*. Johns Hopkins Press, Baltimore, 1935.

4. David, C., in *Degradation of Polymers*, C. H. Bamford and C. F. H. Tipper, Eds. Elsevier, 1975.

5. Schulten, H. R. and Lattimer, R. P., Applications of Mass Spectrometry to Polymers, *Mass Spec. Rev.* 3, 231, 1984.

6. Irwin, W. J., *Analytical Pyrolysis*. Marcel Dekker, New York, 1982.

7. Wiley, R. H., The Mass Spectral Characterization of Oligomers, *Macromol. Revs.*, 14, 379, 1979.

8. Foti, S. and Montaudo, G., Analysis of Polymers by Mass Spectrometry, in *Analysis of Polymers Systems*, L. S. Bark and N. S. Allen, Eds. Applied Science, London, 1982.

9. Chapman, J. R. *Practical Organic Mass Spectrometry*, 2nd Edition. John Wiley, Chichester, UK, 1994.

10. Lyon, P. A. *Desorption Mass Spectrometry*, ACS Symposium Series, Vol. 291. Washington, DC, 1985.

11. Kyranos, J. N. and Vouros, P., The role of mass spectrometry in the analysis of polymers, *J. Appl. Polym. Sci. Applied Polymer Symposia* 43, 211, 1989.

12. Smith, C. G., Nyquist, R. A., Mahle, N. H., Smith, P. B., Martin, S. J., and Pasztor, A. J. Jr., Analysis of synthetic polymers, *Anal. Chem.* 59, 119R, 1987.

13. Smith, C. G., Smith, P. B., Pasztor, A. J. Jr, McKelvy, M. L., Meunier, D. M., Froelicher, S. W., and Ellaboudy, A. S., Analysis of synthetic polymers and rubbers, *Anal. Chem.* 65, 217R, 1993.

14. Montaudo, G., Direct Mass Spectrometry of Polymers, *Brit. Polym. J.*, 18, 231, 1986.

15. Maravigna, P. and Montaudo, G., Cyclic Oligomers, in *Comprehensive Polymer Science* Vol. 5. Pergamon Press, London, 1989.

16. Blazsò, M., Recent Trends in Analytical and Applied Pyrolysis of Polymers, *J. Anal. Appl. Pyrol.*, 39, 1, 1997.

17. Scrivens, J. H., The Characterization of Synthetic Polymer Systems, *Adv. Mass Spectr.* 13, 447, 1995.

18. Lattimer, R. P., Harris, R. E., and Schulten, H. R., Applications of mass spectrometry to synthetic polymers, *Mass. Spec. Rev.*, 4, 369, 1985.

19. Morelli, J. J., Thermal Analysis Using Mass Spectrometry, *J. Anal. Appl. Pyrol.*, 18, 1, 1990.

20. Montaudo, G., Mass Spectral Determination of Cyclic Oligomers Distributions in Polymerization and Degradation Reactions, *Macromolecules*, 24, 5289, 1991.

21. Ballistreri, A., Garozzo, D., Giuffrida, M., Impallomeni, G., and Montaudo, G., Primary Thermal Decomposition Processes in Aliphatic Polyamides, *Polym. Deg. and Stab.*, 23, 25, 1988.

22. Vincenti, M., Pelizzetti, E., Guarini, A., and Costanzi, S., Determination of molecular weight distributions of polymers by desorption chemical ionization mass spectrometry. *Anal. Chem.* 64, 1879, 992.

23. Abate, R., Garozzo, D., Rapisardi, R., Ballistreri, A., and Montaudo, G., Sequence Distribution of β-Hydroxyalkanoate Units in Bacterial Copolyesters Determined by Desorption Chemical Ionization Mass Spectrometry, *Rapid Comm. Mass Spectrom.*, 6, 702, 1992.

24. Hoke, S. H., Cooks, G., Munoz, B., Chang, H., and Green, M. H., Microstructure of Alkyl Isocyanate Copolymers Comprised of Enantiomeric Monomers Determined by Desorption Chemical Ionization MS, *Macromolecules*, 28, 2955, 1995.

25. Boon, J. J., Analytical pyrolysis mass spectrometry: new vistas opened by temperature-resolved in-source PYMS, *Int. J. Mass Spectr. Ion Proc.*, 118, 755, 1992.

26. Ballistreri, A., Garozzo, D., Giuffrida, M., and Montaudo, G., Mechanism of Thermal Decomposition of Nylon 66, *Macromolecules*, 20, 2991, 1987.

27. Schulten, H. R., Plage, B., Ohtani, H., and Tsuge, S., Studies on the Thermal Degradation of Polyamides by Pyrolysis-Field Ionization Mass Spectrometry and Pyrolysis-Gas Chromatography, *Angew. Makromol. Chem.*, 155, 1, 1987.

28. Montaudo, G., Puglisi, C., Blazsò, M., Kishore, K., and Ganesh, K., Thermal Degradation Products of Poly(Styrenesulfides) Investigated by Direct-Pyrolysis MS and Flash-Pyrolysis-GC/MS, *J. Anal. Appl. Pyrolysis*, 29, 207, 1994.

29. Lattimer, R. P., Pyrolysis field ionization mass spectrometry of hydrocarbon polymers, *J. Anal. Appl. Pyrol.*, 39, 115, 1997.

30. Adams, R. E., Pyrolysis mass spectrometry of terephthalate polyesters using negative ionization, *J. Polym. Sci. Polym. Chem. Ed.*, 20, 119, 1982.

31. Adams, R. E., Posititve and negative chemical ionization pyrolysis mass spectrometry of polymers, *Anal. Chem.*, 55, 414, 1983.

32. Garozzo, D., Giuffrida, M., and Montaudo, G., Primary Thermal Decomposition Processes in Aliphatic Polyesters Investigated by CI Mass Spectrometry, *Macromolecules*, 19, 1643, 1986.

33. Garozzo, D., Giuffrida, M., and Montaudo, G., Primary Thermal Fragmentation in Poly(Lactic-Acid) Investigated by Positive and Negative Chemical Ionization Mass Spectrometry, *Polym. Deg. & Stab.*, 15, 143, 1986.

34. Garozzo, D., Giuffrida, M., and Montaudo, G., Mixtures of Cyclic Oligomers of Poly(Lactic-Acid) Analyzed by Negative Chemical Ionization and Thermospray Mass Spectrometry, *Polym. Bull.* 15, 353, 1986.

35. Ballistreri, A., Garozzo, D., Giuffrida, M., Maravigna, P., and Montaudo, G., Thermal Decomposition Processes in Aromatic-Aliphatic Polyamides Investigated by Mass Spectrometry, *J. Polym. Sci. Polym. Chem. Ed.*, 25, 1049, 1987.

36. Montaudo, G., Puglisi, C., Scamporrino, E., and Vitalini, D., Thermal Decomposition Processes in Aliphatic Polysulphides Investigated by Mass Spectrometry, *J. Polym. Sci. Polym. Chem. Ed.*, 25, 475, 1987.

37. Montaudo, G., Puglisi, C., Rapisardi, R., and Samperi, F., Further Studies on the Thermal Decomposition Processes in Polycarbonates, *Polym. Deg. & Stab*, 31, 229, 1991.

38. Carroccio, S., Puglisi, C., Samperi, F., and Montaudo, G., Thermal Decomposition Mechanisms of Polyetherimide Investigated by Direct Pyrolysis Mass-Spectrometry, *Macromol. Chem., Phys*, 200, 2345, 1999.

39. Montaudo, G., Puglisi, C., Scamporrino, E., and Vitalini, D., Mass Spectrometric Analysis of the Thermal Degradation Products of Poly(*o*-, *m*-, and *p*-phenylene sulfide) and of the Oligomers Produced in the Synthesis of these Polymers, *Macromolecules*, 19, 2157, 1986.

40. Montaudo, G., Puglisi, C., Scamporrino, E., and Vitalini, D., Mass Spectrometric Detection of Cyclic Sulfides in the Polycondensation of Dibromoalkanes with Dithiols, *J. Polym. Sci. Polym. Symp.*, 74, 285, 1986.

41. Montaudo, G., Puglisi, C., Scamporrino, E., and Vitalini, D., Thermal Degradation of Aromatic-Aliphatic Polyethers 1. Direct Pyrolysis Mass Spectrometry, *Macromolecules*, 19, 870, 1986.

42. Foti, S., Giuffrida, M., Maravigna, P., and Montaudo, G., Direct Mass Spectrometry of Polymers. Primary Thermal Fragmentation Processes in Totally Aromatic Polyesters, *J. Polym. Sci. Polym. Chem. Ed.*, 22, 1201, 1984.

43. Busch, K. L., Glish, G. L., and McLuckey S. A. *MS/MS: Techniques and Applications of Tandem MS.* VCH, New York, 1988.

44. Foti, S., Liguori, A, Maravigna, P., and Montaudo, G., Characterization of Poly-carboxypiperazine by Mass Analyzed Ion Kinetic Energy Spectrometry, *Anal. Chem.*, 54, 674, 1982.

45. Montaudo, G., Puglisi, C., Scamporrino, E., and Vitalini, D. Identification of Pyrolysis Products of Polysulfides by CAD-Linked Scanning Mass Spectrom-etry, *J. Anal. Appl. Pyrolysis*, 10, 283, 1987.

46. Ballistreri, A., Garozzo, D., Giuffrida, M., and Montaudo, G., Thermal Degra-dation Processes of Polyamides Investigated by Collision Activated MS/MS, *Polym. Deg. & Stab.*, 16, 337, 1986.

47. Ballistreri, A., Garozzo, D., Giuffrida, M., and Montaudo, G., Analysis of Poly-mers by Mass Spectrometry. Metastable Mapping of Pyrolysis Products of an Aromatic Polyamide, *J. Anal. Appl. Pyrolysis*, 12, 3, 1987.

48. Scamporrino, E., Mancino, F., and Vitalini, D., Structural Characterization of Oligomers Formed in the Reaction between Phthaloyl Dichloride and Catecol, *Macromolecules*, 28, 5419, 1995.

49. Abate, R., Ballistreri, A., Montaudo, G., Giuffrida, M., and Impallomeni, G., Separation and Structural Characterization of Cyclic and Open Chain Oligo-mers Produced in the Partial Pyrolysis of Microbial Poly(Hydroxbutyrates), *Macromolecules*, 28, 7911, 1995.

50. Lattimer, R. P., Muenster H., and Budzikiewicz, H., Pyrolysis tandem mass spectrometry (Py-MS/MS) of a segmented polyurethane, *J. Anal. Appl. Pyrol.*, 17, 237, 1990.

51. Montaudo, G., Puglisi, C., and Samperi, F., Primary Thermal Degradation Mech-anism of PET and PBT, *Polym. Deg. and Stab.*, 42, 13, 1993.

52. Lattimer, R. P., Mass spectral analysis of low-temperature pyrolysis products from poly(ethylene glycol), *J. Anal. Apll. Pyrolysis*, 56, 61, 2000.

53. Lattimer, R. P., Mass spectral analysis of low-temperature pyrolysis products from poly(tetrahydrofuran), *J. Anal. Apll. Pyrolysis*, 57, 57, 2001.

54. Qian, K., Killinger, W. E., Casey, M., and Nicol, G. R., Rapid Polymer Identifi-cation by In-Source Direct Pyrolysis Mass Spectrometry and Library Searching Techniques, *Anal. Chem.*, 68, 1019, 1996.

55. Sthatheropoulos, M., Georgakopoulos, C. G., and Montaudo, G., A Method for the Interpretation of Pyrolysis Mass Spectra of Polyamides, *J. Anal. Appl. Py-rolysis*, 20, 15, 1992.

56. Montaudo, G., Georgakopoulos, C. G., Sthatheropoulos, M., and Parissakis, G., A Method for the Interpretation of Pyrolysis Mass Spectra of Polyesters, *J. Anal. Appl. Pyrolysis*, 34, 127, 1995.

57. Georgakopoulos, C. G., Sthatheropoulos, M., and Montaudo, G., An Expert System for the Interpretation of Pyrolysis Mass Spectra of Condensation Poly-mers, *Analytica Chimica Acta*, 359, 213, 1998.

58. Georgakopoulos, C. G., Sthatheropoulos, M., and Montaudo, G., Pyrolysis Path-ways of Polyethers and a Method for the Interpretation of Pyrolysis Mass Spectra of Polyethers, *Polym. Deg. & Stab.* 61, 481, 1998.

59. Puglisi, C., Samperi, F., Carroccio, S., and Montaudo, G., MALDI-TOF Investi-gation of Polymer Degradation. Pyrolysis of Poly(Bisphenol-A-Carbonate), *Macromolecules*, 32, 8821, 1999.

60. O'Mara, M. M., Combustion of PVC, *Pure & Appl. Chem.*, 49, 649, 1977.

61. Starnes, W. H. and Edelson, D. H., Mechanistic aspects of the behavior of molybdenum (IV) oxide as a fire-retardant additive for poly(vinyl chloride). An interpretative review. *Macromolecules*, 12, 1797, 1979.

62. Lattimer, R. W. and Kroenke, W. J., The functional role of molybdenum trioxide as a smoke retarder additive in rigid poly(vinylchloride), *J. Appl. Polym. Sci.*, 26, 1191, 1981.

63. Ballistreri, A., Foti, S., Maravigna, P., Montaudo, G., and Scamporrino, E., Effect of Metal Oxides on the Evolution of Aromatic Hydrocarbons in the thermal decomposition of PVC, *J. Polym. Sci. Polym. Chem. Ed.*, 18, 3101, 1980.

64. Ballistreri, A., Montaudo, G., Puglisi, C., Scamporrino, E., and Vitalini, D., Mechanism of Smoke Suppression by Metal Oxides in PVC, *J. Polym. Sci. Polym. Chem. Ed.*, 19, 1397, 1981.

65. Montaudo, G. and Puglisi, C., Evolution of Aromatics in the Thermal Degradation of Polyvinylchloride: A Mechanicistic Study, *Polym. Deg. & Stab.*, 33, 229, 1991.

66. Marcilla, A. and Beltran, M., Thermogravimetric kinetic study of poly(vinylchloride) pyrolysis, *Polym. Deg. & Stab.*, 48, 219, 1995.

67. Muller, J. and Dongmann, G., Formation of Aromatics during pyrolysis of PVC in the presence of metal chlorides, *J. Anal. Appl. Pyrolysis*, 45, 59, 1998.

68. Miranda, R., Yang, J., Roy, C., and Vasile, C., Vacuum pyrolysis of PVC I. Kinetic study, *Polym. Deg. & Stab.*, 64, 127, 1999.

69. Ballistreri, A., Garozzo, D., Giuffrida M., and Montaudo G., Mechanism of Thermal Decomposition of Nylon 66, *Macromolecules*, 20, 2991, 1987.

70. Ballistreri, A., Garozzo, D., Giuffrida, M., Impallomeni, G., and Montaudo, G., Primary Thermal Decomposition Processes in Aliphatic Polyamides, *Polym. Deg. & Stab.*, 23, 25, 1988.

71. Ballistreri, A., Garozzo, D., Giuffrida, M., Maravigna, P., and Montaudo, G., Thermal decomposition Processes in Aliphatic-Aromatic Polyamides Investigated by Mass Spectrometry, *Macromolecules*, 19, 2693, 1986.

72. Ballistreri, A., Garozzo, D., Giuffrida, M., Montaudo, G., and Pollicino, A., Thermal Decomposition Processes in Polyhydrazides and Polyoxamides Investigated by Mass Spectrometry, *Polymer*, 28, 139, 1987.

73. Buxbaum, L. H., The Degradation of poly(ethylene terephthalate), *Angew Chem. (Int. Ed.)*, 7, 182, 1968.

74. Zimmermann, H., Degradation and Stabilization of polyesters, *Developments in Polymer Degradation*, Grassie, N., Ed. Applied Science Publishers, Vol. 5, London, 1984.

75. Grassie, N. and Murray, E. J., The thermal degradation of Poly(-D)-b-hydroxybutyric acid, Part 1. Identification and quantititative analysis of products, *Polym. Deg & Stab.*, 6, 47, 1984.

76. McNeill, I. C. and Bounekel, M., Thermal degradation studies of therephthalate polyesters: 1. Poly(alkylene terephthalate), *Polym. Deg. & Stab.*, 34, 187, 1991.

77. Othani, H., Kimura, T., and Tsuge, S., Analysis of thermal degradation of terephthalate polyesters by high-resolution pyrolysis-gas chromatography, *Anal. Sci.*, 2, 179, 1986.

78. Luderwald, I., Thermal degradation of polyesters in the mass spectrometer, *Developments in Polymer degradation*, Vol. 2, Grassie, N., Ed. Applied Science Publisher, London, 1979.

79. Ramjit, H. G., The influence of stereochemical structure on the kinetics and mechanism of ester-ester exchange reaction by mass spectrometry, *J. Macromol Sci. Chem. Ed.*, A19, 41, 1983.

80. Tighe, B. J., The thermal degradation of poly-α-esters, *Developments in Polymer Degradation*, Grassie, E., Ed. Applied Science, Vol. 5, London, 1984.

81. Ross, S. D., Coburn, E. R., Leach, W. A., and Robinson, W. B. J., Isolation of a cycle trimer from polyethylene terephthalate film, *J. Polym. Sci.*, 13, 406, 1954.
82. Goodman, I. and Nesbitt, B. F., Polyesters from aromatic and hydroxyaromatic acids, *Polymer*, 1, 384, 1960.
83. Peebles, L. H., Jr., Huffman, M. W., and Ablett, C. T., Isolation and identification of the linear and cyclic oligomers of poly(ethylene terephthalate) and the mechanism of cyclic oligomer formation, *J. Polym. Sci.*, A-1, 7, 479, 1969.
84. Ha, W. S. and Choun, Y. K., Kinetic studies on the formation of cyclic oligomers in poly(ethylene terephthalate) *J. Polym. Sci. Polym. Chem. Ed.*, 17, 2103, 1979.
85. Cooper, D. R. and Semlyen, J. A., Equilibrium ring concentration and the statistical conformations of polymer chains, *Polymer*, 14, 185, 1973.
86. Semlyen, J. A., Cyclic Polymers, Elsevier Ed. Applied Science, London, 1986.
87. Chojinowski, J., Scibiorek, M., and Kowalski, J., Mechanism of the Formation of Macrocycles during the ation Polymerization of Cyclotrisiloxanes. End to End ring closure versus ring expansion, *Makromol. Chem.*, 178, 1351, 1977.
88. Jacobson, H. and Stockmayer, W. H., Intramolecular Reaction in Polycondensate. I. The Theory of Linear Systems, *J. Chem. Phys.*, 18, 1600 1950.
89. Lee, L., Mechanisms of thermal degradation of phenolic condensation polymers. I. Studies on the thermal stability of polycarbonate, *J. Polym. Sci., Part A*, 2, 2859, 1964.
90. Davis, A. and Golden, J. H., Competition between chain scission and crosslinking processes in the thermal degradation of a polycarbonate, *Nature*, 206, 397, 1965.
91. Davis, A. and Golden, J. H., Stability of polycarbonate, *J. Macromol. Sci. Rev. Macromol. Chem.* 1969, C3, 49.
92. Tsuge, S., Okamoto, T., Sujimura, Y., and Tacheuchi, T. *J. Chromat. Sci.* 1969, 7, 253.
93. Wiley, R. H., Mass spectral characteristics of Poly(4,4'-isopropylidendiphenyl carbonate), *Macromolecules*, 4, 264, 1971.
94. Foti, S., Giuffrida, M., Maravigna, P., and Montaudo, G., Primary Thermal Fragmentation Processes in Polycarbonates, *J. Polym. Sci. Polym. Chem. Ed.*, 21, 1567, 1983.
95. Ballistreri, A., Montaudo, G., Puglisi, C., Scamporrino, E., Vitalini, D., and Cucinella, S., Intumescent flame retardant for polymers. The polycarbonatearomatic sulfonate system, *J. Polym. Sci. Polym. Chem. Ed.*, 26, 2113, 1988.
96. Puglisi, C., Sturiale, L., and Montaudo, G., Thermal Decomposition Processes in Aromatic Polycarbonates Investigated by Mass-Spectrometry, *Macromolecules*, 32, 2194, 1999.
97. Ito, Y., Ogasawara, H., Ishida, Y., Ohtani, H., and Tsuge, S. Characterization of end groups in polycarbonate by reactive pyrolysis-gas chromathography, *Polym. J.*, 28, 1090, 1996.
98. Montaudo, G., Puglisi, C., and Samperi, F. Primary Thermal Degradation Processes Occurring in Polyphenylenesulfide Investigated by Direct Pyrolysis Mass-Spectrometry, *J. Polym. Sci. Part A. Polym. Chem.*, 32, 1807, 1994.
99. Montaudo, G., Puglisi, C., and Samperi, F., Primary Thermal Degradation Processes of PEK and PEK/PES Copolymers Investigated by Direct Pyrolysis Mass-Spectrometry, *Makromol. Chem. Phys*, 195, 1241, 1994.
100. Montaudo, G., Puglisi, C., Rapisardi, R., and Samperi, F., Primary Thermal Degradation Processes of PES and PPO Investigated by Direct Pyrolysis Mass-Spectrometry, *Makromol. Chem. Phys*, 195, 1225, 1994.

101. Plage, B. and Schulten, H. R., Pyrolysis-Field Ionization Mass Spectrometry of Aliphatic Polyesters and their Thermal Interactions in Mixtures, *J. Anal. Appl. Pyrol.*, 1989, 15, 1987.

102. Plage, B. and Schulten, H. R., Pyrolysis-Field Ionization Mass Spectrometry of Polyamide Copolymers and Blends, *J. Appl. Polym. Sci.*, 38, 123, 1989.

103. Montaudo, G., Puglisi, C., and Samperi, F., Exchange Reactions Occurring through Active chain Ends Melt Mixing of Nylon 6 and Polycarbonate, *J. Polym. Sci. Part. A. Polym. Chem.*, 32, 15, 1994.

104. Montaudo, G., Puglisi, C., and Samperi, F., Chemical Reactions Occurring in the Thermal Treatment of Polymer Blends Investigated by Direct Pyrolysis Mass Spectrometry. Polycarbonate/Polybuthyleneterephathalate, *J. Polym. Sci. Part A Polym. Chem.*, 31, 13, 1993.

105. Montaudo, G., Puglisi, C., and Samperi, F., Chemical Reactions Occurring in the Thermal Treatment of Polycarbonate/Polyethylene terephathalate Blends, Investigated by Direct Pyrolysis Mass Spectrometry, *Polym. Deg. & Stab.*, 31, 291, 1991.

106. Montaudo, G., Puglisi, C., and Samperi, F., Chemical Reactions Occurring in the Thermal Treatment of PC/PMMA Blends, *J. Polym. Sci. Part A, Polym. Chem.* 36, 1873, 1998.

107. Montaudo, G., Puglisi, C., and Samperi, F. Mechanism of Exchange in BPT/PC and PET/PC Blends. Composition of the Copolymer Formed in the Melt Mixing Process, *Macromolecules.* 31, 650, 1998.

6

Field Ionization (FI-MS) and Field Desorption (FD-MS)

Robert P. Lattimer

CONTENTS

0-8493-3127-7/02/$0.00+$1.50

6.1   Introduction

In the past several years, a number of new ionization methods in mass spectrometry have been introduced. These new techniques have extended mass spectrometric analysis to a wide variety of labile (thermally unstable), highly polar, and higher molecular weight materials. Field ionization (FI) and field desorption (FD) are two of the pioneering techniques in this list of alternative ionization methods. FI-MS, which was introduced for organic molecules in 1954, was the first soft ionization method. (Soft ionization refers to processes that produce high relative abundances of molecular, or quasimolecular, ions.) FD-MS, which was invented in 1969, was the first desorption/ionization method. (Desorption/ionization refers to processes in which the vaporization/ desorption, and ionization steps occur essentially simultaneously.)

FI-MS and FD-MS have a number of useful features, including very high molecular ion abundances, higher mass capability, and applicability to a wide variety of compound types. In this chapter, we will give an overview of FI-MS and FD-MS as techniques for the analysis of polymeric systems. Readers interested in more detailed explanations of FI/FD-MS theory, procedures, and applications are referred to various reviews and monographs on the subject.[1-7]

6.1.1   Development of FI-MS and FD-MS

The invention of field ion microscopy by E. W. Mueller[8] initiated several fields of study, including FI/FD-MS. In 1954, M. G. Inghram and R. Gomer at the University of Chicago described the attachment of an FI microscope to a mass spectrometer.[9] They were able to generate mass spectra from a number of small molecules. In 1957, H. D. Beckey at the University of Bonn began a systematic investigation of FI-MS.[1] His research focused on the mass spectrometric applications of the technique, and he made many fundamental discoveries including the use of FI/FD-MS for organic chemical structure analysis. Beckey and co-workers made several major contributions to our understanding and practice of FI/FD-MS, including the first focusing FI source (1958), field dissociation of organic ions (1961), field ionization kinetics (1961), the development of activated carbon microneedle emitters (1968), and the invention of field desorption (1969).[1]

The first commercial FI/EI ion source was introduced for magnetic sector instruments by MAT (Bremen, Germany) in 1967. The emitter could be retracted for EI operation, but the system had to be vented to change the emitter. The first commercial FD/FI/EI ion source, with the emitter on a sliding pushrod probe, was introduced by Varian MAT in 1973. This provided a marked improvement, because the emitter could easily be replaced or loaded with sample without venting the system. Today all of the major magnetic sector manufacturers offer FI/FD sources as accessories for their instruments.

The number of FI/FD-MS publications rose rapidly in the early to mid-1970s, reaching a peak in the late 1970s.[6] A noticeable drop in FI/FD use occurred after 1983 because of the advent of fast atom bombardment (FAB-MS), and later other desorption/ionization methods. A high percentage of FI/FD-MS articles have come from Germany, which is not surprising in view of the early development of the techniques in Bonn.

6.1.2   Principles and Procedures

The principles involved in FI/FD-MS are fundamentally different from those for methods that rely on beams (electron, ion, atom, particle, laser) or liquid sprays (thermospray, electrospray) for ionization. A simple diagram of a combination FI/FD ion source is shown in Figure 6.1. The field emitter consists of a thin (usually 10 μm) tungsten wire that is supported on two metal posts with a ceramic base. The emitter is on a sliding pushrod probe, which can be removed from the ion source for sample loading or emitter replacement. The emitter wire is covered with microneedles (or dendrites) of pyrolytic carbon or silicon (Figure 6.2). The emitter is held at the accelerating potential, and the wire is situated within a few millimeters of a counterelectrode

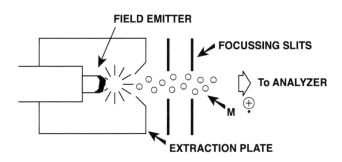

FIGURE 6.1
Diagram of an FI/FD ion source.

FIGURE 6.2
Photomicrograph of a carbon microneedle emitter.

(or extraction plate), which is held at a potential 8–12 kV lower than that of the emitter. The high field strengths necessary for field ionization (10^7–10^8 V/cm) are present near the tips of the microneedles.

Field ionization involves the removal of electrons from a species by quantum mechanical tunneling in a high electric field. In practice, FI-MS refers to the technique in which the sample to be analyzed is introduced as a vapor using a heatable direct probe, heated batch inlet, or GC/MS interface.

Field desorption refers to the technique in which the sample is deposited directly on the emitter before it is inserted into the ion source. This is an ambiguous term, because it implies that it is the electric field that causes desorption and ionization of the analyte from the probe. It is well-known, however, that the field is only one factor in the process; "field ionization" is only one of the ionization processes that may occur. Thus, most practitioners use the term "field desorption" to refer to the sample introduction technique and not necessarily to the method of ionization.

Most practitioners use carbon microneedle emitters, and a few others use silicon emitters.[6] Silicon emitters can be manufactured more rapidly, but carbon emitters are more rugged and can be heated to much higher temperatures (for flash cleaning). Other types of emitters (bare wire, metal microneedles, metal tip, razor blade, or volcano) are only sparingly used. Users either make their own emitters (with commercial or home-built devices) or else purchase them (from mass spectrometer manufacturers or independent vendors). The choice of making or buying emitters usually comes down to cost and time.

For FD-MS analysis, the emitter must be loaded with sample. The first step is to dissolve the material in a suitable solvent (preferably a volatile one like acetone or dichloromethane). If dissolution is not possible, suspension with sonification may be sufficient. The emitter is then dipped into the analyte solution, or the solution is touched onto the wire using a microliter syringe. The dipping technique is a little faster, but the syringe technique should be used when the amount of available sample is very small or when careful control of sample amount and position on the wire is necessary (as in quantitative work). Microsyringe manipulators can be made (or purchased commercially) if the ultimate in sample loading care and convenience is desired. Most practitioners use syringe deposition.[6] Other emitter loading methods (e.g., dipping, aerosol deposition, and freeze loading) are used only infrequently.

6.1.3 Ion Formation Mechanisms

Part of the versatility of the FI-MS and FD-MS methods comes from the variety of ionization mechanisms that can be observed with different analytes. The four most common mechanisms are field ionization, cation attachment, thermal ionization, and proton abstraction.

6.1.3.1 Field Ionization

Field ionization is the first mechanism that mass spectrometrists think of when considering FI/FD-MS, but it is only one of several possibilities. As stated earlier, field ionization is the removal of electrons from a species by quantum mechanical tunneling in a high electric field. This leads to the production of molecular ions ($M^{+\cdot}$ in positive ion mode), and this mechanism of ionization is generally observed for nonpolar or slightly polar organic compounds.

Poly(2,2,4-trimethyl-1,2-dihydroquinoline), an oligomeric antioxidant for rubber, is a typical example for this ionization method. Figure 6.3 shows the FD mass spectrum of a poly-TMDQ sample.[10] Molecular ions for "normal" oligomers ($M^+ = 173n$) are observed, along with minor peaks due to impurities (with differing end-groups) from the synthesis.

Figure 6.4 shows the FD mass spectrum of the TMDQ dimer that was isolated from the polymer by liquid chromatography.[11] This is a typical "single peak" FD mass spectrum that is appealing to many organic chemists. The spectrum immediately confirms (or establishes) the molecular weight of the analyte and tells something about the purity of the material. In the case of mixtures (polymers are always mixtures), one peak is normally observed for each component/oligomer. Therefore, FI-MS or FD-MS can be used as a *screening* technique for complex samples.

In FD-MS analysis the emitter is normally heated resistively to aid in the desorption process. Chemicals that are reasonably volatile and not too polar will desorb with the emitter held at ambient temperature. Most organics, however, will require the application of some *emitter heating current* (EHC) to get them to desorb. The amount of heat needed is much less than that

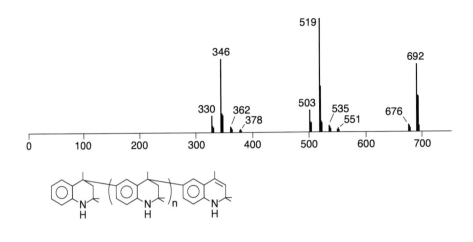

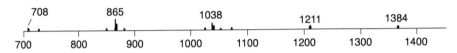

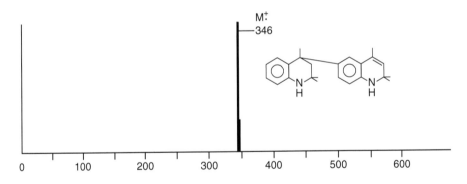

required to vaporize the material from a direct probe. The term *best anode temperature* (BAT) is sometimes used to describe the temperature at which the intensity of the molecular ion is maximal and that of the fragment ions is minimal. As the emitter is heated, the BAT occurs shortly after the point at which molecular ions begin to form. For typical organic molecules, this is in the range of ~10–25 mA EHC, which for 10-μm tungsten wires corresponds to the range ~50–200°C.

6.1.3.2 Cation Attachment

Cation attachment is also called cationization or desolvation. In this process, cations (typically H^+ or Na^+) attach themselves to receptive sites on analyte molecules in the condensed phase. The combination of emitter heating and high field results in the desorption of cation attachment ions (e.g., MNa^+). This mechanism is typically observed for more polar organic molecules (e.g., those with aliphatic hydroxyl or amino groups).

The spectrum of poly(ethylene imine) (Figure 6.5) is a typical example of this ionization method.[12] In this case the principal FD ions are MH^+ for the polymer $H_2N\text{-}(\text{-}C_2H_4\text{-}NH\text{-})_n\text{-}H$. Weak fragment ions and MNa^+ ions (from adventitious sodium in the system) are also observed. By doping the sample with a sodium salt, the intensities of the sodiated ions can be enhanced. Too much salt may result in "sputtering," however; during sputtering the sample bursts irregularly off the emitter, which results in a poorly reproducible spectrum.

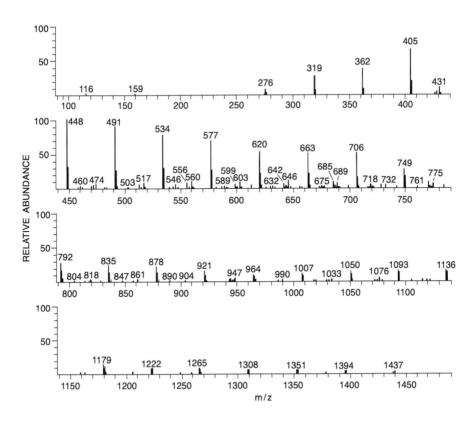

FIGURE 6.5
FD-MS of poly(ethylene imine). (Reprinted from Ref. 12. Copyright 1985, with permission from Elsevier Science.)

6.1.3.3   *Thermal Ionization*

Thermal ionization of preformed ions may be observed for organic and inorganic salts. In this case the field lowers the desorption temperature, facilitates focusing, and enhances the ion current. The emitter is used as a "solid probe" to hold and heat the sample. Thermal ionization is most commonly observed for organic and inorganic salts; it is hardly ever observed for polymers.

6.1.3.4   *Proton Abstraction*

Proton abstraction is a common ion formation mechanism in the negative ion (NI) FD-MS mode. It is not often reported in the literature because little NI-FD-MS work is done. (Field electron emission has to be contended with in NI-FD-MS.) Polar organics in the NI mode will often show $(M - H)^-$ ions, but polymers to date have only been analyzed in the positive ion mode. It is interesting to note that mixtures of poly(ethylene glycol) (PEG) and water are typically used as a viscous solvent in NI-FD-MS studies of organic molecules.[13] In this case, however, no NI-FD signals are observed from the PEG.

6.1.4   Advantages

The principal advantages of FI-MS and FD-MS include fairly high molecular ion abundances, fairly high mass capability, applicability to a wide variety of compound types, and availability on general-purpose organic mass spectrometers.

6.1.4.1   *High Molecular Ion Abundances*

Many low molecular weight organic polymers will produce intense molecular or quasimolecular ions ($M^{+\cdot}$, MH^+, MNa^+) with few fragment ions or none at all. FI-MS and FD-MS are therefore excellent for the determination of oligomer molecular weights. Polymeric mixtures can be screened to assess the number of components (or oligomeric series) and their approximate relative abundances. Chemical structures can often be elucidated from just the molecular weights of the components plus a knowledge of the chemistry and history of the sample.

   FI- and FD-MS are truly "soft" ionization methods, with little excess energy being deposited into the ions that are formed. In a recent study some labile, low molecular weight polyesteramides were analyzed by FD-MS, electrospray (ESI-MS), and MALDI-MS—all of which are techniques that produce high abundances of molecular or quasimolecular ions.[14] In this study ESI-MS was found to be the "softest" method (i.e., the one showing the least ion fragmentation and sample decomposition); FD-MS finished second, and MALDI-MS was third.

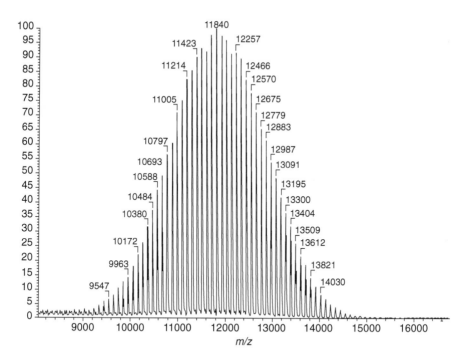

FIGURE 6.6

FD-MS of polystyrene. (Reproduced from Ref. 15 by permission of John Wiley & Sons Limited.)

6.1.4.2 Fairly High Mass Capability

As with most ionization methods, FI and FD mass spectra are most often acquired for compounds of mass <1000 Da (1 kDa). There are many examples in the literature, however, of FI/FD mass spectra for compounds/polymers in the mass range 1–5 kDa. Based on a few literature examples, the maximum practical FD mass range would appear to be 10–15 kDa; thus FD-MS is a technique of intermediate mass range capability. Molecular ions can be obtained at higher masses than have been reported, for example, for EI-MS and CI-MS. However, some desorption/ionization methods, such as MALDI- and ESI-MS, have demonstrated higher mass capabilities.

Figure 6.6 is an example that approaches the upper mass limit for FD analysis of polymers.[15] The sample in this case is a polystyrene standard: $C_4H_9-(-CH_2-CH\phi-)_n-H$. This is the type of soluble, nonpolar polymer that is ideally suited for FD-MS analysis.

6.1.4.3 Applicability to a Wide Variety of Compound Types

FI-MS and FD-MS have been used successfully for both polar and nonpolar materials; organic, inorganic, and organometallic compounds; and neutral molecules and salts. The many compound classes that have been investigated include organic polymers, polymer additives, oils, waxes, surfactants, synthetic organics, natural products, pharmaceuticals, pesticides, organo-metallic complexes, coal liquids, organic salts and dyes, carbohydrates, and polypeptides. The principal reasons for this versatility are the variety of ionization mechanisms that are observed, the soft character of the ionization, and the ability to desorb many thermally labile materials intact. Because FI-MS and FD-MS work best for nonpolar and slightly polar compounds, they often complement other methods (FAB-MS, CI-MS, MALDI-MS, ESI-MS) that are more amenable for use with polar molecules.

6.1.4.4   Availability on General-Purpose Mass Spectrometers

FI/FD-MS ion sources are available from all major magnetic sector instrument manufacturers. Combination FI/FD/EI/CI sources are common. No specialized mass analyzer, such as time-of-flight (TOF-MS) or Fourier transform (FT-MS), is needed. In short, FI-MS and FD-MS are quite compatible (and in fact are optimal) for use with the normal types of double-focusing mass spectrometers typically used for analysis of organic compounds. Since high electric fields are required in the ion source, commercial FI/FD ion sources have not been developed for use with "low voltage" mass spectrometers (such as those with quadrupoles and ion traps).

6.1.5   Disadvantages

Coupled with these features are a number of disadvantages, including difficulty of use, emitter problems, low sensitivity, and ion current stability.

6.1.5.1   Difficulty of Use

Over the years, FI-MS and FD-MS have developed an uneven reputation. Although experienced users practice the techniques routinely, nonusers sometimes have the impression that FI-MS and FD-MS are difficult to master. In our experience, FI-MS and FD-MS are perhaps somewhat more difficult than other methods, but the degree of difficulty is certainly not as severe as some nonusers would imagine.

6.1.5.2   Emitter Problems

Various problems with emitters are often cited. One area of concern is availability; emitters are sometimes said to be too difficult to make and too expensive to buy. In reality, emitters can be readily made with the proper equipment and by properly trained staff, and emitters are no more costly to buy than most other mass spectrometry consumables. Another area that is cited is emitter breakage. The wires are indeed fragile, but with proper care and a reasonably clean ion source (to prevent discharge), breakage should not be a serious problem.

6.1.5.3 Low Sensitivity

It is generally true that absolute FI/FD-MS sensitivities (signal per unit sample) are an order of magnitude or so less than those for EI-MS or CI-MS. Also, FI/FD-MS sensitivities can be highly variable, depending on a number of factors such as quality of the emitter, tuning and cleanliness of the ion source, the chemical class of the compound being analyzed, and matrix effects (such as the presence of inorganic salts). Although these considerations are real, none of these points presents a serious problem for experienced users. Typical FI/FD signal intensities are quite adequate for most applications. Examples can be found in the literature in which FI-MS or FD-MS has been used successfully for high-sensitivity applications such as trace analysis, high-resolution accurate mass measurements, and tandem mass spectrometry (MS/MS).

6.1.5.4 Ion Current Stability

Fluctuations in ion current may occur for certain "difficult" molecules, typically salts or other highly polar organics. However, this is not a problem for organics that yield reasonable FI/FD signal intensities.

6.1.6 Recommendations for New Users

For experienced users, FI-MS and FD-MS techniques are quite manageable and not particularly difficult. For new or potential users, we offer the following suggestions:

- Take time to learn the techniques and their peculiarities. As with almost anything else, there is no substitute for practice.
- Choose a reasonable approach for resupply of emitters. Purchasing emitters may be a better option than making them in-house. The cost can probably be fit into your normal expenses for mass spectrometry consumables.
- Use FI/FD-MS regularly, not just as a "last resort." Our experience is that organic chemists will become enamored of the simple spectra that are produced via FI/FD-MS, and they will begin to request FI/FD results on a regular basis.

6.2 Applications

FI-MS and FD-MS are valuable because they can solve real characterization problems in polymer synthesis and industrial analysis. These techniques often provide information that is unique—i.e., not obtainable by other methods. FI-MS and FD-MS are also complementary to other analytical techniques, both spectroscopic (e.g., IR, NMR) and chromatographic (e.g., LC, GPC).

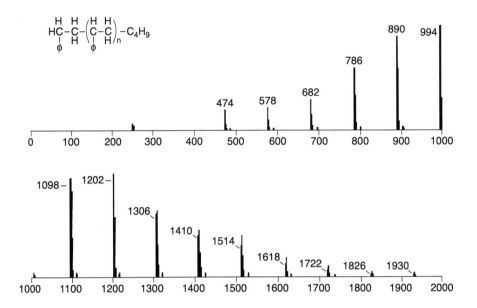

FIGURE 6.7
FD-MS of polystyrene. (Reprinted with permission from Ref. 10. Copyright 1979, American Chemical Society.)

6.2.1 Direct Analysis of Polymeric Mixtures

FI-MS and FD-MS have proven themselves to be excellent techniques for the screening and profiling of complex samples. The number of components, their molecular weights, and their approximate relative abundances can normally be obtained. FI/FD-MS can be a very informative first approach to the analysis of polymeric mixtures.

Often the FI/FD molecular weights alone, combined with a knowledge of the chemistry and history of the sample, are sufficient to obtain the information needed. For example, FI/FD-MS may be used to identify *end-groups* for linear low molecular weight polymers. Figure 6.7 is the FD spectrum of a polystyrene standard.[10] If one subtracts an integral number of monomer units (104 Da) from the observed molecular ions, a "residual" molecular weight of 58 Da is obtained (e.g., MW 474 − 4 × 104 = 58). This is consistent with a butyl group on one end of the chain and a hydrogen on the other: $C_4H_9-(-CH_2-CH\phi-)_n-H$. In this instance, the polymerization initiator was n-butyl lithium.

FI/FD-MS may also help to determine if *cyclic* oligomers are present. Figure 6.8 is the FD spectrum of a toluene extract from a segmented polyurethane.[16] Solvent extraction is often used to remove low molecular weight material from a polymer for analysis. In this case three series of cyclic ester/urethane oligomers could readily be identified from the FD-MS molecular

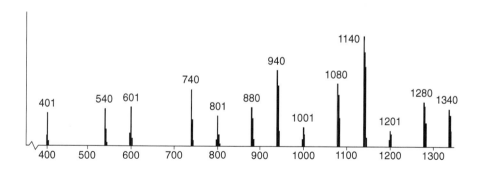

FIGURE 6.8

FD-MS of extract from polyurethane. (Reprinted with permission from Ref. 16. Copyright 1980, Rubber Division, American Chemical Society.)

weights alone:

$$-(-O-CO-C_4H_8-CO-O-C_4H_8-)_n-$$
$$MW = 200n$$

$$-(-O-CO-C_4H_8-CO-O-C_4H_8-)_n-O-CO-NH-\phi-CH_2-\phi-NH-CO-O-$$
$$C_4H_8-MW = 340 + 200n$$

$$-(-O-CO-C_4H_8-CO-O-C_4H_8-)_n-(-O-CO-NH-\phi-CH_2-\phi-NH-CO-O-$$
$$C_4H_8-)_2-MW = 680 + 200n$$

MH^+ ions were dominant for the polyester series, while $M^{+\cdot}$ ions were dominant for the other two series.

In more complex or less well-characterized systems, the FI/FD molecular weights alone may not be sufficient to elucidate the proper chemical structures. In these cases, the FI/FD results can suggest appropriate methods (spectroscopic and/or chromatographic) for further study. For example, other ionization methods can be used to obtain additional mass spectra. Tandem mass spectrometry (MS/MS) can be used to obtain fragmentation patterns, and high resolution accurate mass measurements can be used to obtain atomic compositions (AC-MS) for various components.

There are many examples in the literature of the structural characterization of polymeric systems by FD-MS. Some of these will be briefly mentioned here. Saito and coworkers in Japan have studied a number of polymers by FD-MS.[5] FD spectra were used to identify various poly(ethylene glycol) and poly(propylene glycol) initiators (water, ethyleneimine, glycerol, sorbitol, sucrose).[17] Structures of bisphenol A-based epoxy resins were elucidated.[18,19] The degree of methylation in methylol melamine resins was assessed.[20] Various novalak resins (made from phenol, alkylphenols, and epoxidized phenols) were characterized.[21] Styrene polymerized with various initiators and chain transfer agents was studied; in some cases deuterium labeling was used to help assess the polymerization mechanism.[22–28] Other systems studied included aromatic nitro-containing polymers,[29] blocked urethane prepolymers,[30] polyesters,[31] and acrylics.[32]

Derrick and coworkers have investigated polyglycols (PEG, PPG),[33–36] polybutadiene,[37] polystyrene,[36] and poly(methylmethacrylate)[36] by FD-MS. Prokai[38] and Scrivens et al.[39] have looked at various phenol-formaldehyde (novalak and resole) resins. Wiley and Cook obtained FD spectra of lactone, lactam, and carbonate polymers.[40] Matsuo et al. investigated polystyrene and poly(propylene glycol) as high mass reference compounds.[41] Lattimer and Schulten studied several hydrocarbon polymers (polybutadiene, polyisoprene, polyethylene, polystyrene).[42] Evans et al. used FD-MS to investigate mechanistic aspects of the metal-catalyzed polymerization of ethylene and other olefins.[43–44]

6.2.1.1   Tandem Mass Spectrometry

There are few reports in the literature of FD being used as the ion source in tandem (MS/MS) experiments. The relatively weak ion currents and transient signals make FD-MS/MS spectra (as one paper has put it) "technically challenging to obtain."[45] Most analysts, it would seem, obtain "survey" spectra by FD-MS for screening purposes. Then if more information is needed, MS/MS experiments are performed using other ionization methods (e.g., EI, CI, FAB). In our experience it is almost always preferable to do MS/MS with ionization methods other than FI or FD. Thus the FD-MS/MS data that do appear in the literature seem to have been obtained, for the most part, for demonstration purposes.

In some early FD-MS/MS work, Craig and Derrick studied polystyrene ion dissociation using a double-focusing instrument; both metastable and collisional dissociation experiments were carried out.[46–48] In a recent paper Jackson et al. described the collisional dissociation of polystyrene using a hybrid sector orthogonal time-of-flight (oa-TOF) instrument.[49] This configuration is well-designed for high sensitivity MS/MS experiments, and excellent FD-MS/MS spectra of polystyrene and other chemicals were obtained. Even with this instrument, however, the authors conceded that the applicability is "limited to simple mixtures with relatively strong FD signals."[49]

6.2.1.2   High-Resolution Accurate Mass Measurements

Due to the low ion current and lack of suitable mass reference compounds, high-resolution FI/FD measurements are seldom reported in the literature. In fact, Schulten in Germany is the only mass spectrometrist worldwide who regularly carries out high resolution FI/FD experiments. For high resolution, Schulten uses a double-focusing mass spectrometer with photoplate recording. As an example of its use in polymer analysis, Schulten and Plage recorded high resolution EI and FI spectra of pyrolysis products from nylon 6.10.[50] The accurate mass measurements were used to deduce atomic compositions and chemical structures for various pyrolyzate components. This type of work requires great skill and patience, particularly with the transient nature of FI/FD ionization. Nearly all workers prefer to perform high resolution mass measurements with other ionization methods, usually EI or CI.

6.2.2 Molecular Weight Averages

Molecular weight averages are routinely used by polymer chemists to characterize synthetic macromolecules. The two most common values determined are the *number average* (M_n) and the *weight average* (M_w), defined as follows: $M_n = \Sigma N_i M_i / \Sigma N_i$ and $M_w = \Sigma N_i M_i^2 / \Sigma N_i M_i$. The M_w/M_n ratio, often called the *polydispersity index,* is a measure of the narrowness or broadness of the molecular weight distribution.

In principle it should be possible to determine M_n and M_w by mass spectrometry, since the technique *directly* provides the information needed—oligomer molecular weights (M_i) and relative abundances (or number of molecules, N_i).[51] Field desorption was the first mass spectral method to be used to directly determine molecular weight averages of synthetic polymers (1980).[52]

Since that first paper, several other examples showing the calculation of polymer molecular weight averages by FD-MS have appeared in the literature.[12,15,42,53–55] A summary of most of these results is given in Table 6.1. For the hydrocarbon polymers (PSty, PBd, PIso, PE) and polyglycols (PPG, PEG, PTHF), the FD-MS M_n values agreed well with those determined by "classical" methods (usually vapor pressure osmometry, VPO). The average deviation between the two M_n values is ~5% (Table 6.1). In some cases the FD-MS value is higher than the "reference" value, and in other cases it is lower.

The hydrocarbon polymers and polyglycols in Table 6.1 are all standards used for gel permeation chromatography (GPC) calibration. Thus their polydispersities (M_w/M_n) are all quite low (<1.3). These polymers are also reasonably "well-behaved" when studied by FD-MS. That is, these are "good desorbers" that give relatively intense, long-lasting signals. The polydispersity determined by FD-MS is similar to the "reference" value for each polymer, although in most cases the FD-MS value is slightly lower. This probably represents some mass discrimination at the high end of the oligomer envelope; this would tend to narrow the observed distribution slightly.

Other polymers that are poorly desorbing or have higher polydispersities may not give such good agreement, however. One example from the literature is the FD-MS analysis of *t*-octylphenol/formaldehyde (novalak) resins.[54] For the lower molecular weight resin (M_w/M_n ~1.28), the M_n values determined by FD-MS and liquid chromatographic (LC) analysis agree reasonably well (Table 6.1). The higher mass resin has a broader distribution, however (M_w/M_n ~1.72), and M_n determined via FD-MS is about a third lower than that measured by vapor pressure osmometry (VPO). The polydispersity

TABLE 6.1

Molecular Weight Averages for Synthetic Polymers

Ref.	Polymer[a]	M_n (MS)[b]	M_w/M_n[b]	M_n^c	M_w/M_n^d
52	PSty	1690	1.15	1710	<1.1
52	PSty	2890	1.12	3100	<1.1
15	PSty	5010	—	5100	<1.04
15	PSty	7200	—	7600	<1.04
15	PSty	9350	—	10200	<1.07
15	PSty	11900	—	12500	<1.04
42	PBd	430	1.03	420	1.1
42	PBd	885	1.04	960	1.07
42	PBd	2450	1.02	2350	1.13
42	PIso	931	1.08	940	1.10
42	PE	644	1.05	640	1.10
42	PE	1030	1.08	910	1.10
55	PE	535	—	680[e]	1.18[e]
55	PE	955	—	960[e]	1.20[e]
53	PPG	805	1.04	790	~1.05
53	PPG	1240	1.03	1220	~1.03
53	PPG	1930	1.03	2020	~1.02
53	PEG	1010	1.03	1041	1.05
53	PEG	1360	1.03	1396	1.02
53	PTHF	1110	1.10	1050	1.15
54	OPFR	789	1.23	693[f]	1.28[f]
54	OPFR	822	1.19	1194[g]	1.72[e]

[a] PSty = polystyrene, PBd = polybutadiene, PIso = polyisoprene, PE = polyethylene, PPG = poly(propylene glycol), PEG = poly(ethylene glycol), PTHF = polytetrahydrofuran, OPFR = *t*-octylphenol/formaldehyde resin.
[b] Determined by field desorption (FD-MS).
[c] Value reported by the manufacturer, usually determined by vapor pressure osmometry (VPO).
[d] Value reported by the manufacturer, usually determined by gel permeation chromatography (GPC).
[e] Determined in-house by GPC.
[f] Determined in-house by liquid chromatography (LC).
[g] Determined in-house by VPO.

determined by FD-MS is also much too low. The low FD-MS values may be attributed to "poorer desorption efficiencies for the higher mass oligomers."[54] That is, there tends to be a mass discrimination against the higher mass species.

Evans et al. measured molecular weight averages for three polyethylene samples by FD-MS.[55] Results for two of the samples are given in Table 6.1, and for these the agreement between FD-MS and GPC was reasonably good. The highest mass sample (M_n = 1870 by GPC), however, showed a "loss of sensitivity in the FD-MS data for the higher molecular weight ions of polyethylene."[55] A bimodal distribution was actually observed, and the lower mass envelope was overrepresented in the spectrum. In this case, mass discrimination was a problem even though the polymer itself was of reasonably low polydispersity (M_w/M_n = 1.15 by GPC).

In general, we would expect polymers that are good FD desorbers with low polydispersities (less than ~1.3) to give reliable molecular weight averages via FD-MS analysis. Polymers that are poorly desorbing and/or have broad distributions (M_w/M_n greater than ~1.3) would not be expected to give reliable M_w and M_n values. High polydispersity is not a problem just in FD-MS; other methods (e.g., MALDI-MS) have the same difficulty when it comes to the determination of molecular weight averages.

6.2.3   Direct Polymer Pyrolysis

6.2.3.1   Pyrolysis Field Ionization

Pyrolysis (thermal degradation) is used extensively with mass spectrometry for (i) elucidation of chemical structures of unknown polymers, (ii) assessing the thermal behavior of polymeric materials (e.g., in thermal processing or combustibility studies), and (iii) investigation of polymer decomposition kinetics and mechanisms. Pyrolysis field ionization (Py-FI-MS) turns out to be a very informative method for use in polymer decomposition studies, for several reasons.

First, since the pyrolysis is carried out *in vacuo* using a heated direct probe, the degradation occurs in the ion source very close to the ionization region. Thus, secondary reactions are minimized so that (to a large extent) primary pyrolysis products are observed. Second, since the direct probe is heated slowly (typically 5–20°C/min), pyrolyzates formed at the onset of pyrolysis may be readily detected. These "early pyrolyzates" represent the simplest (lowest energy) pyrolytic reactions, and the highest mass organic pyrolyzates are almost always the first ones that are formed. Third, because of the soft nature of FI-MS, higher mass pyrolysis products can be detected than are observed by other mass spectral methods (such as EI or CI). Fourth, since pyrolyzate mixtures are generally very complex, the very high abundances of molecular (or quasimolecular) ions afforded by FI-MS generally result in the simplest possible spectrum.

A typical Py-FI mass spectrum for a hydrocarbon polymer (isotactic polypropylene) is shown in Figure 6.9.[56] The spectrum shows the initial pyrolyzates that are formed as the polymer is held at 400°C in the direct probe. Volatile pyrolyzates are observed at every carbon number up to ~1400 Da (C_{100}). It is interesting that the principal oligomer series at low mass is simply multiples of the propylene monomer unit (MW = 42n, series **A**). The principal series at higher mass values, however, is an α, ω-diene series (MW = 42n + 12, series **E**). The pyrolysis mechanism may be explained by a free radical degradation pathway.[56]

Hummel and coworkers showed the potential of Py-FI-MS for polymer analysis in some early experiments in the 1970s. For example, in one study polymers containing methylmethacrylate, α-methylstyrene, and acrylonitrile were examined, but only low mass products (<250 Da) could be detected with the instrument used.[57]

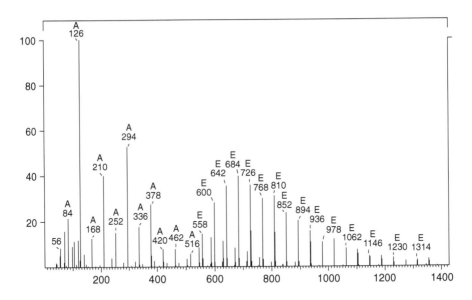

FIGURE 6.9
Pyrolysis FI-MS of isotactic polypropylene. (Reprinted from Ref. 56. Copyright 1995, with permission from Elsevier Science.)

The largest number of Py-FI-MS studies has been carried out by Schulten and coworkers, beginning in 1980.[50,58–73] Most of these studies were conducted with a specially designed "high-temperature, high-sensitivity" direct probe that provides programmed heating up to 800°C.[62]

Schulten's group has studied a large number of polyamides, including aliphatics (nylons),[50,59,65] aromatics (Kevlar, Nomex),[61] and mixed aliphatic/aromatic materials.[58] Other nitrogen-containing polymers studied include polyquinones[66] and acrylonitrile homo- and copolymers.[70] Some hydrocarbon polymers were also investigated, including polyethylene,[62] polystyrene,[67] and copolymers containing butadiene or styrene.[70]

Oxygen-containing polymers studied by Schulten's group included epoxy resins,[63] acrylate and methacrylate polymers,[68,69,71] and polyesters.[60,64] Polyester degradation by Py-FI-MS provides an interesting example.[64] When single polyesters—for example, poly(ethylene succinate) (PES) or poly(butylene adipate) (PBA)—were degraded, the principal pyrolyzates observed were cyclic oligomers. The degradation mechanism involves intramolecular ester exchange, which is a low energy process occurring below 250°C. When physical mixtures of two different polyesters were pyrolyzed at temperatures below ~250°C, the spectra contained the same signals as with single components. At temperatures above ~300°C, intermolecular exchange started to occur, and pyrolyzates were detected that contained fragments from both of the starting polymers. This effect is illustrated in Figure 6.10, which is the Py-FI mass spectrum of mixed PES and PBA. Most of the cyclic oligomer pyrolyzates (**A** = monomers, **B** = dimers, and so forth) contain fragments from both of the starting polymers.[64]

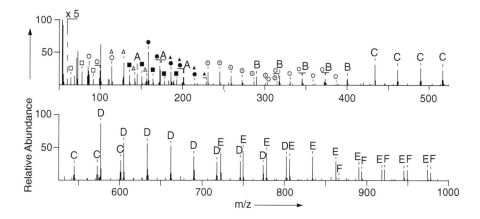

FIGURE 6.10
Pyrolysis FI-MS of a mixture of poly(ethylene succinate) and poly(butylene adipate) (A = monomers, B = dimers, etc.). (Reprinted from Ref. 64. Copyright 1989, with permission from Elsevier Science.)

Lattimer and coworkers have studied a number of saturated and unsaturated hydrocarbon polymers by Py-FI-MS.[56,72–75] These include cured and uncured diene rubbers,[72,73,75] and a number of polyolefins (polyethylene, polypropylene, polyisobutylene, et al.).[56,74]

6.2.3.2 Pyrolysis Field Desorption

Most of the pyrolysis FI/FD studies in the literature have been done in the Py-FI-MS mode. An alternative approach is to place the polymer (from solution) directly on the field emitter, and then heat the wire to induce pyrolysis. Some of the earlier studies from Schulten's laboratory used this method (Py-FD-MS). Polymers examined were polyamides[58,59,63] and a polyester.[60] Derrick et al. reported the Py-FD-MS analysis of a poly(olefin sulfone).[76]

Py-FD-MS is somewhat more difficult to carry out experimentally as compared to Py-FI-MS. It may be noted that Schulten used Py-FD only in his earlier polymer degradation studies; all the reports after 1988 give Py-FI results only. There are a number of reasons why Py-FD is more challenging. First, the polymer needs to be dissolved (or at least suspended in a solvent) for deposition on the field emitter. Py-FI-MS, on the other hand, can easily be run with intractable solids. Second, ion "sputtering" can occur as the polymer degrades and desorbs on the emitter. This leads to transient (intermittent) signals and, in the worst case, to emitter breakage. Third, temperature control while heating the emitter is inexact. There is, in fact, a significant temperature gradient along the length of the wire (with the center being the hottest). A programmed direct probe, on the other hand, provides precise temperature control in Py-FI-MS.

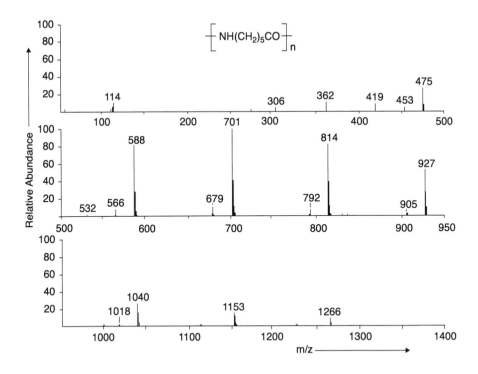

FIGURE 6.11
Pyrolysis FD-MS of nylon 6. (Reprinted with permission from Ref. 59.)

Nevertheless, Py-FD-MS does have the advantage that there is essentially no time separation between the degradation and ionization events. Thus primary products are most likely to be observed. An excellent example is shown in Figure 6.11. This is the Py-FD mass spectrum of nylon 6, which was dissolved in 1,1,1,3,3,3-hexafluoro-2-propanol for deposition on the emitter.[59] A very clean spectrum of the oligomeric pyrolyzates is seen; the major peaks are MNa^+, and the minor peaks are MH^+.

6.2.4 Polymer Additives

Commercial rubbers and plastics are often very complex materials. In addition to various polymers, commercial formulations contain a number of compounding ingredients (additives) that are included to give particular physical and/or chemical properties. These additives include plasticizers, processing/extender oils, waxes, carbon black, inorganic fillers, antioxidants, antiozonants, antifatigue agents, heat and light stabilizers, tackifying resins, processing aids, crosslinking agents, accelerators, retarders, adhesives, pigments, smoke and flame retardants, and others. In this section we will be concerned exclusively with organic additives, since these are the ones that are most readily analyzed by FI/FD mass spectrometry. All of the examples in this section are of field *desorption* applications, although there is no inherent reason why field ionization could not be used for the analysis of more volatile additives as well.

We will consider direct FD-MS analysis and off-line LC/FD-MS analysis separately. This section considers the analysis of additives as "raw materials." The next section considers the analysis of additives that are found in polymer formulations/compounds.

6.2.4.1 Direct FD-MS Analysis

It has been known for a number of years that FD-MS is an effective analytical method for direct analysis of many rubber and plastic additives. Major components and impurities in commercial additives can be assessed quickly, and the FD-MS data can be used to help determine what (if any) additional analytical characterization is needed. Lattimer and Welch showed that FD-MS gives excellent molecular ion spectra for a number of polymer additives, including rubber accelerators (dithiocarbamates, guanidines, benzothiazyl, and thiuram derivatives)[16,77] antioxidants (hindered phenols, aromatic amines)[16,77] *p*-phenylenediamine-based antiozonants,[16,77] processing oils,[77] and phthalate plasticizers.[77] Zhu and Su characterized alkylphenol ethoxylate surfactants by FD-MS.[78] Jackson et al. analyzed some plastic additives (hindered phenol antioxidants and a benzotriazole UV stabilizer) by FD-MS.[79]

Lattimer et al. have used direct FD-MS in some mechanistic model compound studies. Reaction products of a *p*-phenylenediamine antiozonant and *cis*-9-tricosene (a model olefin) were assessed by FD-MS.[80] Several products from a model compound study of phenolic resin vulcanization were characterized by FD-MS.[81]

6.2.4.2 Liquid Chromatography and FD-MS

Schulten has reviewed applications combining liquid chromatography (LC or HPLC) with FD-MS.[82] In considering whether an *on-line* LC/FD-MS system was feasible, Schulten concluded that "there is no practical and efficient way of transferring the HPLC eluents onto the emitter without increasing the technical complexity considerably. Hence these methods should be used in the off-line mode...."[82]

Lattimer and coworkers have used off-line LC/FD-MS in a number of studies.[10,11,54,83–86] In a typical example, an oligomeric antioxidant—poly (2,2,4-trimethyl-1,2-dihydroquinoline), or poly-TMDQ—was separated by LC, giving the chromatogram in Figure 6.12.[10] LC fractions were collected for individual peaks, and the solvent (tetrahydrofuran/water) was evaporated with gentle heating under a stream of nitrogen. The residues were then taken up in tetrahydrofuran for deposition on the field emitter. FD spectra for three fractions are shown in Figure 6.13.[10] It may be noted that the FD spectra clearly indicate when there are multiple components present from unresolved or adjacent LC peaks. This is especially noticeable in the spectrum

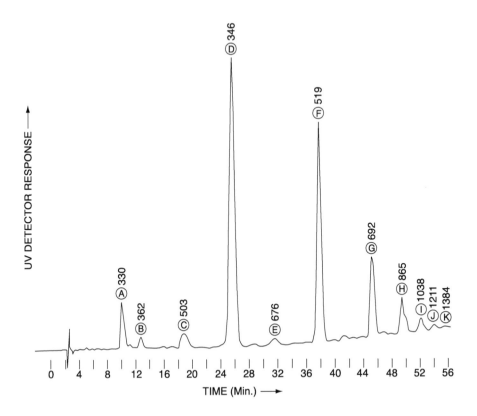

FIGURE 6.12
Liquid chromatogram of poly-TMDQ. (Reprinted with permission from Ref. 10. Copyright 1979, American Chemical Society.)

for Peak I (Figure 6.13c). In addition to the main component at M^+ 1038 (TMDQ hexamer), there is a small amount of a second component at M^+ 1022 (TMDQ hexamer minus methane).

In later work, Lattimer et al. identified eight oligomeric series in a commercial version of poly-TMDQ by using the LC/FD-MS combination.[11] Components of *t*-octylphenol/formaldehyde resins were also characterized using this method.[54] In a series of studies, ozonation products of several *p*-phenylenediamine compounds (rubber antiozonants) were separated by LC and identified by using FD-MS.[83–85] In another study, model "efficient vulcanization" products were separated by column chromatography and characterized by FD-MS.[86]

6.2.5 Polymer Compound Analysis

The identification of the ingredients in a compounded polymer can be a difficult task for the analytical chemist. A wide variety of components is involved—polymers, fillers, solvents, organic and inorganic additives.

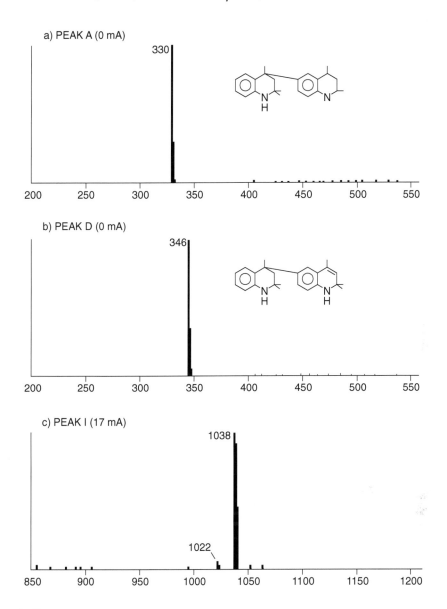

FIGURE 6.13
FD-MS of poly-TMDQ LC fractions. (a) Peak A, (b) Peak D, (c) Peak I. (Reprinted with permission from Ref. 10. Copyright 1979, American Chemical Society.)

Various methods have been developed to separate, identify, and quantify the numerous ingredients. In recent years there has been an increasing interest in *direct* methods of analysis, i.e., examining the compounded polymer with no or minimal pretreatment of the material. It has been demonstrated that FI-MS and FD-MS can play an important role in this type of analysis.[87]

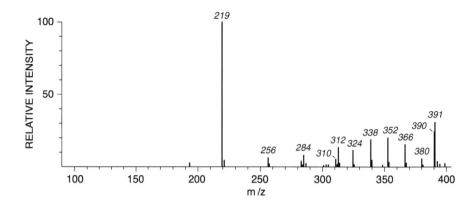

FIGURE 6.14

FD-MS of extract from an EPDM vulcanizate. (Reprinted with permission from Ref. 88. Copyright 1984, Rubber Division, American Chemical Society.)

There are two basic approaches to mass spectral analysis of a compounded rubber or plastic material: *extract* analysis and *direct polymer* analysis. Comparing these approaches, analysis of a *solvent extract* has these advantages: (i) the organic additives are isolated, which eliminates mass spectral interferences that may arise from the rubber and filler components; (ii) isolation of the additives facilitates further chromatographic or spectroscopic analysis; and (iii) higher mass, less volatile components (e.g., oligomeric antioxidants) can generally be identified more readily. On the other hand, *direct polymer* analysis has these advantages: (i) this is a more rapid approach; (ii) one avoids problems due to variable extraction efficiencies of different solvents; (iii) the more volatile additives (e.g., solvents, accelerator fragments) may be detected more readily; and (iv) one has the possibility of identifying both the organic additives and the polymer components in the same experiment.

Lattimer and coworkers have published several reports on polymer compound analysis by mass spectrometry.[74,87–91] In earlier studies, the emphasis was on field *desorption* analysis of rubber *extracts*.[88–90] A typical example is shown in Figure 6.14, which is the FD mass spectrum of the acetone extract from an EPDM vulcanizate.[88] The spectrum shows the presence of several ingredients: phenyl-β-napthylamine antioxidant (MW 219), fatty acid (MW 256, 282, 284), dioctylphthalate plasticizer (MW 390), and a paraffin wax (MW 324, 338, et al.).

In later papers in this series, there was more emphasis on *direct* analysis of the rubber or plastic material by field *ionization*.[74,89–91] Figure 6.15 is the total ion current (TIC) vs. time (or temperature) profile for a diene rubber compound.[90] The sample was heated in the direct probe from 50–750°C, with FI-MS. There are two distinct regions in which TIC maxima are observed. The first occurs between 50–400°C and largely represents the evaporation of organic additives from the rubber (Figure 6.16). Additives in the rubber include fatty acid (MW 256, 284), a p-phenylenediamine antiozonant (MW 332),

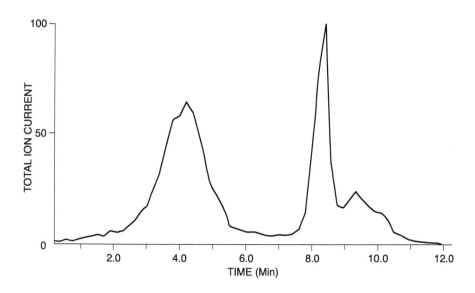

FIGURE 6.15

FI-MS total ion current vs. time (temperature) profile for diene rubber compound. (Reprinted with permission from Ref. 90. Copyright 1988, Rubber Division, American Chemical Society.)

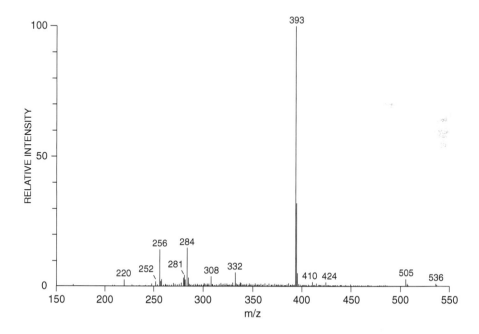

FIGURE 6.16

FI-MS of volatile components in diene rubber compound. (Reprinted with permission from Ref. 90. Copyright 1988, Rubber Division, American Chemical Society.)

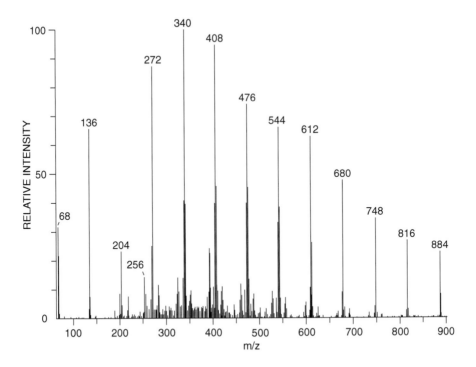

FIGURE 6.17
Pyrolysis FI-MS of diene rubber compound. (Reprinted with permission from Ref. 90. Copyright 1988, Rubber Division, American Chemical Society.)

a *t*-octylated diphenylamine antioxidant (MW 393, 505), a *t*-octylphenol/ formaldehyde tackifying resin (MW 424, 536), and a processing oil (weaker ions in the "background").

The second TIC maximum occurs at higher temperatures (400–750°C) and represents the evolution of rubber thermal decomposition products (pyrolyzates). The major FI-MS peaks in this region are isoprene oligomers (MW = 68n, Figure 6.17).[90] Note that the two envelopes in Figure 6.15 are well-separated in time (or temperature); thus one can obtain separate mass spectra for the organic additives and the rubber pyrolyzates.

6.3   Conclusion

FI-MS and FD-MS are only used in a handful of laboratories worldwide. Those of us who have used the methods for several years still rely upon them heavily. Despite the large number of other ionization methods that have become available, it is clear that FI-MS and FD-MS still possess some capabilities that have not been duplicated by newer methods. In particular, essentially all of the newer methods (e.g., MALDI and ESI) work best for polar molecules, while FI/FD-MS is optimal for nonpolar and slightly polar compounds. FI-MS and FD-MS are superb techniques for the "survey" analysis of complex polymeric mixtures, as long as the molecular weight range needed is not too high.

Perhaps an encouraging sign is the recent commercialization by MicroMass of a new benchtop orthogonal acceleration time-of-flight (oa-TOF) mass spectrometer that offers field ionization as an accessory. This is the first commercial FI-MS system that is not connected to a magnetic sector instrument. Hopefully this may open up the FI-MS technique to a new generation of users.

At any length, those who think of FI-MS and FD-MS as pioneering techniques that are no longer useful should rethink their position. It is premature to put these methods on the endangered species list. FI-MS and FD-MS have some attractive features that make them both complementary to other desorption/ionization methods and unique in their applications.

References

1. Beckey, H. D., *Principles of Field Ionization and Field Desorption Mass Spectrometry*, Pergamon Press, Oxford, 1977.
2. Schulten, H.-R., Advances in field desorption mass spectrometry, in *Soft Ionization Biological Mass Spectrometry*, Morris, H. R., Ed., Heyden, London, 1981, p. 6.
3. Wood, G. W., Field desorption mass spectrometry: Applications, *Mass Spectrom. Rev.*, 1, 63, 1982.
4. Schulten, H.-R., Analytical application of field desorption mass spectrometry, in *Ion Formation from Organic Solids*, Benninghoven, A., Ed., Springer-Verlag, Berlin, 1983, Ch. 1.2.
5. Saito, J., Waki, H., Teramae, N., and Tanaka, S., Application of field desorption mass spectrometry to polymer and oligomer analysis, *Progress Org. Coatings*, 15, 311, 1988.
6. Lattimer, R. P. and Schulten, H.-R., Field ionization and field desorption mass spectrometry: Past, present, and future, *Anal. Chem.*, 61, 1201A, 1989.
7. Prokai, L., *Field Desorption Mass Spectrometry*, Marcel Dekker, New York, 1990.
8. Mueller, E. W. and Tsong, T. T., *Field Ion Microscope*, Elsevier, New York, 1969.
9. Inghram, M. G. and Gomer, R., Mass spectrometric analysis of ions from the field microscope, *J. Chem. Phys.*, 22, 1279, 1954.
10. Lattimer, R. P., Harmon, D. J., and Welch, K. R., Characterization of low molecular weight polymers by liquid chromatography and mass spectroscopy, *Anal. Chem.*, 51, 1293, 1979.
11. Lattimer, R. P., Hooser, E. R., and Zakriski, P. M., Characterization of aniline-acetone condensation products by liquid chromatography and mass spectroscopy, *Rubber Chem. Technol.*, 53, 346, 1980.
12. Lattimer R. P. and Schulten, H.-R., Field desorption and fast atom bombardment mass spectrometry of poly(ethylene imine), *Int. J. Mass Spectrom. Ion Proc.*, 67, 277, 1985.

13. Ott, K. H., Roellgen, F. W., Zwinselman, J. J., Fokkens, R. H., and Nibbering, N. M. M., Negative ion field desorption mass spectra of some inorganic and organic compounds, *Org. Mass Spectrom.*, 15, 419, 1980.

14. Muscat, D., Henderickx, H., Kwakkenbros, G., van Benthem, R., de Koster, C. G., Fokkens, R., and Nibbering, N. M. M., In-source decay of hyperbranched polyesteramides in matrix-assisted laser desorption/ionization time-of-flight mass spectrometry, *J. Am. Soc. Mass Spectrom.*, 11, 218, 2000.

15. Rollins, K., Scrivens, J. H., Taylor, M. J., and Major, H., The characterization of polystyrene oligomers by field-desorption mass spectrometry, *Rapid Commun. Mass Spectrom.*, 4, 355, 1990.

16. Lattimer, R. P. and Welch, K. R., Direct analysis of polymer chemical mixtures by field desorption mass spectroscopy, *Rubber Chem. Technol.*, 53, 151, 1980.

17. Saito, J., Toda, S., and Tanaka, S., Analysis of polyalkylene oxides by field desorption mass spectrometry, *Shitsuryo Bunseki*, 28, 175, 1980.

18. Saito, J., Toda, S., and Tanaka, S., Field desorption mass spectrometry of epoxy resin prepolymer, *Bunseki Kagaku*, 29, 462, 1980.

19. Saito, J., Toda, S., and Tanaka, S., Analysis of various epoxy resins by field desorption mass spectrometry, *Netsu Kokasei Jushi*, 1, 79, 1980.

20. Saito, J., Toda, S., and Tanaka, S., Chemical structural investigation of methylated methylol melamine resins by field desorption mass spectrometry, *Netsu Kokasei Jushi*, 1, 18, 1980.

21. Saito, J., Toda, S., and Tanaka, S., Analysis of phenolic novalak resins by field desorption mass spectrometry, *Netsu Kokasei Jushi*, 2, 72, 1981.

22. Saito, J., Toda, S., and Tanaka, S., Molecular species analysis of styrene oligomer by field desorption mass spectrometry, *Bunseki Kagaku*, 30, 706, 1981.

23. Saito, J., Hara, J., Toda, S., and Tanaka, S., Analysis of radical oligomerization of styrene by field desorption mass spectrometry, *Chem. Lett. (Japan)*, 1982, 311.

24. Saito, J., Teramae, N., Hara, J., Toda, S., and Tanaka, S., Analysis of thermal behavior of styrene oligomers by field desorption mass spectrometry, *Kobunshi Ronbunshu*, 40, 485, 1983.

25. Saito, J., Hara, J., Toda, S., and Tanaka, S., Analysis for the oligomerization mechanism of styrene by field desorption mass spectrometry, *Bull. Chem. Soc. Japan*, 56, 748, 1983.

26. Saito, J., Teramae, N., Hara, J., Toda, S., and Tanaka, S., Analysis of thermal oligomerization mechanism of styrene by field desorption mass spectrometry, *J. Appl. Polym. Sci.*, 28, 2303, 1983.

27. Hara, J., Teramae, N., Saito, J, and Tanaka, S., Analysis of initiation mechanism in thermal polymerization of styrenes by field desorption mass spectrometry, *Kobunshi Ronbunshu*, 41, 453, 1984.

28. Hara, J., Teramae, N., Saito, J., and Tanaka, S., Evidence for the presence of cage reaction in the initiation step of thermal polymerization of styrenes by field desorption mass spectrometry, *J. Appl. Polym. Sci.*, 30, 1461, 1985.

29. Saito, J., Teramae, N., Hara, J., Toda, S., and Tanaka, S., Analysis of inhibition reaction of 2,2-diphenyl-1-picrylhydrazyl in polymerization by field desorption mass spectrometry, *Kobunshi Ronbunshu*, 40, 531, 1983.

30. Saito, J., Teramae, N., Toda, S., and Tanaka, S., Analysis of thermal behavior of blocked isocyanates by field desorption mass spectrometry, *Bunseki Kagaku*, 32, 637, 1983.

31. Hara, J., Teramae, N., Saito, J., and Tanaka, S., Analysis of curing mechanism of unsaturated polyester resin. (1) Field desorption mass spectrometric study on termination reaction in copolymerization of styrene and diethylfumarate, *Netsu Kokasei Jushi*, 5, 142, 1984.

32. Saito, J., Teramae, N., Hara, J., and Tanaka, S., Analysis of acrylic oligomers by field desorption mass spectrometry, *Kobunshi Ronbunshu*, 41, 623, 1984.

33. Neumann, G. M., Cullis, P. G., and Derrick, P. J., Mass spectrometry of polymers: Polypropylene glycol, *Z. Naturforsch., Teil A*, 35, 1090, 1980.

34. McCrae, C. E. and Derrick, P. J., The role of the field in field desorption fragmentation of polyethylene glycol, *Org. Mass Spectrom.*, 18, 321, 1983.

35. Davis, S. C., Neumann, G. M., and Derrick, P. J., Field desorption mass spectrometry with suppression of the high field, *Anal. Chem.*, 59, 1360, 1987.

36. Tottszer, A. I., Neumann, G. M., Derrick, P. J., and Willett, G. D., Laser heating versus resistive heating in the field-desorption mass spectrometry of organic polymers, *J. Phys. D: Appl. Phys.*, 21, 1713, 1988.

37. Craig, A. G., Cullis, P. G., and Derrick, P. J., Field desorption of polymers: Polybutadiene, *Int. J. Mass Spectrom. Ion Phys.*, 38, 297, 1981.

38. Prokai, L., Investigation of phenol-formaldehyde condensates by field desorption mass spectrometry, *J. Polym. Sci.: Part C: Polym. Lett.*, 24, 223, 1986.

39. Blease, T. G., Paterson, G. A., and Scrivens, J. H., Thermal characterisation of polymeric systems by mass spectrometry, *Br. Polym. J.*, 21, 37, 1989.

40. Wiley, R. H., and Cook Jr., J. C., Field desorption mass spectral data for oligomers up to 2400 amu, *J. Macromol. Sci. – Chem.*, A10, 811, 1976.

41. Matsuo, T., Matsuda, H., and Katakuse, I., Use of field desorption mass spectra of polystyrene and polypropylene glycol as mass references up to mass 10000, *Anal. Chem.*, 51, 1329, 1979.

42. Lattimer, R. P. and Schulten, H.-R., Field desorption of hydrocarbon polymers, *Int. J. Mass Spectrom. Ion Phys.*, 52, 105, 1983.

43. Evans, W. J., DeCoster, D. M., and Greaves, J., Field desorption mass spectrometry studies of the samarium-catalyzed polymerization of ethylene under hydrogen, *Macromolecules*, 28, 7929, 1995.

44. Evans, W. J., DeCoster, D. M., and Greaves, J., Metalation as a termination step in polymerization reactions involving α-olefins and ethylene as detected by field desorption mass spectrometry, *Organometallics*, 15, 3210, 1996.

45. Jackson, A. T., Jennings, K. R., and Scrivens, J. H., The effects of internal energy deposition during ionization on the collision-induced decomposition spectra of an organic polymer additive, *Rapid Commun. Mass Spectrom.*, 10, 1459, 1996.

46. Craig, A. G. and Derrick, P. J., Collision-induced decomposition of cationic radical polystyrene chains, *J. Chem. Soc., Chem. Commun.*, 1985, 891.

47. Craig, A. G. and Derrick, P. J., Spontaneous fragmentation of cationic polystyrene chains, *J. Am. Chem. Soc.*, 107, 6707, 1985.

48. Craig, A. G. and Derrick, P. J., Production and characterization of beams of polystyrene ions, *Aust. J. Chem.*, 39, 1421, 1986.

49. Jackson, A. T., Jennings, R. C. K., Scrivens, J. H., Green, M. R., and Bateman, R. H., The characterization of complex mixtures by field desorption-tandem mass spectrometry, *Rapid Commun. Mass Spectrom.*, 12, 1914, 1998.

50. Schulten, H.-R. and Plage, B., Thermal degradation of aliphatic polyamides studied by field ionization and field desorption mass spectrometry, *J. Polym. Sci.: Part A: Polym. Chem.*, 26, 2381, 1988.

51. Lattimer, R. P., Harris, R. E., and Schulten, H.-R., Mass spectrometry, in *Determination of Molecular Weight*, Cooper, A. R., Ed., John Wiley & Sons, New York, 1989, Ch. 14.

52. Lattimer, R. P., Harmon, D. J., and Hansen, G. E., Determination of molecular weight distributions of polystyrene oligomers by field desorption mass spectrometry, *Anal. Chem.*, 52, 1808, 1980.

53. Lattimer, R. P. and Hansen, G. E., Determination of molecular weight distributions of polyglycol oligomers by field desorption mass spectrometry, *Macromolecules*, 14, 776, 1981.

54. Lattimer, R. P., Hooser, E. R., Diem, H. E., and Rhee, C. K., Analytical characterization of tackifying resins, *Rubber Chem. Technol.*, 55, 442, 1982.

55. Evans, W. J., DeCoster, D. M., and Greaves, J., Evaluation of field desorption mass spectrometry for the analysis of polyethylene, *J. Am. Soc. Mass Spectrom.*, 7, 1070, 1996.

56. Lattimer, R. P., Pyrolysis field ionization mass spectrometry of polyolefins, *J. Anal. Appl. Pyrolysis*, 31, 203, 1995.

57. Hummel, D. O. and Duessel, H.-J., Field ionization- and electron impact-mass spectrometry of polymers and copolymers. 3. Copolymers of α-methylstyrene with methylmethacrylate and acrylonitrile, *Makromol. Chem.*, 175, 655, 1974.

58. Schulten, H.-R. and Duessel, H.-J., Pyrolysis field desorption mass spectrometry of polymers. II. Pyrolysis field ionization and field desorption mass spectrometry of aliphatic and aromatic poly(4,4'-dipiperidylamides), *J. Anal. Appl. Pyrolysis*, 2, 293, 1980/1981.

59. Bahr, U., Luederwald, I., Mueller, R., and Schulten, H.-R., Pyrolysis field desorption mass spectrometry of polymers. III. Aliphatic polyamides, *Angew. Makromol. Chem.*, 120, 163, 1984.

60. Doerr, M., Luederwald, I., and Schulten, H.-R., Characterization of polymers by field desorption and fast atom bombardment mass spectrometry, *Fresenius Z. Anal. Chem.*, 318, 339, 1984.

61. Schulten, H.-R., Plage, B., Ohtani, H., and Tsuge, S., Studies on the thermal degradation of aromatic polyamides by pyrolysis-field ionization mass spectrometry and pyrolysis-gas chromatography, *Angew. Makromol. Chem.*, 155, 1, 1987.

62. Schulten, H.-R., Simmleit, N., and Mueller, R., High-temperature, high-sensitivity pyrolysis field ionization mass spectrometry, *Anal. Chem.*, 59, 2903, 1987.

63. Plage, B. and Schulten, H.-R., Pyrolysis-field ionization mass spectrometry of epoxy resins, *Macromolecules*, 21, 2018, 1988.

64. Plage, B. and Schulten, H.-R., Pyrolysis-field ionization mass spectrometry of aliphatic polyesters and their thermal interactions in mixtures, *J. Anal. Appl. Pyrolysis*, 15, 197, 1989.

65. Plage, B. and Schulten, H.-R., Pyrolysis-field ionization mass spectrometry of polyamide copolymers and blends, *J. Appl. Polym. Sci.*, 38, 123, 1989.

66. Blazso, M., Jakab, E., Szekely, T., Plage, B., and Schulten, H.-R., Pyrolysis-gas chromatography mass spectrometry and field ionization mass spectrometry of polyquinones, *J. Polym. Sci.: Part A: Polym. Chem.*, 27, 1027, 1989.

67. Ohtani, H., Yuyama, T., Tsuge, S., Plage, B., and Schulten, H.-R., Study on thermal degradation of polystyrenes by pyrolysis-gas chromatography and pyrolysis-field ionization mass spectrometry, *Eur. Polym. J.*, 26, 893, 1990.

68. Plage, B., Schulten, H.-R., Schneider, J., and Ringsdorf, H., Bulk reactions in amphiphilic acrylic copolymers studied by time/temperature-resolved pyrolysis-field ionization mass spectrometry, *Macromolecules*, 23, 3417, 1990.

69. Plage, B. and Schulten, H.-R., Thermal degradation mechanisms of amphiphilic acrylic copolymers studied by temperature-resolved pyrolysis-field ionization mass spectrometry, *J. Anal. Appl. Pyrolysis*, 19, 285, 1991.

70. Plage, B. and Schulten, H.-R., Sequences in copolymers studied by high-mass oligomers in pyrolysis-field ionization mass spectrometry, *Angew. Makromol. Chem.*, 184, 133, 1991.

71. Plage, B., Schulten, H.-R., Ringsdorf, H., and Schuster, A., Thermal reactions of amphotropic copolymers studied by thermogravimetry and temperature-resolved pyrolysis-field ionization mass spectrometry, *Makromol. Chem.* 192, 1567, 1991.

72. Lattimer, R. P., Harris, R. E., Rhee, C. K., and Schulten, H.-R., Identification of organic components in uncured rubber compounds using mass spectrometry, *Rubber Chem. Technol.*, 61, 639, 1988.

73. Schulten, H.-R., Plage, B., and Lattimer, R. P., Pyrolysis-field ionization mass spectrometry of rubber vulcanizates, *Rubber Chem. Technol.*, 62, 698, 1989.

74. Lattimer, R. P., Direct analysis of polypropylene compounds by thermal desorption and pyrolysis-mass spectrometry, *J. Anal. Appl. Pyrolysis*, 26, 65, 1993.

75. Lattimer, R. P., Pyrolysis field ionization mass spectrometry of hydrocarbon polymers, *J. Anal. Appl. Pyrolysis*, 39, 115, 1997.

76. Jardine, D. R., Nekula, S., Than-trong, N., Haddad, P. R., Derrick, P. J., Grespos, E., and O'Donnell, J. H., Field desorption mass spectrometry of poly(olefin sulfones), *Macromolecules*, 19, 1770, 1986.

77. Lattimer, R. P. and Welch, K. R., Field desorption mass spectra of polymer chemicals, *Rubber Chem. Technol.*, 51, 925, 1978.

78. Zhu, P. and Su, K., Field desorption mass spectrometric analysis of mixtures of surfactants and inorganic salts, *Org. Mass Spectrom.*, 25, 260, 1990.

79. Jackson, A. T., Jennings, K. R., and Scrivens, J. H., Analysis of a five-component mixture of polymer additives by means of high energy mass spectrometry and tandem mass spectrometry, *Rapid Commun. Mass Spectrom.*, 10, 1449, 1996.

80. Lattimer, R. P., Layer, R. W., and Rhee, C. K., Mechanisms of antiozonant protection: Antiozonant-rubber reactions during ozone exposure, *Rubber Chem. Technol.*, 57, 1023, 1984.

81. Lattimer, R. P., Kinsey, R. A., Layer, R. W., and Rhee, C. K., The mechanism of phenolic resin vulcanization of unsaturated elastomers, *Rubber Chem. Technol.*, 62, 107, 1989.

82. Schulten, H.-R., Off-line combination of liquid chromatography and field desorption mass spectrometry: Principles and environmental, medical and pharmaceutical applications, *J. Chromatogr.*, 251, 105, 1982.

83. Lattimer, R. P., Hooser, E. R., Diem, H. E., Layer, R. W., and Rhee, C. K., Mechanisms of ozonation of N,N'-di-(1-methylheptyl)-*p*-phenylenediamine, *Rubber Chem. Technol.*, 53, 1170, 1980.

84. Lattimer, R. P., Hooser, E. R., Layer, R. W., and Rhee, C. K., Mechanisms of ozonation of N-(1,3-dimethylbutyl)-N'-phenyl-*p*-phenylenediamine, *Rubber Chem. Technol.*, 56, 431, 1983.

85. Lattimer, R. P., Layer, R. W., Hooser, E. R., and Rhee, C. K., The ozonation of N,N'-di-*n*-octyl-*p*-phenylenediamine and N,N'-di-(1,1-dimethylethyl)-*p*-phenylenediamine, *Rubber Chem. Technol.*, 64, 780, 1991.

86. Gregg J., E. C., Jr. and Lattimer, R. P., Polybutadiene vulcanization. Chemical structures from sulfur-donor vulcanization of an accurate model, *Rubber Chem. Technol.*, 57, 1056, 1984.

87. Lattimer, R. P. and Harris, R. E., Mass spectrometry for analysis of additives in polymers, *Mass Spectrom. Rev.*, 4, 369, 1985.
88. Lattimer, R. P., Harris, R. E., Ross, D. B. and Diem, H. E., Identification of rubber additives by field desorption and fast atom bombardment mass spectrometry, *Rubber Chem. Technol.*, 57, 1013, 1984.
89. Lattimer, R. P., Harris, R. E., Rhee, C. K., and Schulten, H.-R., Identification of organic additives in rubber vulcanizates using mass spectrometry, *Anal. Chem.*, 58, 3188, 1986.
90. Lattimer, R. P., Harris, R. E., Rhee, C. K., and Schulten, H.-R., Identification of organic compounds in uncured rubber compounds using mass spectrometry, *Rubber Chem. Technol.*, 61, 639, 1988.
91. Lattimer, R. P., Direct analysis of elastomer compounds by soft ionization, tandem (MS/MS) and high resolution (AC-MS) mass spectrometry, *Rubber Chem. Technol.*, 68, 783, 1995.

7

Fast Atom Bombardment of Polymers

Giorgio Montaudo and Filippo Samperi

CONTENTS

7.1 Introduction

In this chapter we discuss the role of Fast Atom Bombardment Mass Spectrometry (FAB-MS) in the structural characterization of synthetic polymeric materials.

The FAB technique can be used for the chemical structural characterization of some samples not amenable to conventional ionization methods such as electron impact or chemical ionization mass spectrometry. It permits the analysis of polar and ionic compounds, often without the need for purification, isolation, and derivatization.

0-8493-3127-7/02/$0.00+$1.50

Historically, FAB has had a great impact on the use of MS in the biological sciences, and it has provided the bioanalyst with the ability to obtain mass-specific detection of individual compounds in complex mixtures.[1-4]

FAB belongs to the wide range of desorption/ionization techniques (Chapter 1) that avoid evaporating molecules into the gas phase prior to the formation of ions.

Fast atom bombardment of a chemical compound dissolved in a viscous organic liquid matrix (usually an alcohol) induces desorption phenomena that lead to the detection of quasimolecular ions in the mass spectra.[1-12]

Intense "molecular ions" (M^+, M^-) and/or "quasimolecular ions" (e.g, MH^+, MLi^+, MNa^+, MK^+, $[M-H]^-$) corresponding to low mass oligomers are detected in FAB mode (some appear in positive mode, others in negative mode, some in both modes).

For a few years after the introduction of FAB (in 1982), there was a limited interest in applying it to synthetic polymers,[8] and FAB was mostly used for proteins and other biomolecules.[1-12]

Later on, however, the importance of the FAB technique for polymer analysis grew considerably. For instance, the chain statistics method for the determination of copolymer sequence by MS (Chapter 2) was originally developed using the FAB technique.

Reasons for the delay in the application of FAB to the analysis of synthetic polymers are related to the understanding of the ionization mechanism on which FAB is based.

Initial uncertainties in the interpretation of FAB spectra of polymers indeed existed and were debated.[13-16] Peaks obtained in earlier work on the FAB spectra of crude aliphatic polyesters were interpreted as corresponding to the products originating from thermal degradation or from ion fragmentation reactions induced by atomic bombardment of the macromolecules investigated.[13,14]

Further work showed instead that the peaks appearing in the FAB mass spectra of the crude polyesters were not due to degradation or fragmentation processes.[15] They were identified as protonated molecular ions (MH^+) of pre-formed low mass species (cyclic and linear oligomers) contained in the polymer samples, which desorb intact from the liquid matrix under FAB conditions.[15]

Liquid Chromatography (both SEC and HPLC) has conclusively shown that condensation polymers (polyesters, polyamides, and the like) often contain sizeable amounts of low molar mass components. Even a small amount of oligomers may produce intense peaks in the FAB spectra of a high polymer sample. Assuming a degree of polymerization of about 100–200, a content of only 1% (on a weight basis) of oligomers in the polymer would yield a 1:1 mixture on a molar basis. Lower mass components generally desorb/ionize much more efficiently.

When the crude polyester samples were accurately purified from the lower mass oligomers, no significant peaks were observed. On the contrary, FAB spectra of the material extracted from the polyesters were found to be very similar to those obtained for the crude polymers.[15]

The exact mechanism of ionization occurring in the FAB experiment remains uncertain. Whether it is exclusively a surface phenomenon, or whether there is a contribution of sputtered ions from the bulk, remains unclear. Instead, studies on the mechanism of formation of ions in FAB-MS agree that the species desorbed from the liquid matrix are not those that collide directly with the fast atom beam.[7-11]

The molecules hit by the fast atoms are destroyed in the collision and do not appear in the spectrum, whereas the molecules not directly hit (matrix and analyte) acquire enough energy to ionize and desorb from the target surface. Therefore, fast atom bombardment in a liquid matrix should allow desorption of the intact analyte molecule from the condensed phase.

The earlier contention that FAB is a "soft" ionization method with no ion fragmentation of the desorbing molecules is not tenable today, but it is true that the ion fragmentation level is usually modest in many cases.[1-16]

In general all desorption/ionization methods, including FAB, are intended to produce ions of high mass and to minimize the fragmentation processes.

In fact, even for the simplest homopolymers one often experiences several mass peaks due to oligomer series having the same repeat unit but different end-groups.

Ion fragmentation complicates the mass spectra and may prevent quantitative analysis. However, the ion fragmentation level of specific FAB adducts has often been found to be nearly constant (i.e., independent of the molar mass of the oligomers, within the mass range accessible to FAB), thus yielding useful structural information, especially in copolymer sequence analysis (Chapter 2).

Together with SIMS (Chapter 8), and prior to the advent of MALDI technique (Chapter 10), FAB was the most important ionization method that allowed the analysis of intact polymer molecules, and therefore the MS literature on synthetic polymers in the last fifteen years is in sizeable part based on FAB analysis.

Today FAB has to compete with the formidable analytical power of the most recent MS soft ionization techniques such as MALDI (Chapter 10); however, it is still used in current polymer work because it allows one to obtain mass spectra with excellent resolution in the region up to 1000–2000 Da, whereas MALDI spectra are sometimes obscured by the matrix clusters in the mass range above 1000 Da.

Due to the relatively low mass range accessible to FAB (Chapter 1), FAB spectra do not yield direct structural information on large macromolecules.

However, the polymer structure can be inferred from the analysis of the relatively low molar mass species (oligomers) that often present in high polymer samples, if it is assumed that these low molar mass oligomers have the same structure as the undetected larger ones.[15-17]

As an alternative approach, polymer degradation can be used to produce low molar mass species. Partial degradation of high polymers, followed by FAB-MS analysis, provides a suitable method for the identification of the oligomeric species formed and affords the structural characterization of homopolymers and copolymers.[16,18]

Direct MS analysis of oligomer mixtures by FAB can be combined with the Collision Induced Dissociation (CID) technique, which is used to induce the fragmentation of selected molecular ions by means of collision with an inert gas (MS/MS). This allows one to obtain daughter ions: classical MS/MS techniques utilize the linked scanning of two or more mass analyzers to obtain product (daughter) and/or precursor (parent) ion spectra; see Chapters 1 and 4. Systematic application of tandem Mass Spectrometry (MS/MS) has produced important results in the analysis of polymeric materials by FAB.[16–22]

Specific examples and significant applications of FAB-MS to the analysis of polymeric materials are illustrated in the following sections.

7.2   Structure of Low Molar Mass Oligomers Contained in Polymers

The production of high molar mass polymers is often accompained by the formation of sizeable amounts of linear and/or cyclic low molar mass oligomers, and the identification of their structures in crude homopolymer and copolymer samples is a task in contemporary polymer characterization work.

The formation of low mass oligomers during polymerization has been ascertained in the synthesis of polyesters, polysulfides, polyureas, polyamides, polyethers, polysiloxanes, and in several other condensation polymers.[15–18]

Therefore, the knowledge of the structure and amount of oligomers present in a crude polymer sample is often desirable for evaluating the synthetic process, for the characterization of the sample, and for the evaluation of the polymer properties.

Traditional procedures of detecting oligomers contained in polymer samples are based on gas, liquid, and size exclusion chromatography (SEC), combined with several structure identification methods. These techniques are indeed powerful, but sometimes low volatility of samples, low solubility in suitable organic solvent, or low resolution in chromatography make alternative and rapid methods of detection and direct identification of mixtures highly desirable.

Mass Spectrometry is particulary suitable to the detection of these materials. Numerous reports have appeared where GC/MS and DPMS (using conventional EI, CI, DCI, and FI ionization modes), were coupled to detect cyclic

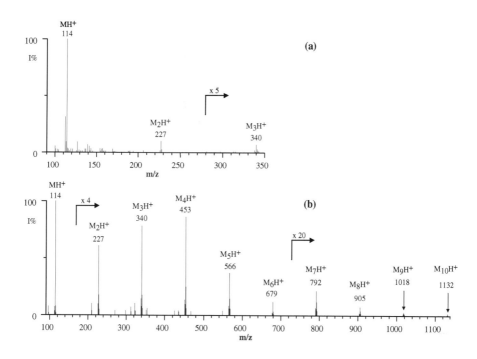

FIGURE 7.1
Positive ions FAB mass spectra of the crude Ny6 **(a)** and of the oligomers extracted from Ny6 with CH_3OH **(b)**. (Reprinted with permission from Refs. 15 and 17, Copyright 1987, American Chemical Society.)

oligomers present in a variety of condensation polymer samples. However, if the oligomers are thermally labile or if they are not stable under EI or CI modes, the identification of higher molar mass species becomes increasingly difficult.

FAB-MS does not in general have this limitation, and it permits the direct analysis of a polymer sample without prior clean-up (purification) procedures.

The FAB-mass spectrum of the crude commercial polycaprolactam (Ny6) consists (Figure 7.1a) of three peaks at m/z 114, 227, and 340. These correspond to the protonated cyclic monomer (m/z 114), dimer (m/z 227), and trimer (m/z 340), respectively.[15] The FAB mass spectrum of the oligomers extracted from Ny6 with methanol (Figure 7.1b) shows also the presence of peaks due to cyclic oligomers from tetramer up to decamer.[17] These cyclics were identified by comparison with the collision-induced decomposition (CID) mass spectra (i.e., by the product ion (B/E) spectra) of the authentic cyclic dimer and cyclic trimer.[17] Furthermore, the B/E mass spectra revealed that the lower cyclic oligomers of Ny6 (monomer, dimer, and trimer) are also generated by ion fragmentation processes occurring in the cyclic trimer and tetramer. Therefore, it is not possible to estimate the real distribution of cyclic oligomers in crude Ny6 by FAB-MS analysis.[17]

When Nylon 6 was carefully purified, removing the low molar mass compunds, no significant peaks were observed in the FAB mass spectra of the polymer.[15]

Similar work was performed on several polyesters (polycaprolactone, polyethyleneadipate, polybutylene isophthalate) and on Nylon 6,6, allowing the identification of the mixtures of cyclic and linear low mass oligomers contained in all these polymers.[15,17]

In other investigations, several cyclic sulfides were found to be present in the corresponding sulfur-containing polymers, and FAB-MS allowed the identification of all the oligomers resolved in the SEC traces.[23, 24]

An FAB-MS study on several aromatic, aliphatic, and aliphatic-aromatic polysulfides doped with heavy metal salts allowed the detection of adducts of the cyclic sulfides with heavy metal ions Ag^+, Hg^+, and Cu^+.[24] The molecular ions of the cyclic sulfides and those corresponding to the metal adducts appeared in the spectra with widely differing intensities, providing a tool to investigate the selectivity of metals toward macrocyclic sulfides of different structure and size.[24]

The positive ion FAB mass spectra of the mixtures of cyclic sulfides extracted from poly(hexamethylene sulfide), and that of the same sample doped with $AgNO_3$ or CuCl are reported in Figure 7.2. Only protonated molecular ions are present in the FAB spectrum of the undoped mixture

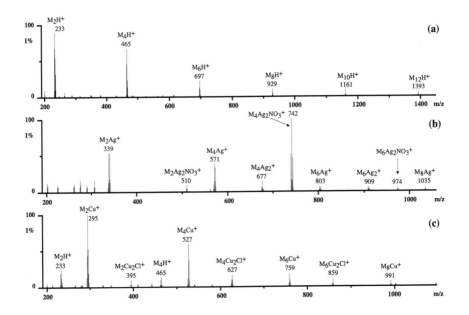

FIGURE 7.2
FAB mass spectra of the cyclic oligomers extracted from poly(examethylene sulfide) (PHMS): (a) crude PHMS, (b) PHMS doped with $AgNO_3$, and (c) PHMS doped with CuCl. (Reprinted with permission from Ref. 24, Copyright 1988, American Chemical Society.)

(Figure 7.2a). These peaks have been replaced by peaks due to cyclics containing Ag^+, Ag_2^+, or $Ag_2NO_3^+$ in the FAB mass spectrum of the cyclic oligomers doped with $AgNO_3$ (Figure 7.2b). Furthermore, the intensity of the peak at m/z 742 ($M_4Ag_2NO_3^+$) is much higher than that at m/z 510 ($M_2Ag_2NO_3^+$), indicating that the formation of complexes with the Ag salt in these cyclic sulfides is a size-dependent process. In fact, the relative abundance of dimer (M_2H^+) and tetramer (M_4H^+) are reversed in the FAB mass spectrum of the pure mixture (Figure 7.2a).[24]

The FAB mass spectrum of the cyclic sulfide doped with CuCl (Figure 7.2c), shows the peaks due to the M_nCu^+ and $M_nC_2Cl^+$ species, together with the protonated molecular ions (MH^+) of the cyclic dimer and trimer. In this case the relative intensity of the M_2Cu^+ and M_4Cu^+ is similar to that of the M_2H^+ and M_4H^+ in the FAB mass spectrum of the undoped mixture (Figure 7.2a), indicating a lack of selectivity of copper salt toward the cyclic sulfide dimer and trimer linked with the Cu^+ ion.[24] Similar results have been obtained for the mixture of cyclic oligomers extracted from poly(hexamethylene-co-*m*-phenylene sulfides).[24]

FAB-MS analysis also has been used to characterize low mass commercial polyethylene- and polypropylene-glycols, by detecting the oligomer ion abundances in the mass spectra of the samples.[20] However, the relative peak intensities of the lower mass oligomers are distorted by the fragmentation processes of the oligomers at higher molar masses. Therefore it is not possible to estimate molar mass distributions by FAB-MS.[19, 20]

The complex fragmentation pathways leading to the ions observed in the FAB spectra of polyglycols were also studied by the MS/MS technique.[20–22,25] In these studies it was observed that [M+Li]$^+$ adducts provide a number of important advantages as precursors for the MS/MS study of polyglycols as compared to the use of [M+H]$^+$ or [M+Na]$^+$.[21,25] In fact, [M+Li]$^+$ ions are generally much more intense than the corresponding [M+H]$^+$ adducts. Also, [M+Li]$^+$ product ions are formed over the entire mass range for low molar mass polyglycols.

[M+Li]$^+$ ions yield excellent MS/MS spectra for polyglycols because the lithium cation is more tightly bound to oxygen in organic compounds than are the other alkali metals. Furthermore, for sodiated ions, the metal ion (Na^+) is released preferentially upon collisional activation, leaving the organic molecule as a neutral fragment. For lithiated ions, the metal has a greater tendency to stay attached to the organic molecule. This allows cleavages of the organic portion of the molecule to proceed more readily, leading to lithiated organic product ions.[21,25] Using [M+Li]$^+$ both charge-site initiated and charge-remote decomposition reactions were observed; while using [M+H]$^+$ a charge-site initiated decomposition occurred. This behavior is useful for the chemical structure elucidation (i.e., monomer sequence and end-group characterization) of unknown polyglycols.[21,25]

Also studied by FAB-MS/MS were the proton- and deuteron-attachment ions in several ethylene glycol polymers (PEG).[22] Considerable H/D exchange

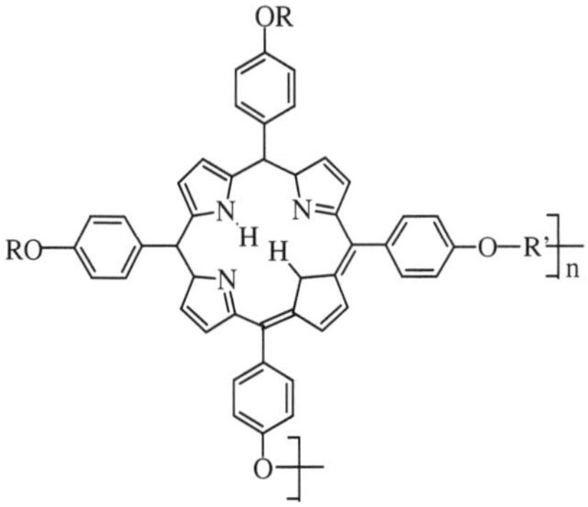

(R: -CH$_3$;   R': -CH$_2$-);   Polymer I

(R: -CH$_3$;   R': -(CH$_2$)$_6$-);   Polymer II

(R: -CH$_3$;   R': -(CH$_2$)$_8$-);   Polymer III

(R: -CH$_2$-CH$_3$;   R': -(CH$_2$)$_6$-);   Polymer IV

SCHEME 7.1

was observed, which occurs during the formation of hydroxyl-terminated carbonium product ions from [M+D]$^+$ precursors. These carbonium ions can be formed from either cyclic or linear PEG [M+D]$^+$ precursors, and it was proposed that reversible H/D exchange occurs between the deuterated hydroxyl end-group and the carbonium center (CH$_2^+$) of this intermediate. The decomposition of this intermediate via backbiting leads to elimination of one or more ethylene oxide units, with some loss of deuterium.[22] Fragmentation processes induced by the fast atom beam were also observed in the FAB-MS analysis of two poly(ethylene imine) samples.[26]

The structure of polyethers containing mesotetraarylporphyrin units in the backbone of the polymer chain has been investigated by FAB-MS. These polyethers contain linear aliphatic linkages of different lengths (methylene, hexamethylene, or octamethylene) between the porphyrin units (Scheme 7.1).[27]

The SEC curves of these polyethers show low molar masses, calculated by using polystyrene standards.

The FAB mass spectrum of the oligomers present in the polyether I (Scheme 7.1) shows essentially three clusters of peaks due to cyclic (species C; Figure 7.3a) and open-chain oligomers (species A and B; Figure 7.3a).

The peaks at *m/z* 1437 and 2155 correspond to the protonated molecular ions of cyclic dimer and trimer, whereas the peak at *m/z* 719 is a fragment ion formed from oligomers having higher molar mass, as confirmed by the product ion mass spectrum (B/E) of the cyclic dimer (*m/z* 1437). The formation of the monomeric cyclic structure (*m/z* 719) is sterically hindered. The other peaks marked in the spectrum correspond to open-chain oligomers having hydroxy end-groups.[27]

The positive ion FAB mass spectrum of polyether II (Figure 7.3b, Scheme 7.1) reveals, in addition to the above features, that many peaks correspond to unexpected oligomers having one (peaks at *m/z* 762 and 1550; 863 and 1551; 881 and 1669) or two butyl end-groups (peaks at *m/z* 819 and 1607). These products were formed by reaction between the butyl group of the tetrabutylammoniumbromide (TBAB), used as a phase transfer agent in the synthesis, and the porphyrin hydroxy groups. In fact the oligomers containing butyl groups disappear in the FAB spectrum of the same polymer prepared using 18-crown-6 ether (instead of TBAB) as a phase transfer agent.[27]

For polyethers III and IV (Scheme 7.1), inspection of the FAB-mass spectra (Figures 7.3c,d) reveals that cyclic oligomers having an aliphatic ether connection between two porphyrin units were also formed (species G).[27]

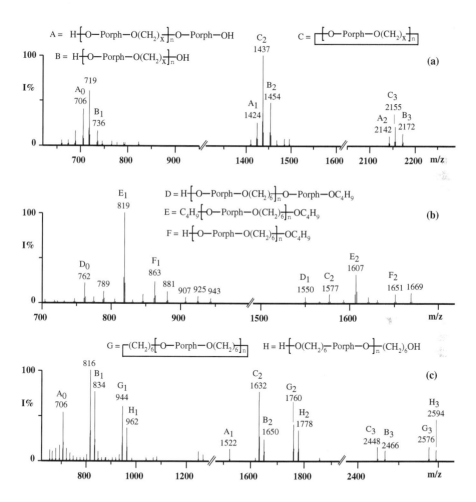

FIGURE 7.3

Positive ions FAB mass spectra of porphyrin-polyethers I-IV (Scheme 1): **(a)** polyether I, **(b)** polyther II, **(c)** polyether III, **(d)** polyether IV, and **(e)** polyether IV reacted with cobalt(II) acetate. For the oligomers having the A, B, and C structures, X = 1, 6, 8, 6 indicate the methylene chain in polymer I, II, III, and IV, respectively (Scheme 1). (Reprinted with permission from Ref. 27, Copyright 1992, American Chemical Society.)

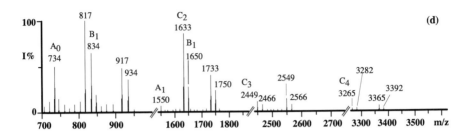

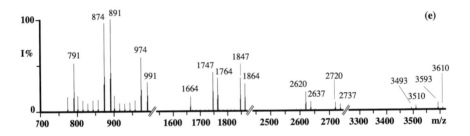

FIGURE 7.3
(Continued.)

The formation of metaloporphyrin derivatives could also be ascertained for these polyethers.[27,28] In fact, the positive ion FAB spectra of these polyethers, treated with the acetate of bivalent transition metals (Co, Mn, Cu, Ni, Fe, and Zn), show the presence of the peaks corresponding to compounds in which the two NH hydrogen atoms of each porphyrin ring have been replaced by one metal atom. Figure 7.3e shows the effect of cobalt on the spectrum of copolyether IV.[27,28] A similar behavior was observed in copolyethers containing different amounts of porphyrin and bisphenol-A units.[28]

The FAB-MS technique has been used to identify oligomers with different end-groups obtained either by carbonate interchange reaction of bisphenol-A with dimethyl carbonate or by partial methanolysis of poly(bisphenol-A carbonate) (PC).[29] The molecular peaks due to these oligomers appear as lithiated ions (M+7) in the FAB spectra, since they were recorded using 3-nitrobenzyl alcohol as matrix doped with LiCl. The FAB-mass spectra show that two series of oligomers (A and B) are present in the case of carbonate interchange between bisphenol-A with dimethyl carbonate (Figure 7.4a), whereas three species of oligomers (A, B, and C) were obtained in the methanolysis of a PC (Figure 7.4b). Oligomers with two hydroxy end-groups (species A), and with one hydroxy and one methyl carbonate at the chain ends (species B), were detected in both cases. A third series of oligomers with two methyl carbonate end-groups (species C), was obtained only by methanolysis of PC. This interesting difference between the two sets of experiments reflects the different mechanisms operating in the two synthetic methods. Methanolysis, which involves a random scission of the carbonate linkages,

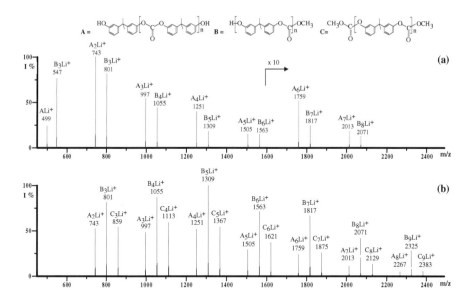

FIGURE 7.4
Positive ions FAB-MS spectra of poly(bisphenol-A carbonate) oligomers obtained: by **(a)** synthesis and **(b)** by methanolysis of high molar mass poly(bisphenol-A carbonate). (Reprinted with permission from Ref. 29.)

produces all the three possible series of oligomers. The complete absence of oligomers bearing two methyl carbonate end-groups (series C) in the carbonate-carbonate exchange raction indicates the high reactivity of methyl carbonate groups, which prevents their accumulation among the reaction products.[29]

FAB-MS has been applied also to the characterization of the oligomers present in copolymers, in order to obtain detailed information on the comonomer sequence.[30,31] For instance, an exactly alternating copolyester (PET/PETx) containing ethylene terephthalate (PET) and ethylene truxillate (PETx) units

was analyzed in comparison with an equimolar random copolyester corresponding to the same structural formula, and the differences observed in their FAB spectra were used to elucidate the microstructure of the two copolymers.[30,31] In the FAB spectrum of the alternating copolymer were detected open-chain oligomers containing ethylene terephthalate (A) and ethylene truxillate (B) comonomer units in 1:1 ratio, together with oligomers containing a number of A units differing by only one from the number of

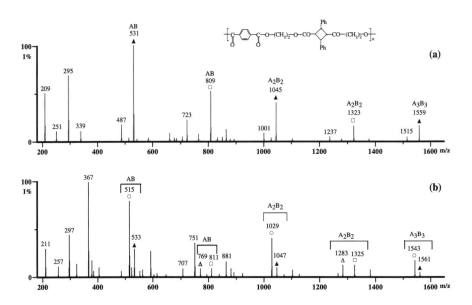

FIGURE 7.5

Negative **(a)** and positive **(b)** ions FAB mass spectra of alternating PET/PETx copolyester. (Reprinted with permission from Ref. 31, Copyright 1989, American Chemical Society.)

B units. Cyclic oligomers were also present, and these had an even number of both comonomer A and B.[30]

Figure 7.5 reports both negative and positive ions FAB mass spectra of this alternating copolyester. In the negative ion FAB mass spectrum (Figure 7.5a) are observed only the linear oligomers bearing both truxillic acid (species □) and either hydroxy and truxillic acid end-groups (species ▲). Their intensity decreases in the positive ion FAB mass spectrum (Figure 7.5b), in which the cyclic oligomers (species ○) give the most intense peaks. Linear oligomers with two hydroxy end-groups (species △) are also present in Figure 7.5b, although with low intensity.[31]

For the random PET/PETx copolyester, the FAB spectrum (Figure 7.6b) reveals peaks corresponding to oligomers containing the comonomers A and B in different ratios, as expected for a random copolymer. Oligomers containing only ethylene terephthalate units (A) and oligomers containing only ethylene truxillate units (B) were also detected. These results (Figure 7.6a,b) show the different sequence distributions in the two isomeric copolyesters.[31] The relative abundance of oligomers containing different amounts of terephthalic (A) and truxillic (B) units is subject to change as a function of the copolyester composition, and this fact is observed in their FAB mass spectra.[31]

Telechelic prepolymers having epoxide end-groups as well as telechelics containing amino end-groups and cyclooligomers were found in the FAB mass spectra of epoxy-amine addition polymers.[32,33] An FAB-MS study characterized in detail the oligomeric products formed in the synthesis of Novalak-type resin.[34]

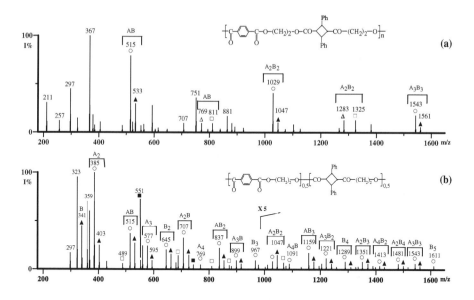

FIGURE 7.6

Positive ions FAB mass spectra of **(a)** alternating PET/PETx copolyester and **(b)** random PET/PETx (50.50 mol/mol) copolyester. (Reprinted with permission from Ref. 31, Copyright 1989 American Chemical Society.)

Application of FAB-MS to the identification and quantification of additives contained in polymer samples has been reported. In fact, the amount of phthalate contained in plasticized PVC could be determined by analyzing directly on the FAB probe the solid pieces of PVC suspended in the glycerol liquid matrix.[35]

7.3 Characterization of Cyclic Oligomers Formed in Polymerization Reactions

As mentioned above, the production of synthetic polymers in polycondensation and ring-opening reactions is often accompanied by the formation of sizeable amounts of cyclic oligomers.[36–48] Cyclic oligomers of different sizes may be produced in the polymerization or in a variety of degradation processes, and it is of great interest to determine their relative abundances (distributions), which are related to the mechanism of formation of the cyclic species.[36–39]

The molar mass values resulting from step-growth or ring-opening polymerization processes depend solely on the extent of reaction (p) and, since they follow a Flory distribution at any p, low mass oligomers should be absent for values of p close to unity.

If low molar mass oligomers are present at the end of these polymerizations, they might originate: (i) from an intramolecular cyclization reaction competing with the polymerization process; (ii) from a depolymerization reaction involving the linear polymer that is formed.

Depolymerization of a high molar mass polymer to cyclic oligomers may also be induced by certain catalysts, or it may occur during thermal degradation.[36]

Macrocyclization equilibria are able to describe the behavior of linear polymer chains when they are allowed to reach thermodynamic equilibrium and generate cyclic oligomers.[36–44]

According to Jacobson and Stockmayer (JS),[44] the cyclization probability is related to the mean separation of the reaction sites, and the equilibrium concentration of each cyclic oligomer is predicted to decrease according to Equation 7.1:

$$C_n = A\, n^{-2.5} \tag{7.1}$$

where C_n is the concentration of the given cycle with n repeating units; n is the ring size expressed in the number of repeat units present and A is a constant. Systems that depend on ring-chain equilibrium have been identified, and the theory has been verified in detail.[38]

Sometimes the distribution of cyclic oligomers may deviate from the thermodynamic equilibrium because of kinetic factors. In fact, the end-groups may be capable of reacting at a higher rate compared to the reaction rate of the functional groups attached to the inner portion of the chain molecule. When this happens, an end-biting process takes place (negating the equal reactivity principle), and a kinetically controlled distribution occurs. The concentration of the cyclic oligomers formed by end-biting is predicted to decrease proportionally to $n^{-1.5}$.[43,44]

Consequently, determining the experimental distributions of cyclic oligomers allows one to distinguish between thermodynamically and kinetically controlled cyclization processes. In the former case the cyclic oligomer concentration should decrease with $n^{-2.5}$, and in the latter with $n^{-1.5}$.[43]

In a comparative study, the distributions of cyclic oligomers formed in the polymerization of cyclic stannoxanes such as 2,2-dibutyl-1,3,2-dioxastannolane with diacylchlorides (Scheme 7.2), were determined by SEC and by FAB.[45]

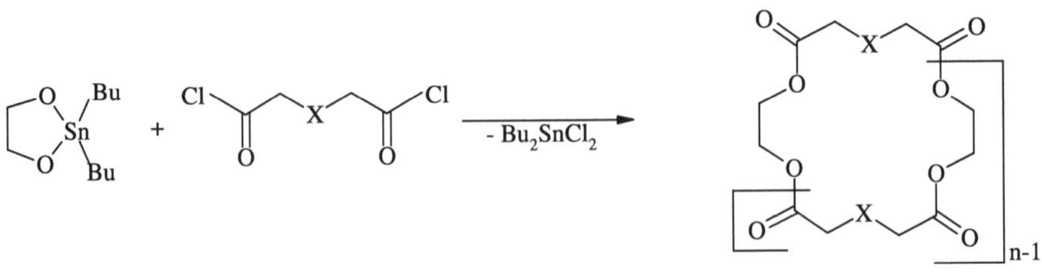

SCHEME 7.2

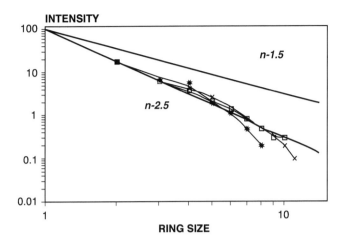

FIGURE 7.7
Experimental cycle concentrations for polymerization data reported in Table 7.1: (∗) Polymer 1;
(□) polymer 2; (x) polymer 3; compared with theoretical distribution laws. (Reprinted with
permission from Ref. 43, Copyright 1991, American Chemical Society.)

The relative abundances of cyclic oligomers formed in the polymerization
of cyclic stannoxanes such as 2,2-dibutyl-1,3,2-dioxastannolane with diacylchlorides, determined by SEC, provided unambigous evidence that the product mixtures are well-behaved eqilibrium distributions, showing that the
experimental data are in essential agreement with the $n^{-2.5}$ law predicted by
the JS theory.[45]

The relative abundances of the cyclic esters formed in the polymerization reactions were obtained on the basis of the "molecular ion" (M^+, M^-)
and/or "quasimolecular ion" (e.g., MH^+, MNa^+, $[M–H]^-$) intensities recorded
in the FAB spectra. When these experimental FAB data were compared with
the distributions calculated according to the JS theory, a very reasonable
agreement is found in all three cases (Figure 7.7, Table 7.1, polymers I-III).[45]
This result suggests that cyclic oligomer distributions generated in polymerization reactions can be quickly estimated, when FAB-MS data are available.

A number of polycondensation systems have been investigated by FAB-
MS. Table 7.1 reports the relative abundances, estimated from FAB, for the
cycles formed in the polycondensation of bis(2-hydroxyethylene)terephtha-
late with truxylloyl chloride.[31] These appear to follow a distribution law of the
type $n^{-1.5}$, indicating a kinetic control in the ring formation.[43]

Table 7.1 also reports the relative FAB abundances for the cycles contained in
polycaprolactone obtained by ring-opening polymerization. Their concentration decreases in proportion to $n^{-2.5}$, following therefore the thermodynamic
equilibration process, a result in agreement with SEC estimates.[43,46]

The same result is obtained analyzing the FAB data for polycaprolactam,
reported in Table 7.1, polymer 6. In this case, however, there are significant

TABLE 7.1

Distribution of Cyclic Oligomers Generated in Polymerization Reactions[a]

	Structure of Cyclic Oligomer	n	m/z[b]	FAB-MS Abundance obsd	FAB-MS Abundance calcd
1	$[COCH_2OCH_2COOCH_2CH_2O]_n$	2	321	17.7[c]	17.7[d]
		3	481	7.1	6.4
		4	641	5.8	3.1
		5	901	1.9	1.8
		6	961	1.1	1.1
		7	1121	0.5	0.8
		8	1281	0.2	0.5
2	$[CO(CH_2)_4COO(CH_2)_2O]_n$	2	345	17.7[c]	17.7[d]
		3	517	6.0	6.4
		4	689	3.7	3.1
		5	851	2.1	1.8
		6	1033	1.4	1.1
		7	1205	0.8	0.8
		8	1377	0.5	0.5
		9	159	0.3	0.4
		10	1721	0.3	0.3
3	$[CO(CH_2)_3COO(CH_2)_2O]_n$	3	475	6.4[c]	6.4[d]
		4	633	4.3	3.1
		5	791	2.8	1.8
		6	949	1.5	1.1
		7	1107	0.9	0.8
		8	1265	0.5	0.5
		9	1423	0.35	0.4
		10	1581	0.2	0.3
		11	1739	0.1	0.25
4	Ph — $[CO{-}\langle\rangle{-}COO(CH_2)_2OCO{-}\langle\rangle{-}COO(CH_2)_2O]_n$ — Ph	2	515	35.3[c]	35.3[e]
		4	1029	17.6	12.5
		6	1543	7.60	6.8
5	$[CO(CH_2)_6O]_n$	3	343	6.4[c]	6.4[d]
		4	457	3.2	3.1
		5	571	2.2	1.8
		6	685	2.6	1.1
6	$[CO(CH_2)_5NH]_n$	1	114	100[c]	100[d]
		2	227	20.0	17.7
		3	340	11.0	6.4
7	$[CO(CH_2)_4CONH(CH_2)_6NH]_n$	1	227	100[c]	100[e]
		2	453	42.0	35.3
		3	679	11.0	12.5
8	$[SCH_2CH_2CH_2]_n$	4	297	3.1[c]	3.1[d]
		6	445	1.2	1.1
		8	593	0.5	0.5
		10	741	0.5	0.3
		12	889	0.25	0.2
		14	1037	0.19	0.13
9	$[S{-}(CH_2)_6]_n$	2	233	35.3[c]	35.3[e]
		4	465	23.0	12.5
		6	697	7.7	6.8
		8	929	3.4	4.4
		10	1161	2.5	3.2
		12	1393	1.5	2.4
10	$[\langle\rangle{-}S]_n$	3	324	6.4[c]	6.4[d]
		4	432	3.1	3.1
		5	540	2.6	1.8
		6	648	0.7	1.1
		7	756	1.4	0.8
		8	864	1.3	0.5
		9	972	0.8	0.4
		10	1080	0.5	0.3

[a] Data from step-growth and ring-opening polymerization are reported here.
[b] FAB-MS data are from Ref. 43.
[c] The relative intensity of the first MS peak has been taken equal to that calculated theoretically for the corresponding cyclic oligomer.
[d] Theoretical values calculated according to the $n^{-2.5}$ distribution law (see text).
[e] Theoretical values calculated according to the $n^{-1.5}$ distribution law (see text).

deviations from equilibrium values, and this fact has been already observed using chromatographic methods.[39,43]

FAB-MS data showing the relative abundances for the cyclic oligomers contained in a solid sample of Nylon 6,6 are reported in Table 7.1 (polymer 7). Their concentration decreases in proportion to $n^{-1.5}$, indicating the presence of kinetic factors in this cyclization process.[43]

The last three examples in Table 7.1 are concerned with the distributions of some cyclic sulfides formed in polycondensation reactions of dithiols with dibromides. The FAB spectra have a sufficient number of peaks, and therefore the distribution laws can be derived with some confidence. Poly(trimethylene sulfide) and poly(m-phenylene sulfide) appear to yield mixtures of cyclic oligomers corresponding to JS equilibrium conditions. However, MS data for poly(hexamethylene sulfide) show a better fit for the $n^{-1.5}$ distribution law, indicating the possible control of an end-biting reaction.[43]

FAB analysis, together with the application of Eq. 7.1, has been also performed in order to detect the distribution of cyclic oligomers formed in the polycondensation of isophthalaldehyde with ethylenediamine.[47] The FAB-MS spectrum of the crude polymer (Figure 7.8) shows only peaks due to cyclic Schiff bases from dimer up to heptamer. This indicates that the reaction proceeds essentially without the concomitant formation of linear oligomers. The relative abundance of the macrocycles formed decreases in proportion to a factor close to $n^{-2.5}$, consistent with a thermodynamic equilibrium.[47]

FAB-MS analysis established the cyclic nature of the aliphatic polyester synthesized by polymer-supported reactions.[48] The cyclic polyester was prepared by intramolecular alkylation of an ω-bromo-carboxylic acid using an anion-exchange resin. This method offers an advantage over classical preparative procedures, because the excess and spent reagent may easily be removed from the final reaction mixtures. This method also makes it possible to perform the reaction on columns, where the reagents can be generated. The FAB analysis of the reaction products showed that the specimens were predominantly cyclic, with little evidence for bromine-terminated species.[48] Cyclic aliphatic and aliphatic-aromatic oligo-esters (such as PET and PBT), obtained

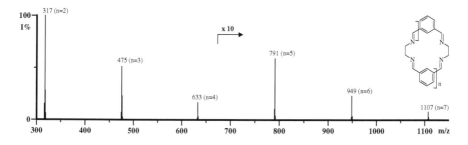

FIGURE 7.8
Positive ions FAB mass spectrum of macrocyclic Schiff bases formed in the condensation of isophthaldehyde with ethylenediamine.

by ring-chain equilibration in dilute solutions or extracted from the crude polymers, were characterized by FAB-MS.[49–51]

Ring-chain equilibrations in solution were carried out in appropriate solvents, using a variety of catalysts, for poly(decamethylene terephthalate) (PDT), poly(ethylene terephthalate) (PET), and poly(butylene terephthalate) (PBT). The reactions were monitored up to equilibrium and the molar cyclization equilibrium constant (K_x) for the individual oligomers was deduced using Flory and Jacobson-Stockmayer theories. The K_x values were calculated from SEC traces by analyzing peak areas of individual resolved cyclic species.[48–51]

In the case of polydecamethyleneterephthalate (PDT), the FAB-MS spectra showed that a large portion of cyclic species uncontaminated by linear oligomers was obtained. The FAB spectra show also that no monomeric cycles are formed, and that the products consist of dimer and higher cyclic oligomers.[49]

FAB-MS analysis and liquid chromatography tandem mass spectrometry (LC-MS/MS) of the oligomers obtained from PET, show that two series of cyclic oligomers are formed.[51] The FAB spectrum, together with the peaks corresponding to the cyclic oligomers containing the repeat unit of PET, shows also a series of molecular peaks due to the cyclic oligomers containing one diethylene glycol residue. Cyclic oligomers from trimer up to heptamer of both species were identified as protonated and sodiated ions in the FAB-MS spectrum. The structure of these oligomers was confirmed by LC-MS/MS.[50]

A similar study was carried out for the cyclic oligomers of PBT, obtained both from the extract of a commercial polymer and by solution ring-chain equilibration.[51] Besides the peaks due to the cyclic compounds, corresponding to multiples of the repeat unit (*m/z* 220.2), peaks were observed due to the ion fragmentation of the cyclic trimer of PBT.[51] This fragmentation process was confirmed by the detailed results obtained by negative ion chemical ionization MS/MS of the authentic cyclic trimer of PBT.[52]

Large cyclic ether-esters from oxyethyleneglycol succinate, prepared by dilute solution transesterification, were also characterized by FAB-MS.[53,54] As for the PDT cyclic oligomer, the molar cyclization equilibrium constant K_x was found in agreement with the Jacobson-Stockmayer theory.

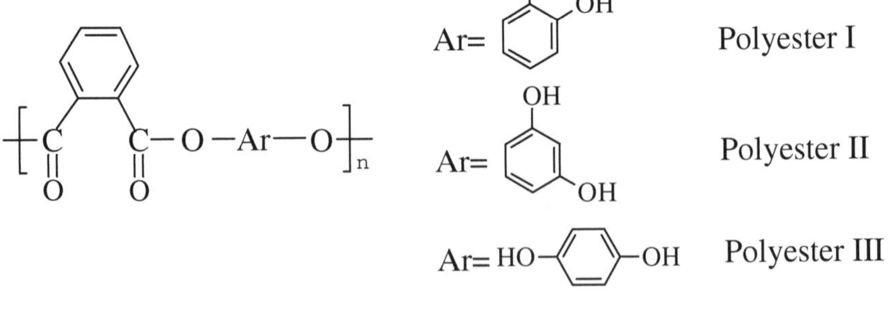

SCHEME 7.3

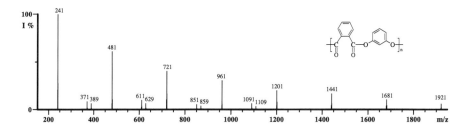

FIGURE 7.9
Positive ions FAB mass spectrum of poly(resorcinol phathalate) (polyester II). (Reprinted with permission from Ref. 55, Copyright 1994, American Chemical Society.)

Cyclic oligomers formed in the synthesis of isomeric polyesters obtained by condensation of phthaloyl chloride and catechol (Polyester I), resorcinol (Polyester II), or hydroquinone (Polyester III) (Scheme 7.3), respectively, were characterized by FAB-MS.[55]

The positive ion FAB mass spectra of these polyesters are very similar, since these polymers differ only in the isomeric structure of the dihydroxybenzene unit present in their repeating unit. Figure 7.9 shows the FAB-mass spectrum of Polyester II, which indicates, essentially, a series of intense peaks at *m/z* 241, 481, 721, 961, 1201, 1441, 1681, and 1921, corresponding to protonated molecular ions of cyclic oligomers (from monomer up to octamer). The other two mass series, which appear with low intensity at *m/z* 371 + n240 and 389 + n240, are due to secondary ion fragmentation phenomena.[55]

Poly(1,2–dihydroxybenzene phthalate) and poly(1,2–dihydroxy-4-methyl-benzene phthalate) were prepared by ring-opening bulk polymerization of spiro compounds I and II without the addition of a catalyst (Scheme 7.4).[56]

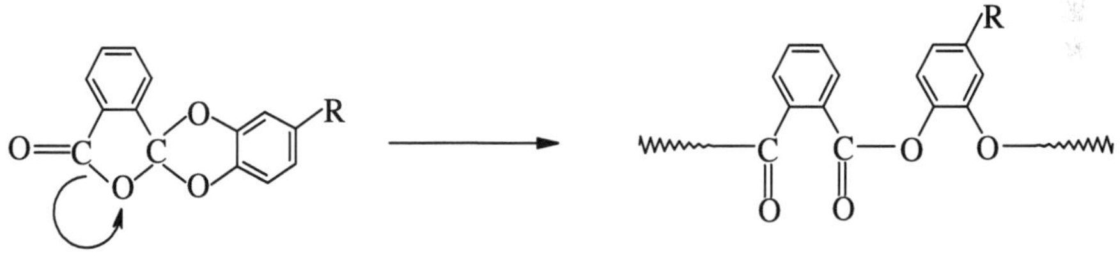

R=H, CH$_3$

SCHEME 7.4

FAB and MALDI mass spectrometric analyses of both polymers reveals that only macromolecules having hydroxy end-groups are formed by this inter-esting synthetic procedure. Figure 7.10 reports the FAB spectrum of the oligomers present in the unfractionated poly(1,2–dihydroxy-4-methylbenzene phthalate). Remarkably, the spectrum shows peaks up to *m/z* values of 6000,

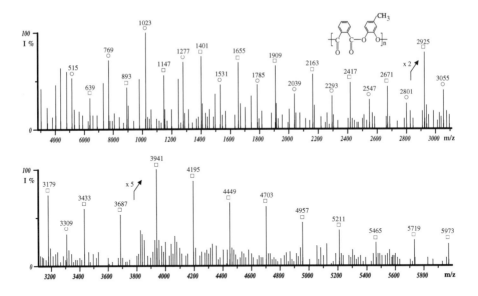

FIGURE 7.10
Positive ions FAB mass spectrum of poly(1,2-dihydroxy-4-methylbenzene phthalate). (Reprinted
with permission from Ref. 56, Copyright 1996, American Chemical Society.)

and is constituted mainly by two families of peaks. The most intense series
consists of peaks at m/z 131 + n254, with n = 2-25, corresponding to linear
oligomers detected as MLi^+ ions, having both dihydroxy-4-methylbenzene
end-groups.

The other series of peaks corresponds to the lithiated ions of the cyclic
oligomers (species $\bigcirc$), due to secondary fragmentation processes which occur
in the FAB-MS source.[56] In fact, the MALDI-TOF spectrum of this polymer
sample shows only one series of peaks due to open-chain macromolecules.[56]

FAB and MALDI-TOF techniques revealed that macrocyclic oligoesters
were the main condensation products of catechol or O,O'-Bistrimethylsilyl
catechol (BTSC) with adipoyl chloride, sebaroyl chloride, and sebacoyl
chloride, respectively.[57] Macrocyclic polyesters were also obtained when the
noncyclic 1,3-bis(tributylstannoyl)butane was used as monomer.[58]

FAB-MS was also used for the characterization of tin-containing macrocyclic oligo- and polylactones prepared by ring expansion polymerization.[59]
Molecular peaks due to these macrocyclics were not detected, because the
alcohols (i.e., 3-nitro-benzyl alcohol) used as matrix cleave the Sn-O bonds
and the FAB-spectra showed only the signals of OH-terminated telechelic
polylactones. However, the macrocyclic polylactone containing the Sn-S bond
yields the corresponding molecular peaks in the FAB spectra. This result,
confirmed also by MALDI-TOF analysis, suggested that Sn compounds containing the more stable Sn-S bond survive the interaction with the matrix
much better than compounds containing Sn-O bonds.[59]

FAB-MS analysis revealed that a carbazole cyclic tetramer was obtained in a high yield in the synthesis of carbazole main chain polymers by the Knoevenagel polycondensation of 3,6-diformyl-9-heptylcarbarzole with 3,6-bis (cyanoacetoxymethyl)-9-heptylcarbazole.[60]

Fast Atom Bombardment-Fourier Transform-Mass Spectrometry (FAB-FT-MS) and low-energy collisional activation experiments proved that the oligomers of β-ketolactone generally have a macrocyclic ring structure rather than a catenane ring.[61]

The FAB-MS technique has been used for quantitative studies of complex formation between alkali metal cations and macrocyclic ligands such as 18-crown 6 and cyclogentiotetraose peracetate. A method to calculate the stability constant from the FAB spectra was proposed, and the stability constants obtained were in good agreement with those calculated by calorimetric techniques.[62]

7.4 Sequence of Copolymers. Partial Degradation of Polymers Coupled with FAB-MS Analysis

The characterization of sequence arrangement in multicomponent condensation polymers having large comonomer subunits sometimes cannot be easily achieved by current NMR methods, that otherwise have proven to be of general utility in the case of vinyl, olefin, and diene copolymers.[63]

Fast atom bombardment mass spectrometry (FAB-MS), being able to detect the masses of the individual molecules in a mixture of homologues, is particularly suitable for the detection of a series of oligomers in copolymer samples. The assumption on which the sequential analysis of a copolymer by mass spectrometry is based is that ions detected in FAB-mass spectra originate from species already present in the polymeric sample. Furthermore if FAB ionization produces the fragmentation of macromolecules, it is assumed that this is independent of chain length.[18,64–66]

MS peak intensities reflect the relative abundance of the oligomers present in the spectrum and are directly related to the composition and sequence distribution of copolymers.

FAB spectra do not yield direct structural information on higher macromolecules, but this can be deduced from the structural analysis of relatively low molar mass oligomers, since the sequence present in the oligomers reflects the comonomer sequence present in the copolymer.[30,31,64]

Chain statistics modeling of the mass spectral intensities of copolymers has been used to derive information on the distribution of monomers along the copolymer chain (Chapter 2), and an automated procedure to find the composition and the sequence of copolymers has been developed.

Chain statistics (Bernoullian, first- or second-order Markoffian), allows one to generate any arrangement of comonomer units along the chain.

Starting from any sequence, a theoretical mass spectrum can be generated, based on the assignment of each spectroscopic peak to a set of sequential arrangements of monomers. The intensity of each peak is related to the relative abundances of the sequential arrangements present in the copolymer, and their relative intensities reflect the comonomer distribution.[39,40]

This means that one has the possibility to build a theoretical mass spectrum for any given copolymer sequence, and that this can be compared with the experimental mass spectrum corresponding to the copolymer sample being investigated. By this modeling it is also possible to distinguish a pure copolymer from a binary mixture of copolymers.[66,67]

Because of the relatively low mass range limit of FAB, the high mass copolymer chains may not be seen in FAB spectra. This problem can be solved by partial degradation of the copolymer before FAB analysis. Several thermal and chemical degradation processes, which produce compounds of low molecular mass without altering the original sequence of the monomer units along the chains (such as hydrolysis, methanolysis, ammonolysis, ozonolysis, photolysis, and pyrolysis), have been used for the partial degradation of many copolymers before FAB analysis.

7.4.1 Partial Photolysis and Partial Hydrolysis

Photolytic degradation was used for the partial degradation of two copolyamides containing truxillic units (Scheme 7.5) that are highly photodegradable materials, since the cyclobutane ring may be easily cleaved by UV irradiation.[68] By increasing the exposure time, a marked lowering of the copolymer molar mass was obtained.

The photolytic cleavage of the cyclobutane rings generates oligomers containing cinnamoyl end-groups. The lower molar mass fraction of the irradiated sample (up to about 2000 Daltons) can be detected by direct FAB-MS analysis.[68] The inspection of the FAB mass spectrum in Figure 7.11a reveals that in the case of an equimolar adipic/truxillic copolyamide, oligomers containing both adipic and truxillic units and having cinnamoyl end-groups are formed in the photolysis.

The random structure of the copolymer was deduced from the relative abundance of the two comonomeric units in the photolyzed compounds.[68]

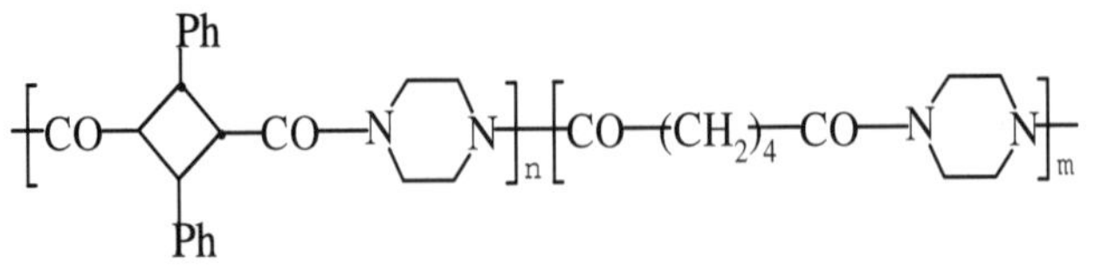

I: n = 0.8, m = 0.2
II: n = 0.5, m = 0.5

SCHEME 7.5

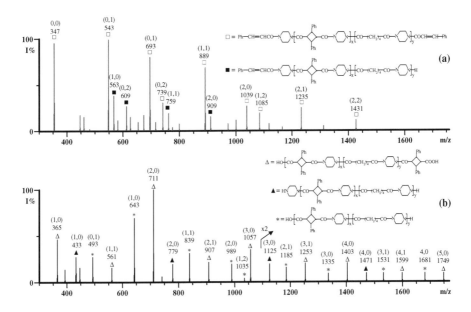

FIGURE 7.11

Positive ions FAB mass spectra of oligomers generated by **(a)** photolysis and **(b)** by hydrolysis of random copolyamide containing equimolar truxilloylpiperazine and adipoyl piperazine units. (Reprinted with permission from Ref. 68, Copyright 1989, American Chemical Society.)

Acid-catalyzed hydrolysis was also used to partially degrade these copolyamides and the FAB spectrum (Figure 7.11b) confirmed the random structure of the equimolar adipic/truxillic copolymers. It was also possible to distinguish the different structure of the end-groups formed in the two partial degradation experiments. In fact, it can be observed that the hydrolysis reaction produces oligomers with the piperazine, truxillic and/or adipic end-groups, whereas the oligomers with cinnamoyl end-groups are absent.[68]

7.4.2 Partial Ammonolysis

Polycarbonates are known to undergo ammonolysis quite easily at room temperature.

FAB-MS has been used to identify the low molar mass oligomers produced in the partial ammonolysis of an essentially alternating copolycarbonate containing resorcinol and bisphenol-A (BPA) units.[69]

The positive ion FAB mass spectrum of the crude mixture obtained by ammonolysis shows protonated molecular ions (MH+) that can be unequivocally assigned to sequences with an alternating structure. However, several peaks present in the FAB spectrum do correspond to sequences containing blocks of two, three, or four consecutive units of BPA. This indicates that a

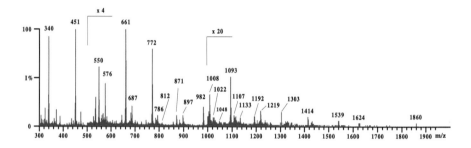

SCHEME 7.6

FIGURE 7.12
Positive ions FAB mass spectrum of the aminolyzed residue of poly(bisphenol-A carbonate) sample heated at 400°C for 1 hour. (Reprinted with permission from Ref. 70, Copyright 1999, American Chemical Society.)

sizeable portion of bisphenol-A-bischloroformate, used in the interfacial synthesis of the copolymer, underwent hydrolysis during the course of the polymerization reaction, yielding BPA and/or BPA chloroformate, which reacted further, producing blocks of bisphenol-A carbonate units that were included in the copolymer chains.[69] Thus, FAB analysis revealed the presence of minor side-reaction products in this copolymer.

In another study, ammonolysis of the main-chain carbonate groups (Scheme 7.6), followed by FAB-MS analysis, was performed to characterize the pyrolysis residue of poly(bisphenol-A carbonate) (PC).[70]

The FAB-mass spectra of the aminolyzed residue of the PC sample heated at 400°C (Figure 7.12), showed the presence of PC oligomers containing several consecutive xanthone and ether units (Table 7.2).

The mass spectra indicate that the isomerization and condensation processes leading to these structures are quite extensive during the thermal degradation of PC.[70]

TABLE 7.2
Molecular Ions Detected by FAB-MS Analysis of the Aminolyzed Residue Obtained After Heating PC at 400°C for 1 h

Structures / X	N–CO–O–X–OH MH⁺ (+1 Pip, +2 Pip, +3 Pip, +4 Pip)[a]	N–CO–O–X–O–CO–N MH⁺ (+1 Pip, +2 Pip, +3 Pip)[a]
[–O–CO–O–X–]n with X = –C(CH₃)₂– diphenyl (BPA carbonate)	340	n=0: 451, 536 n=1: 790
X = –C(CH₃)₂– diphenyl ether (BPA ether)	550	661
Xanthone-containing polymer	n=1: 576 (661) n=2: 812 (897) (982) n=3: 1048 (1133) (1218) (1303) n=4: – – (1539) (1624)	n=1: 687 (772) n=2: – (1008) (1093) n=3: – – (1414)
Xanthone-containing polymer (ether)	n=1: 786 (871) n=2: 1022 (1107) (1192)	n=1: 897 (982) n=2: 1133 (1218) (1303) n=3: – – (1540) (1624) n=4: – – – (1860)

[a] The protonated molecular ions are desorbed as adducts with piperidine. In presence of xanthone units one, two, three, and four molecules of piperidine are added, depending on the number of these units.

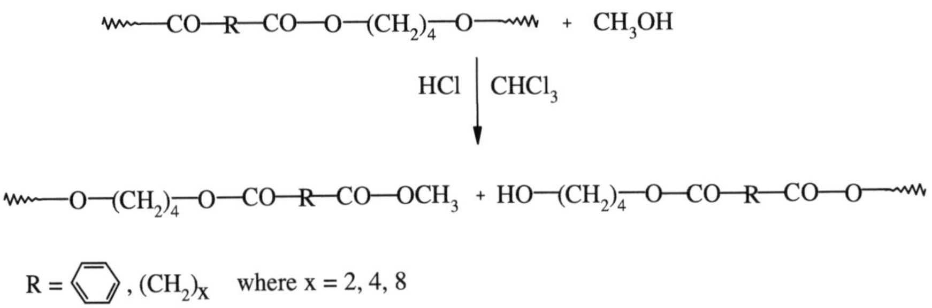

$$\text{ww——CO—R—CO—O—(CH}_2)_{\overline{4}}\text{—O——ww} \ + \ \text{CH}_3\text{OH}$$

$$\text{HCl} \ \Big| \ \text{CHCl}_3$$

$$\text{ww——O—(CH}_2)_{\overline{4}}\text{—O—CO—R—CO—OCH}_3 \ + \ \text{HO—(CH}_2)_{\overline{4}}\text{—O—CO—R—CO—O——ww}$$

$$R = \bigcirc \ , (\text{CH}_2)_x \quad \text{where } x = 2, 4, 8$$

SCHEME 7.7

FAB-MS analysis of the partial ammonolysis products has been also used to characterize the random structure of copolyester containing 20 mol% of ethylenetruxillate and 80 mol% of ethyleneterephthalate units.[71]

7.4.3   Partial Methanolysis and Partial Methoxidation

The composition and sequence distribution of equimolar multicomponent copolyesters, obtained from 1,4-butandiol (B) and mixtures of succinic (Su), adipic (A), sebacic (Se), and terephthalic acid (T), were determined by analysis of the FAB-MS spectra of the oligomers obtained in the controlled methanolysis reaction of these copolymers (Scheme 7.7).[72,73]

Figure 7.13 reports the experimental and calculated FAB mass spectra of the partial methanolyzed random P(BA-co-BT-co-BSe) terpolyester. The majority of peaks can be assigned to lithiated molecular ions of oligomers (dimer up to hexamer) bearing two types of end-groups, namely, HO/.../ OCH₃ (species □) and CH₃O/.../ OCH₃ (species ■) (Figure 7.13a, Table 7.2). The intensity of the whole spectrum was used to compute the composition of the copolymer through a computer program (MACO4), which performs a least-squares minimization in which the experimental intensities of the oligomers are compared with the theoretical mass spectra (Figure 7.13b) generated by chain statistics (Chapter 2).[65,72]

The partial methanolysis was also applied for the characterization of poly (β-hydroxybutyrate) (PHB) and poly(β-hydroxybutyrate-co-β-hydroxyvalerate) P(HB-co-HV), as will be discussed in Section 7.6.

The sequence distribution of four poly(ether-sulfone)/poly(ether-ketone) (PES/PEK) copolymer samples were determined by FAB-MS analysis of their oligomers obtained by controlled partial degradation with sodium methoxide in dimethyl sulfoxide solution.[74] As for the aliphatic copolyesters discussed above, the sequential arrangements of ether-sulfone/ether-ketone units in these copolymers were estimated by a best-fit minimization method using the MACO4 algorithm.[65]

The methoxidation occurs selectively at the ether bonds and the negative ion FAB spectra of the methoxidated PES/PEK copolymer showed peaks

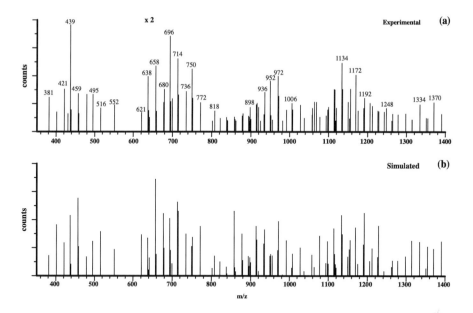

FIGURE 7.13
Mass spectra of the partial methanolyzed PBA/PBT/PBSe terpolyester: **(a)** experimental FAB mass spectrum; **(b)** simulated mass spectrum. (Peak assigments are reported in Table 7.2.) (Reprinted with permission from Ref. 72, Copyright American Chemical Society.)

due to the oligomers with OH and OH/OCH$_3$ end-groups.[74] The FAB-mass spectra of the methoxidation products show peaks only up to trimers due to the difficult desorption of the oligomers from the matrix solution.

Random PES/PEK copolymers yielded oligomers having the sequence arrangements expected from the monomer-to-feed ratios used in the synthesis.

Instead, a PES/PEK copolymer sample expected to be exactly alternating (from the synthesis procedure), showed the presence of 44% random sequences, owing to transetherification rearrangements which occur during the synthesis.[74]

7.4.4 Partial Ozonolysis

It is well-known that the cleavage of olefinic double bonds by ozone attack produces compounds having aldehyde, ketone and/or carboxyl end-groups (Scheme 7.8).

Partial degradation induced by an ozonolysis process was used to produce low molar mass oligomers, and FAB-MS was applied for the identification of the ozonolysis-formed oligomers, from polyisoprene,[75] poly(chloroprene),[75] and butadiene/acrylonitrile and butadiene/styrene copolymer samples.[76,77]

It was ascertained that poly(chloroprene) is degraded by the ozone more rapidly than poly(isoprene) and poly(butadiene).[75]

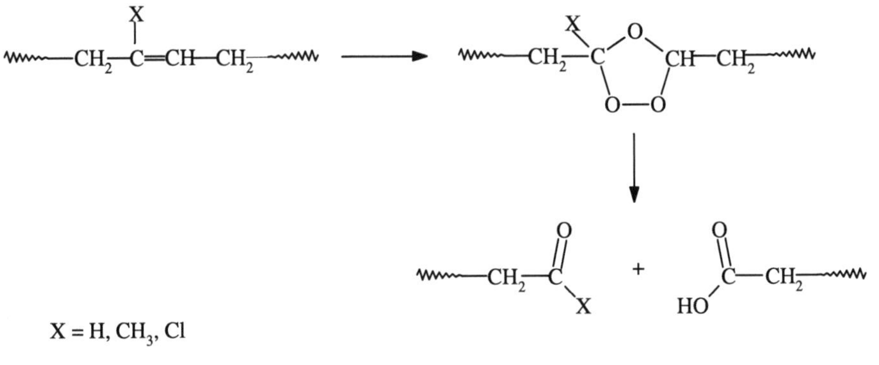

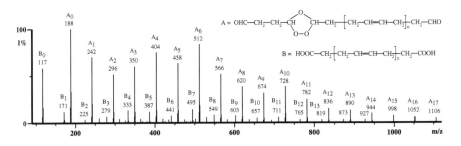

X = H, CH$_3$, Cl

SCHEME 7.8

FIGURE 7.14
Negative ions FAB mass spectrum of poly(butadiene) ozonolysis products. (Reprinted with permission from Ref. 76, Copyright 1999, American Chemical Society.)

According to Scheme 7.8, the negative ion FAB-mass spectra of the ozonized poly(isoprene) and poly(chloroprene) consist mainly of a family of peaks corresponding to molecular ions (M–H$^-$) of compounds having carboxyl and ketone end-groups.[75]

In the case of poly(chloroprene), the ozonized mixtures were treated with piperidine before the FAB-MS analysis to transform the reactive acyl chloride groups, formed by cleavage of the double bounds along the main chain,[75] to stable amide end-groups.

The negative ion FAB mass spectra of the oligomers formed by partial ozonolysis of poly(butadiene) (Figure 7.14) show that two families of peaks corresponding to the oligomers having aldehyde (species A) or carboxyl end-groups (species B). The structure of compounds having aldehyde end-groups also present an ozonide unit in the molecule. Oligomers containing the same unit have been found among the ozonolysis products of random styrene/butadiene (St/But) and butadiene/acrylonitrile copolymers. By FAB-MS analysis it has also been possible to compute the molar composition of these copolymers.[76–77]

Furthermore, it was found that the content of butadiene units in the ozonolyzed oligomers decreases with ozonolyis time, since they are destroyed by the ozone attack. It is therefore possible to isolate the blocks of styrene units present in the copolymers, and to estimate their chain length by FAB analysis.[76]

7.5 Sequence of Copolymers from Reactive Polymer Blends

Blends of condensation polymers containing functional groups internally or at chain ends (such as polycarbonates, polyesters, polyamides, and the like) may undergo intermolecular exchange reactions when mixed in the molten state.[78–80]

The exchange reactions may be induced by the presence of catalysts used in the homopolymer synthesis or added to the blends, and/or caused by reactive terminal groups (i.e., OH, COOH, NH_2). Block, random, or graft copolymers, which may affect the blend properties, may be formed in various amounts as a function of the mixing time.[78,80]

The sequence distribution and composition of the copolymers generated by melt mixing of blends—such as poly(ethylene terephthalate)/poly(ethylene adipate) (PET/PEA) or poly(ethylene terephthalate)/poly(ethylene truxillate) (PET/PETx)—were determined by analysis of the FAB mass spectra of the oligomers present in the crude blends or else formed by appropriate partial degradation (hydrolysis or aminolysis) of the mixtures.[78,79]

Figure 7.15(a–c) shows the positive ion FAB spectra of the oligomers present in the crude PEA/PET blend melt-mixed at 290°C for increasing times (0, 20, and 270 min., respectively).[78]

The FAB spectrum of the initial mixture (M0, Figure 7.15a) is quite similar to that of pure PEA and does not show peaks corresponding to PET oligomers. In fact, the PET sample used for the blend had been completely freed from all the low molar mass oligomers before mixing it with the PEA.[78]

The mass spectra of the samples obtained after 20 min. (M20, Figure 7.15b) and after 270 min. mixing (M270, Figure 7.15c) show a series of peaks corresponding to protonated molar ions of cyclic homo- and co-oligomers. Peaks due to open-chain oligomers, not observed in the positive ion FAB spectra, were detected in the negative ion FAB mode.[78]

The MS peak intensities of the PET/PEA samples reacted from 10 up to 270 min. were used to estimate the copolymer composition, the extent of exchange and the number average block lengths by the chain statistics modeling of the mass spectra of copolymers (Chapter 2).[78]

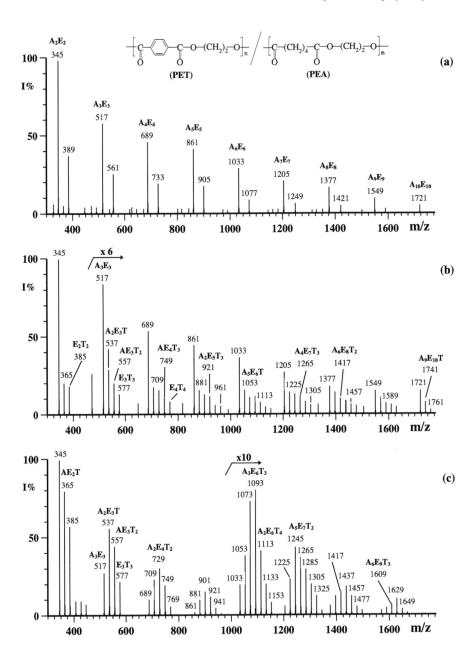

FIGURE 7.15
Positive ions FAB mass spectra of an equimolar PET/PEA blend reacted at 290°C for: **a)** 0 min; **b)** 20 min; and **c)** 270 min. The symbols: **A, E,** and **T** indicate the adipic, ethylene, and terephthalic units, respectively. (Reprinted with permission from Ref. 78, Copyright 1992, American Chemical Society.)

7.6 Sequence of Copolymers of Microbial Origin: Polyhydroxyalkanoates

Poly(3-hydroxyalkanoates) (PHAs) are optically active polyesters that are produced as an intracellular energy source and carbon storage materials by a wide variety of bacteria, and have great importance as biocompatible and biodegradable thermoplastic materials.[81–100]

These biopolymers, which are produced in the form of small granules within the cell, have the general structure:

$$\left[OCHCH_2\overset{\overset{\displaystyle O}{\|}}{C} \right]_n$$

$$\underset{R}{|}$$

in which R is an n-alkyl group varying in size from C_1 to C_{12} and even higher, depending on the bacterium and on the carbon substrate used for the micro-organism growth. In many cases the bacteria produce copolyesters, and the copolymers containing 3-hydroxybutyrate (HB) and 3-hydroxyvalerate (HV) units are often found in nature.

The structures of poly(3-hydroxybutyrate) P(HV), of poly(3-hydroxybutyrate-co-3-hydroxyvalerate) P(HB-co-HV), and of several other PHAs, have been intensively studied.[81–100]

Since the mechanical, physical, and processing properties of the P(HB-co-HV) copolymers vary systematically with composition and presumably with sequence distributions, it is important to have accurate estimates of these parameters.

The sequence distribution of copolymers P(HB-co-HV) with differing composition has been deduced from dyad and triad analysis by 13C-NMR spectrometry, and it has been concluded that some of the bacterially synthesized P(HB-co-HV) samples are random copolymers of HB and HV units, but other materials are mixtures of P(HB-co-HV) random copolymers with different compositions.[84] However, problems emerge in the NMR characterization of the 3-alkylsubstituted copolyesters that are produced when the biosynthesis is carried out by using sodium salts of n-alkanoic acids as carbon sources.[94,97,98,100] In fact, it is not possible to determine by ^{13}C-NMR the sequence distribution of the monomer units present in the polyesters with alkyl chains larger than propyl, because the chemical shifts of the carbon atoms in the higher alkanoates are essentially identical. Furthermore, the patterns of their ^{1}H-NMR spectra are too complex to be used for such determination.[94,97,98,100] One cannot therefore distinguish if the material produced in the bacterial fermentation is a mixture of homopolymers or of copolymers, or a real multicomponent copolymer.

The resolving power of FAB mass spectrometry is adequate to deal with the problem, and large oligomers differing in monomer composition can be easily identified using the FAB-MS technique. Because the FAB-MS analysis is generally limited to lower masses, a partial methanolysis of the high molar mass PHAs is necessary (Scheme 7.9).

FAB-MS was used to identify the oligomers formed in the partial methanolysis, followed by HPLC fractionation, of PHB and P(HB-co-HV) copolymers.[86,87,93,95] From these data it was possible to calculate the copolymer composition and to determine the sequence distribution of comonomer units in these materials.

Figure 7.16 reports the positive ion FAB mass spectrum of the one HPLC fraction of the methanolysis products of the P(HB-co-HV) copolymer

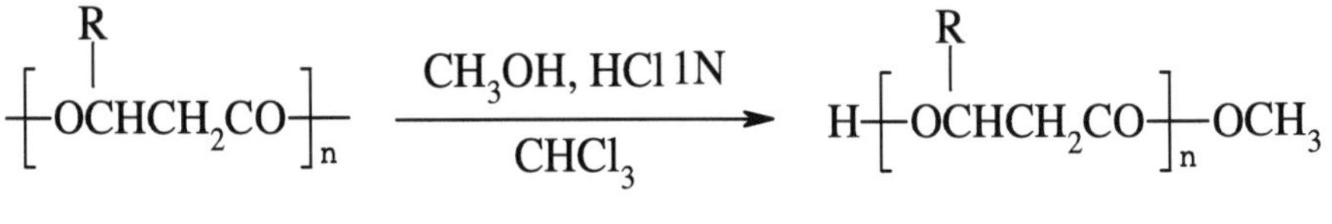

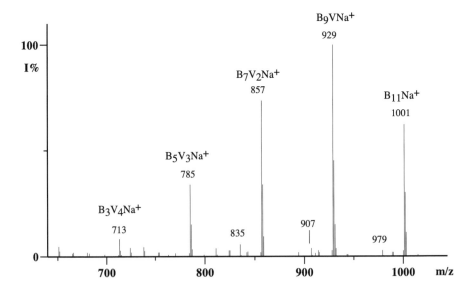

SCHEME 7.9

FIGURE 7.16
Positive ions FAB mass spectrum of a HPLC fraction collected from methanolysis products of the copolyester (PHB-co-15%HV). 3-Nitrobenzyl alcohol doped with NaCl is used as a matrix.

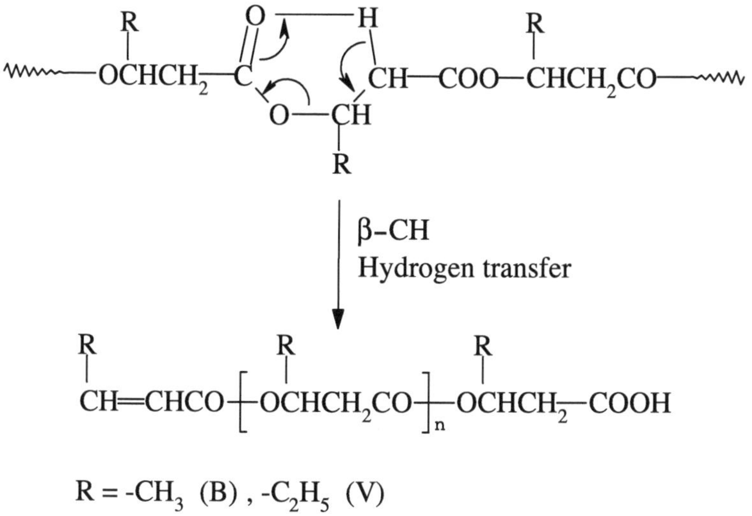

The reaction scheme shows:

$$R = -CH_3 \text{ (B) }, -C_2H_5 \text{ (V)}$$

SCHEME 7.10

(85/15 mol/mol ratio).[86,93,95] The spectra show peaks corresponding to sodiated pseudomolecular ions of BxVy oligomers ($B_xV_yNa^+$).

Partial pyrolysis, as an alternative to a partial methanolysis, has been used for the partial degradation of the P(HB-co-HV) microbial copolyesters.[93,95,96]

The partial pyrolysis of PHB and of P(HB-co-HV) samples was performed in a TG apparatus to monitor the pyrolysis and to stop it at predetermined weight loss levels.[93,95,96] The pyrolysis residue consists of a mixture of oligomers with carboxyl end-groups (Scheme 7.10).

Because these acid compounds can easily lose one proton, the FAB spectra were recorded in the negative ion mode, revealing them as carboxylate anions. The negative ion FAB mass spectrum of the pyrolysate of a P(HB-co-HV) copolymer consists essentially of the $(M-H)^-$ ions corresponding to oligomers, and was used to estimate the copolymer compositions and sequence distribution.[93,95,96] The negative ion FAB mass spectrum of the partial pyrolyzate (PHB-co-15%-HV) copolyester (Figure 7.17) exhibits the $(M_n-H)^-$ ions corresponding to oligomers of the structure shown in Scheme 7.10. Assuming a random distribution of the HB and HV units in the copolymers and using the normalized intensities of the peaks corresponding to dimers, trimers, tetramers, pentamers, and hexamers, a molar composition of 85/15 in HB and HV units was calculated. This is similar to that calculated by the methanolysis-HPLC-FAB-MS method and is close to the value 87/13 calculated by ^{1}H-NMR.

The FAB mass spectra obtained for some (PHB-co-HV) copolymers allow the identification of oligomer series up to hexads. A comparison between the experimental and the calculated peak intensities shows a clear distinction

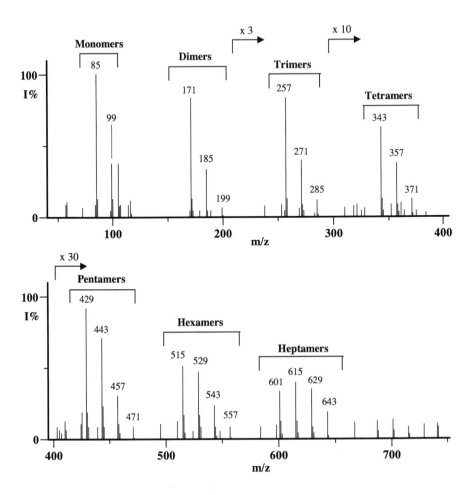

FIGURE 7.17
Negative ions FAB mass spectrum of the TG residue from the partial pyrolysis of P(HB-co-15%HV) copolyester. (Reprinted with permission from Ref. 96, Copyright 1991, American Chemical Society.)

between pure copolymers and mixtures of homo- or copolymers with different composition.[93,96]

The FAB-MS analysis is not limited to any particular type of PHA; it is therefore suitable to sequence the complex multicomponent materials obtained when bacteria are fed with a mixture of long-chain n-alkanoates.[94,97,98,100]

The two analysis methodologies illustrated above (HPLC-FAB and TG-FAB) have been applied to the complex PHA samples, i.e., multicomponent copolymers (containing from 3 to 7 comonomers). Chain statistics calculations showed a random distribution of monomeric units within these polymeric materials, allowing one to calculate their composition.[94,97,98,100]

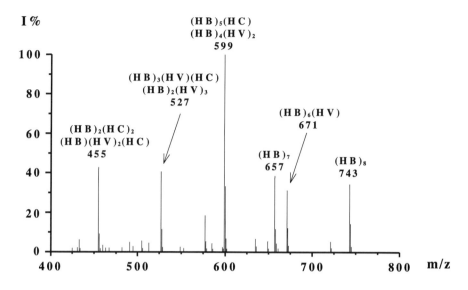

FIGURE 7.18
FAB mass spectrum corresponding to the HPLC fraction 5 of the methanolysis products from microbial terpolyester. The symbols HB, HV, and HC refer to the 3-hydroxybutyrate, 3-hydroxyvalerate, and 3-hydroxyhexanoate units, respectively. (Reprinted with permission from Ref. 100, copyright 1999, Elsevier Science.)

Figure 7.18 reports the FAB mass spectrum of one fraction collected in the HPLC run of the methanolysis products of a microbial terpolyester obtained by fermentation of *Rhodospirillum rubrum,* grown with 3-hydroxyhexanoic acid.[100] The peaks correspond to the protonated (MH^+) and sodiated (MNa^+) ions of the octamers having OH/OCH_3 end-groups.[100]

The negative ion FAB mass spectrum of the partially pyrolyzed copolyester, at 20% weight loss in an isothermal TG run, is shown in Figure 7.19. The spectrum essentially consists of the $(M-H)^-$ pseudomolecular ions corresponding to oligomers having olefinic and carboxylic end-groups formed according to Scheme 7.9. Peaks corresponding to the monomers—3-hydroxybutyrate (HB), 3-hydroxyvalerate (HV), and 3-hydroxyhexanoate (HC), at m/z 85, m/z 99, and m/z 113—were observed together with those from higher oligomers. Statistical analysis, applied to the relative intensities of the peaks present in the FAB mass spectra of the partial pyrolyzate or in the FAB spectra of the HPLC fractions of the partial methanolyzate, give a molar copolymer composition close to 77/8/15 in HB, HV, and HC units, respectively. It also revealed that the sequence distribution in this copolyester follows Bernoullian statistics, indicating that it is a random terpolyester.[100]

These data confirming the high discrimination power of the TG-FAB and HPLC-FAB methods applied to identification of the microbial polyester microstructure, because by NMR analysis it was not possible to establish

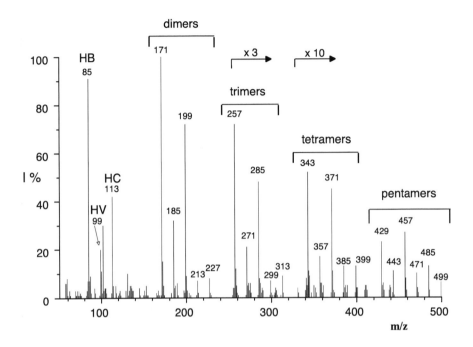

FIGURE 7.19
Negative ions FAB mass spectrum of the TG residue from the partial pyrolysis of the microbial terpolyester. The symbols HB, HV, and HC refer to the 3-hydroxybutyrate, 3-hydroxyvalerate, and 3-hydroxyhexanoate units, respectively. (Reprinted with permission from Ref. 100, copyright 1999, Elsevier Science.)

whether a single terpolymer, a mixture of copolymers, or a mixture of homopolymers had been obtained by microbiological synthesis.

References

1. Barber, M., Bordoli, R.S., Sedgwick, R.D., and Tyler, A.N., Fast Atom Bombardment of Solids (FAB): A New Ion Source for Mass Spectrometry, *J. Chem. Soc. Chem. Comm.* 325, 1981.
2. Barber, M., Bordoli, R.S., Elliot, G.J., Sedgwick, R.D., and Tyler, A.N., Fast Atom Bombardment Mass Spectrometry, *Anal. Chem.*, 54, 645A 1982.
3. Mass Spectroscopy in the Health and Life Science; Proceedings of an International Symposium, San Francisco, California, Sept. 1984, Burlingame, A.L. and Castagnoli, N. Jr., Eds., Elsevier: Amsterdam, 1985.
4. Barber, M., Bordoli, R.S., Elliot, G.J., Sedgwick, R.D., and Tyler, A.N., Fast Atom Bombardment Mass Spectrometry (FABMS)—A Study of Surface Coverage Effects in FABMS, *J. Chem. Soc., Faraday Trans. I*, 79, 1249, 1983.
5. Lehmann, W.D., Kessler, M., and König, W.A., Investigations on Basic Aspects of Fast Atom Bombardment Mass Spectrometry: Matrix Effects, Sample Effects, Sensitivity and Quantification, *Biomed. Mass Spectrom.*, 11, 217, 1984.

6. Desorption Mass Spectrometry, Lyon, P.A., Ed. ACS Symposium Series, American Chemical Society: Washington, DC, 1985.

7. Chapman, J.R., Practical Organic Mass Spectrometry, Wiley: Chichester, UK, 1985.

8. Schulten, H.R. and Lattimer. R.P., Applications of Mass Spectrometry to Polymers, *Mass Spectrom. Rev.* 3, 231, 1984.

9. Cooks, R.G. and Busch, K.L., Matrix Effects, Internal Energies and MS/MS Spectra of Molecular Ions Sputtered from Surfaces, *Int. J. Mass Spectrom. Ion Phys.*, 53, 111, 1983.

10. Wong, S.S., Rollgen, F.W., and Przybylski, M., Evidence for a Surface Self-cleaning Sputtering Mechanism in Fast Atom Bombardment Mass Spectrometry, *Biomed. Mass Spectrom.*, 12, 43 1985.

11. Gower, J.L., Matrix Compounds for Fast Atom Bombardment Mass Spectrometry, *Biomed. Mass Spectrom.*, 12, 191, 1985.

12. Williams, D.H., Bradley, C, Bjesen, G., Santikarn, S., and Taylor, L.C.E., Fast Atom Bombardment Mass Spectrometry: A Powerful Technique for the Study of Polar Molecules, *J. Am. Chem. Soc.* 103, 5700, 1981.

13. Doerr, M., Luderwald, I., and Schulten, H.R., Characterization of Polymers by Field Desorption and Fast Atom Bombardment Mass Spectrometry, *Fresenius' Z. Anal. Chem.*, 318, 339, 1984.

14. Doerr, M., Luderwald, I., and Schulten, H.R., Investigations of Polymers by Field Desorption and Fast Atom Bombardment Mass Spectrometry, *J. Anal. Appl. Pyrol.*, 8, 109, 1985.

15. Ballistreri, A., Garozzo, D., Giuffrida, M., and Montaudo, G., Identification of the Ions Produced by Fast Atom Bombardment in Some Polyesters and Polyamides, *Anal. Chem.*, 59, 2024, 1987.

16. Montaudo, G., Scamporrino, E., and Vitalini, D., Structural Characterization in Synthetic Polymers and Copolymers by Fast Atom Bombardment Mass Spectrometry, in *Applied Polymer Analysis and Characterization*, Mitchell, J. Jr., Ed., Oxford University Press: New York, Vol. II, p. 79, 1992.

17. Ballistreri, A., Garozzo, D., Giuffrida, M., Montaudo, G., Filippi, A., Guaita, C., Manaresi, P., and Pilati, F., Fast Atom Bombardment Mass Spectrometry Identification of Oligomers Contained in Poly(ε-caprolactam) and Poly(butyleneisophthalate), *Macromolecules*, 20, 1029, 1987.

18. Montaudo, G., Microstructure of Co-polymers by Fast-atom Bombardment Mass Spectrometry, *Rapid Comm. Mass Spectrom.*, 5, 95, 1991.

19. Lattimer, R.P., Fast Atom Bombardment Mass Spectrometry of Polyglycols, *Int. J. Mass Spectrom. Ion Processes*, 55, 221, 1984.

20. Lattimer, R.P., Münster, H., and Budizikiewicz, H., Tandem Mass Spectrometry of Polyglycols, *Int. J. Mass Spectrom. Ion Processes*, 90, 119, 1989.

21. Lattimer, R.P., Tandem Mass Spectrometry of Lithium-Attachment Ions from Polyglycols, *J. Am. Soc. Mass Spectrom.*, 3, 225, 1992.

22. Lattimer, R.P., Tandem mass spectrometry of poly(ethylene glycols) proton- and deuteron-attachment ions, *Int. J. Mass Spectrom. Ion Processes*, 116, 23, 1992.

23. Montaudo, G., Puglisi, C., Scamporrino, E., and Vitalini, D., Mass Spectrometry Analysis of the Thermal Degradation Products of Poly(o-,m-, and p-phenylene sulfide) and of the Oligomers Produced in the Synthesis of These Polymers, *Macromolecules*, 19, 2157, 1986.

24. Montaudo, G., Puglisi, C., Scamporrino, E., and Vitalini, D., Fast Atom Bombardment Mass Spectrometry Identification of Oligomers Contained in Polysulfides and Their Complexes with Heavy Metals, *Macromolecules*, 21, 1594, 1988.

25. Lattimer, R.P., Tandem Mass Spectrometry of Poly(ethylene glycol) Lithium-Attachment Ions, *J. Am. Soc. Mass Spectrom.*, 5, 1072, 1994.

26. Lattimer, R.P. and Schulten, H.-R., Field Desorption and Fast Atom Bombardment Mass Spectrometry of Poly(ethylene imine), *Int. J. Mass Spectrom. Ion Processes*, 67, 277, 1985.

27. Scamporrino, E. and Vitalini, D., Main-Chain Porphyrin Polymers. 1. Synthesis and Characterization of Polyethers Containing Porphyrin Units and Their Metal Derivatives, *Macromolecules*, 25, 1625, 1992.

28. Scamporrino, E. and Vitalini, D., Main Chain Porphyrin Polymers. 2. Synthesis and Characterization of Some Copolyethers and Their Metal Derivatives, *Macromolecules*, 25, 6605, 1992.

29. Shaikh, A.G., Sivaram, S., Puglisi, C., Samperi, F., and Montaudo, G., Poly(arylenecarbonate)s oligomers by carbonate interchange reaction of dimethyl carbonate with Bisphenol-A, *Polymer Bulletin*, 32, 427, 1994.

30. Montaudo, G., Scamporrino, E., and Vitalini, D., Synthesis and Structural Characterization by Fast-Atom Bombardment Mass Spectrometry of Exactly Alternating Copolyesters Containing Photolabile Units in the Main Chain, *Polymer*, 30, 297, 1989.

31. Montaudo, G., Scamporrino, E., and Vitalini, D., Characterization of Copolymer Sequences by Fast Atom Bombardment Mass Spectrometry. 2. Identification of Oligomers Contained in Alternating and Random Copolyesters with Photolabile Units in the Main Chain, *Macromolecules*, 22, 627, 1989.

32. Klee, J.E., Hägelle, K., and Przybylski, M., Mass Spectrometry Characterization of a 2,2-bis-4-(2,3-epoxy)phenylpropane-aniline Addition Polymer and its Telechelic Prepolymers, *Macromol. Chem. Phys.*, 196, 937, 1995.

33. Klee, J.E., Hägelle, K., and Przybylski, M., Mass Spectrometry Characterization of Telechelic Prepolymers and Addition Polymers of DGEBA-Aliphatic Amines, *J. Polym. Sci. Polym. Chem.*, 34, 2791, 1996.

34. Prokay, L. and Simonsick W.J., Jr., Direct Mass Spectrometry Analysis of Phenol-Formaldehyde Oligocondensates: A Comparative Desorption Ionization Study, *Macromolecules*, 25, 6532, 1992.

35. Loy, J.O., Jr., and Miller, B.J., Plasticizers in Pacifiers: Direct Determination by FAB-MS, *Anal. Chem.*, 59, 1323A, 1987.

36. Maravigna, P. and Montaudo, G., Formation of Cyclic Oligomers, in *Comprehensive Polymer Science*, Allen, G., Bevington, J.C., Eds., Pergamon Press: Oxford 1989, Vol. 5, p. 63.

37. Penczek, S., Kubisa, P., and Matyiaszewski, K., Macromolecular Monomers-Macromonomers Synthesis and Applications. *Adv. Polym. Sci.*, 35, 68/69, 1985.

38. Semlyen, J.A., Ed., *Cyclic Polymers, Adv. Polym. Sci.*, 21, 41, 1976. Elsevier Applied Science Publishers: London, 1986.

39. Chojinowski, J., Scibiorek, M., and Kowalski, J., Mechanism of the Formation of Macrocycles During the Cation Polymerization of Cyclotrisiloxanes. End to End Ring Closure Versus Ring Expansion, *Makromol. Chem.*, 178, 1351, 1977.

40. Spanagel, E.W. and Carothers, W.H., Macrocyclic Esters, *J. Am. Chem. Soc.*, 57, 929, 1935.

41. Flory, P.J., Principles of Polymer Chemistry, Cornell University Press: Ithaca, New York 1953.

42. Montaudo, G. and Puglisi, C. in *Developments in Polymer Degradation-7*, Grassie, N., Ed. Elsiever: New York, 1987, p. 35.

43. Montaudo, G., Mass Spectral Determination of Cyclic Oligomer Distributions in Polymerization and Degradation Reactions, *Macromolecules*, 24, 5829, 1991.

44. Jacobson, H. and Stockmayer, W.H., Intramolecular Reaction in Polycondensations. I. The Theory of Linear Systems, *J. Chem. Phys.*, 18, 1600, 1950.

45. Mandolini, L., Montaudo, G., Scamporrino, E., Roelens, S., and Vitalini, D., Organotin-Mediated Synthesis of Macrocyclic Tetraesters. A Combined ^{1}H-NMR Spectroscopy, Gel Permeation Chromatography, and Fast Atom Bombardment Mass Spectrometry Approach to Complete Product Analysis, *Macromolecules*, 22, 3275, 1989.

46. Ito, K., Hashizuka, Y., and Yamashita, Y., Equilibrium Cyclic Oligomer Formation in the Anionic Polymerization of ε-Caprolactone, *Macromolecules*, 10, 821, 1977.

47. Failla, S. and Finocchiaro, P., Mass Spectral Determination of Azomethyne Macrocycle Distribution Formed in Polycondensation Reactions, *J. Chem. Soc. Perkin Trans.*, 2, 701, 1992.

48. Wood, B.R., Hodge, P., and Semlyen, J.A., Cyclic polyesters: 1. Preparation by a New Synthetic Method, Using Polymer-Supported Reagents, *Polymer*, 34, 3052, 1993.

49. Hamilton, S.C. and Semlyen, J.A., Cyclic Polyesters: 5. Cyclic Prepared by Poly(decamethylene terephthalate) Ring-Chain Reactions, *Polymer*, 38, 1685, 1997.

50. Bryant, J.J. and Semlyen, J.A., Cyclic Polyesters: 6. Preparation and Characterization of Two Series of Cyclic Oligomers from Solution Ring-Chain Reactions of Poly(ethylene terephthalate), *Polymer*, 38, 2475, 1997.

51. Bryant, J.J. and Semlyen, J.A., Cyclic polyesters: 7. Preparation and Characterization of Cyclic Oligomers from Solution Ring-Chain Reactions of Poly(butylene terephthalate), *Polymer*, 38, 4531, 1997.

52. Montaudo, G., Puglisi, C., and Samperi, F., Primary Thermal Degradation Mechanism of PET and PBT, *Polym. Deg. & Stab.*, 42, 13, 1993.

53. Wood, B.R., Hamilton, S.C., and Semlyen, J.A., Preparation of Some Large Cyclic Oxyethylene Succinate Ether-Esters, *Polymer Int.*, 44, 397, 1997.

54. Hamilton, S.C., Semlyen, J.A., and Haddleton, D.M., Part 8. Preparation and Characterization of Cyclic Oligomers in Six Aromatic Ester and Ether-Ester Systems, *Polymer*, 39, 3241, 1998.

55. Vitalini, D., Maravigna, P., and Scamporrino, E., Selective Formation of Cyclic Monomer, Dimer, or Trimer in the Pyrolysis of Isomeric Dihydroxybenzene Phthalate Polyesters, *Macromolecules*, 27, 2291, 1994.

56. Scamporrino, E., Mineo, P., and Vitalini, D., Synthesis and MALDI-TOF MS Characterization of High Molecular Weight Poly(1,2-dihydroxybenzene phthalates) Obtained by Uncatalyzed Bulk Polymerization of O,O'-Phthalid-3-ylidenecatechol or 4-Methyl-O,O'-phthalid-3-ylidenecatechol, *Macromolecules*, 29, 5520, 1996.

57. Kricheldorf, H.R., Lorenc, A., Spickermann, J., and Maskos, M., Macrocycles: 11. Polycondensation of Aliphatic Dicarboxylic Acid Dichlorides with Catechol or Bis-trimethylsylil Catechol, *J. Polym. Sci. A: Polym. Chem.*, 37, 3861, 1999.

58. Kricheldorf, H.R., Langanke, D., Spickermann, J., and Schmidt, M., Macrocycles: 10. Macrocyclic Poly (1,4-butanediol-ester)s by Polycondensation of 2-stanna-1,3-dioxepane with Dicarboxylic Acid Chlorides, *Macromolecules*, 32, 3559, 1999.

59. Kricheldorf, H.R. and Eggerstedt, S., MALDI-TOF Mass Spectrometry of Tin-Initiated Macrocyclic Polylactones in Comparison to Classical Mass Spectroscopic Methods, *Macromol. Chem. Phys.*, 200, 1284, 1999.

60. Zhang, Y., Wada, T., and Sasabe, H., Synthesis and Characterization of Novel Carbazole Cyclic Oligomer and Main Chain Polymer, *J. Polym. Sci. A: Polym. Chem.* 35, 2041, 1997.

61. Wu, J., Chen, C., Kurth, M.J., and Lebrilla, C.B., Mass Spectrometry Analyses of β-Ketolactone Oligomers, Macrocyclic or Catenane Structures, *Anal. Chem.*, 68, 38, 1996.

62. Bonas, G., Bosso, C., and Vignon, M.R., Determination of the Stability Constants of Macrocyclic Ligand-Alkali Cation Complexes by Fast Atom Bombardment Mass Spectrometry, *J. Incl. Phenomena Mol. Rec. Chem.*, 7, 637, 1989.

63. Randall, J.C., Polymer Sequences Determination. Carbon-13 NMR Method, Academic Press: New York, 1977.

64. Montaudo, M.S., Ballistreri, A., and Montaudo, G., Determination of Microstructure in Co-polymers. Statistical Modeling of Computer Simulation Mass Spectra, *Macromolecules*, 24, 5051, 1991.

65. Montaudo, M.S. and Montaudo, G., Further Studies on the Composition and Microstructure of Copolymers by Statistical Modeling of Their Mass Spectra, *Macromolecules*, 25, 4264, 1992.

66. Montaudo, G., Sequence of Polymers by Mass Spectrometry, *Macromol. Symp.*, 98, 899, 1995.

67. Montaudo, M.S., Monte-Carlo Simulation of Copolymer Mass Spectra, *Makromol. Chem., Theory Simul.*, 2, 735, 1993.

68. Montaudo, G., Scamporrino, E., and Vitalini, D., Characterization of Copolymer Sequences by Fast Atom Bombardment Mass Spectrometry. 1. Identification of Oligomers Produced in the Hydrolysis and Photolysis of Random Copolyamides Containing Photolabile Units in the Main Chain, *Macromolecules*, 22, 623, 1989.

69. Montaudo, G., Puglisi, C., and Samperi, F., Sequencing an Aromatic Copolycarbonate by Partial Ammonolysis and FAB-MS Analysis of the Products, *Polym. Bulletin*, 21, 483, 1989.

70. Puglisi, C., Sturiale, L., and Montaudo, G., Thermal Decomposition Processes in Aromatic Polycarbonates Investigated by Mass Spectrometry, *Macromolecules*, 32, 2194, 1999.

71. Montaudo, G., Scamporrino, E., and Vitalini, D., Structural characterization of copolyester by fast atom bombardment mass spectrometry analysis of the partial ammonolysis products, *Makromol. Chem., Rapid Commun.*, 10, 411, 1989.

72. Montaudo, M.S., Puglisi, C., Samperi, F., and Montaudo, G., Structural Characterization of Multicomponent Copolyesters by Mass Spectrometry, *Macromolecules*, 31, 8666, 1998.

73. Montaudo, M.S., Puglisi, C., Samperi, F., and Montaudo, G., Partially Selective Methanolysis of Sebacic Units in Biodegradable Multicomponent Copolyesters, *Macromol. Rapid Commun.*, 19, 445, 1998.

74. Montaudo, G., Montaudo, M.S., Puglisi, C., and Samperi, F., Sequence distribution of Poly(ether-sulfone)/poly(ether-ketone) Copolymers by Mass Spectrometry and ^{13}C NMR, *Makromol. Chem. Phys.*, 196, 499, 1995.

75. Montaudo, G., Scamporrino, E., Vitalini, D., and Rapisardi, R., Fast Atom Bombardment Mass Spectrometry Analysis of the Partial Ozonolysis Products of Poly(isoprene) and Poly(chloroprene), *J. Polym. Sci, Polym. Chem.*, 30, 525, 1992.

76. Montaudo, G., Scamporrino, E., and Vitalini, D., Structural Characterization of Butadiene/Styrene Copolymers by Fast Atom Bombardment Mass Spectrometry Analysis of the Partial Ozonolysis Products, *Macromolecules*, 24, 376, 1991.

77. Scamporrino, E. and Vitalini, D., Structural characterization of butadiene/ acrylonitrile copolymers by fast atom bombardment mass spectrometry analysis of the partial ozonolysis products, *Polymer*, 33, 4597, 1992.

78. Montaudo, G., Montaudo, M.S., Scamporrino, E., and Vitalini, D., Mechanism of Exchange in Polyesters. Composition and Microstructure of Copolymers Formed in the melt Mixing Process of Poly(ethylene terephthalate) and Poly(ethylene adipate), *Macromolecules*, 25, 5099, 1992.

79. Montaudo, G., Montaudo, M.S, Scamporrino, E., and Vitalini, D., Composition and microstructure of copolyesters formed in the melt mixing of poly(ethylene terephthalate) and poly(ethylene truxillate), *Makromol. Chem.*, 194, 993, 1993.

80. Montaudo, G., Puglisi, C., and Samperi, F., Copolymer Composition: a Key to the Mechanism of Exchange in Reactive Polymer Blending, in *Transreactions in Condensation Polymers*, Kakirov, S., Ed., Wiley-VCH, Weinheim, 1999.

81. Bloembergen, S., Holden, D.A., Hamer, G.K., Bluhm, T.L., and Marchessault, R.H., Studies of Composition and Crystallinity of Bacterial Poly(β-hydroxybutyrate-co-β-hydroxyvalerate), *Macromolecules*, 19, 2865, 1986.

82. Bluhm, T.L., Hamer, G.K., Marchessault, R.H., Fyfe, C.A., and Vergin, R.P., *Macromolecules*, 19, 2871, 1986.

83. Doi, Y., Kunioka, M., Nakamura, Y., and Soga, K., Nuclear magnetic resonance studies on poly(β-hydroxybutyrate) and a copolyester of β-hydroxybutyrate and β-hydroxyvalerate isolated from Alcaligenes eutrophus H16, *Macromolecules*, 19, 2860, 1986.

84. Kamiya, N., Yamamoto, Y., Inoue, Y., Chujo, R., and Doi, Y., Microstructure of bacterially synthesized poly(3-hydroxybutyrate-co-3-hydroxyvalerate), *Macromolecules*, 22, 1676, 1989.

85. Holmes, P. A., Application of PHB a Microbially Produced Biodegradable Thermoplastic, *Phys. Technol.*, 16, 32, 1985.

86. Ballistreri, A., Garozzo, D., Giuffrida, M., Impallomeni, G., and Montaudo, G., Sequencing Bacterial Poly(β-hydroxybutyrate-co-β-hydroxyvalerate) by Partial Methanolysis, High-Performance Liquid Chromatography Fractionation, and Fast Atom Bombardment Mass Spectrometry Analysis, *Macromolecules*, 22, 2107, 1989.

87. Nedea, M.E., Morin, F.G., and Marchessault, R.H., Microstructure of bacterial poly(β-hydroxybutyrate-co-β-hydroxyvalerate) by FAB mass-spectrometry, *Polymer Bulletin*, 26, 549, 1991.

88. Morikawa, H. and Marchessault, R.H., Pyrolysis of Bacterial Polyalkanoates, *Can. J. Chem.*, 59, 2306, 1981.

89. Grassie, N., Murray, E.J., Holmes, P.A., The Thermal Degradation of Poly(d-β-hydroxybutyric acid): part 3—The Reaction Mechanism, *Polym. Degr. & Stab.*, 6, 127, 1981.

90. Kunioka, M. and Doi, Y., Thermal Degradation of Microbial Copolyesters: Poly(3-hydroxybutyrate-co-3-hydroxyvalerate) and Poly(3-hydroxybutyrate-co-4-hydroxyvalerate), *Macromolecules*, 23, 1933, 1990.

91. Lehrle, R.S. and Williams, R.J., Thermal Degradation of Bacterial Poly(hydroxybutyric acid): Mechanism from the Dependence of Pyrolysis Yields on Sample Thickness, *Macromolecules*, 27, 3782, 1994.

92. Abate, R., Ballistreri, A., Impallomeni, G., and Montaudo, G., Thermal Degradation of Microbial Poly(4-hydroxybutyrate), *Macromolecules*, 27, 332, 1994.

93. Ballistreri, A., Garozzo, D., Giuffrida, M., and Montaudo, G., Microstructure of Bacterial Poly(β-hydroxybutyrate-co-β-hydroxyvalerate) by Fast Atom

Bombardment Mass Spectrometry Analysis of Their Partial Degradation Products, *Novel Biodegradable Microbial Polymers*, 49, 1990.

94. Ballistreri, A. Montaudo, G., Impallomeni, G., Lenz, R.W., Kim, Y. B., and Fuller, C.R., Sequence Distribution of β-Hydroxyalkanoate Units with Higher Alkyl Groups in Bacterial Copolyesters, *Macromolecules*, 23, 5059, 1990.

95. Ballistreri, A., Garozzo, D., Giuffrida, M., Impallomeni, G., and Montaudo, G., Analytical Degradation: An Approach to the Structural Analysis of Microbial Polyesters by Different Methods, *J. Anal. App. Pyrolysis*, 16, 239, 1989.

96. Ballistreri, A., Garozzo, D., Giuffrida, M., Montaudo, G., and Montaudo, M.S., Microstructure of Bacterial Poly(β-hydroxybutyrate-co-β-hydroxyvalerate) by Fast Atom Bombardment Mass Spectrometry Analysis of the Partial Pyrolysis Products, *Macromolecules*, 24, 1231, 1991.

97. Ballistreri, A., Giuffrida, M., Lenz, R.W., Fuller, C.R., Kim, Y.B., and Montaudo, G., Determination of Sequence Distributions in Bacterial Copolyesters Containing Higher Alkyl and Alkenyl Pendant Groups, *Macromolecules*, 25, 1845, 1992.

98. Ballistreri, A., Impallomeni, G., Lenz, R.W., Fuller, C.R., Montaudo, G., and Ulmer, H.W., Synthesis and Characterization of Polyesters Produced by *Rhodospirillum rubrum* from Pentenoic Acid, *Macromolecules*, 28, 3664, 1995.

99. Abate, R., Ballistreri, A., Giuffrida, M., Impallomeni, G., and Montaudo, G., Separation and Structural Characterization of Cyclic and Open Chain Oligomers Produced in the Partial Pyrolysis of Microbial Poly(hydroxybutyrates), *Macromolecules*, 28, 7911, 1995.

100. Ballistreri, A., Giuffrida, M., Impallomeni, G., Lenz, R.W., and Fuller, C.R., Characterization by mass spectrometry of poly(3-hydroxyalkanoates) produced by *Rhodospirillum rubrum* from 3-hydroxyacids, *Int. J. Biol. Macromol.*, 26, 201, 1999.

8

Time-of-Flight Secondary Ion Mass Spectrometry (TOF-SIMS)

David M. Hercules

CONTENTS

0-8493-3127-7/02/$0.00+$1.50

8.1   Introduction

Conventional mass spectrometry has been limited in its application to polymers with regard to both ion formation and instrumentation. Polymers are large molecules having low volatility and are thermally unstable when heated; conventional electron impact sources are "hard," tending to cause extensive fragmentation. Conventional sector-field instrumentation is limited in mass range and the response of detectors is often poor for large molecules. Recent developments in ionization sources, mass analyzers, and detectors have expanded the scope of mass spectrometry making it possible to obtain spectra of nonvolatile compounds. A logical outgrowth has been to take advantage of these developments to obtain mass spectra from polymers.

A variety of volatilization/ionization methods have been applied to polymers; a recent review or key paper is cited here for each. Extensive reviews that include mass spectrometry of polymers can be found in *Analytical Chemistry.*[1] Other topical reviews are: field desorption,[2] laser desorption,[3] plasma desorption,[4] fast-atom bombardment,[5] pyrolysis,[6] and electrospray ionization.[7] The present review will focus on polymer characterization using secondary-ion mass spectrometry (SIMS) in the high mass range; comparison with other methods will be presented where appropriate.

Perhaps the first logical question to ask is what kind of information can be obtained about polymers by SIMS. Table 8.1 answers this question. Information about repeat units, functional groups, and terminal groups is useful for identification. Oligomer distributions can be measured. Studying copolymers with SIMS provides information about subunits, whether they are random or ordered, and information about crosslinking and branching. Varied information can be obtained about polymer films. Quantitative analysis is possible as is obtaining information about stereoregularity.

TABLE 8.1

Important Information about Polymers Obtained from SIMS

1. Identification
   - Which Polymer?
   - Polymer Repeat Units
   - Functional Groups Present
   - Terminal Groups

2. Oligomer Distribution
   - Number-Average Molecular Weights
   - Weight-Average Molecular Weights

3. Copolymers
   - Subunit Identification
   - Lengths of Segments
   - Random vs. Ordered
   - Crosslinking and Branching

4. Polymer Films
   - Surface Impurities
   - Surface Segregation
   - Inhomogeneities
   - Maps and Images of Polymer Blends

5. Quantitative Analysis
   - Ratios of Monomers
   - Degree of Branching

6. Steroregularity

Because polymers represent a special challenge for mass spectrometry, their analysis imposes certain constraints on instrumentation; these are summarized in Table 8.2. The ion source must be able to handle an intact sample, provide soft ionization, and be both reproducible and efficient. The analyzer must have high transmission, high mass range, and high resolution. The detection system should have high efficiency, a large dynamic range, and low noise. Instrumentation should be capable of imaging and be able to compensate for sample charging. A time-of-flight (TOF) SIMS instrument qualifies on all counts. Thus this chapter will deal primarily with studies which have been performed using TOF-SIMS.

From the standpoint of mass spectrometry, a polymer is a series of repeat units separating two terminal groups: T_1-R_n-T_2. There is a similarity between the mass spectra of polymers and small molecules. Both show parent ions. In small molecules the parent is a molecular ion characteristic of a single molecular species, in polymers there will be multiple parent ion peaks representing the distribution of oligomers. Charge on small molecules is usually caused by loss of an electron, but in polymers by gain of a proton or other cation (e.g., Na^+). In polymers, fragmentation and charge formation may be either simultaneous or decoupled. For example, in the low mass region

TABLE 8.2

Characteristics of a Mass Spectrometer for Polymer Studies by SIMS

1. Ion Source
   • Handle Intact Sample
   • Provide Soft Ionization
   • Do Both Surface and Bulk Studies
   • High Ion-Formation Efficiency
   • Stable and Reproducible

2. Analyzer
   • High Transmission (>10%)
   • High Mass Range (>10^5)
   • High Resolution (M/ΔM > 10^4)
   • Simultaneous Detection of All Masses

3. Detection System
   • High Detection Efficiency
   • Large Dynamic Range (>10^6)
   • Low Noise
   • Simultaneous Detection of All Masses

4. Other Features
   • Charge Compensation
   • Imaging

polymers show ions which probably result from larger charged species. In the high mass region, cationization can occur either before fragmentation or afterward with neutral fragments.

A SIMS spectrum illustrating oligomer and fragment ion peaks for poly-isoprene is shown in Figure 8.1. The peaks labeled (•) are intact oligomers cationized by silver, and the peaks labeled (o) are cationized fragment ions.

8.1.1   Nature of the SIMS Process

Sputtering and ionization in SIMS are due to events caused by the impact of a high velocity ion on a surface. The process is depicted schematically in Figure 8.2. A primary ion (typically Ar$^+$) of approximately 10 keV energy strikes a surface, usually a metal substrate having an organic (polymer) overlayer. The impact of the primary ion causes disruption of the surface and sputtering of atoms and molecules both from the metal foil and the organic overlayer. The sputtered particles include electrons, positive ions, negative ions, and neutrals; neutrals constitute the vast majority of the sputtered material (>99%). The secondary ions sputtered from the surface are collected by a mass spectrometer and mass analyzed. Secondary ions include atomic ions and cluster ions from the substrate, and molecular and fragment ions from the overlayer.

The SIMS experiment is relatively simple in principle. A specially prepared Ag foil is produced by a pretreatment involving preoxidation in 20% nitric

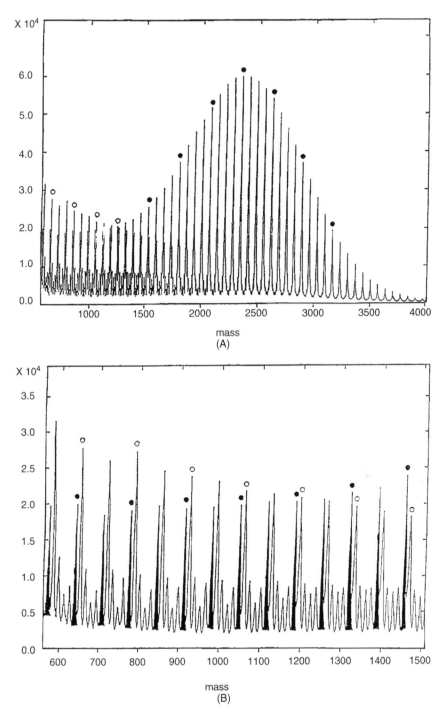

FIGURE 8.1
TOF-SIMS spectrum of polyisoprene (MW = 2300). A. Oligomer distribution. B. Region showing overlap between oligomer and fragment-ion peaks. ● - Oligomer peaks, o - Fragment-ion peaks.

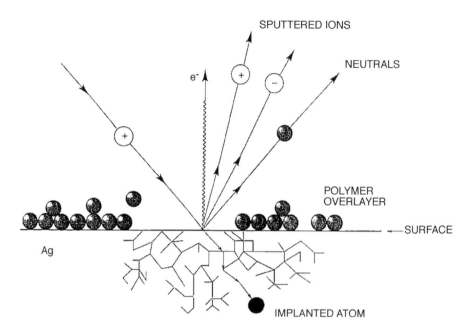

FIGURE 8.2
Diagram of SIMS sputtering process.

acid, washing with distilled water in an ultrasonic cleaner, and air drying.[8] A dilute polymer solution in a volatile solvent is added to the foil from a micropipet and allowed to evaporate, producing a thin polymer film on the metal. A SIMS spectrum is then obtained directly from the foil.

It is also possible to measure thick polymer films directly in SIMS. These differ from the model shown in Figure 8.2 in that the Ag foil is no longer involved. In this case only polymeric material will be sputtered and fragment ions from the polymer will be produced. The major difference between the two experiments is that only small fragment ions (<500 Da) are generally observed from sputtered polymer films, whereas much larger polymer fragments are sputtered from metal foils. A typical TOF-SIMS film spectrum is shown in Figure 8.3.

There are two modes in which SIMS can operate: dynamic and static. In dynamic SIMS a high primary ion current is used, typically 10^{-6} A/cm^2, resulting in multiple sputtering events at the same site on the surface. For an organic material this mode produces only very small fragments, generally not larger than C_3. This is the mode used for depth profiling with an ion beam and is *not* the mode used for SIMS of polymers. In the static SIMS mode, the primary ion dose is kept low so that the probability of multiple sputtering events from the same site is vanishingly small. A monolayer contains about 10^{15} atoms or molecules per cm^2 and a total dose of 10^{11}–10^{12} ions/cm^2 is typically used in a static SIMS experiment. This is the equivalent of sputtering as little as 10^{-4} of

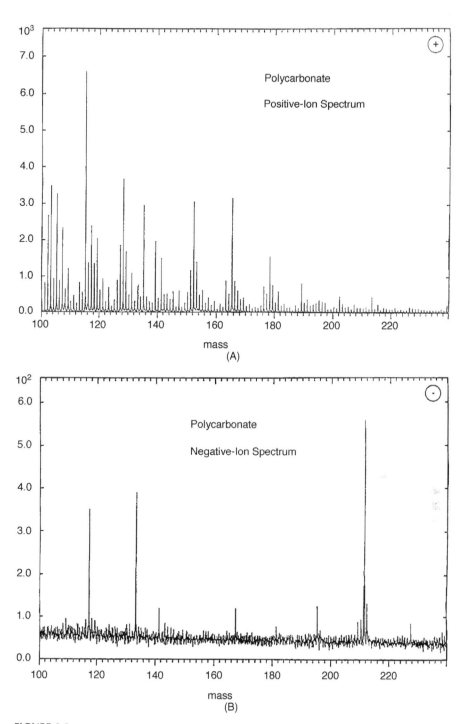

FIGURE 8.3
TOF-SIMS spectra of a polycarbonate in the low mass range. (A) Positive-ion spectrum.
(B) Negative-ion spectrum.

a monolayer. The equivalent primary ion current is $<10^{-10} A/cm^2$. When operating in the static mode one is dealing with isolated events; primary ions, like lightning, do not strike the same spot twice. The static mode is capable of producing very large ions from organic overlayers.

The SIMS process, as it relates to polymers, is characterized by three important parameters: ion yield, transformation probability, and disappearance cross section. The ion yield (Y) is the ratio of the number of secondary ions produced divided by the number of primary ions incident on the sample: $Y = N(si)/N(pi)$. SIMS bombardment of a surface covered by a monolayer or less results in an exponential decay of the measured signal intensity: $I(t) = I^{\circ} * exp[-\sigma(x)j_p t/e]$ where j_p is the primary ion current density, x the species measured, and σ the disappearance cross section. Thus σ can be calculated from SIMS intensity decay curves. The disappearance cross section is the damage area caused by the primary ion which leads to desorption of a polymer molecule.

The integrated area under the SIMS intensity decay curve is a measure of the total number of ions detected in a SIMS experiment. The total number of ions detected for the species x can be written as: $N_d(x) = P_t * N_m(x) * T(x) * D(x)$ where $N_m(x)$ is the number of molecules of x in the area analyzed, $T(x)$ is instrument transmission for x, $D(x)$ is the detector efficiency for x, and P_t is the transformation probability. P_t represents the probability that a molecule, M, sputtered from the surface will be detected as an ionic species, x.

8.1.2 Ion Formation Processes in SIMS

All of the ion formation processes in SIMS are known for other forms of mass spectrometry. The fundamental difference is that several ion formation mechanisms may occur simultaneously in competition with each other; often one does not have control over which ion formation process will dominate. Also, the dominant ionization process may vary with the particular type of polymer involved. One great advantage of SIMS is that both positive and negative ions are formed, often in comparable yields.

Table 8.3 summarizes the ion formation processes observed in SIMS. The first of these is gain or loss of electrons, much like in conventional electron-impact mass spectrometry. Molecules having low ionization potentials will tend to form positive ions; those with high electron affinities will form negative ions. However, gain or loss of electrons is an uncommon process in SIMS, particularly in polymers. The equivalent of odd-electron ions is encountered only in the low mass range for most polymers. Organic compounds that are salts exist as pre-formed ions, and sputtering only needs to overcome lattice forces to form ions. Ion yields from such materials are high, but few polymers exist as salts. Thus, this is a minor process in polymer SIMS.

Gain or loss of protons is frequently observed when ionizable groups are present in the molecule, much like in conventional chemical ionization (CI) mass spectrometry. Whereas in conventional CI mass spectrometry the

TABLE 8.3

Ion Formation Processes in SIMS

(M = Polymer Molecule)

Gain or Loss of Electrons

$$M \rightarrow M^+ + e^-$$
$$M + e^- \rightarrow M^-$$

Gain or Loss of Protons

$$M + BH \rightarrow MH+ B$$
$$MH + B \rightarrow M^- + BH^+$$

Ionization of Salts

$$M^+ X^- \rightarrow M^+ + X^-$$

Ion Attachment

$$M + C^+ \rightarrow (MC^+)$$
$$M + A^- \rightarrow (MA^-)$$

proton donor (or acceptor) is known, generally the source of protons is ill-defined in SIMS. However, gain or loss of protons is a major ionization process in polymer SIMS. Multiply charged ions are the exception rather than the rule in SIMS, unlike in electrospray.

Attachment of cations to polymer molecules is a major and often dominant ion formation process in SIMS. Attachment of anions is known, but not common. The cation may be derived from the substrate, such as Ag^+, or may be present as an impurity or intentionally added to the substrate, such as Na^+ or K^+. In some cases peaks from impurity Na^+ attachment may be more intense than those from the substrate ion, Ag^+. Even though polymers may have multiple sites that could readily attach metal ions, the usual ionization process is attachment of a single ion to a large molecule, even for masses as large as 10,000 Da. Some doubly charged polymer ions have been observed, but this is the exception.

SIMS spectra of polymers generally contain three types of ions: intact oligomer ions, large fragment ions, and small fragment ions. These are summarized in Table 8.4. Each type of ion has its own characteristics with regard to mass range and the type of information that can be derived from it. Intact oligomer ions are generally seen in the mass range below 10,000 Da and correspond to cationized polymer molecules. They can be used to measure oligomer distributions, as will be discussed later.

Large fragment ions are observed in the range 500–5000 Da. Cationized neutral fragments are observed both with and without end-groups as indicated by 1a and 1b in Table 8.4. The relative intensities of peaks for fragments

TABLE 8.4

Types of Ions Observed in SIMS Polymer Spectra

Oligomers

Mass Range 500–15,000
$[T_1 - R_n - T_2 + Ag]^+$

Large Fragments

Mass Range 500–7,000

1. *Neutral Fragments*
   a. With end groups: $[T - R_n \pm H + Ag]^+$
   b. Without end groups: $[R_n \pm H + Ag]^+$

2. *Charged Fragments (Perfluororethers)*
   a. With end groups: $T - R_n CF_2^+$
   $$T - R_n CF_2 O^-$$
   b. Without end groups: $R_n CF_2^+$
   $$R_n CF_2 O^-$$

Small Fragments

Mass Range Below 500
1. Multiples of Repeat Unit: $H \pm R_n \pm \bullet$
2. Fragments: $= Frag \pm \bullet$

containing end-groups decrease as polymer molecular weight increases. Small fragments having intrinsic charge are observed for all polymers below 500 Da; they are singly charged. The nature of the fragments ranges from multiples of repeat units (with or without added hydrogen), some number of repeat units plus a fragment of the repeat unit, or fragments from one repeat unit. In general, fragments in this range are not larger than two or at most three polymer repeat units.

Large fragments having intrinsic charge (mass range 500–7000 Da) have been observed for fluoropolymers, the perfluoropolyethers[9] are the best example. As shown in Table 8.4, 2a and 2b, they may occur with or without end-groups; both positive and negative ions have been observed. These polymers are unique in this respect because no other polymer system has shown fragment ions of this size having intrinsic charge. Another interesting aspect of the perfluoropolyethers is that metastable ions are observed in their spectra. These polymers will be discussed in detail later.

8.1.3   Important Factors for SIMS of Polymers

A number of important factors must be considered when studying TOF-SIMS of polymers. Included are the nature of the substrate, substrate coverage, substrate pretreatment, the primary ion, polymer molecular weight, and the solvent used in sample preparation.

8.1.3.1 Nature of the Substrate

The yield of oligomer or high-mass fragment ions strongly depends on the nature of the substrate material. Studies with polystyrene showed that silver surfaces that had been etched with nitric acid gave the maximum yield of cationized secondary ions.[10] Other metals were studied, including Au and Pt. Although they gave measurable yields of secondary ions, they were about an order of magnitude lower than from silver. Reactive metals, such as Cu or Ni, showed ion yields about two orders of magnitude lower than Ag. So far it seems that the silver foil used initially for SIMS experiments[8] remains the optimum substrate.

Although etching Ag substrates generally improves ion yields, it is possible to obtain perfectly good TOF-SIMS spectra from flat Ag foils, or from substrates prepared by depositing a layer of Ag on a substrate such as a Si wafer or mica. It is also possible to obtain TOF-SIMS spectra from a chemically treated Si wafer that has on its surface a small amount of a cation such as Cu^{+2} or Ag^{+}.[11]

8.1.3.2 Substrate Coverage

The amount of polymer on a substrate surface has a dramatic effect on the ion yield. Figure 8.4a shows a plot of ion yield for oligomer ions of a 3000 MW polystyrene as a function of amount deposited on the substrate.[12] Deposition was 10 μL of solution on a 150 mm^2 substrate. Based on the cross-section of a polystyrene molecule, deposition of 50 μg/mL corresponds to full monolayer coverage, which is just below the vertical line separating Region I and Region II in the diagram. Figure 8.4b shows the polystyrene ion yield as a function of surface coverage measured by ion-scattering spectroscopy. Surface coverage is only about a quarter monolayer at the line. Further, surface coverage is only about half a monolayer at the line between Regions II and III even though the concentration has been increased by three orders of magnitude over the range.

The explanation relates to the surface roughness of the silver foils. Scanning electron microscopy revealed that the surface consists of conical pores about 2 μm in radius and 9 μm deep. As the polymer concentration is increased, the polymer preferentially fills the bottoms of the pores and has only a thin coating on the remainder of the surface. The thin layer of polymer on the external surface has a high ion yield while that inside the pores has a low ion yield. Thus, as the concentration increases the yield increases. After half a monolayer, the pores are nearly filled, and thick layers of polymer begin to form on the outer surface. This decreases the Ag^{+} ion yield, depriving the polymer of necessary ions for cationization.

8.1.3.3 Substrate Modification

Alternatives to the Ag substrate have been tried, some modifying the Ag substrate, others using a different material. These have resulted in enhancement of signals for organic compounds, including polymers. To date, none

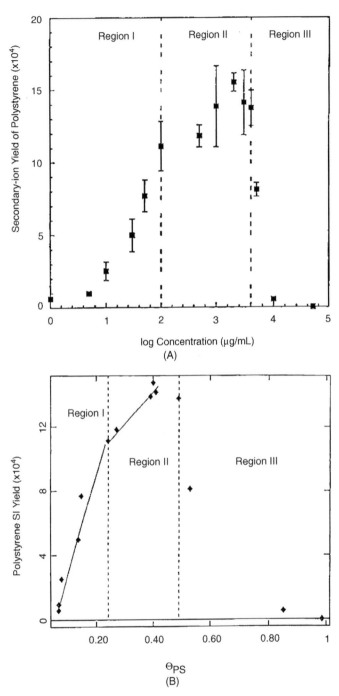

FIGURE 8.4
TOF-SIMS ion yields for polystyrene deposited on a silver substrate. A. Secondary-ion yield as a function of polystyrene concentration. B. Secondary-ion yield as a function of surface coverage. (Reprinted with permission from Ref. 12, Copyright 1994 American Chemical Society)

has become widely accepted, and in general, the reason for the enhancement is a matter of conjecture.

The first of these methods was by Cook's group[13] and involved use of an ammonium chloride matrix. An increase in molecular ion intensity was observed along with reduction in fragmentation for sugars, polypeptides, nucleotides, and vitamins. The mechanism proposed was that large clusters of ions are sputtered initially, consisting of analyte ions surrounded by matrix ions. It is the formation of these clusters that results in increased yield of protonated species (by proton transfer from the ammonium ion) and softer ionization because of relaxation processes in the clusters.

Another method involving a different matrix was reported by Wu and Odom.[14] They deposited organic materials on the surfaces of organic crystals used as MALDI matrices, such as 2,5-dihydroxybenzoic acid. Enhanced molecular ion signals were observed for peptides and oligonucleotides up to masses of 10,000 Da, a mass range not normally accessible by SIMS for these materials. Detection limits in the subpicomole range were reported.

Our group reported three different modifications of matrices that resulted in increased ion yields. The first of these was the observation that ion yields of materials varied significantly from a Si substrate, depending on whether the method of pretreatment rendered the surface hydrophobic or hydrophilic.[11] Polymers studied were polyethylene, polypropylene, poly(propylene oxide) and polystyrene. The yields of oligomer ions were consistently higher from the hydrophilic surface; oligomer ions were either protonated or cationized with sodium. A second, and most interesting observation, was that traces of Cu^{+2} ions on the *hydrophobic* Si surface gave significant yields of cationized polystyrene oligomers, in preference to the hydrophilic surface;[11] an example is shown in Figure 8.5. This effect was caused by 6 ppm of Cu in the HF solution used to treat the Si surface.

Enhanced ion emission was also observed for polymers and large biomolecules from Ag surfaces that had been treated with a sub-monolayer coverage of cocaine hydrochloride.[15] Several possible mechanisms were proposed but it was not possible to distinguish between them. A recent discovery in our laboratory has been that halide modification of the SIMS substrate surface enhanced signal intensity for compounds such as cyclosporin A, angiotensin II, and a small nucleoside (dA-dA).[16] Improvements ranged from 2–30 times for cationized species and 10–1000 times for protonated species. An interesting effect is that treatment of Ag surfaces with HBr and HI caused enhancement, while treatment with HCl did not.

8.1.3.4 *Primary Ion*

The nature of the primary ion can have a significant impact on SIMS ion yield. While the effect of primary ion on yields in dynamic SIMS has been well-documented, the effect in static SIMS has become of interest only recently, other than by comparing yields for different rare gas ions. Several groups have become interested in the interaction of polyatomic ions with surfaces and their potential as primary ions in SIMS.

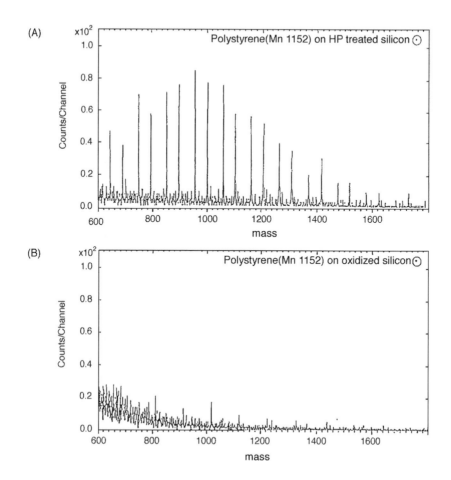

FIGURE 8.5

TOF-SIMS spectra obtained from silicon surfaces (from Ref. 11). A. Silicon treated with HF solution before polymer deposition. B. Oxidized silicon surface - no HF treatment.

Cooks' group has published experiments dealing with soft landing of ions on surfaces[17] and studies of reactive collisions with them. Schweikert and coworkers have pioneered the use of polyatomic projectiles as primary ions in SIMS and related methods. They have reported nonlinear effects caused by clusters[18] and have compared yields for cluster ions with those from monatomic ions.[19] Most recently, Benninghoven's group has compared the use of SF_5^+ as a primary ion for obtaining spectra from polymers with monatomic ions.[20] They observed enhancements in the higher mass range on the order of 100–500 times for spin-coated samples of poly(ethylene terphthalate) and polystyrene. They were also able to observe structurally significant peaks up to 2000 Da from thick polymer films using SF_5^+ as the primary ion for cases where the maximum mass seen by atomic ions was 1000 Da.

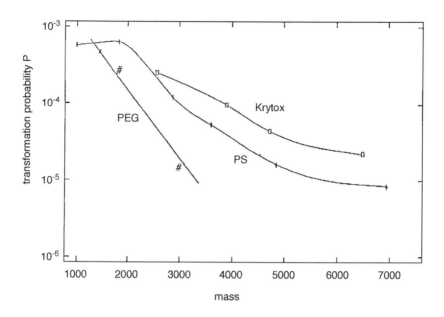

FIGURE 8.6

Transformation probabilities of polymer oligomers. (Reprinted with permission from Ref. 10, Copyright 1989 American Vacuum Society)

8.1.3.5 Molecular Size

The size of a polymer molecule has a significant effect on the nature of the SIMS spectrum. Low-molecular-weight oligomers generally are desorbed intact, while large oligomers are fragmented. Figure 8.6 shows the transformation probability (P_t) of polystyrene oligomers as a function of their molecular weight.[10] Note that P_t remains constant up to about mass 2000, and then decreases substantially as a function of increasing mass in the range 2000–7000 Da; the equivalent change in P_t is from 3×10^{-4} to 2×10^{-6}. Extrapolating this trend to about mass 10,000, one obtains a P_t value of approximately 10^{-7}. By inserting reasonable numbers into the equation for P_t, one calculates that, for a reasonable signal-to-noise ratio (3:1), the limit of detection would be about 10,000 Da. This correlates very closely with observations. The polarity of the polymer also seems to affect P_t for a given mass; PEG has a lower value than polystyrene, which is lower than the value for a fluoroether.

The disappearance cross sections for polymers correspond to a circle about 40–50 Å in diameter. If one considers that the length of a polystyrene monomer is about 3 Å, a stretched out polymer having 18 monomer units would be just about 50 Å. This and end-groups correspond to 1930 Da, amazingly close to 2000 where the "fall off" begins in Figure 8.6. Thus we can conclude that efficient sputtering of a polymer molecule is limited by the "action area" of the sputtering process.

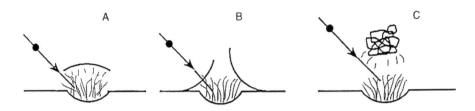

FIGURE 8.7
Simple sputtering model for polymers.

The data of Figure 8.6 are consistent with a rather simple model of sputtering. Figure 8.7 shows such a model, relating the action cross section of the primary ion to the size of a polymer ion which can be emitted intact.[21] Figure 8.7A shows a small polymer molecule overlying the action area; the molecular dimensions are approximately the same as that of the action area. In this situation the intact polymer will be sputtered by the primary ion. In Figure 8.7B, the polymer molecule is much longer, part of it overlying the action area and part lying outside of it. When sputtering occurs, the part of the molecule over the action area will be given a thrust upward, while the other part remains weakly bonded to the surface. If the latter forces are strong enough, bond breaking will occur and a fragment will be emitted. This fragment can be either a section cut from the chain or containing an end-group.

8.1.3.6   Sample Preparation

The method used to prepare samples for TOF-SIMS analysis can have a dramatic effect on the spectrum obtained. For example, the solvent used to deposit the polymer on the Ag foil can be critical. In Figure 8.7C a large polymer is on the surface in a coiled configuration. Because the size of the polymer coil is about that of the interaction area, the polymer is sputtered intact and an oligomer ion will be observed. If the same polymer were stretched out over the surface, only fragment ions would be seen. The effect on SIMS spectra of varying the solvent used for deposition is seen experimentally. For example, the SIMS fragmentation pattern changed considerably for a 410K MW polyisoprene, depending on whether it was deposited from hexane or toluene.[22]

8.2   Intact Oligomer Spectra

8.2.1   Introduction

Because TOF-SIMS can obtain spectra of intact oligomers, it is potentially useful for determining molecular weight (MW) distributions of polymers. Many methods currently used to measure polymer molecular weights give only average values. Because TOF-SIMS records peaks for individual oligomers,

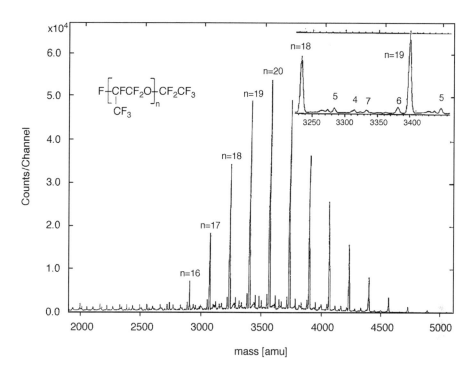

FIGURE 8.8

Positive-ion TOF-SIMS spectrum of a perfluoro polyether. The inset shows the presence of other terminal groups. (Reprinted with permission from Ref. 23, Copyright 1990 American Chemical Society)

it can measure distributions and give absolute molecular weights. The technique has been used successfully to measure M_n values for polymers below 5000, and a few above, but generally, it is not applicable to higher MW polymers, for reasons that will be discussed below.

Figure 8.8 shows a TOF-SIMS spectrum obtained for a ca. 3400 MW Krytox; the polymer structure is shown at the top of the figure.[23] The spectrum is typical of those obtained in this mass range. The intense peaks correspond to individual oligomers (some are labeled) cationized with Ag; the inset shows the spectrum in greater detail for the region of $n = 18$ and 19. Oligomer peaks can be identified positively by their mass, because the TOF-SIMS can measure masses to an accuracy of 0.03 Da. The spacing between the oligomer peaks is the mass of the polymer repeat unit.

In some cases, polymer fragment-ion peaks occur in the same region as the oligomer peaks, the spectra of poly(isoprene) shown in Figure 8.1 are a case in point. However, the weak peaks in the Krytox spectrum of Figure 8.8, between the oligomer peaks, are not fragment ions, but come from oligomers having end-groups other than those shown in the structure, or from Na-cationized oligomers. The peak labeled "4" in the inset is for the $n = 19$ oligomer cationized with sodium. Typically sodium impurities are present on the silver surface or in the polymer; therefore Na-cationized peaks may appear weakly or strongly depending on the Na level and the relative affinities of the polymer for Na and Ag ions. The peaks labeled "5" in Figure 8.7 are for $n = 18$ and $n = 19$ oligomers having a perfluoroisopropyl terminal group (CF_3CFCF_3) instead of the perfluoroethyl group. The peak labeled "6" is for oligomer $n = 19$ with a CH_3CHF- terminal group instead of CF_3CF_2-, and the peak labeled "7" corresponds to an additional CF_3CF_2- terminal group on the left-hand side for oligomer $n = 18$ instead of F. Careful analysis of polymer spectra in the oligomer region has been used to identify impurities in polymer samples arising from the presence of different end-groups.[23]

8.2.2 Molecular Weight Distributions

Because conventional methods used for polymer molecular weight analysis are largely averaging methods, polymer molecular weights are represented either as number-average (M_n) or weight average (M_w) molecular weights. These numbers themselves do not provide information about molecular weight distributions, although the ratio of M_w/M_n is often used of an indicator of such.

M_n and M_w for polymers can be calculated from the mass spectrum by measuring the relative peak intensities of individual oligomers and calculating M_n and M_w as follows:

$$M_n = \frac{\Sigma_i I_i M_i}{\Sigma_i I_i} \qquad M_w = \frac{\Sigma_i I_i M_i^2}{\Sigma_i I_i M_i}$$

where I_i is the intensity for a given oligomer peak, i, in the molecular weight distribution, and M_i is the mass of the ith oligomer. For a narrow, symmetrical molecular weight distribution, the ratio M_w/M_n will be slightly larger than 1.0. As the distribution becomes more skewed and becomes broader, the ratio will increase.

Certain assumptions are inherent in calculating molecular weight averages from a TOF-SIMS oligomer distribution; these are summarized in Table 8.5. The first assumption is that fragmentation of a polymer is not a function of chain length. From the preceding discussion, this is clearly not valid. Large oligomers have a greater tendency to fragment than small oligomers. This effect will tend to shift the measured values for M_n or M_w to lower values. The second assumption is that the sputtering cross section of an oligomer is independent of chain length. The disappearance cross sections of cationized oligomers of polystyrene showed a linear increase over the range of mass 1000–7000,[10] so this assumption lacks validity. This effect will also tend to lower measured values of M_n and M_w. A third assumption is that cationization efficiency is the same for all oligomers. This seems unlikely, given that there are more points for attachment of an ion in a larger oligomer. This effect will tend

TABLE 8.5

Assumptions for Calculating M_n from TOF-SIMS Spectra

Assumption	Valid?	Favors
1. Fragmentation is not a function of chain length	No	Low molecular weights
2. Sputtering cross section independent of chain length	No	Low molecular weights
3. Cationization efficiency the same for all oligomers	No	High molecular weights
4. Detection efficiency not mass-dependent	No	Low molecular weights
5. Analyzer transmission not mass-dependent	Yes	Neither
6. Volatilization not mass-dependent	No	High molecular weights

to skew the molecular weight distribution toward higher values. Fourth, it is assumed that detector efficiency is not mass-dependent. This assumption is also invalid and will tend to shift the measured distribution to lower values. This effect will be discussed in some detail below. The fifth assumption is that the TOF analyzer transmission does not vary with molecular size and this is probably a valid assumption. Sixth, there is the assumption that there will be no preferential evaporation of oligomers from the probe in the SIMS vacuum. This is clearly not valid, given that small oligomers will have lower surface affinities and higher vapor pressures than large ones. The importance of this effect will be highly dependent on the specific polymer. The seemingly low intensity of the $n = 16$ and $n = 17$ oligomers in Figure 8.8 is probably due to evaporation; the absence of oligomers smaller than $n = 16$ is also probably due to this effect.

The efficiency of detectors used in TOF-SIMS is strongly mass-dependent for large ions, because detector efficiency varies with ion kinetic energy. In a TOF instrument, ion velocity (and therefore kinetic energy) is lower for large ions. Theoretically calculated curves for detection efficiency as a function of mass are shown in Figure 8.9.[24] The ion extraction voltage used in the TOF-SIMS is 3 keV, and ions emerging from the flight tube are typically postaccelerated by 10 keV before they strike the detector; the sum of the two determines the kinetic energy of an ion striking the detector. If one uses an average of the 15 and 20 keV curves of Figure 8.9 as typical for the TOF-SIMS, and assumes that an oligomer distribution will have a range of 1000 Da on either side of the center, for a polymer having $M_n = 6000$, the difference in detector efficiency between the highest and lowest MW oligomers would be ca. 25%.

The net effect of all of the parameters listed in Table 8.5 implies that TOF-SIMS measurement of M_n and M_w should be accurate for narrow MW distributions, but for polymers having broad MW distributions this would

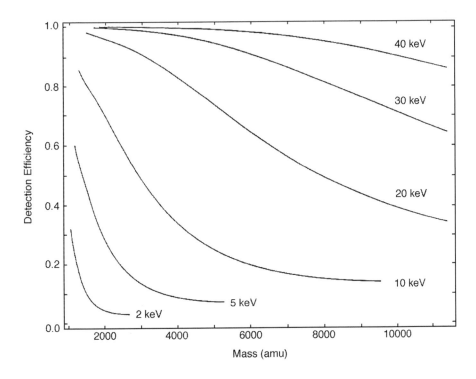

FIGURE 8.9
Detection efficiencies as a function of ion mass (from Ref. 24). Ion kinetic energies at the detector
are indicated.

not be the case. To illustrate this point, Figure 8.10 shows a TOF-SIMS
spectrum obtained for a mixture of two polystyrenes having M_n values of
2500 and 5000;[25] the two were mixed in a 1:1 mole ratio. Clearly the intensity
of the lower MW polymer (centered around 2600 Da) is much greater than
the higher MW polymer (centered around 5100 Da). Based on the data
from Figure 8.6, one would anticipate about an order of magnitude
decrease due to the difference in transformation probabilities, and from
Figure 8.9, the decrease in detector efficiency should be about 35% for the
higher MW polymer.

Another problem for TOF-SIMS measurements is the possible dependence
of M_n and M_w on sputtering time during acquisition of a spectrum. Low MW
oligomers may be sputtered preferentially to high mass oligomers introducing a time-dependent bias into the MW distribution measurement. The effect
of sputtering time on MW distributions was studied for polystyrenes having
M_n values of 1000, 3000, and 5000.[26] A decrease of 19% in the measured value
of M_n over a period of 60 min. was observed for the 1000 MW polymer, 6%
for 3000, and 1% for 1000. However, these changes occurred for sputtering
times much longer than are normally used for TOF-SIMS data acquisition;
1–5 min. data acquisition times are typical. Thus the effect of sputtering time

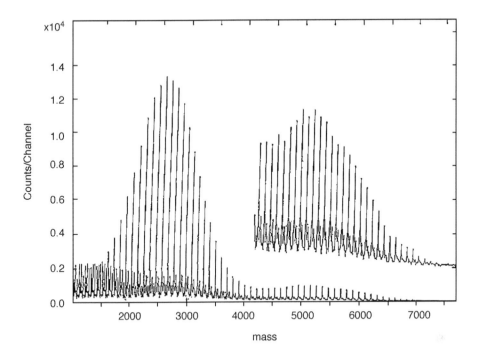

FIGURE 8.10
TOF-SIMS spectrum for a mixture of two polystyrenes (from Ref. 25).

on measurement of M_n and M_w for normal polymer samples would be extremely small.

8.2.3 Average Molecular Weight Measurements

Molecular weight distributions have been measured for a number of polymers using TOF-SIMS. Table 8.6 shows representative measurements (but not an exhaustive list) and compares them with results using conventional techniques like gel permeation chromatography (GPC), supercritical fluid chromatography (SCF), NMR, and end-group analysis. Measurements of M_n and M_w by TOF-SIMS shown in Table 8.6 agree quite well with results from conventional methods. These data indicate that TOF-SIMS can be used to measure MW distributions for polymers having low polydispersities and M_n values and for calculating their M_n/M_w ratios.

A more detailed comparison of TOF-SIMS with conventional methods is in order. GPC is probably the most widely used technique for MW measurements because it measures distributions of oligomers and is convenient to use. Generally GPC can resolve individual oligomers in the low mass (3000) range, but distributions merge into a broad band for higher molecular weights. Also, GPC is not applicable to very low MW polymers.[27] TOF-SIMS is very competitive with GPC for measuring MW distributions to M_n values

TABLE 8.6

Comparison of M_n Measurements for Polymers

Polymer	TOF-SIMS		Conventional		Ref.
	M_n	M_w/M_n	M_n	M_w/M_n	
Polybutadienes					
PB-900	861	1.05	982[a]	1.07	28
PB-2000	2490	1.01	2760[a]	1.08	28
Polydimethylsiloxanes					
SF-1	1650	1.10	1542[b]	1.09	29
SF-2	2748	1.1	2278[b]	1.07	29
Polystyrenes					
PS-1000	1020	1.1	930[a]	1.2	26
PS-5000	4620	1.03	4970[a]	1.04	26
PS-7000	6890	1.02	7390[a]	1.03	26
Perfluoropolyethers					
K-AA	2560	1.01	2450[c]		26
K-1525	3110	1.00	3420[c]		26
Polypropyleneglycols					
P-600	651	1.04	440[d]	1.07	30
P-1000	1004	1.03	877[d]	1.30	30
P-2000	1962	1.02	2074[d]	1.12	30
P-3000	2935	1.03	2197[d]	1.19	30
Polyethyleneglycol diesters dimethacrylates					
DM-200	409	1.01	350[c]		30
DM-1000	1105	1.02	1100[c]		30

[a] Gel permeation chromatography (GPC)
[b] Supercritical fluid chromatography (SFC)
[c] NMR
[d] Hydroxyl end-group determination

around 5000 because of its ability to resolve oligomers. A major advantage over GPC is that TOF-SIMS is an absolute method and GPC requires use of carefully selected standards.

Non-chromatographic techniques such as NMR, light scattering, boiling point elevation, and end-group analysis give values for M_n and/or M_w, but do not provide oligomer distributions. However, they are applicable to polymers of higher MW than is TOF-SIMS. A chromatographic or mass spectrometric method is preferred, assuming that the distribution measured is accurate, because it provides information about oligomer distributions. TOF-SIMS qualifies for polymers having $M_n < 6000$ and low polydispersities, and in some cases for larger polymers. An alternative TOF-SIMS approach has also been successful for measuring M_n.[31] By comparing negative-ion intensity ratios of fragment-ion peaks having terminal groups to those without, it was possible to accurately measure M_n up to 10,000 Da for perfluoropolyethers.

It is interesting that, even with all of the invalid assumptions as listed in Table 8.5, the results for TOF-SIMS and GPC agree quite well. The reason is that the comparison is only for low MW polymers of low polydispersity. Many low MW polymers have narrow MW distributions and fragmentation of low MW oligomers is not extensive. Fall-off in detector efficiency is not bad in the low-mass region and the effect of changes in sputtering cross-section will not be severe for narrow MW distributions.

Two other mass spectral methods are used to measure oligomer distributions, field-desorption mass spectrometry (FDMS) and matrix-assisted laser desorption/ionization (MALDI). FDMS has been applied effectively to polymers in the same MW range as TOF-SIMS.[2] The major disadvantage of FDMS relative to TOF-SIMS is in the ease of obtaining data. Although in the hands of experts FDMS is a powerful tool (see Chapter 6), many laboratories find it difficult to use and therefore do not use it routinely. MALDI was developed originally to measure the molecular weights of biopolymers; applications to synthetic polymer characterization have become prevalent only recently. In the low mass range MALDI can resolve oligomers as well as TOF-SIMS. However, MALDI provides superior measurement of oligomer distributions for higher MW polymers.[32] Polymers with molecular weights as high as 170,000 Da have been measured by MALDI.[33] (See Chapter 11.)

8.3 Fragment Ions

8.3.1 Introduction

When high molecular weight (MW of ca. 50,000) polymers deposited on a metal surface are bombarded with keV ions, the SIMS spectra observed differ significantly from spectra obtained by pyrolysis mass spectrometry (PyMS) for the same polymers. In conventional PyMS only small molecular fragments are observed (<300 Da) whereas fragments can be observed in SIMS spectra at masses greater than 2000 Da. Although there is a difference in the types of ions formed in SIMS and conventional PyMS, SIMS ions are cationized neutral fragments and PyMS ions have intrinsic charge, one would hope that there is a relationship between the two. The processes leading to the two differ also. PyMS is basically an isothermal process in that the molecules are heated uniformly, so when one bond has the energy to break, others can fragment similarly. SIMS is adiabatic in that the part of the chain over the action area (see Figure 8.7) receives a pulse of energy while the rest of the chain is at ambient temperature. However, species generated by SIMS fragmentation (presented below) are characteristic of processes involving thermal energies; thus it is reasonable that there would be a relationship between SIMS and PyMS. The time domains of the two measurements are different, picoseconds for SIMS and milliseconds for PyMS. Therefore the SIMS process may be looking at thermal chain fracture closer to the initial event, making it a useful method to probe processes fundamental to thermal degradation of polymers.

Another important question is the mechanism of chain fracture in SIMS. Much of the work on SIMS of polymers has been interpreted on the basis of free radical mechanisms. One reason for this is that most of the polymers that have been studied are formed by radical polymerization, and the tacit assumption has been that chain fracture is related to the depolymerization process. A second reason is that PyMS and low-mass SIMS data have generally been interpreted successfully on the basis of radical chain decompositions. However, consistent with the depolymerization argument, simple radical chain cleavage cannot explain the TOF-SIMS spectra of polymers formed by condensation reactions. This is consistent with proposals by Montaudo[59] for PyMS where other mechanisms provide better explanations for certain systems.

Charged species are observed in the low mass region of TOF-SIMS spectra (<500 Da), comparable to those in PyMS. They are characteristic of species commonly observed in EI mass spectrometry, a mixture of odd- and even-electron ions. It is possible that both ionic and radical mechanisms can be operative in SIMS; smaller fragments would be formed by radical mechanisms and larger fragments would result from both radical cleavage and intra-chain reactions such as hydrogen transfer. It also appears in some cases that functional group migrations occur, a phenomenon more common for ionic reactions than for radical processes. These issues will be discussed in greater detail for specific polymers.

As was illustrated in Figure 8.7, formation of polymer chain fragments in SIMS results from "reverse thrust" by surface disruption induced by impact of the primary ion. Chemically, this is the equivalent of giving a thermal pulse to the part of the chain overlying the action area. Therefore, a reasonable approach to interpretation is the premise that "the weakest bond breaks first." Accurate data on bond energies in polymers should allow semi-quantitative interpretation of fragmentation pathways. Unfortunately, such data are not available. At best one has only qualitative approximations of polymer bond energies computed by analogy with bond energies for a limited data set of small molecules.[41]

In the general sense, $C-C$, $C-O$, and $C-N$ single bonds will be the weakest in most polymers. Typically $C-C$ bond energies are 80–85 Kcal/mole, $C-O$ bonds are in the same range, and $C-N$ bonds are roughly 65–70 Kcal/mole. The $C-H$ bond energies are generally higher, normally 90–100 Kcal/mole, although in some cases they are in the range of $C-C$ bonds. Above these are multiple bonds, $C=C$, $C=O$, and $C\equiv N$, which have energies in the 120–140 Kcal/mole range. To date, successful interpretation of polymer fragmentation has considered that multiple bonds remain intact as do aromatic rings.

One should always remember that in SIMS one is looking at what is a minority event. Generally ion yields are 10^{-3} or lower which means that the majority process may differ from that reflected in the SIMS spectrum. One must assume, therefore, that the ions seen in SIMS do reflect the majority process, and that the low yields result from low-efficiency ionization processes (e.g., cationization). Another important assumption is that the metal surface plays no role in the fragmentation process per se. In general this last assumption seems to be valid. For example, although ion yields are strongly substrate-dependent,

there is no substrate influence on disappearance cross sections.[10] Also, cationization occurs similarly on metal surfaces and on non-metal surfaces doped
with the same metal ions.[11]

8.3.2   Nature of Fragment Ion Spectra

The nature of ions produced in SIMS spectra of polymers have been illustrated in Table 8.4. In this section we will deal with fragments which contribute to SIMS spectra in the mass range 500–7000 Da. We will consider primarily
spectra from polymers having molecular weights in the range of 20,000 or
more. Such polymers produce large neutral fragments without end-groups;
the fragments are usually composed of a number of monomer units, with or
without additional parts of a monomer unit. This section will deal with how
fragmentation occurs, and how these processes affect SIMS spectra.

A typical fragment ion spectrum from a high molecular weight polymer
(MW = 24,000) is shown for polybutadiene in Figure 8.11. As one can see, a series
of intense peaks is observed in the mass range 700–2500 Da, monotonically
decreasing in intensity with increasing mass; this spectrum approached the
noise limit for masses greater than 4000 Da. The spacing between the largest
peaks is 54 Da, the mass of the repeat unit of the polymer. One can also see

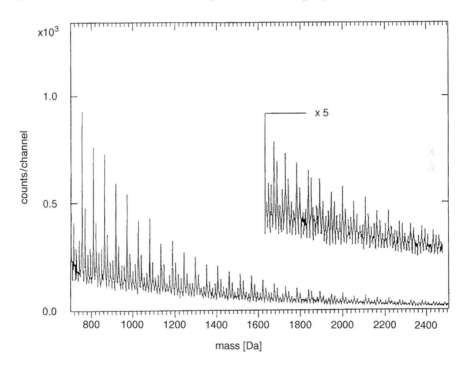

FIGURE 8.11
TOF-SIMS spectrum of polybutadiene, MW = 24,000. (Reprinted with permission from Ref. 28,
Copyright 1994 J. Wiley and Associates)

weaker series of peaks having the same spacing; these are due to fragments having additional parts of the repeat unit, e.g., 14 Da. For homopolymers the spectra are reasonably simple, but they become more complex as substitution occurs on the chain.

As polymers increase in complexity and the monomer units increase in mass, the appearance of the SIMS spectra change from that like Figure 8.11 to one containing a series of repeating patterns such as those shown in Figure 8.12 for substituted polystyrenes (PS). The PS spectrum consists of a very definite pattern, repeating itself to 2000–3000 Da. Fragmentation in the polystyrenes will be discussed in detail later in this chapter. The reason that the spectrum is shown here is to introduce the terminology used to

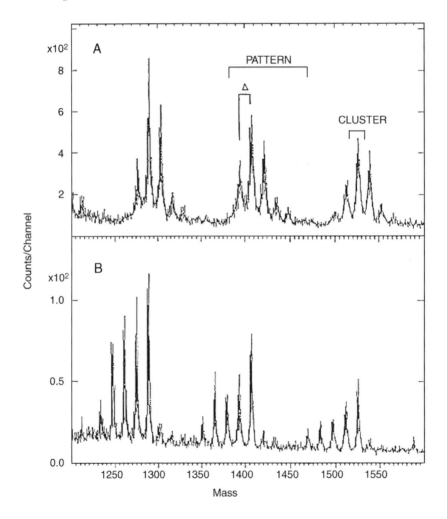

FIGURE 8.12

Patterns in TOF-SIMS spectra of polystyrenes. A. Poly(4-methylstyrene), B. Poly(α-methylstyrene). (Reprinted with permission from Ref. 35, Copyright 1992 American Chemical Society)

describe polymer fragment-ion spectra. The spectra consist of a repeating *pattern*, which occurs throughout the entire range; Figure 8.12 shows three of these. Repeat patterns have been observed in all polymers studied to date. The patterns consist of a series of *clusters*, also indicated in Figure 8.12. Each cluster is comprised of a series of individual peaks, generally separated by one mass unit. The spacing between clusters is given the symbol Δ, which is approximately 14 Da, depending on the exact distribution of intensities within the cluster. The typical cluster structure for PS is shown in Figure 8.13.

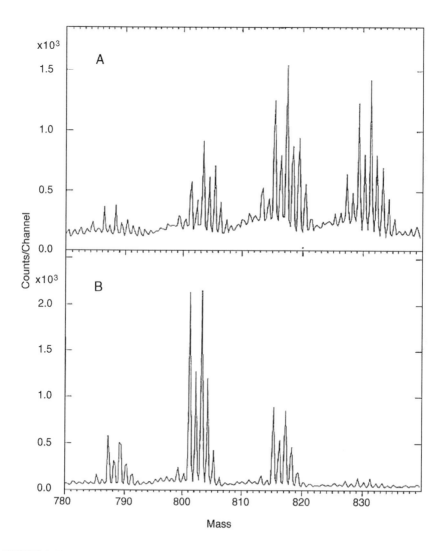

FIGURE 8.13

Detailed cluster structure for polystyrene TOF-SIMS spectra. A. Poly(4-methylstyrene), B. Poly(α-methylstyrene). (Reprinted with permission from Ref. 35, Copyright 1992 American Chemical Society)

TABLE 8.7

Assumptions for Treating Polymer Fragmentation

1. Semi-infinite chains (MW > 50,000)
   Terminal groups will not influence spectrum
2. Consider limited chain fracture
   A. Primary ion does *not* strike the polymer chain
   B. Chain fracture is caused by collision cascade
3. Chain fracture is an isolated event
   A. Low primary ion dose and low surface coverage
   B. *Inter*molecular phenomena not important
4. Chain fracture is a thermal event
   A. Single initial break
   B. Subsequent break for each part
   C. Neutral fragments are produced which cationize

It is important to consider the assumptions and formalism used to interpret polymer fragment-ion spectra. These are summarized in Table 8.7. First, it is assumed that the polymers exist as semi-infinite chains. This means that the molecular weight is large enough that terminal groups do not influence the SIMS spectrum. This assumption is generally valid for polymers having molecular weights greater than 20,000. The second assumption is that chain fracture is an isolated thermal event. This means that the primary ion does not strike the polymer directly, but penetrates the substrate and causes fragmentation by a collision cascade as discussed earlier. Chain fracture, therefore, is caused by the cascade and *not* by the primary ion itself. Therefore limited chain fracture occurs, which is characteristic of thermally induced chemical processes. A third assumption is that intermolecular phenomena are not important. The reason for this assumption is that polymers are deposited on Ag foils at low surface coverage and therefore one can ignore interactions between polymer chains. This also means that ions produced from two "hits" at the same point on the surface are unlikely. The ramification is that polymer fragmentation can be interpreted as an *intra*molecular phenomenon.

Because chain fracture is caused by breakup of the surface, fragmentation processes resemble thermal chemical reactions rather than resulting from the high- energy processes expected for collision between a keV ion and a polymer chain. The formalism used considers that a polymer chain undergoes an initial break randomly along the chain and away from the end-groups. This produces two radical ends, each of which can fragment farther down the chain. The fourth assumption is that hydrogen transfer produces neutral fragments that subsequently cationize by attachment of a silver ion, presumably in the "gas" phase over the substrate surface. This last assumption has validity, given that multiply charged ions are rarely observed in TOF-SIMS spectra.

On the basis of the above, one can interpret polymer fragmentation mechanisms with a statistical model which assumes that bond breaking along the chain is random. It also assumes that sufficient energy is available to break any bond along the chain. Consider a simple linear polymer consisting of two segments, A and B; the polymer would be .....ABAB(AB)$_n$..... where

(AB) is the polymer repeat unit. For example, in polystyrene, A would be the CH_2 group and B the $CH-C_6H_5$ group. A polystyrene chain is shown at the top of Figure 8.14; the white circle is the CH_2 group and the dark circle is the $CH-C_6H_5$ group. When the chain fragments initially, two different radical chain ends are produced, one terminated by $CH_2\bullet$ and the other by $\bullet CH-C_6H_5$. Each of these segments can fragment further as shown by Paths A and B in Figure 8.14, which are of equal probability. Only two types of chain fragments can be produced by Path A, one having an equal

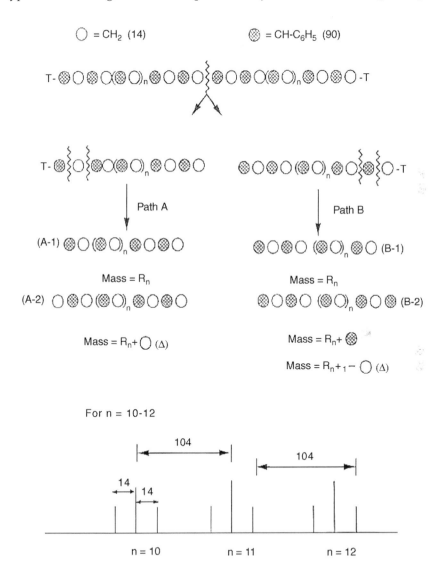

FIGURE 8.14
Simple statistical model of polymer fragmentation.

number of CH_2 and $CH-C_6H_5$ groups, and one having an extra CH_2 group. Path B also produces a fragment with an equal number of CH_2 and $CH-C_6H_5$ groups and one with an extra $CH-C_6H_5$ group (which is the equivalent of having one too few CH_2 groups). Therefore, the probability of producing a fragment with an equal number of CH_2 and $CH-C_6H_5$ groups is twice that of having either of the groups present in excess. This simple statistical model predicts a three-cluster pattern, with 14 Da internal spacing, and 104 Da between patterns as shown at the bottom of Figure 8.14. This closely resembles the cluster structure of the PS spectrum in Figure 8.12.

If one also considers chemical effects, weaker bonds would have a higher probability of breaking and therefore processes which require breaking of weaker bonds would be more likely. Also, radicals produced by the initial chain fracture will have different activities for hydrogen atom extraction, and therefore will also influence the pattern. Thus, if chemical effects are operative, one would not necessarily expect to observe the relative intensities of the peaks within a cluster predicted by statistical cleavage. Also, as one introduces more bonds into the repeat unit, chemical effects will become more likely simply because more radical reactions are possible. The reality of the situation is that both chemical and statistical effects are important. To some extent the spectra can be interpreted statistically by considering only fracture of single bonds in the polymer. However, it will become clear that in some cases this treatment is inadequate and more complicated mechanisms must be considered.

8.4   Homopolymer Spectra

This section will treat the TOF-SIMS spectra of homopolymers taken in the high mass range, >600 Da. Emphasis will be on interpreting the spectra using the statistical model presented above, and indicating where it fails and where chemical considerations become important. The research surveyed will be mainly from our research group along with Benninghoven's because we have collaborated extensively in this area for more than a decade. The specific polymer systems to be surveyed are simple linear polymers (butadiene, isoprene, isobutylene, ethylene glycol, propylene glycol), substituted vinyl polymers (styrenes, acrylates, methacrylates), polyesters, polyamides, and polysiloxanes. Not all details will be presented; the original references should be cited for greater detail.

8.4.1   Linear Homopolymers

Linear homopolymers represent the simplest case for studying polymer fragmentation in SIMS. They have only a limited number of fragmentation possibilities because all bonds (except for $C-H$) are backbone bonds.

The simplest polymer is polyethylene in which all C—C bonds are equivalent and the C—H bond energy is greater than that of C—C. Thus, one would expect a monotonically decreasing series of peaks separated by 14 Da, and such is observed.

8.4.1.1   Polybutadiene

The C—C bonds are much weaker than the C=C bonds, so the spectra can be interpreted by fracture of only C—C single bonds. One complication is that there are two types of C—C single bonds, CH_2—CH_2 and CH_2—CH=CH. The spectrum consists of a series of peaks having unequal intensity, but occurring in a regular pattern as shown in Figure 8.11. Figure 8.15 shows the two possible pathways for chain fracture of polybutadiene involving only C—C single bonds.[28] In Path A the CH_2—CH_2 bond breaks, initially producing two identical fragments. Each of these can subsequently break as shown, producing one of three fragments: an even number of repeat units (R_n); an even number of repeat units plus a CH_2 group ($+\Delta$), and an even number of repeat units minus a CH_2 group ($-\Delta$). In Path B two different radicals are produced by the initial chain fragmentation and they can each fragment to produce different chain segments. Path B1 yields R_n, $+\Delta$ and $+2\Delta$ (two extra CH_2 groups), while Path B2 yields R_n, $-\Delta$ and $+2\Delta$. If one assumes equal probabilities for Paths A and B and equal probabilities for the branching of Path B, the relative intensities of the peaks within a cluster predicted statistically for R_n:$+\Delta$:$+2\Delta$:$-\Delta$ would be 33:25:17:25. Experimental values measured for high molecular weight polymers to avoid influence of end-groups were R_n:$+\Delta$:$+2\Delta$:$-\Delta$ equal to 35:25:20:20. Agreement is quite good, so one can conclude that fragmentation of polybutadiene is essentially statistical and involves only fracture of C—C single bonds.

8.4.1.2   Poly(ethylene glycol)

The C—C and C—O bonds in poly(ethylene glycol)s (PEG) have similar energies, with the C—C bond energies being slightly larger. Thus one would expect a PEG spectrum to resemble that of polyethylene, a series of monotonically decreasing peaks approximately 14–16 Da apart. A PEG spectrum (MW = 100,000) is shown in Figure 8.16. What one actually sees is a pattern consisting of two intense clusters and a third about half the intensity of the other two; this repeat pattern continues for the entire spectrum. The relative intensities of the fragment clusters for PEGs of different molecular weights are shown in Figure 8.17.

There are only two possible initial chain fragmentation events, scission of a C—O or a C—C bond; these are shown as Paths A and B, respectively, in Figure 8.18. Each radical can fragment to yield the same three fragments, R_n, $+\Delta$, and $-\Delta$. If only statistical effects are considered, the intensity ratios of the three peaks should be 1:1:1. However, if fracture of the C—O bond is

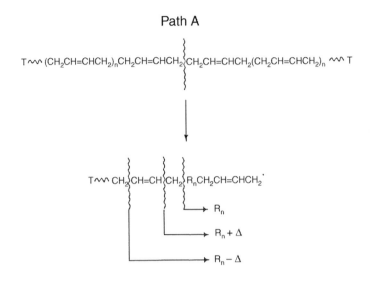

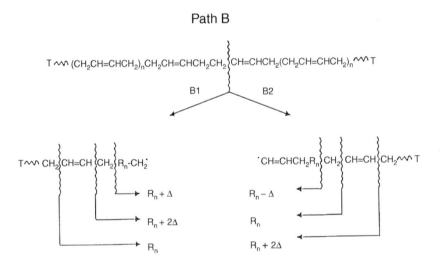

FIGURE 8.15
Fragmentation pathways for polybutadiene.

more probable than the C—C bond by a factor of three, the mechanism is consistent with the observed 2:2:1 ratio.[30] However, it seems unlikely that the observed difference in relative intensities could result from the small difference in C—C and C—O bond energies (ca. 5 kcal/mole). A more likely reason is that the difference arises from different reactivities of carbon- and oxygen-terminated radicals.

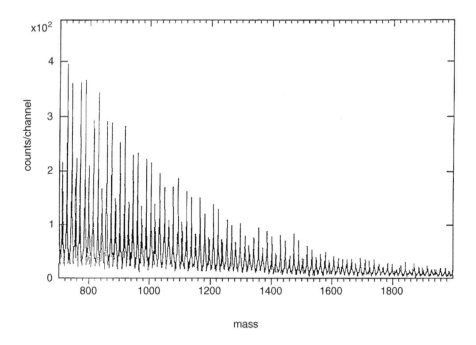

FIGURE 8.16

TOF-SIMS spectrum of poly(ethylene glycol), MW = 100,000. (Reprinted with permission from Ref. 30, Copyright 1994 American Chemical Society)

8.4.1.3   Methyl Substituted Linear Polymers

Methyl substitution of PEG or polybutadiene has a dramatic effect on the high-mass TOF-SIMS spectra. There are two effects: an increase in the number of peaks in a pattern, and a change in relative cluster intensities. The reasons for this can be seen from the scheme for fragmentation of poly(propylene glycol) (PPG) shown in Figure 8.19. Compared with the comparable scheme for PEG, there are four primary paths that must be considered. Paths A and B are comparable to those for PEG in Figure 8.18. Path C is for C—O bond cleavage on the other side of the C—O bond producing radicals different from those in Path A. Path D corresponds to initial loss of the methyl group and subsequent chain fracture to yield a terminal olefin (D2). Four clusters are observed for PPG patterns corresponding to R_n, $+\Delta$, $+2\Delta$, and $-\Delta$. Simple statistical fragmentation with all paths having equal probability predicts intensities of R_n:$+\Delta$:$+2\Delta$:$-\Delta$ of 8:5:5:6; the observed intensities for high molecular weight PPGs (MW = 100,000) are 8:5:4:5.[30]

A similar situation is observed for comparison of poly(butadiene) (PBD) and poly(isoprene) (PIP).[34] Whereas four clusters are observed for a PBD pattern, five are seen for PIP. The predicted ratio for statistical cleavage of the PIP chain, R_n:$+\Delta$:$+2\Delta$:$+3\Delta$:$-\Delta$, is 8:5:2:3:6; the observed ratio is 8:5:3:3:4 for a 410,000 MW polymer. The relative intensities of clusters within the patterns vary significantly depending on the number of repeat units in the pattern.

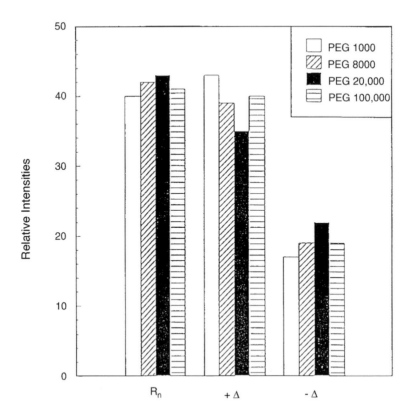

FIGURE 8.17
Relative intensities of clusters for poly(ethylene glycols). (Reprinted with permission from Ref. 30, Copyright 1994 American Chemical Society)

For example, for a MW = 410,000 PIP, the R_n cluster accounts for about 35% of the total pattern intensity for $n = 10$, but about 50% for $n = 17$.

8.4.1.4 Cluster Structure

The distributions of the individual peak intensities within the clusters provide information about the nature of the species produced by polymer fragmentation. Consider, for example, the cluster structure for a PIP with MW = 3000, shown in Figure 8.20. The mass spectrum of each species will have contributions from the isotopic distributions of carbon (^{12}C, ^{13}C) and silver (^{107}Ag, ^{109}Ag). The single species spectrum in Figure 8.20C is that expected for R_n having $n = 5$. Contrast this with the spectrum in Figure 8.20A which is observed for PIP 3000. Clearly there are more peaks in Figure 8.20A than in 20C with the "extra peaks" being at higher mass. This means that there are species with a higher level of saturation present. By use of linear least-squares regression analysis, one can demonstrate that three species are necessary to describe the spectrum in Figure 8.20A, having isotope structures

PATH A

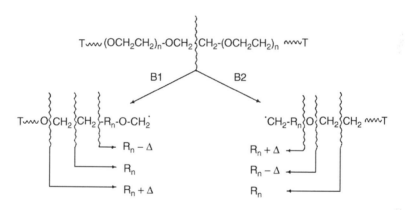

PATH B

FIGURE 8.18
Fragmentation pathways for poly(ethylene glycol). (Reprinted with permission from Ref. 30, Copyright 1994 American Chemical Society)

like that of Figure 8.20C. By considering three species such as those shown below

$$CH_3-CMe=CHCH_2-Rx-CH_2CMe=CHCH_3 \tag{8.1}$$
(No extra double bonds)

$$CH_2=CMeCH=CH-Rx-CH_2CMe=CHCH_3 \tag{8.2}$$
(One extra double bond)

$$CH_2=CMe-CH=CH-Rx-CH=CMeCH=CH_2 \tag{8.3}$$
(Two extra double bonds)

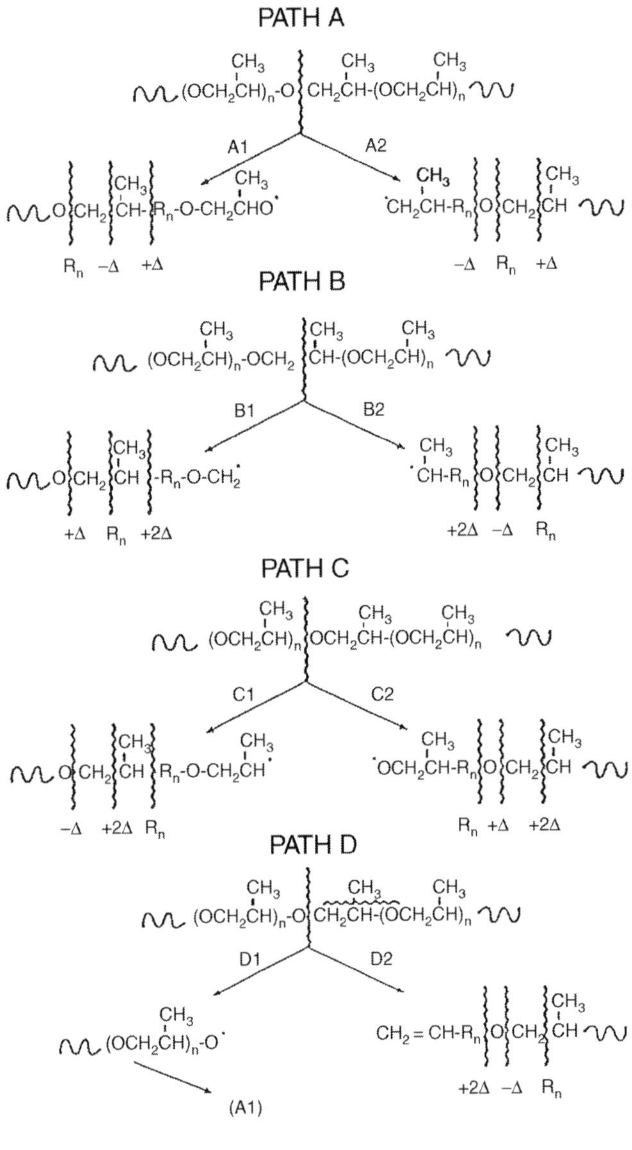

FIGURE 8.19
Fragmentation pathways for poly(propylene glycol). (Reprinted with permission from Ref. 30, Copyright 1994 American Chemical Society)

one can construct a theoretical spectrum (Figure 8.20B) which matches the observed spectrum quite well.[34] The spectrum in Figure 8.20B was calculated by assuming 38% of a species like (8.1), 53% like (8.2), and 9% like (8.3).

This approach was used to demonstrate that the same levels of unsaturation can explain the cluster structure of PBD TOF-SIMS spectra.[28] It also demonstrated that the degrees of unsaturation vary with the cluster, as

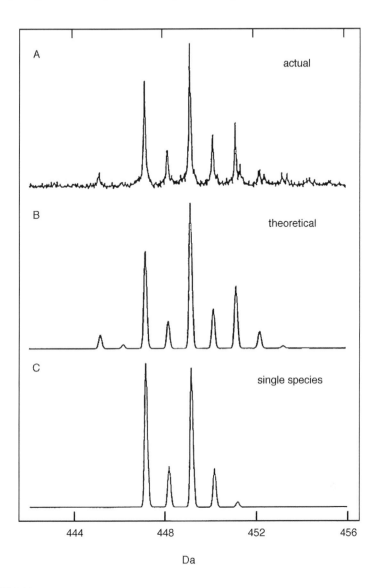

FIGURE 8.20
Cluster structure for polyisoprene. A. Observed spectrum, B. Calculated spectrum, C. Cluster for a single species. (Reprinted with permission from Ref. 34, Copyright 1996 Springer-Verlag)

shown in Table 8.8. The theoretical ratios shown in the table were calculated by assuming that secondary radicals formed in PBD chain fracture would only abstract a hydrogen atom, but that primary radicals could either abstract a hydrogen or form a double bond. The agreement is far from perfect (except for the $+\Delta$ species), but the potential of the method is illustrated. Formation of cyclic species or double bond rearrangements were not accounted for, nor was the problem of multiple oligomer fragmentation addressed.

TABLE 8.8

Linear Least-squares Regression Analysis of Species
Within Fragment Clusters

Cluster	No. of Unsaturations			Theoretical Ratios
	2	1	0	
R_n	18 ± 5	60 ± 6	22 ± 5	1:2:1
$R_n + \Delta$	35 ± 4	34 ± 4	31 ± 3	1:1:1
$R_n + 2\Delta$	52 ± 3	20 ± 4	29 ± 4	2:2:1
$R_n - \Delta$	77 ± 12	23 ± 12	—	1:1

8.4.2 Polystyrenes

The polystyrenes have been studied more extensively than other homopolymer systems and have provided insights into SIMS polymer fragmentation
mechanisms. Therefore, even though they are simple monosubstituted linear
polymers, they will be discussed as a separate class. Segments of the high-
mass spectra of poly(4-methylstyrene) (P4MS) and poly(α-methylstyrene)
(PAMS) are shown in Figure 8.12. The P4MS spectrum is characteristic of
polystyrene (PS) and other styrenes in which methyl groups are substituted
on the phenyl ring; poly(vinylpyridine) has a similar spectrum.[42] Clearly,
PAMS is different. P4MS will be discussed as the "typical" example.[35]

8.4.2.1 Poly(4-methylstyrene)

The pattern of P4MS consists of five clusters, three strong and two weak.
The three strong ones correspond to R_n, $-\Delta$ and $+\Delta$; the weaker to $+2\Delta$ and
$+3\Delta$. The relative intensities of the clusters within a pattern vary with the
size of the pattern, but a typical value for PS with $n = 15$, for $-\Delta:R_n:+\Delta:+2\Delta:+3\Delta$
is 24:30:23:13:10. Methoxy and di-methyl substituted polystyrenes show an
additional peak at -2Δ.

The patterns for PS and P4MS can be explained semi-quantitatively by
considering two initial chain-breaking events, fracture of the polymer chain
and loss of a phenyl group, as shown in Figure 8.21.[35] Path A produces both
primary and secondary radicals which cleave subsequently to yield only R_n,
$+\Delta$ and $-\Delta$ fragments in a ratio, $-\Delta:R_n:+\Delta$, of 1:2:1. Path B accounts for the
other two peaks (B-1); path B-2 is redundant with path A-1. Thus the intensities
of the five peaks will be determined by the relative importance of paths A
and B. One should also consider loss of the allylic hydrogen as an initial
chain-breaking event because it is weaker than the other C—H bonds. This
would result in breaking the chain C—C bond, the equivalent to path A. The
nature of the chain ends would be different, but would affect only the
intensity distributions within the clusters, not the relative cluster intensities.

The best way to evaluate the relative cluster intensities for PS and P4MS
is to compare the coefficients for the individual fragments in Figure 8.21,
calculated statistically, with those calculated from observed spectral intensities.

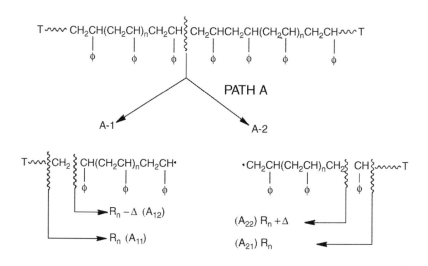

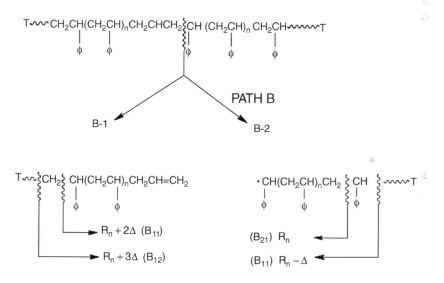

FIGURE 8.21

Fragmentation pathways for polystyrenes. (Reprinted with permission from Ref. 35, Copyright 1992 American Chemical Society)

It is possible to solve a set of simultaneous equations relating the various pathways to peak intensities in the SIMS spectra. These are compared in Table 8.9; the theoretical values are based on equal probabilities of chain fracture and phenyl loss as the primary event. Remember that A_{11} and A_{12} will be redundant with B_{21} and B_{22}. The theoretical and experimental values agree well except for A_{21} and A_{22}; the former is too small and the latter too large.

TABLE 8.9

Branching Ratios for Polystyrene Fragmentation

	Coefficient					
Polystyrene	A_{11}	A_{12}	A_{21}	A_{22}	B_{11}	B_{12}
Polystyrene	26	24	3.5	23	13	11
4-Methyl	22	28	1.3	28	12	9
4-Methoxy	29	20	3.2	27	14	5.5
2,4-Dimethyl	31	18	0.0	30	13	7.4
2,5-Dimethyl	30	18	0.0	22	22	8.5
2,6-Dimethyl-4-*tert*-butyl	29	21	3.0	27	11	8.4

This is reflected in the spectrum as a larger than expected intensity for the +Δ peak. Similar behavior is observed for the substituted polystyrenes. This represented a puzzle in the original paper;[35] however, there is a logical explanation. Pathway A_{22} is the only one in the scheme that produces a radical having a terminal $CH_2 \bullet$ group. There is considerable evidence (*vide infra*) that the second step in the chain cleavage process involves H-atom extraction by the primary radical. It is quite likely that abstraction of the allylic hydrogen would be quite efficient, and this would cause cleavage one CH_2 group beyond that shown in Figure 8.21 for A_{21}. The net effect would be to change the product from an R_n fragment to a +Δ fragment. The modified theoretical coefficients shown in Table 8.9 were calculated using this idea; they closely match the observed values. The relative cluster intensities do not vary significantly for PS and P4MS as a function of fragment size. For chain segments having from 5 to 20 repeat units, there is a small increase in the relative intensity of R_n at the expense primarily of the +Δ fragment.

The polystyrene data provide evidence that a cyclic intermediate is involved as part of the chain-breaking mechanism. If one assumes that the kinetics of polymerization and depolymerization are the same, and that solution and gas-phase kinetics are governed by the same molecular factors, then one can apply the relationship:

$$k_{cy} = AM^{-\gamma} \tag{8.4}$$

where k_{cy} is the cyclization rate constant, M is the monomer number, and A and Y are constants. Normalized plots of cyclization rates vs. monomer number were found to have slopes between −1.3 and −2.0, depending on the solvent.[36] Similar least-squares plots for PS and P4MS SIMS data gave slopes of −1.62 and −2.32, respectively, in the same range as the solution cyclization data. These results support the idea that a cyclic intermediate is involved in the second chain-breaking step rather than simply chain fracture. This is consistent with allylic hydrogen abstraction as discussed above. Thus it is quite likely, at least for the polystyrenes, that initial chain fracture produces radicals that then attack the polymer chain to which they are attached. This interpretation is also supported by the lack of intermolecular hydrogen abstraction (*vide infra*) and the observation that chain fragments having masses lower than 600–800 Da are seldom observed for any polymer in TOF-SIMS.

The cluster structure of P4MS can be explained by the presence of three species having 0, 1, or 2 double bonds, as was the case for the simple linear polymers. The relative amounts of the three vary somewhat among the polystyrenes. The ratios of the number of unsaturations, 2:1:0, averaged for clusters with $n = 5$–7, are 26:45:29 for PS, 26:37:37 for P4MS, 39:39:22 for 4-methoxy PS, and 21:37:42 for 2,5-dimethyl PS. The distributions also vary with fragment size for the different polymers, for example, the ratio changes to 22:66:12 for $n = 12$ for P4MS.

8.4.2.2   Poly(α-methylstyrene)

The "oddball" of the polystyrenes is PAMS as can be seen from the spectra in Figure 8.12. Whereas the reaction schemes shown in Figure 8.21 explain many aspects of the SIMS spectra of the other polystyrenes, this scheme is unable to do so for PAMS. The most striking feature of the PAMS spectrum is that the $+\Delta$ peak is greatly reduced in intensity, and significant peaks appear at -2Δ, -3Δ, and -4Δ. Also, the detailed structure of the clusters indicates that the species produced mostly have one degree of unsaturation; the 2:1:0 ratio is 9:86:5. A further complication is that the relative cluster intensities vary significantly with size for fragments with fewer than 12 monomer units. For example, for $n = 5$ the most intense peak is $-\Delta$, for $n = 8$ it is -2Δ, and for $n = 12$ it is R_n.

It is not possible to explain the spectra of PAMS using the scheme shown in Figure 8.21, even considering loss of the α-methyl group as an initial event. This process produces only R_n and $-\Delta$ fragments. By considering loss of the methyl group as a major initiating event, and rearrangements encountered in pulse radiolysis studies,[37] it is possible to explain the intense R_n and $-\Delta$ peaks along with loss of the $+\Delta$, $+2\Delta$, and $+3\Delta$ peaks.[35] A little creative organic chemistry can define possible pathways for producing -2Δ and -3Δ, but the -4Δ peak cannot be accounted for.

8.4.2.3   Hydrogen-Deuterium Exchange

One important assumption made for interpreting high-mass fragment TOF-SIMS spectra is that they result from isolated, unimolecular processes. This requires that reactions between polymer chains be small compared to reactions within a chain. This premise was tested by studying H-D exchange between perdeutero-PS (DPS) and perhydro-PS (HPS).[38] A 1:1 mixture of DPS and HPS was used; the polymers had $M_n = 4500$.

Specifically, *inter*molecular H-D exchange was studied for the $n = 6$–10 R_n and $-\Delta$ fragments.[38] The kinds of structures considered and the isotopic patterns of their SIMS spectra are shown in Figure 8.22. Varying degrees of

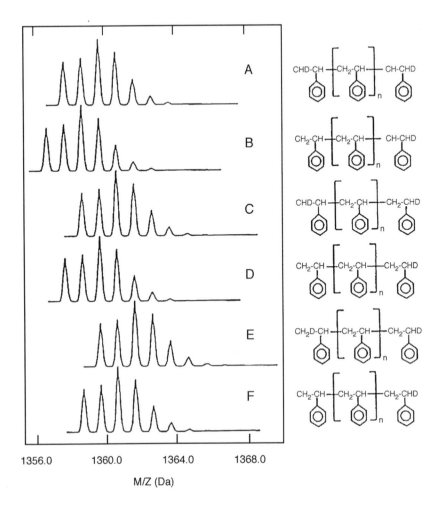

FIGURE 8.22

Cluster structures for polystyrene from H-D exchange studies. (Reprinted with permission from Ref. 38, Copyright 1994 American Chemical Society)

unsaturation and different levels of deuterium exchange both affect the spectrum and must be considered. Clearly there will be redundancies, e.g., structures C and F and A and D. Hence, different degrees of deuterium substitution and unsaturation will affect relative intensities of the peaks within a given cluster. The R_n cluster for $n = 12$, shown in Figure 8.22, will be a composite of the six spectra shown.

Obviously, determining whether or not the relative intensities of peaks in a given cluster have been affected by H-D exchange is a complex problem. Two different methods were used for this study of fragment ions. First, the relative peak intensities were measured for a given cluster for both HPS and DPS alone and the statistical limits for each peak intensity were established. For example, Figure 8.23 shows the $n = 7$ R_n cluster pattern for HPS; the

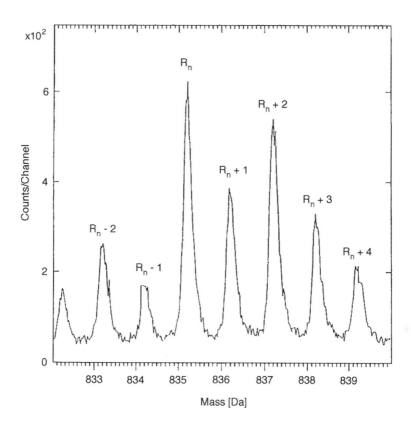

FIGURE 8.23

Polystyrene H-D exchange studies. Cluster pattern for R_n having $n = 7$. (Reprinted with permission from Ref. 38, Copyright 1994 American Chemical Society)

same series was measured for DPS. The HPS and DPS data are shown in Table 8.10, marked "HPS alone" and "DPS alone," respectively. Next to these columns are the relative intensity data for a 50:50 mixture of HPS and DPS measured at their respective masses, along with their standard deviations. In the next columns marked "pooled STD" are the pooled variances of HPS, DPS, and their 50:50 mixtures. The columns labeled "HPS-mixture" and "DPS-mixture" are the absolute values of the differences between the individual polymers and the 50:50 mixture. When the HPS- and DPS-mixture values were t-tested against their pooled STDs at the 95% confidence level, no significant difference could be found for any case.

A second method was used to determine if there was a significant difference between HPS, DPS, and the 50:50 mixture. Linear least-squares regression analysis was used to correlate the spectra of HPS and DPS with the 50:50 mixture. The point of maximum correlation was calculated as 60 ± 4% HPS and 40 ± 7% DPS. The residuals showed no systematic series of peaks indicating that the deviations from 50:50 were due to random chance. Therefore, based on two data analysis methods, one can conclude that there is not a significant amount

TABLE 8.10

Comparison of R_n Clusters for Individual Components and Mixture

Peak	HPS Alone[a]	50/50 Mixture[a]	σ_p (Pooled STD)	Δ_{actual} (HPS-Mixture)	DPS Alone[a]	50/50 Mixture[a]	σ_p (Pooled STD)	Δ_{actual} (DPS - Mixture)
$R_n - 2$	12.0 ± 3.7	8.7 ± 1.7	2.9	3.3	14.6 ± 1.2	13.3 ± 6.6	5.1	1.3
$R_n - 1$	8.0 ± 2.6	6.8 ± 1.8	2.2	1.2	18.3 ± 1.2	15.5 ± 3.5	2.8	1.8
R_n	21.0 ± 3.0	21.6 ± 2.5	2.8	0.6	17.2 ± 2.7	17.1 ± 3.8	3.4	0.1
$R_n + 1$	14.2 ± 2.2	15.1 ± 3.3	2.8	0.9	18.2 ± 2.8	18.3 ± 3.7	3.4	0.1
$R_n + 2$	20.4 ± 1.9	21.6 ± 2.7	2.3	0.8	14.9 ± 2.0	15.9 ± 4.0	3.3	1.0
$R_n + 3$	13.5 ± 2.3	15.7 ± 3.9	3.2	2.2	9.8 ± 2.6	11.5 ± 2.5	2.6	1.7
$R_n + 4$	10.8 ± 2.1	10.3 ± 2.4	2.2	0.5	7.0 ± 1.7	8.4 ± 2.7	2.3	0.6

[a] Relative peak intensities.

of intermolecular hydrogen transfer in polystyrene chain fragmentation. It is possible that small amounts of H-D exchange could occur and be missed by the data analysis methods, but this would not amount to more than 5–10%.

Partially deuterated polystyrenes (PS–d_3 and PS–d_5) were studied by TOF-SIMS, using regression analysis to determine if *intra*molecular H-D exchange occurs in PS.[39] PS–d_3 has the carbon atoms on the chain deuterated and PS–d_5 has its D atoms on the phenyl rings. The exchange of interest is between the chain and the phenyl ring. This is a complex problem, given that one is dealing with 0, 1, or 2 unsaturations and a number of possible H-D exchanges. Analysis of the PS–d_5 +Δ cluster structure for $n = 7$ showed the presence of six components; these were consistent with the occurrence of at least one H-D exchange. It was not possible to be more definitive because of redundancies between 1U + 1E and 2U + 2E, and 0U + 0E and 1U + 2E, where U = unsaturation and E = exchange. This was confirmed by studies on the R_n, +Δ and −Δ clusters of PS–d_3, with evidence supporting an additional exchange, especially for the species having zero unsaturations. These results are particularly significant in that they support the idea that intramolecular hydrogen abstraction plays an important role in polymer chain fragmentation.

8.4.3 Acrylic Polymers

Acrylics are unsymmetrically substituted polyethylenes which are widely used and of technological importance. The high mass range spectra of the acrylics provide valuable insight into the SIMS fragmentation process, so they will be treated as a separate category. Both methacrylates and acrylates will be discussed.

8.4.3.1 Methacrylates

The poly(alkyl methacrylate)s (PAMA) show repeating patterns in the spectral region 800–4000 Da. Poly(cyclohexyl methacrylate) is a typical example; the mass region of the TOF-SIMS spectrum from 1000–1600 Da is shown in Figure 8.24.[40] The repeating pattern is complex, consisting of 12 clusters; as

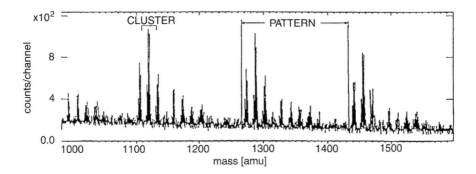

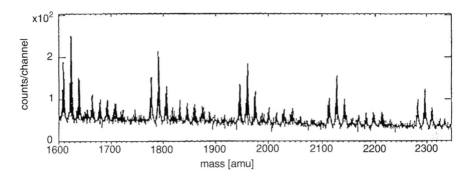

FIGURE 8.24

TOF-SIMS spectrum of poly(cyclohexyl methacrylate). (Reprinted with permission from Ref. 40, Copyright 1993 American Chemical Society)

for other PAMAs the TOF-SIMS spectrum is a specific fingerprint. The discussion here will focus on generalities.

All PAMA spectral patterns show clusters at R_n, $+\Delta$, $+2\Delta$, $+3\Delta$, $+4\Delta$, and $-\Delta$; these clusters will be referred to as "main-chain fragments" because they are formed largely independent of the ester group. An illustration of the relative intensities for PMMAs with small ester R groups is shown in Figure 8.25 for the methyl (PMMA), ethyl (PEMA), and n-propyl (PnPMA) methacrylates. The spectra in the figure are arranged so that the peaks from their main-chain fragments are aligned. The relative intensities of these peaks, normalized to the R_n peak, are amazingly constant as shown in Figure 8.26. Each fragment intensity was averaged over five different sets of clusters. This consistency indicates that these chain fragments are formed largely independent of the ester group. The major effect of the ester group is to add additional small clusters between the $-\Delta$ and $+4\Delta$ peaks as seen in Figure 8.25. The number of peaks in this region reflects the size of the ester R chain.

One must consider at least three initial bond-breaking events when formulating a mechanistic pathway for PAMA chain fracture: loss of the ester group, loss of the main chain methyl group, and fracture of the main chain C—C bond. These are shown as processes A, B, and C, respectively, in Figure 8.27.

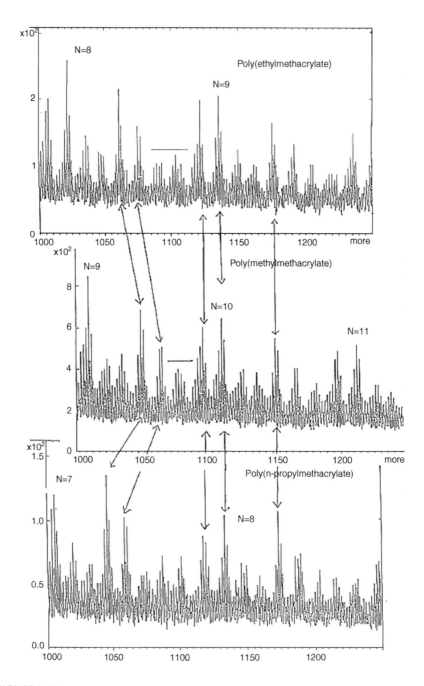

FIGURE 8.25
Poly(methacrylate) spectra showing main chain fragments. (Reprinted with permission from Ref. 40, Copyright 1993 American Chemical Society)

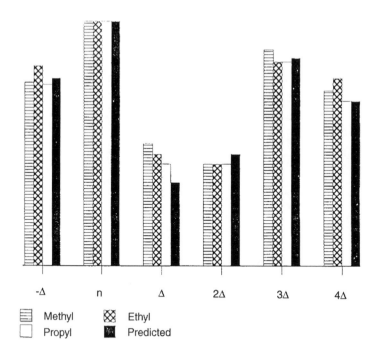

FIGURE 8.26
Comparison of cluster intensities for poly(methactylates). (Reprinted with permission from Ref. 40, Copyright 1993 American Chemical Society)

Path A-1 leads to a terminal olefin and subsequent fragmentation will give $+4\Delta$ and $+3\Delta$ fragments. Path A-2 will yield R_n and $-\Delta$ fragments. Loss of a methyl group will also yield a terminal olefin, via B-1, and a radical via B-2 redundant with A-2; both pathways yield R_n and $-\Delta$ fragments. Initial fracture of the main chain will give the usual $-\Delta$, R_n, and $+\Delta$ fragments; path C-2 is also redundant with A-2. The three pathways shown account for all of the main-chain fragments except for $+2\Delta$. This fragment can be introduced into the scheme by considering loss of a formate ester from A-2 and subsequent fragmentation to yield $+2\Delta$ and $+3\Delta$.

The scheme in Figure 8.27 can account for the clusters qualitatively, but it is not possible to use the same semi-quantitative treatment as was done for the polystyrenes (too many unknowns). However, if one makes assumptions about the relative importance of the three paths, it is possible to approximate this treatment. Assume that the major mode of initial scission is path A and that it accounts for 60% of the intensity. Similarly, assume that path B accounts for 10% and C for 30%. Assume that for radical A-2, fragments R_n and $-\Delta$, and loss of HCOOR have equal probabilities and that other branches (e.g., branching between B-1 and B-2) are 50:50. This would predict that the ratio $-\Delta$:R_n:$+2\Delta$:$+3\Delta$:$+4\Delta$ would be 17.5:25:7.5:10:22:18.[40] These values are shown as the black (predicted) lines in Figure 8.26. Agreement is quite good indicating

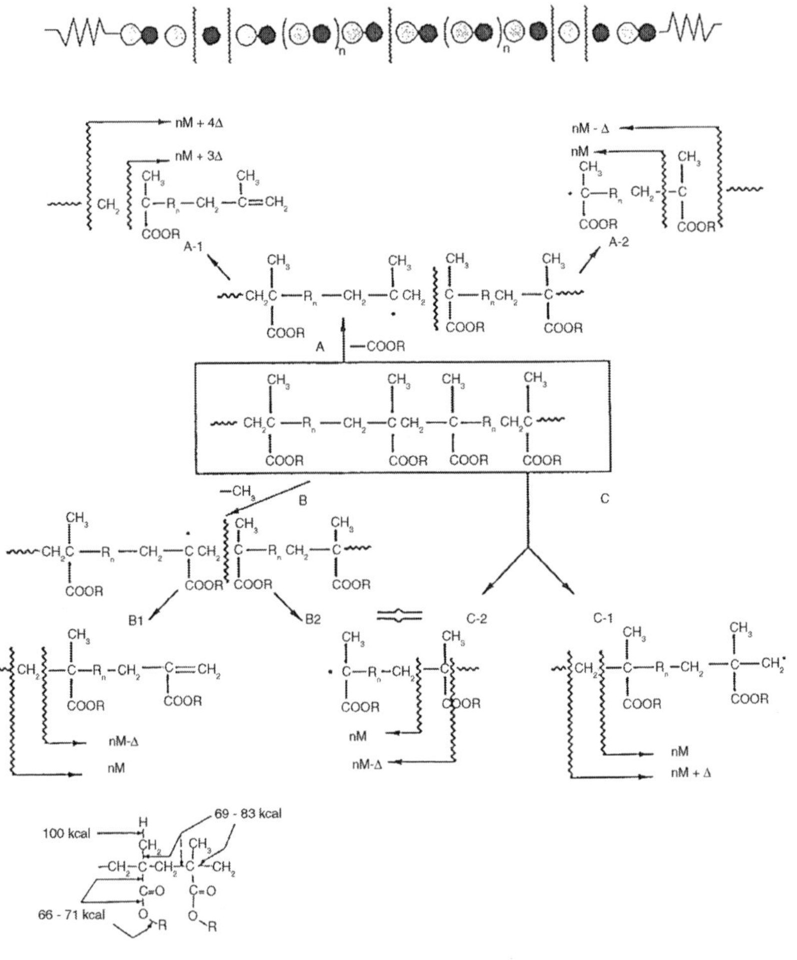

FIGURE 8.27
Fragmentation pathways for poly(alkylmethacrylates). (Reprinted with permission from Ref. 40, Copyright 1993 American Chemical Society)

that all three processes initiate polymer chain fragmentation, although initial loss of the ester group is the major one.

The relative intensities of the individual peaks within a cluster are nearly identical for a given polymer, independent of fragment size. Least-squares curve fitting showed that four is the minimum number of species contributing to each cluster, corresponding to completely saturated species, and those containing one, two, and three rings and/or double bonds. The assumption was made that all unsaturation in a fragment is on the main chain and not in the ester group. The relative intensities for the species contributing to a cluster vary considerably from one polymer to the other, despite the consistency for a given polymer. The variation could not be correlated with simple structural changes.[40]

TABLE 8.11

Relative Intensities of the Isometric C-6 Poly(acrylate) Clusters

	-5Δ	$-\Delta$	$n\mathrm{M}$	Δ	2Δ	3Δ	4Δ
Poly(cyclohexyl methacrylate)	39	76	100	56	38	59	48
Poly(benzyl methacrylate)	36	87	100	49	34	98	71
Poly(phenyl methacrylate)	142	114	100	71	73	150	104
Poly(n-hexyl methacrylate)	121	126	100	76	81	134	93

Two sets of isomeric methacrylates were studied to investigate changes in spectra caused by variation of the ester R group: the isomeric butyl, and the C—6 and C—7 methacrylates. Data for four separate patterns were analyzed. For the butyl polymers, the main-chain fragments, $-\Delta$, $+\Delta$, $+2\Delta$, $+3\Delta$, and $+4\Delta$, decreased relative to R_n in the order: *n*-butyl>*i*-butyl>*s*-butyl>*t*-butyl. This order is what would be expected qualitatively based on the availability of hydrogens for secondary cleavage. Thus, as the ester R group becomes longer, one will see more fragments not having the R group intact. The second series compared the *n*-hexyl, cyclohexyl, phenyl, and benzyl esters. Curiously, the benzyl and cyclohexyl esters were similar and the phenyl and *n*-hexyl were similar, but different from the other two. The data are shown in Table 8.11. What is particularly striking is the decrease in the $+3\Delta$ peak for the cyclohexyl ester, relative to the others, and the large -5Δ peaks for the *n*-hexyl and phenyl compounds. Thus it would appear that initial loss of the ester group is less likely for the cyclohexyl derivative.

8.4.3.2 Acrylates

As might be expected, because of their similarities in structure, the SIMS spectra of the poly(alkyl acrylates) (PAA)s, are quite similar to those of the methacrylates. Figure 8.28 compares segments of the spectra of poly(*n*-butyl acrylate) and poly(*n*-butyl methacrylate). One similarity is that both types of polymers show repeat patterns extending beyond 3500 Da[43] with spacings corresponding to the monomer unit. Second, they both have four species contributing to cluster structure, corresponding to 0, 1, 2, and 3 unsaturations and/or double bonds. Third, changes in the ester R group affect relative cluster intensities.

The relative intensities of the main-chain fragments are different for the PAAs and PAMAs, as can be seen in Figure 8.28. For example, the $+3\Delta$ cluster is of greater intensity than the $+2\Delta$ cluster for all PAMAs, but not for any of the PAAs. Fragmentation of the PAAs cannot be interpreted as well as the PAMAs using the scheme in Figure 8.27, with adaptations for the missing methyl group.[43] The scheme would predict five main-chain fragments for PAAs: $-\Delta$, R_n, $+\Delta$, $+2\Delta$, and $+3\Delta$. The other peaks should then be weaker. Although this may appear to be the case for the *n*-butyl derivative shown in Figure 8.28, the relative intensity of the $+5\Delta$ cluster is greater than the $+4\Delta$ for all other polymers studied.

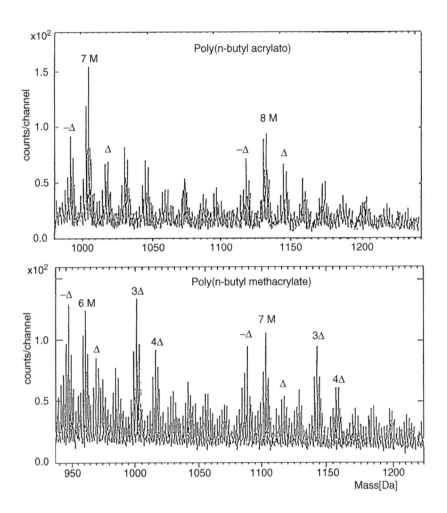

FIGURE 8.28
Comparison of poly(acrylate) and poly(methacrylate) spectra. (Reprinted with permission from
Ref. 43, Copyright 1994 Elsevier)

8.4.3.3   *Stereoregularity*

Polymer stereoregularity was observed to have a significant effect on TOF-SIMS
spectra, the best example of which is poly(methyl methacrylate) (PMMA).[44]
Figure 8.29 shows a comparison of the spectra of isotactic and atactic PMMA.
Clearly the spectra differ; the spectrum of the syndiotactic polymer resembles
that of atactic PMMA. The spectrum of the atactic polymer shows a repeat
pattern of seven clusters per pattern, while the isotactic polymer has only
two clusters per pattern. Even more interesting is that the masses of the
isotactic clusters correspond to an integral number of repeat units plus an
odd mass that is not simply related to chain structure. For example, the more
intense cluster corresponds to $R_n + C_2H_6O$, cationized with silver. The pattern

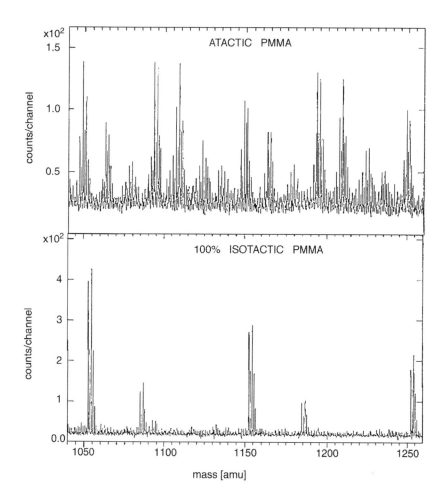

FIGURE 8.29

TOF-SIMS spectra of atactic and isotactic poly(methyl methacrylate). (Reprinted with permission from Ref. 44, Copyright 1994 Society of Applied Spectrometry)

is repeated to high masses as for the atactic polymer. Also, the isotactic PMMA clusters are composed of mainly one species, while the atactic clusters correspond to four species as discussed above.

It was shown subsequently[45] that the effect of polymer stereoregularity is related to the double helical structure of PMMA. Thus, while the simple statistical model is effective for interpreting TOF-SIMS spectra of polymers having isolated chains, it is not for a polymer having the structure of isotactic PMMA. The crystal structure of iso-PMMA has acrylate groups rotated toward the center of the double helix in close proximity to each other. Also, LB films of iso-PMMA show the double helical structure. Therefore there are many possibilities for cross-chain reactions in iso-PMMA films deposited on a surface. A combinatorial spreadsheet analysis was done to determine

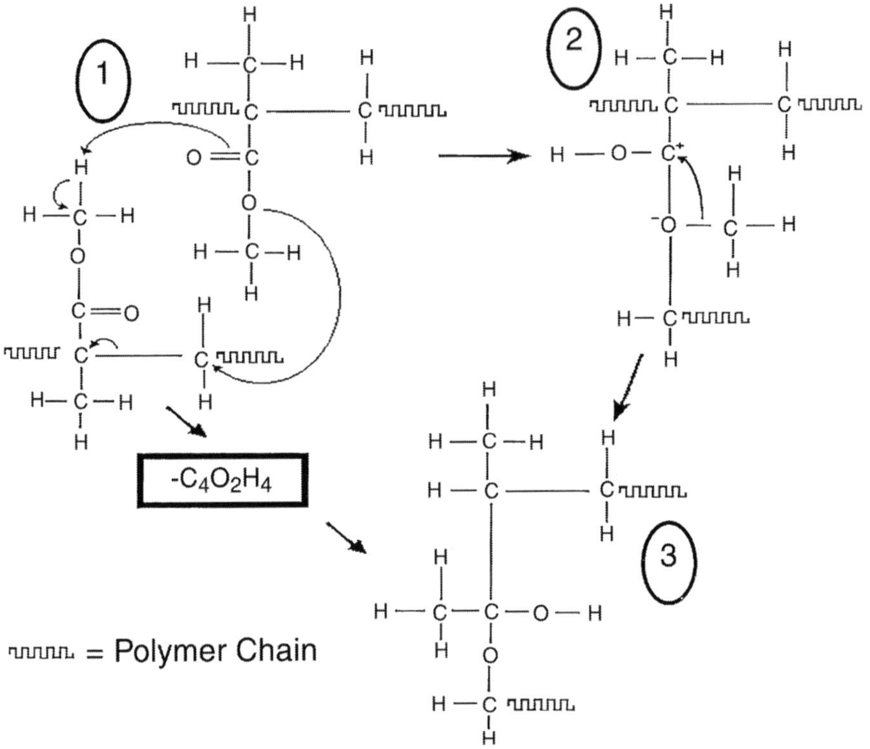

FIGURE 8.30
Proposed interchain ion formation mechanism. See text for details.

which polymer group transfers could account for the observed iso-PMMA spectrum. It was determined that transfer of a hydrogen from an ester methyl group to a carbonyl oxygen initiated the process. Subsequent reactions joined the two chains which then eliminated fragments to give the final product. This is illustrated in Figure 8.30 for the 1185 Da peak.

These results for PMMA are particularly important because they demonstrate that TOF-SIMS spectra can be sensitive to secondary structure. PMMA is not an isolated case. Similar effects were observed for poly(propylene) and poly(propylene oxide).[44]

8.4.4 Other Polymers

The discussion above has laid out the general principles of polymer chain fragmentation as derived from TOF-SIMS studies, and presented two important, illustrative examples: polystyrenes and acrylics. The present section will deal with studies of other polymer systems, in less detail, to provide an overview of other systems that have been studied. Specifically, nylons, polyesters, and poly(dimethyl siloxanes) will be considered. In addition, polyesters will be discussed to illustrate the type of information that can be obtained by combining TOF-SIMS with limited chemical degradation.

8.4.4.1 Nylons

This system is included primarily for historical purposes. Nylons were the first series of polymers studied using TOF-SIMS in the high mass range.[46] The specific nylons studied were nylon-6 (N–6), N–8, N–66, N–69, and N–66(α6). The major series of clusters observed for all nylons were R_n fragments, presumably cyclic; peaks were observed to 2500–3000 Da. Weaker peaks were seen at every ca. 14 mass units. Spacing between major peaks corresponded to the repeat unit, and side chains on the nylons remained intact. No studies were done to compare different molecular weights; this system is worth revisiting.

8.4.4.2 Polyesters

Two detailed studies have been reported on TOF-SIMS of polyesters. The first dealt with a series of seven polyesters having approximately the same molecular weight,[47] and the second studied poly(butylene adipate) (PBA) in greater detail.[48]

Three series of peaks were observed for polyesters having MWs in the range 1000–3000: oligomers, repeat unit, and fragment ion series.[47] The high-mass range (600–2000 Da) spectrum of poly(ethylene adipate) is shown in Figure 8.31. Peaks due to the oligomer and repeat unit series are indicated as O_{EA} and R_{EA}, respectively; R_{EA} is the equivalent of R_n in our present terminology, corresponding to an integral number of repeat units. As can be seen in Figure 8.31B, weaker series of peaks are also evident. Both O_{EA} and R_{EA} series consist of a single species, based on comparison of calculated and observed cluster structures. These data are shown for PBA in Figure 8.32.

It is noteworthy that fragmentation patterns for polyesters differ considerably from those for ethylene-based polymers; namely, the pattern consists of only two clusters, and each cluster is due primarily to a single species. This argues for a very different fragmentation mechanism, as was argued by Montaudo.[49] A mechanism which is consistent with the observed spectra is that fragmentation of polyesters occurs primarily by intramolecular ester exchange.[48] This is shown in Figure 8.33. Such a mechanism is consistent with FAB studies on polyesters[50] and poly(dimethylsiloxanes).[51] The reaction shown in Figure 8.33 produces a cyclic fragment and a smaller linear oligomer; there is no net energy change for this process. This is consistent with the observation in Figure 8.31 that the lower mass oligomers are in greater abundance than the higher mass oligomers and that there is a monotonic decrease in the R_{EA} ions. Another argument for the mechanism is that the ratio R_{EA}/O_{EA} increases with increasing polymer MW over the range 14,000–58,000. As the chain length of the oligomer increases, cyclization can proceed in a stepwise fashion, increasing the ratio of cyclics to oligomers for a given molecular size.

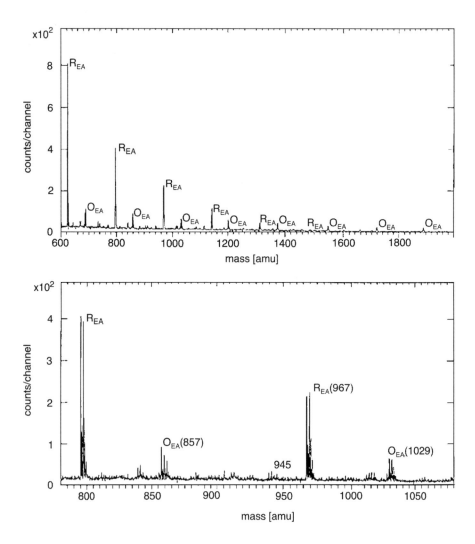

FIGURE 8.31
TOF-SIMS spectrum of poly(ethylene adipate). (Reprinted with permission from Ref. 47, Copyright 1994 American Chemical Society)

The weaker peaks in the polyester spectra can be explained largely by hydrogen-transfer reactions via a McLafferty rearrangement.[48] Depending on which side of the ester group contributes the hydrogen, species will occur at $O_{EA} - 18(y_2)$, $R_{EA} + 18(x_2)$, $O_{EA} + 68(y_1)$, and $R_{EA} - 68(x_1)$. These will account for the chain cleavages shown in Figure 8.34.

8.4.4.3   Polysiloxanes

High mass range TOF-SIMS spectra have been studied for three series of polysiloxanes: poly(dimethylsiloxane)s (PDMS) as a function of molecular weight, PDMS with different end-groups, and polysiloxanes with different

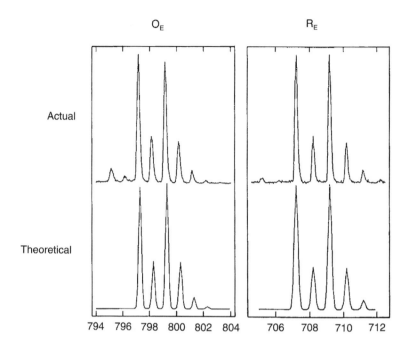

FIGURE 8.32

Comparison of calculated and observed cluster structures for poly(ethylene adipate). O_E is the oligomer series, R_E is a fragment having an integral number of repeat units. (Reprinted with permission from Ref. 47, Copyright 1994 American Chemical Society)

chain substituents. The high mass range spectra of PDMS are amazingly similar to those of the polyesters. They consist of two intense clusters per pattern corresponding to O_{EA} and R_{EA}.[51] O_{EA} is a linear fragment referred to as $[nR + 14]$ in the original paper because the end-groups are $-CH_3$ and $-(CH_3)_3Si$, which add up to 14 Da more than the repeat unit. Formation of cyclic species is consistent with the PDMS thermal degradation mechanism.[52] The ratio of the cyclic fragment increases relative to the linear fragment with increasing molecular weight, as for the polyesters. The main difference is that the increase is linear in $(MW)^2$ instead of (MW). Different end-groups do not dramatically affect the TOF-SIMS spectra in the high-mass range. Those studied (*vide infra*) showed the same pair of peaks in a cluster.

There are two unique aspects of the polysiloxane spectra seen in the mass range 500–1000. First, polymer fragments cationized with two silver ions to produce a doubly charged fragment are observed. Figure 8.35A shows details of the spectrum from PDMS ($M_n = 620$) showing the R_{EA} species along with the doubly charged cluster and one due to $[R_{EA} + 16]$. The latter is produced by hydrogen transfer from a terminal silylmethyl group.[51] Doubly charged ions such as these are rare events in TOF-SIMS spectra.

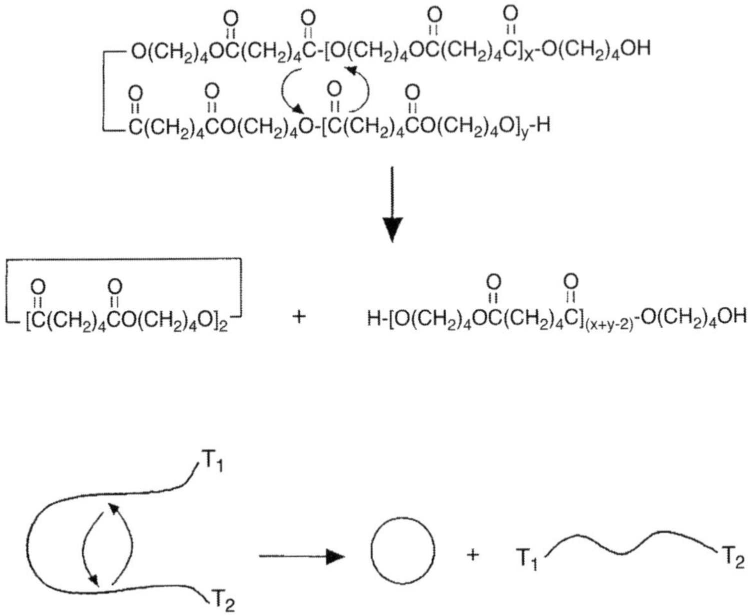

FIGURE 8.33
Fragmentation of polyesters via intramolecular ester exchange. (Reprinted with permission from Ref. 48, Copyright 1995 American Chemical Society)

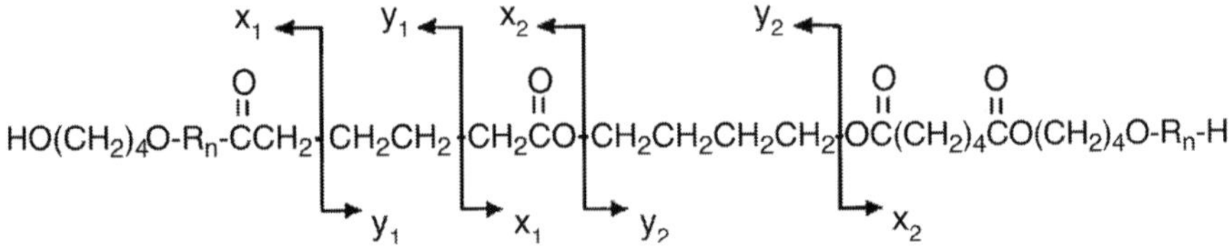

FIGURE 8.34
Fragmentation of polyesters via mcLafferty rearrangement. (Reprinted with permission from Ref. 48, Copyright 1995 American Chemical Society)

A second "oddity" occurs in the spectra of PDMS derivatives substituted with aminopropyl (APD) or amidopropyl (MPD) terminal groups on both ends of the polymer chain. A segment of the MPD-terminated PDMS spectrum is shown in Figure 8.35B. There are two intense clusters present. One corresponds to the Ag-cationized linear oligomer, and the other to a linear fragment formed by loss of the amidopropyl group from an oligomer, but *not* cationized by silver. In other words, it has an intrinsic charge. The cluster at 783 Da is due to an impurity, but the clusters at 799 and 835 correspond to $(R_{EA}-CH_3)^+$ and $(O_{EA}-CH_3)^+$, respectively. It is rare to see such a high-mass fragment having intrinsic charge for a nonfluorinated polymer.

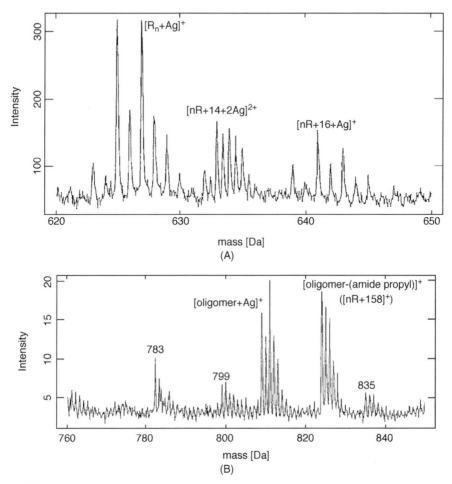

FIGURE 8.35

TOF-SIMS spectra of poly(dimethylsiloxane)s having different terminal groups. (A) Trimethyl-silyl terminated, (B) Amide-propyl terminated. (Reprinted with permission from Ref. 51, Copyright 1997 American Chemical Society)

PDMS was studied along with polyhydromethylsiloxane (PHMS) and polymethyl-phenylsiloxane (PMPhS) to investigate the effect of functional changes, on the siloxane chain, on polysiloxane fragmentation mechanisms.[53] Cyclic fragments are observed in the spectra of PDMS and PHMS, but not for PMPhS; the spectrum of PMPhS shows only linear species. The mechanism proposed for the formation of cyclic fragments in PDMS involves a four-membered cyclic intermediate. The flexibility of the PDMS chain allows formation of this intermediate; the situation is similar for PHMS. However, the chain of PMPhS is more rigid, making it difficult to form the cyclic intermediate. As a result, loss of a phenyl group and cleavage of the siloxane chain become the major fragmentation pathways.

8.4.4.4   Limited Chemical Reaction

Fragmentation of polymers in TOF-SIMS is complicated by the presence of multiple oligomers and multiple fragmentation pathways. Also, if polymers are insoluble or highly crosslinked it is often difficult to characterize them with TOF-SIMS because of solubility limitations, since it is necessary to deposit polymers as a thin layer on the Ag substrate to obtain a high-mass spectrum. Therefore it is desirable to combine limited chain cleavage by a controlled chemical reaction with derivatization to put markers on the polymer at the place where chain cleavage has occurred. The potential value of such an approach for polymer analysis is clear from the diagram shown in Figure 8.36. The structures on the left correspond to polymers

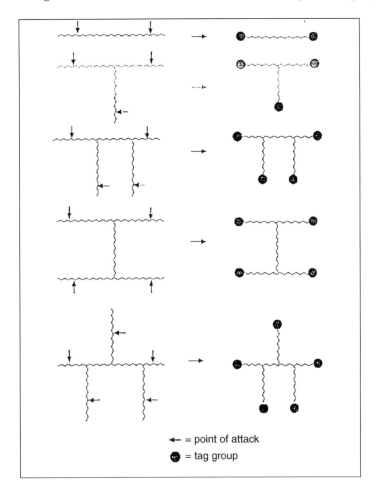

FIGURE 8.36
Schematic diagram for fragments generated from branched polymers. (Reprinted with permission from Ref. 47, Copyright 1994 American Chemical Society)

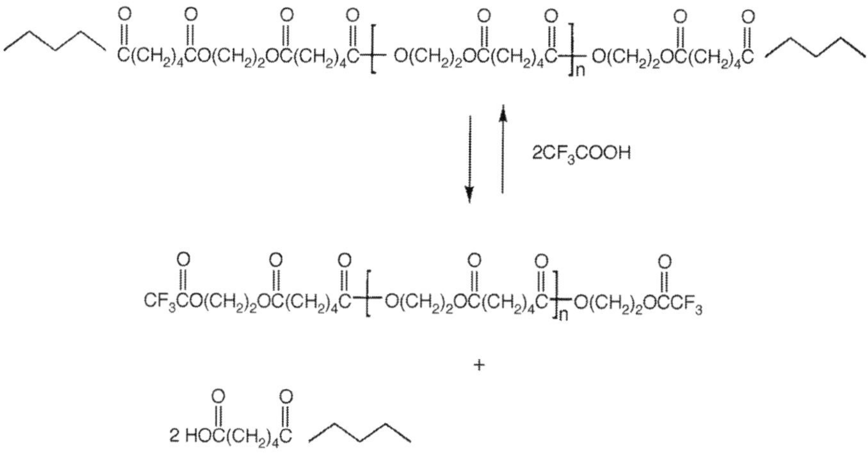

FIGURE 8.37
Reaction of trifluoroacetic acid with a polyester. (Reprinted with permission from Ref. 47, Copyright 1994 American Chemical Society)

having different degrees of branching and points of attack of a cleaving/ derivatizing reagent. The structures on the right represent segments of the chain with the markers added. By using either mass markers or those with isotopic patterns, it is possible to distinguish between fragments having 2, 3, 4, or 5 "tags." This approach was used initially to characterize polyurethanes.[54]

A detailed study has been carried out on the reaction of polyesters with two reaction/derivatizing agents, trifluoroacetic acid (TFA) and chlorodifluoroacetic acid (CFA);[47] the first is a mass marker, the second an isotopic marker. The reaction of these reagents with a polyester is shown in Figure 8.37. The reaction was referred to in the original papers as a transesterification, although technically it is not. The species produced will have an integral number of repeat units plus an extra diol; the mass marker is provided by the fluorines (3,6,9) due to the 19 Da mass of the F atom. The spectrum produced by the reaction of TFA with poly(ethylene adipate) is shown in Figure 8.38. The peaks labeled T_{EA} are from chain fragments having two tagged groups (R_{EA} is R_n).

Use of CFA is better for more complex polymers (crosslinked) because it is easier to spot the isotopic clusters for different levels of chain substitution. Poly(1,3-butylene adipate) was studied as an example of reaction of CFA with a branched polymer;[47] a segment of the spectrum is shown in Figure 8.39. The labels C_{2t}, C_{3t}, and C_{4t} correspond to fragments having 2, 3, and 4 CFA tags, respectively. The calculated isotopic patterns for these three species are shown at the bottom of Figure 8.39; the measured patterns agreed well. On this basis, the species are easily distinguishable.

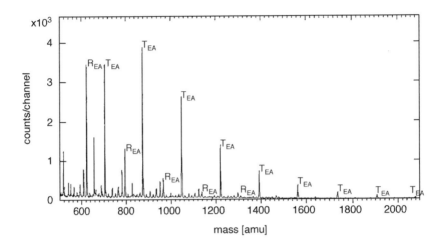

FIGURE 8.38
TOF-SIMS spectrum of poly(ethylene adipate) after treatment with trifluoroacetic acid. (Reprinted with permission from Ref. 47, Copyright 1994 American Chemical Society)

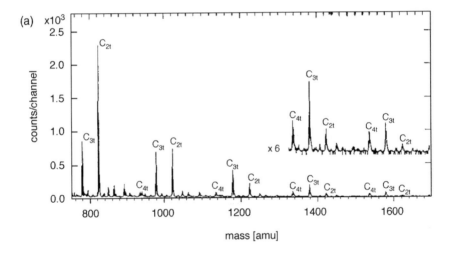

FIGURE 8.39
TOF-SIMS spectra of poly(1,3-butylene adipate) treated with chlorodifluoroacetic acid. (A) Mass range 750–1700 Da. (B) Theoretical patterns for species containing 2, 3, and 4 chlorines. (Reprinted with permission from Ref. 47, Copyright 1994 American Chemical Society) (Continued)

Reaction with TFA has also been applied to polyurethanes.[48,54] Two types of fragments are produced, a segment derived only from the ester portion of the chain, and one containing the polymer hard block (original diisocyanate). The reaction scheme is shown in Figure 8.40. This makes TOF-SIMS a powerful tool for polymer and copolymer characterization because one can control the level of "fragmentation" chemically. It is also possible to use

(b)   Theoretical Spectra.

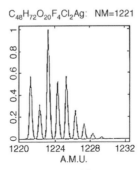

C$_{48}$H$_{72}$O$_{20}$F$_4$Cl$_2$Ag: NM=1221

Theoretical Spectra.

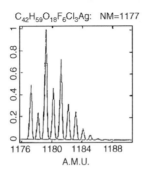

C$_{42}$H$_{59}$O$_{18}$F$_6$Cl$_3$Ag: NM=1177

Theoretical Spectra.

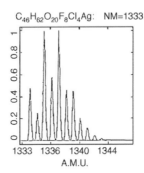

C$_{46}$H$_{62}$O$_{20}$F$_8$Cl$_4$Ag: NM=1333

FIGURE 8.39
(Continued)

reaction with TFA to obtain high-mass fragments from solid polymer samples. Figure 8.41 shows a TOF-SIMS spectrum obtained from a plastic beverage bottle by treatment with TFA and subsequent deposition of the solution on an Ag foil. From the spacing between the peaks, it can be concluded that the bottle was made from poly(ethylene terphthalate).

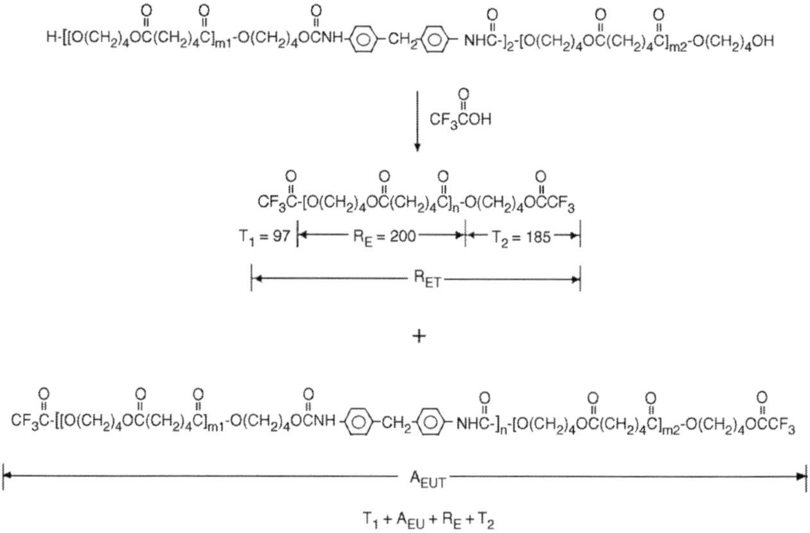

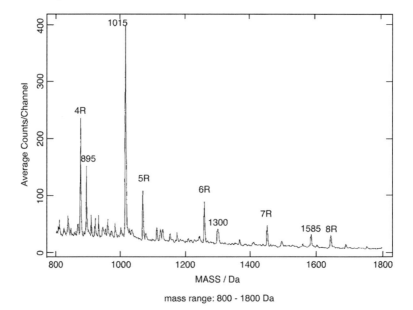

FIGURE 8.40
Reaction of polyurethane[4,4,1] with trifluoroacetic acid. (Reprinted with permission from
Ref. 48, Copyright 1995 American Chemical Society)

FIGURE 8.41
TOF-SIMS spectrum from a plastic beverage bottle. Reacted with trifluoroacetic acid. (From
Dong, X., Characterization of Polymers by Time of Flight Mass Spectrometry, Ph.D. Thesis,
University of Pittsburgh, 1999.)

8.5 Copolymer Spectra

As might be expected, high mass range TOF-SIMS spectra of copolymers are more complicated than those of homopolymers. The level of complexity will be determined by how many monomers comprise the polymer, their relative percentages, and whether one is dealing with a random or block copolymer. In the latter case, the block size will be important; for a 5000 MW polymer, it is different if one has many small (5 monomer) blocks or a diblock copolymer. Studies in the high mass range have been reported for only a few copolymer systems. The two most extensively studied are the perfluoropolyethers and the polyurethanes. Another important point to consider involves the relative ion yields of the components of a copolymer system. If they are vastly different, quantification will be difficult. To date, there has been little work reported in this important area.

8.5.1 Polyester Polyurethanes

Two extensive studies of polyester polyurethanes have been reported[48,56] along with an earlier study of model polyurethanes made from diols and diisocyanates.[54] TOF-SIMS spectra obtained for all three showed the same characteristics, but only one study[48] was performed on a high resolution instrument. The results of that study will be discussed in some detail.

The polyester polyurethanes (PE-PUR) were prepared from poly(butylene adipate) (PBA) and 4,4′-diphenylmethane diisocyanate (MDI), a TOF-SIMS spectrum for the $M_n = 40{,}000$ polymer is shown in Figure 8.42, along with the polymer structure and fragment masses. Two intense patterns are observed composed of two clusters which correspond to a fragment from the polyester chain, R_E, and a fragment containing the MDI functionality, A_{EU}. The spacing between both sets is the mass of the polymer repeat unit, and the mass difference between the two is the mass of the diisocyanate. Oligomer peaks, O_{EA}, were also observed but only from low MW (<15,000) materials. The cluster structures of O_{EA}, R_{EA}, and R_E showed a singly unsaturated species, indicating a probable cyclic structure. As was observed for the polyesters, the oligomer series decreased in intensity relative to the fragment series as the MW of the polymer increased.

It is not surprising that fragmentation of the PE-PURs is similar to that of the polyesters. Cyclization by intramolecular ester exchange is the major pathway for fragmentation. Thermal degradation of PURs is known to occur by cyclization and ester exchange.[57] The earlier study of PURs[56] showed the same two major series for a collection of 24 PURs along with lower intensity fragment-ion peaks which corresponded to $R_E + R_U + CO$, $R_E + R_U + O$, and $R_E + R_U + CO + O$ for some PURs. These simply represent cleavage of different bonds in the urethane functionality. Because the spectra in that study were

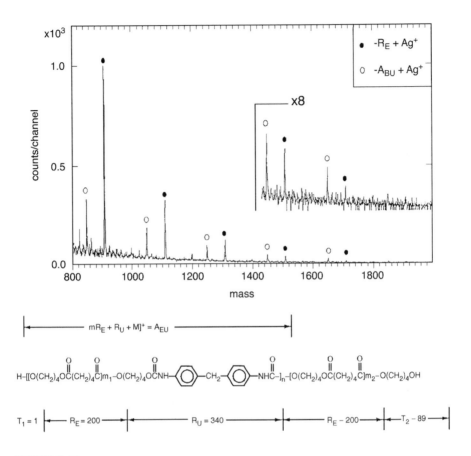

FIGURE 8.42
TOF-SIMS spectrum of polyurethane[4,4,1]. Polyurethane structure is shown below the spectrum. (Reprinted with permission from Ref. 48, Copyright 1995 American Chemical Society)

not high resolution, it is not possible to tell the number of contributing species. Studies on 12 model PURs made from diols and diisocyanates showed similar behavior; the major series consisted of peaks due to oligomers and repeat units.[54] Peaks due to fragmentation were observed, but at lower intensity.

An interesting sidelight of one of the earlier studies concerned identification of unknown extenders in PURs based on polycaprolactones.[56] PURs having unknown extender alcohols and different number-average molecular weights were examined. The unknown alcohols are part of the B_{EU} fragments, so their masses can be calculated from these peaks after accounting for the known parts of the B_{EU} fragments. For example, one PUR showed a Na-cationized peak at 749 Da. Subtracting the masses of the urethane (252), caprolactone (114), and sodium (23) gave a remainder of 132 which corresponds to the mass of trimethylolpropane.

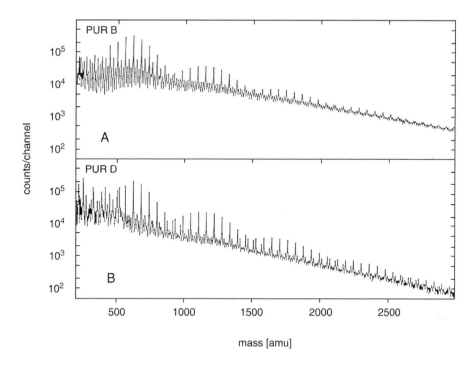

FIGURE 8.43
TOF-SIMS spectra of polyether polyurethanes having poly(propylene glycol) soft blocks. A. 1,6 Hexamethylenediisocyanate hard block, B. Toluene-2,4-diisocyanate hard block.

8.5.2 Polyether Polyurethanes

Materials studied were prepared from polyethers and diisocyanates; both branched and linear polyethers were studied.[58] The important characteristic of the TOF-SIMS spectra of these materials is that the oligomer distribution of the polyether is reflected in the spectra of the PURs. Figure 8.43 shows the TOF-SIMS spectra of two PURs. Both polymers have poly(propylene glycol) (PPG) as the soft block; PUR B was prepared with 1,6 hexamethylenediisocyanate (HX) and PUR D was prepared using toluene-2,4-diisocyanate (TDI). The patterns of peak distributions can be seen in the spectra and are related to the oligomer distribution of the original PPG. Fragment ion spectra indicate that fragmentation in polyether PURs is not random, but that bond cleavage occurs such that the glycol molecular weight distribution is maintained after chain fragmentation. This is consistent with earlier studies which indicate that cleavage in PURs occurs primarily around the urethane bond. This will be discussed in detail below.

Figure 8.44 shows spectra of PUR A, prepared from a PPG having $M_n =$ 425 and 4,4'-diphenylmethane diisoocyanate (MDI). The PUR structure is

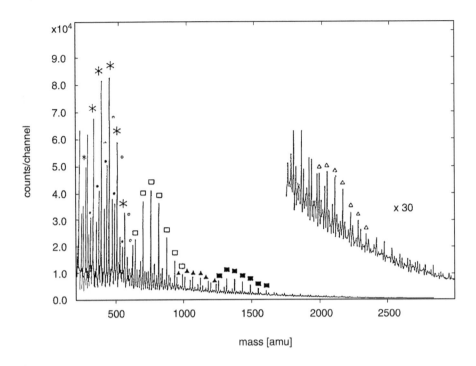

FIGURE 8.44

TOF-SIMS spectrum of a polyether polyurethane. See text for details.

shown below:

If fragmentation involves only the urethane group, one would expect sets of peaks for various combinations of n and m. The smallest of these should be due to cleavage of the PPG units alone from either end of the molecule; these would correspond to $m = 0$. This is observed in the region below 500 Da where three sets of peaks occur, all due to PPG species cationized with Na. The peaks labeled (o) are due to PPG oligomers with $n = 4$–17 and $n_{max} = 7$. The original PPG also had $n_{max} = 7$. The most intense set of peaks ($*$) is for $n = 4$–13 and $n_{max} = 7$ and corresponds to PPG minus two hydrogens. Because the spectra were run under low resolution, no additional structural information is available. The third set ($\bullet$) is presumably due to formation of cyclic ethers and has $n = 5$–12 and $n_{max} = 7$.

There are four additional sets of regularly occurring patterns in the region from 500–3000 Da, corresponding to different combinations of PPG and MDI blocks. These are summarized in Table 8.12. The most intense set in Figure 8.42, centered at 755 Da ($\square$) is due to fragments consisting of one MDI and

TABLE 8.12

Origin of Patterns in PPG-MDI Polyurethane Spectra

Blocks	m	n	n_{max}	Mass Max
PPG-MDI	1	2–17	8	755
$(PPG)_2$-MDI	1	10–22	14	1121
$(PPG)_2(MDI)_2$	2	7–25	13	1313
$(PPG)_3$-$(MDI)_2$	2	15–30	19	1929

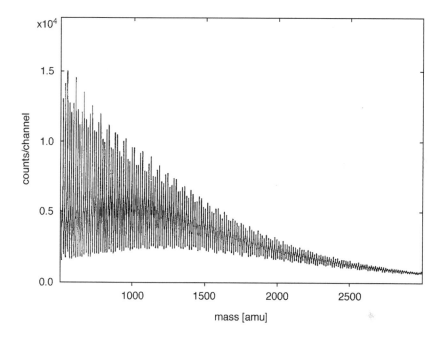

FIGURE 8.45

TOF-SIMS spectrum of a copolyglycol polyurethane. Derived from poly(ethylene glycol), poly(propylene glycol), and MDI.

different numbers of PPG oligomers. The two sets, centered at 1121 (▲) and 1313 Da (■), contain two PPG blocks and different numbers of MDIs. This is shown by the same range of n values and the same n_{max}. The series centered around 1929 Da (△) must contain three PPG blocks, consistent with the n values observed. Using higher molecular weight PPGs and MDI gave similar spectra, although they are more strung out because of the larger oligomer distributions. All of the series listed in Table 8.12 are not seen because of mass limitations of the SIMS experiment.

Copolyglycol PURs also show patterns in their mass spectra, but they are not as readily apparent. Figure 8.45 shows the TOF-SIMS spectrum of the copolyglycol derived from PEG, PPG, and MDI. Although not readily apparent, oligomer peaks are seen for both PEG and PPG. Series for the PUR correspond to species $(PPG)_n$ $(PEG)_m(MDI)_x$; those observed were for $x = 0, 1$ and n and m equal to one or two times the normal oligomer distributions.

8.5.3 Perfluorinated Polyethers

This interesting series of polymers has been studied by a number of workers. Although most of the work has been reported for the Krytoxes (DuPont), some work has been done on perfluorinated polyether (PFPE) copolymers. Thus they are included in this section. The SIMS behavior of PFPEs is unique in several ways. The PFPEs have high yields of both positive and negative ions and have fragment ion peaks extending to higher mass than for most polymers, to more than ca. 5000 Da. Another interesting feature is that the spectra are derived from species having intrinsic charge, i.e., cationization is not essential to observe fragment ions. However, when oligomer distributions are observed, the spectra correspond to Ag-cationized species. Additionally, the PFPEs show high yields of metastable ions, particularly in their positive-ion spectra.

Most of the work reported for PFPEs has been done on the Krytoxes, their structure is shown below (**1**). A typical Fomblin (Montedison) copolymer is also shown (**2**). Other PFPEs, derived from common polyethers, have also been studied.[60]

$$F\left[\begin{array}{c} CF_3 \\ | \\ CF\ CF_2O \end{array}\right]_n CF_2CF_3 \qquad\qquad CF_3\left[OCF_2\right]_m\left[OCF_2\right]_n OCF_3$$

1   **2**

Figure 8.46 shows a segment of the high-mass, positive-ion, TOF SIMS spectrum of Krytox AD.[23] Similar spectra were reported later;[9] in addition, SIMS and PDMS measurements have been compared for PFPEs and other polymers.[61] The earliest study[23] reported more peaks in the Krytox spectra than later results, due primarily to differences in experimental conditions. To further complicate matters, the nomenclature used also changed. In all studies, however, the same major peaks were observed and the origin of these peaks is indicated in Table 8.13. We will adopt the nomenclature used in the latest study.[9]

The intense peaks in the Krytox spectra are derived from carbonium ions produced either by single or double cleavage of the polymer chain. The major peak corresponds to cleavage and charge-retention at the CF_2—O bond (C) which is preferred to breaking the CF—CF_2 bond which produces fragment D. The intensity ratio of the fragments, C/D, is greater than unity, but decreases with increasing mass due to metastable ion formation (*vide infra*). Ions A and B are probably formed by double chain cleavage, similarly to the fragments observed for most polymers. A significant number of weaker ions were reported in the first study;[23] the greater degree of fragmentation reflects the use of a different type of TOF-SIMS instrument.

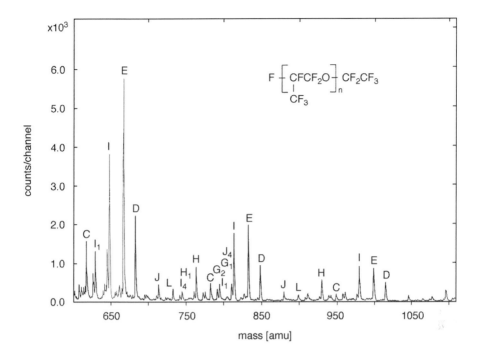

FIGURE 8.46

Positive-ion TOF-SIMS spectrum of Krytox AD. (Reprinted with permission from Ref. 23, Copyright 1990 American Chemical Society)

Figure 8.47 shows a section of the negative-ion spectrum of Krytox AD;[23] this spectrum is essentially the same as those reported in later studies.[9,31,60] Table 8.13 summarizes the interpretation of major peaks in the negative-ion spectra; the various nomenclatures are reviewed. The peaks all correspond to species which are oxyanions, mostly derived from a single chain cleavage. As with the positive-ion spectra, chain cleavage and resulting ion formation is unsymmetrical. Retention of charge on an oxygen adjacent to a primary carbon occurs at greater probability than for a secondary carbon. This effect has been studied in some detail[60] and is related to the mechanism of negative ion formation. The authors present a compelling argument that anion formation in the PFPEs results from one-step dissociative electron capture at the ether oxygens to produce the major peaks (E,F) in the SIMS spectra. They demonstrate that for PFPEs, which are symmetrical about the ether bond, the peak intensities are comparable, but they are not unsymmetrical ethers (Krytoxes). They also indicate that minor peaks can be derived from dissociative electron capture to produce fluoride ions with subsequent ionization caused by secondary electron capture. An example is that peak G for Krytox AD would be derived from an acyl fluoride; the structure is shown in Table 8.13.

TABLE 8.13

Structures of Major Ions in Krytox Spectra

$$F - \left[\begin{array}{c} CF_3 \\ | \\ CF\ CF_2\ O \end{array} \right] - CF_2\ CF_3$$

Krytox
AD

		Positive Ions		
			Nomenclature	
Formula	Ion Structure		Ref. (9)	Ref. (23)
$R_nC_3F_7O$	$^{\oplus}CF_2O$-R_n-CF_2CF_3		D	D-1
$R_nC_3F_7$	$F-R_n-\overset{\overset{\textstyle CF_3}{\textstyle \|}}{C}FCF_2^{\oplus}$		C	E
$R_nC_3F_6$	$R_n\overset{\overset{\textstyle CF_3}{\textstyle \|}}{C}F\,CF^{\oplus}$		B	I
$R_nC_2F_4$	$R_n-\overset{\overset{\textstyle CF_3}{\textstyle \|}}{C}F^{\oplus}$		A	H

		Negative Ions		
			Nomenclature	
Formula	Ion Structure	Ref. (9)	Ref. (23)	Ref. (31)
R_nF	$F-Rn-\overset{\overset{\textstyle CF_3}{\textstyle \|}}{C}F\ CF_2O^{\ominus}$	E	D-2	R_0^-
$R_nC_2F_5O$	$^{\ominus}O\,\overset{\overset{\textstyle CF_3}{\textstyle \|}}{C}F\ CF_2O-R_n CF_2CF_3$	F	N	R_1^-
$R_nC_3F_6O$	$^{\ominus}O-R_n\overset{\overset{\textstyle CF_3}{\textstyle \|}}{C}F\ CF_2$	G	G-3	I_2^-

Figure 8.48 shows the negative-ion spectrum obtained from Fomblin Z. Fomblin Z is an AB random copolymer (2) with m/n ca. 4/6. Needless to say, the spectrum is more complex than those of the Krytoxes. It is possible to interpret these spectra based on the assumption of complete randomness of the blocks and totally random cleavage of the C—O bonds.[23,60] A supporting argument is that in the more recent study,[60] it was possible to simulate the Fomblin Z spectrum using a statistical model. It is particularly interesting to contrast results for the Fomblin Z with those for PFP dioxolane, which has similar units: $CF_3-O-(CF_2CF_2-O-CF_2-O)_n-CF_3$; they are entirely different.

The relationships between high mass fragments, the number average molecular weight, and composition of PFPEs has been studied by TOF-MS.[31]

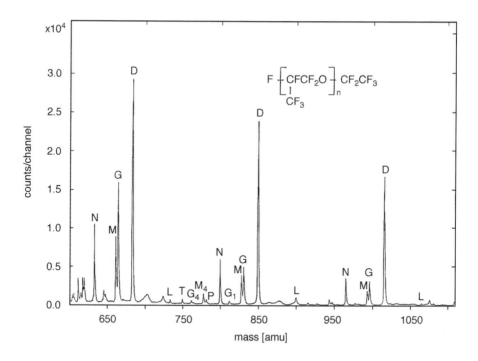

FIGURE 8.47
Negative-ion TOF-SIMS spectrum of Krytox AD. (Reprinted with permission from Ref. 23, Copyright 1990 American Chemical Society)

A linear, quantitative relationship exists between relative peak intensities for the PFPEs and their molecular weights. This method appears to be superior to measuring oligomer distributions. High mass fragments in the negative-ion spectrum were used because the quantitative relationships are better for negative-ion spectra, there are fewer peaks for each pattern, and there is no interference from cationized oligomer spectra. The peaks chosen were E and G in Table 8.13. Peak E results from a single-chain cleavage and therefore contains an end-group; peak G does not contain an end-group and represents a true chain segment. A linear relationship was observed between the G/E intensity ratio and the number-average of monomer units. The PFPEs are a particularly good choice for this type of analysis and may represent a special case because they give large fragment ions without cationization.

Both positive- and negative-ion Krytox fragment-ion spectra show peaks due to metastable ions when recorded in the reflectron mode.[9] The greater stability of the negative ions is indicated by their occurrence to more than 5000 Da, whereas positive ions are observed to only mass 2500. The relative ion stabilities are most apparent for spectra recorded in the linear mode. Figure 8.49 shows spectra of Krytox AD recorded in the range of 800–1200 Da. In the linear mode the parent and metastable ions are separated by different flight times caused by the postacceleration gap. Because the negative ions are derived from a single

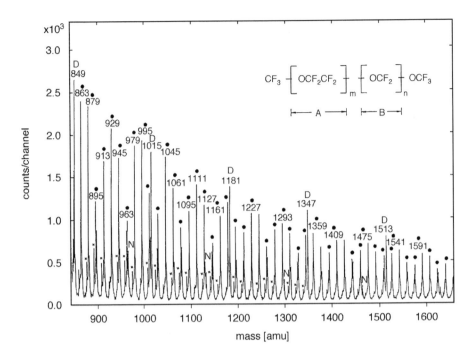

FIGURE 8.48
Positive-ion TOF-SIMS spectrum of Fomblin Z copolymer. (Reprinted with permission from
Ref. 23, Copyright 1990 American Chemical Society)

fragmentation step, the parent and daughter ions are not much different in mass
and therefore the major peaks in the negative-ion spectrum (Figure 8.49a)
are due only to parent ions.

However, in the positive-ion spectrum (Figure 8.49b) the metastable ions
appear as a broad peak (shown with shading) ahead of the parent ion. The
shape of the metastable peaks is Gaussian, indicating that many different
fragment ions are formed, some small and others large, characteristic of
multiple fragmentation processes. Another interesting feature is that the
intensity ratio of parent to daughter ions decreases as the mass of the parent
ion increases. This effect is shown by the top curve (•) in Figure 8.50, which
plots the relative ion yield as a function of mass. From this, one can calculate
the lifetimes of the parent ions; the calculated lifetimes for the different
parent ions are shown at the bottom in Figure 8.50 (o). It is clear that the
lifetimes are independent of the mass of the parent ion. The measured
lifetimes were 17 μsec for the positive ions and 79 μsec for the negative
ions. The shorter lifetimes for the positive ions explain why the intensity
ratios of parent to daughter change with mass. The lifetimes of the metastable
ions are on the order of the time required to traverse the flight tube. Because
the positive ions have shorter half-lives, at any given mass they will show

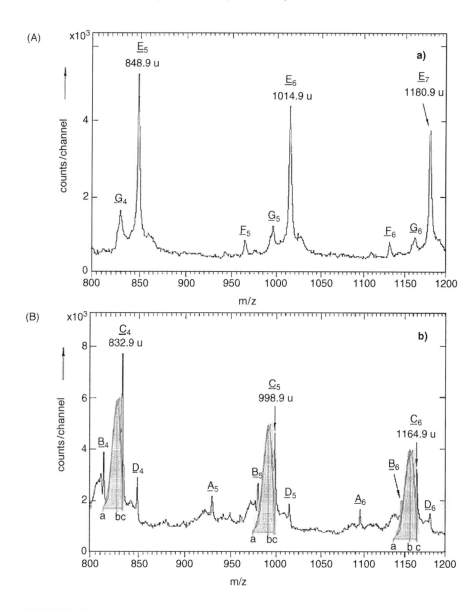

FIGURE 8.49

TOF-SIMS spectra of Krytox taken in the linear mode. A. Negative-ion spectrum, B. Positive-ion spectrum. (Reprinted with permission from Ref. 9, Copyright 1993 American Chemical Society)

more decay than the negative ions. In a way, the positive ion lifetimes are optimum for observing metastable ions in TOF-SIMS. If the half-life is much shorter, decay will occur before extraction into the flight tube. If ions are more stable, they will pass through the flight tube before significant decomposition occurs.

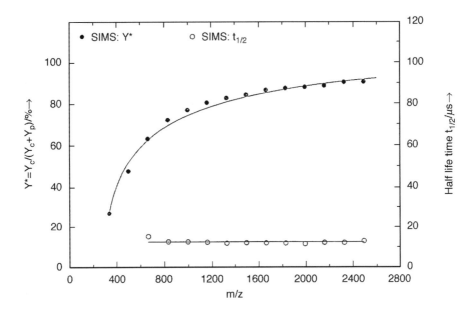

FIGURE 8.50
Reaction yields and half-lives of ions from Krytox recorded as a function of mass. A. Ion yields, B. Half-lives. (Reprinted with permission from Ref. 9, Copyright 1993 American Chemical Society)

Acknowledgments

Many thanks are due to the graduate students and postdocs who have contributed to our TOF-SIMS polymer program: I.V. Bletsos, P. D. Zimmerman, M. P. Chiarelli, Y. L. Kim, L. R. Cohen (Hittle), J. T. Mehl, B. K. Choi, K. Xu, D. C. Muddiman, A. J. Nicola, A. M. Cognetti, and X. Dong. Also thanks are due to my collaborators at Bayer USA, C. G. Karakatsanis and J. M. Rieck, and to Prof. A. Benninghoven (Münster) and his co-workers. Special thanks are due to my senior research collaborators, M. Houalla, A. Proctor, and A. I. Gusev. Support for my polymer research by the National Science Foundation is gratefully acknowledged.

References

1. (a) Burlingame, A.L., Boyrd, R.K., and Gaskell, S.J., Mass spectrometry, *Anal. Chem.*, 68, 599R, 1996. (b) Smith, P.B., Pasztor, A.J., Jr., McKelvy, M. L., Meunier, D.M., Froelicher, S.W., and Wang, F.C.Y., Analysis of synthetic polymers and rubbers, *Anal. Chem.*, 69, 95R, 1997.

2. (a) Lattimer, R.P. and Schulten, H.-R., Field ionization and field desorption mass spectrometry: past, present, and future, *Anal. Chem.*, 61, 1201a, 1989. (b) Saito, J., Waki, H., Teramae, N., and Tanaka, S., Application of field desorption mass spectrometry to polymer and oligomer analysis, *Prog. Org. Coatings*, 15, 311, 1988.

3. (a) Karas, M., Bahr, U., Ingendoh, A., and Hillenkamp, F., Laser desorption/ ionization mass spectrometry of proteins of mass 1000 to 250000 Dalton, *Angew. Chem. Int. Ed. Engl.*, 28, 760, 1989. (b) Fenselau, C., Maldi MS and strategies for protein analysis, *Anal. Chem.*, 69, 661A, 1997.

4. Cotter, R. J., Plasma desorption mass spectrometry: coming of age, *Anal. Chem.*, 60, 781A, 1988.

5. (a) Ballisteri, A., Garozzo, D., Giuffrida, M., and Montaudo, G., Identification of the ions produced by fast atom bombardment mass spectrometry in some polyesters and polyamides, *Anal. Chem.* 59, 2024, 1987; (b) Lattimer, R. P., Fast atom bombardment mass spectrometry of polyglycols, *Int. J. Mass. Spectrom. Ion Processes*, 55, 221, 1984.

6. (a) Schulten, H.-R. and Lattimer, R. P., Physical properties of synthetic high polymers, *Mass Spectrom Rev.*, 3, 231, 1984, (b) Turecek, F. and Longevialle, P., eds., *Advances in Mass Spectrometry*, 1988; Wiley, New York (1989) 1079. (c) Plage, B. and Schulten, H.-R., Thermal degradation mechanisms of amphiphillic acrylic copolymer studied by temperature-resolved pyrolysis-field ionization mass spectrometry, *Anal. Appl. Pyrolysis* (1988).

7. Smith, R.D., Loo, J.A., Loo, R.R.O., Busman, M., and Udseth, H.R., Principles and practice of electrospray ionization—mass spectrometry for large polypeptides and proteins, *Mass Spectrom. Rev.*, 10, 359, 1991.

8. Benninghoven, A., Jaspers, D., and Sichterman, W., Secondary-ion emission of amino acids, *Appl. Phys.*, 11, 35, 1976.

9. Feld, H., Leute, A., Rading, D., Benninghoven, A., Chiarelli, M.P., and Hercules, D.M., Secondary ion emission from perfluorinated polyethers using megaelectronvolt and kiloelectronvolt ion bombardment, *Anal. Chem.*, 65, 1947, 1993.

10. vanLeyen, D., Hagenhoff, B., Niehuis, E., Benninghoven, A., Bletsos, I.V., and Hercules, D.M., Time-of-flight secondary ion mass spectrometry of polymer materials, *J. Vac. Sci. Technol.*, A7, 1790, 1989.

11. Zimmerman, P.D., Practical applications of time-of-flight secondary ion mass spectrometry, *Ph.D. Thesis*, Univ. of Pittsburgh, Ch. 4, 1994.

12. Muddiman, D.C., Brockman, A.H., Proctor, A., Houalla, M., and Hercules, D.M., Characterization of polystyrene on etched silver using ion scattering and x-ray photoelectron spectroscopy: correlation of secondary ion yield and time-of-flight sims with surface coverage. *J. Phys. Chem.*, 98, 11570, 1994.

13. (a) Liu, L.K., Busch, K.L, and Cooks, R.G., Matrix-assisted secondary ion mass spectra of biological compounds, *Anal. Chem.*, 53, 109, 1981. (b) Unger, S. E., Day, R. J., and Cooks, R.G., Positive and negative secondary ion mass spectrum and mass analyzed ion kinetic energy spectra of some amides, amines and related compounds; mechanisms in molecular SIMS, *Int. J. Mass Spectrom. Ion Processes*, 39, 231, 1981. (c) Busch, K. L., Hsu, B. H., Xie, Y.W., and Cooks, R. G., Matrix effects in secondary ion mass spectrometry, *Anal. Chem.*, 55, 157, 1983.

14. Wu, K.J. and Odom, R.W., Matrix-enhanced secondary ion mass spectrometry: a method of molecular analysis of solid surfaces, *Anal. Chem.*, 68, 873, 1996.

15. Nicola, A., Muddiman, D.C., and Hercules, D.M., Enhancement of ion intensity of time-of-flight secondary-ionization mass spectrometry, *J. Am. Soc. Mass Spectrom.*, 7, 467, 1996.

16. Gusev, A.I., Choi, B.K., and Hercules, D.M., Improvement of signal intensities, in static secondary-ion mass spectrometry using halide additives and substrate modification, *J. Mass. Spectrom.*, 33, 480, 1998.

17. Cooks, R. G., Ast, T. A., and Mabud, M.A., Collisions of polyatomic ions with surfaces, *Int. J. Mass Spectrom. Ion Processes*, 100, 209, 1990.

18. Benguerba, M., Brunelle, A., Della-Negra, S., Depauw, J., Joret, H., LeBeyec, Y., Blain, M.G., Schweikert, E. A., Assayag, G. B., and Sudreau, P., Impact of slow gold clusters on various solids: nonlinear effects in secondary ion emission. *Nucl. Instr. Meth. Phys. Res.*, B62, 8, 1991.

19. Stipdonk, M.J.V., Harris, R.D., and Schweikert, E. A., A comparison of desorption yields from C_{60}^+ to atomic and polyatomic projectiles at keV energies, *Rapid Commun. Mass Spectrom.*, 10, 1987, 1996.

20. Kötter, F., Niehuis, E., and Benninghoven, A., Secondary ion emission from polymer surfaces under Ar^+, Xe^+ and SF_5^+ primary ion bombardment, *SIMS XI-Eleventh International Conference on Secondary Ion Mass Spectrometry*, Orlando, FL, 1997.

21. Hercules, D.M., Surface analysis of organic materials with TOF-SIMS, *J. Mol. Structure*, 292, 49, 1993.

22. Xu, K., unpublished studies, University of Pittsburgh, 1993.

23. Bletsos, I.V., Hercules, D. M., Fowler, D., vanLeyen D., and Benninghoven, A., Time-of-flight secondary ion mass spectrometry of perfluorinated and poly-ethers, *Anal. Chem.*, 62, 1275, 1990.

24. Niehuis, E., Inaugural-Dissertation zür Erlangung des Doctorgrades, Universität Münster, 1988.

25. Benninghoven A., unpublished data, University of Münster.

26. Bletsos, I.V., Hercules, D. M., vanLeyen, D., Hagenhoff, B., Niehuis, E., and Benninghoven, A., Molecular weight distributions of polymers using time-of-flight secondary-ion spectrometry, *Anal. Chem.*, 63, 1953, 1991.

27. Elais, H.G., *Macromolecules*, 2nd ed. Plenum Publishing Co., New York, 1984.

28. Hittle, L.R. and Hercules, D.M., Time-of-flight secondary ion mass spectrometry of polybutadienes, *Surf. Interface Anal.*, 21, 217, 1994.

29. Hagenhoff, B., Benninghoven, A., Barthel, H., and Zoller, W., Supercritical fluid chromatography and time-of-flight secondary ion mass spectrometry of poly(dimethylsiloxane) oligomers in the mass range 1000–10,000 Da. *Anal. Chem.*, 63, 2466, 1991.

30. Hittle, L.R., Altland, D.E., Proctor, A., and Hercules, D.M., Investigation of molecular weight and terminal group effects on the time-of-flight secondary ion mass spectra of polyglycols, *Anal. Chem.*, 66, 2302, 1994.

31. Fowler, D.E., Johnson, R. D., vanLeyen D., and Benninghoven, A., Determination of molecular weight and composition of a perfluorinated polymer from fragment intensities in time-of-flight secondary ion mass spectrometry, *Anal. Chem.*, 62, 2088, 1990.

32. (a) Montaudo, G., Scamporrino, E., Vitalini, D., and Mineo P., Novel procedure for molecular weight average measurement of polydisperse polymers directly from matrix-assisted laser desorption/ionization time-of-flight mass spectra, *Rapid Commun. Mass Spectrom.*, 10, 1551, 1996. (b) Danis, P. O., Karr, D. E., Xiong, Y., and Owens, K.G., Methods for the analysis of hydrocarbon polymers by matrix assisted laser desorption/ionization time-of-flight mass spectrometry. *Rapid Commun. Mass Spectrom.*, 10, 862 1996.

33. Danis, P. O., Karr, D.E., Mayer, F., Holle, A., and Watson, C.H., The analysis of water-soluble polymers by matrix-assisted laser desorption time-of-flight mass spectrometry, *Org. Mass. Spectrom.*, 27, 843, 1992.

34. Xu, K., Proctor, A., and Hercules, D.M., Time-of-flight secondary ion mass spectrometry (TOF-SIMS) of polyisoprenes, *Mikrochim, Acta.*, 122, 15, 1996.

35. Chairelli, M.P., Proctor A., Bletsos I.V., Hercules, D.M., Feld, H., Leute, A., and Benninghoven, A., High resolution TOF-SIMS studies of substituted polystyrenes, *Macromolecules*, 25, 6970, 1992.

36. Winnik, M.A., Redpath, J., and Richards, D.H., The dynamics of end-to-end cyclization in polystyrene probed by pyrene excimer formation, *Macromolecules*, 13, 328, 1980.

37. Pesson DeAmorim, S., Bouster, C., Vermande, P, and Veron, J, Evolution of the product yield with temperature and molecular weight in the pyrolysis of polystyrene. *J. Anal. Appl. Pyrol.*, 3, 19, 1981.

38. Hittle, L.R., Proctor, A., and Hercules, D.M., Hydrogen-deuterium exchange in time-of-flight secondary ion mass spectra of polystyrene, *Anal. Chem.*, 66, 108, 1994.

39. Hittle, L.R., Proctor, A., and Hercules, D.M., Investigation of intramolecular hydrogen-deuterium exchange in the time-of-flight secondary-ion mass spectra of polystyrene, *Macromolecules*, 28, 6238, 1995.

40. Zimmerman, P.A., Hercules, D.M., and Benninghoven, A., Time-of-flight secondary ion mass spectrometry of poly(alkylmethacrylates), *Anal. Chem.*, 65, 983, 1993.

41. (a) Wayner, D.D.M. and Griller, D., Free radical thermochemistry, *Advances in Free Radical Chemistry*, 1, 159, 1990. (b) McMillen, D.F. and Golden, D.M., Hydrocarbon bond dissociation energies, *Ann. Rev. Phys. Chem.*, 33, 493, 1982. (c) Berkowitz, J., Ellison, G. B., and Gutman, D., Three methods to measure RH bond energies, *J. Phys. Chem.*, 98, 2744, 1994.

42. Cognetti, A.M., Characterization of polyvinylpyridines and poly(4-vinylpyridine-co-butyl methacrylate)s by time-of-flight secondary ion mass spectrometry, *MS Thesis*, University of Pittsburgh, 1995.

43. Zimmerman, P.A. and Hercules, D.M., Time-of-flight seccondary ion mass spectrometry of poly(alkyl acrylates): comparison of poly(alkyl methacrylates), *Anal. Chim. Acta*, 297, 301, 1994.

44. Zimmerman, P. A. and Hercules, D.M., Effect of stereoregularity on TOF-SIMS spectra of polymers, *Applied Spectrosc.*, 48, 620, 1994.

45. Nowak, R.W., Gardella Jr., J.A., Wood, T.D., Zimmerman, P.A., and Hercules, D.M., The double helical structure of isotactic poly (methylmethacrylate) in adsorbed monolayers leads to inter-chain ion formation in secondary ion mass spectrometry, *Anal. Chem.*, 72, 4585, 2000.

46. Bletsos, I.V., Hercules, D.M., Greifendorf, D., and Benninghoven, A., Time-of-flight secondary ion mass spectrometry of nylons: detection of high mass fragments, *Anal. Chem.*, 57, 2384, 1985.

47. Kim, Y. L. and Hercules, D. M., Structural characterization of polyesters by transesterification and time-of-flight secondary ion mass spectrometry, *Macromolecules*, 27, 7855, 1994.

48. Cohen, L.R.H., Hercules, D. M., Karakatsanis, C.G., and Rieck, J. N., Characterization of polyester polyurethanes by time-of-flight secondary-ion mass spectrometry, *Macromolecules*, 28, 5601, 1995.

49. Montaudo, G., Microstructure of co-polymers by fast-atom bombardment mass spectrometry. *Rapid Comm. Mass Spectrom.*, 5, 3, 1991.

50. Ballisteri, A., Garozzo, D., Guiffrida, M. and Montaudo, G., Identification of the ions produced by fast atom bombardment mass spectrometry in some polyesters and polyamides, *Anal. Chem.*, 59, 2024, 1987.

51. Dong, X., Proctor, A., and Hercules, D.M., Characterization of poly-(dimethylsiloxane)s by time-of-flight secondary ion mass spectrometry, *Macromolecules*, 30, 63, 1997.

52. Clarson, S.J. and Semlyen, J.A., *Siloxane Polymer*, Prentice Hall; Englewood Cliffs, NJ, 1993.

53. Dong, X., Gusev, A., and Hercules, D.M., Characterization of polysiloxanes with different functional groups by time-of-flight secondary ion mass spectrometry, *J. Am. Soc. Mass Spectrom.*

54. Bletsos, I.V., Hercules, D.M., vanLeyen, D., Benninghoven, A., Karakatsanis, C.G., and Rieck, J.N., Structural characterization of model polyurethanes using time-of-flight secondary ion mass spectrometry, *Anal. Chem.*, 61, 2142, 1989.

55. Dong, X., Characterization of Polymers by Time of Flight Mass Spectrometry, Ph.D. Thesis, University of Pittsburgh, 1999.

56. Bletsos, I. V., Hercules, D.M., vanLeyen, D., Benninghoven, A., Karakatsanis, C.G., and Rieck, J.N., Structural characterization of model polyester polyurethanes using time-of-flight secondary ion mass spectrometry, *Macromolecules*, 23, 4157, 1990.

57. Foti, S., Guiffrida, M., Maravigna, P., and Montaudo, G. Direct mass spectrometry of polymers. VIII. Primary thermal fragmentation processes in polyurethanes, *J. Polym. Sci., Polym. Chem. Ed.*, 21, 1583, 1983.

58. Bletsos, I.V., Time-of-flight secondary ion mass spectrometry of polymers, *Ph.D. Thesis*, University of Pittsburgh, 1990.

59. Ballistreri, A., Garozzo, D., Giuffrida, M., and Montaudo, G., Microstructure of bacterial poly(β-hydroxybutyrate-co-β-hydroxyvalerate) by fast atom bombardment mass spectrometry analysis of their partial degradation products, in *Novel Biodegradable Microbial Polymers*, 49,E. Dawes, E.A., ed., Kluwer, The Netherlands, 1990.

60. Spool, A.M. and Kasai, P.H., Perfluoropolyethers: analysis by TOF-SIMS, *Macromolecules*, 29, 1691, 1996.

61. Feld, H., Leute, A., Zurmühlen R., and Benninghoven, A., Comparative and complementary plasma desorption mass spectrometry/secondary ion mass spectrometry investigations of polymer materials, *Anal. Chem.*, 63, 903, 1991.

9

Laser Fourier Transform Mass Spectrometry (FT-MS)

Salvador J. Pastor and Charles L. Wilkins

CONTENTS

9.1 Introduction

Lasers have revolutionized mass spectrometry. Nowhere is this more true than in Fourier transform mass spectrometry (FTMS), also commonly known as Fourier transform ion cyclotron resonance (FTICR). A recent comprehensive review[1] includes, among other topics, discussion of the uses of pulsed lasers with FTMS. A Fourier transform mass spectrometer, unlike most other mass spectrometers, traps ions within the confines of a magnetic (or electromagnetic) field. Ions undergo a cyclotron motion that keeps them orbiting in small circular paths that are perpendicular to the magnetic field (Figure 9.1). Without some type of constraint, the ions would eventually spiral out of the ends of the magnet. To prevent this, electrostatic potentials are applied to trapping

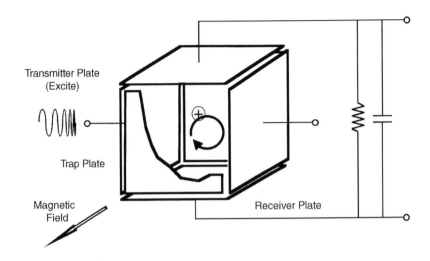

FIGURE 9.1
Diagram of FTMS cubic cell showing electrodes and their function.

plates on opposite ends of the analysis cell—positive potentials to trap positive ions and negative potentials to trap negative ions (like charges repel one another). For analysis, a frequency-swept excitation is applied to transmission plates on opposite ends of the cell. As the orbiting ions become in phase with the applied excitation, they absorb the energy and accelerate to higher overall radii. By careful application of the applied excitation, ion ensembles can be made to orbit near the inner edges of the cell without contacting cell plates and subsequently being neutralized. The phase-coherent ion packets then induce an image current onto all plates, including oppositely positioned detection plates. A sinusoidal image current is produced that exactly matches the frequency characteristics of the ions. For complex samples such as polymers, a complicated transient response results, which damps out as a function of collisions that leads to loss of phase-coherence. The image current signal is amplified and digitized, followed by Fourier transformation to yield the individual ion frequencies that comprise the signal. When combined with calibration, using known standards relating frequency-to-mass, the result is the simultaneous accurate detection of the masses of the types of ions present. The exceptionally high resolving capabilities of this method result from the ability to trap and observe ions for long enough periods to measure their frequencies with extremely high accuracy. An additional advantage of the method is that it is nondestructive.

Lasers have provided a convenient means to create gas phase ions from nonvolatile substances. In this chapter, application of lasers to FTMS for polymer analysis will be considered. The earliest combination of direct laser desorption and laser ablation techniques with FTMS will be discussed first. Moving on from that topic, the impact of matrix-assisted laser desorption/ ionization (MALDI) on FTMS will be addressed, with particular emphasis on its capability to permit unprecedented high mass analysis with minimal fragmentation owing to its "soft-ionization" nature. Next, the complementary high mass analysis method of electrospray ionization will be briefly considered along with its impact on polymer analysis by FTMS. Because FTMS, in common with all mass spectrometric analysis methods, is subject to possible mass discrimination effects, this issue is examined in a separate section. Following this, the use of quadrupolar excitation to allow very efficient mass selection for FTMS analysis of polymers is discussed. Applications of laser FTMS methods to polymer and polymer additives analysis are also considered. Finally, the current status of techniques for end-group determination and other polymer structure analysis questions is addressed. In this regard, the future of polymer analysis by FTMS is very exciting. Because FTMS works by separating ions in time rather than in space such as quadrupole, sector, or TOF instruments, there is theoretically no limit to the number of individual ion manipulation events that may be employed for structure analysis by multiple stage mass spectrometry experiments (MS^n).

9.2 Laser Desorption (LD)

Laser desorption Fourier transform mass spectrometry (LD-FTMS) yielded some impressive early results, following its introduction by McCrery et al.[2] Direct laser desorption using a pulsed CO_2 laser and a 3-Tesla superconducting magnet became the *de facto* standard for early polymer analysis applications. Wilkins and co-workers were one of the first research groups to see the potential of FTMS for high mass applications. In one of the earliest papers on the subject, Wilkins and co-workers[3] combined a Nicolet FTMS-1000 instrument with a Tachisto 215G pulsed TEA CO_2 laser using a 40-ns pulse at 10.6 μm. Their first paper showed a LD-FTMS spectrum of poly(ethylene glycol) with an average molecular weight of 3350. The use of negative ion modes was also demonstrated with a spectrum of a polyperfluorinated ether, Krytox 16140, where oligomer ions were observed up to m/z 7000, the highest mass molecular ion ever observed by FTMS. Figure 9.2 is the original Krytox 16140 negative ion spectrum. These encouraging results revealed the potential of FTMS for high mass analysis and also the need for "soft" ionization sources capable of delivering intact molecular ions into the gas phase.

The following year, 1986, Brown et al.[4] published a follow-up paper on polymer analysis. They found that the majority of undoped polymers readily attached sodium or potassium cations, presumably introduced during sample synthesis or preparation, and that these molecular ion species could be produced with little fragmentation. Performance was investigated using a series of poly(ethylene glycol) samples. The results of investigating PEG 1450, PEG 3350, and PEG 6000 lead to some general conclusions. First, it was necessary

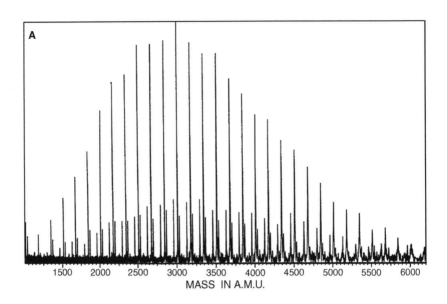

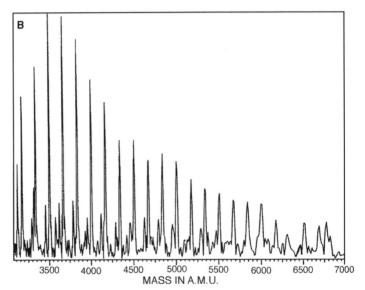

FIGURE 9.2

(A) Negative ion laser desorption FT mass spectrum of Krytox 16140. (B) High mass region Krytox 16140 taken with narrower bandwidth to show ions extending to m/z 7000. (Reproduced from ref. 3 with permission of the copyright holder)

to modify excitation and detection electronics to bypass the 3000 Da mass limit of the commercial system. Second, as theory dictates, resolution is inversely proportional to mass, leading to decreased mass resolving power at higher m/z. In fact, unit mass resolution was only achievable to m/z 2000 with the

3-T system. Other polymers examined were poly(ethylenimine) (PEI) 600 and 1200, poly(ethylene glycol methyl ether) (PEGME) 5000, poly(caprolactone diol) 2000, poly(propylene glycol) (PPG) 4000, and poly(styrene) 2000. Because of the inherent decrease in resolving power as m/z increases, integration of the spectral peak area for all isotopes for each oligomer species was done, as well as mass correcting for the attached cation. Both number and weight averages were calculated and shown to be in reasonable agreement with both manufacturers' specifications and also the samples analyzed by other mass spectral techniques.

In a chapter partially devoted to polymer work, Brown and Wilkins[5] optimized the analysis of a poly(ethylene glycol) sample of PEG 3350. Unit mass resolution was extended to almost m/z 4000, and resolution of 7,500 was achieved for m/z 3000. The spectrum of a combined mixture of PEG 1450 and PEG 3350 demonstrated resolving power of 15,000 for ions with m/z 1500. A plot of resolution vs. mass showed the predicted linear loss of resolving power with increasing mass.

At about the same time, results of several other LD-FTMS studies done in collaboration with scientists at Nicolet Analytical Instruments (the first commercial producers of FTMS systems) began to appear. Brown et al.[6] examined poly(p-phenylene) (PPP), a polymer that is popular because of its qualities of thermal stability and resistance to oxidation and radiation. Because of the synthetic routes employed, different procedures produced PPPs with somewhat different physical properties and colors. LD-FTMS was chosen as the technique for analysis based upon its ability to produce molecular ions of intact polymer chains with better than unit mass resolution. Four preparations yielded different average molecular weights as well as varying repeat units. Halogenation of these materials was assessed, as well as the proposed synthetic mechanisms. A follow-up paper on the subject by Brown et al.[7] examined in more detail the formation of PPPs by polymerization of benzene, by polymerization of biphenyl, and by polymerization of terphenyl. These authors also attempted to correlate stabilization of charge with the appearance of negative or positive ion spectra of the PPPs. Furthering their previous work on PPPs, Brown et al.[8] investigated analysis of several heterocyclic aromatic polymers such as poly(1-methyl-2,5-pyrrolylene), poly(1-phenyl-2,5-pyrrolylene), poly(2,5-thienylene), and poly(2,5-selenienylene). These polymers would provide meaningful comparisons of electrical conductivity.[9] The investigation of several related polymers of PPPs was continued by Brown et al.,[10] extending their investigations to poly(phenylene sulfide), polyaniline, poly(vinyl phenol), polypyrene, and several others. Among their findings were evidence that carbon clusters with high mass formed during infrared laser desorption, as well as experimental evidence that some polymers fragment during the data acquisition period. Later, So and Wilkins established that singly charged, even-numbered positive cluster ions up to C-600 could be observed by pulsed infrared laser desorption of benzene soot.[11]

There was still considerable interest in coupling alternative ionization sources to FTMS. Infrared laser desorption was an emerging technique having been compared to other techniques such as field desorption (FD), plasma desorption (PD), electrohydrodynamic ionization (EH), fast-atom bombardment (FAB), and secondary ion mass spectrometry (SIMS).[12,13] The comparisons were done by sector-based instrumentation. Nuwaysir and Wilkins[14] examined alkoxylated pyrazole and hydrazine polymers in the weight range between 600 and 1300. Using LD-FTMS, spectra were found to contain more regular polymer distributions, fewer fragment ions, and less mass discrimination when compared with analyses by FAB and SIMS ionization published earlier.[15]

A significant advancement in performance occurred by the introduction of a higher field (7.2 T) superconducting magnet.[16] A Nicolet FTMS-2000 mass spectrometer equipped with dual cubic cells, an autoprobe, and a Tachisto 215 CO_2 laser demonstrated the improved results that could be obtained by such a system. When spectra of PEG 8000 were obtained by averaging fifty individual laser shots acquired in direct mode, they showed oligomer ions extending up to m/z 9700, setting a new high mass record for direct LD-FTMS of polymers. In other experiments where ion populations were controlled by ejecting ions, resolving power of 160,000 was achieved for m/z 3200 ions from PEG 3350, and resolving power of 60,000 for m/z 5922 ion from PPG 4000 (Figure 9.3). These results are impressive even by today's standards.

Partially as a result of these early studies, there was growing interest from the polymer community to see the analysis of complex examples of polymer systems rather than the standard polar homopolymers such as poly(ethylene glycol), which had been investigated first. This was partially addressed through collaborative efforts of academic scientists with industry (see the industrial applications section later in the chapter). Nuwaysir et al.[17] examined copolymer systems of methylmethacrylate/butylacrylate (MMA/BA), methylmethacrylate/styrene (MMA/STY), and poly(ethylene glycol)/poly(propylene glycol) (PEG/PPG). In addition to calculating number and weight averages, mass spectra were analyzed to estimate copolymer monomer contributions and compared with manufacturer feed ratios. All results were run in parallel with gel permeation chromatography (GPC) and the results compared. Also, the work discussed possible discrimination effects of ion transfer to the analyzer cell through the conductance limiting plate separating the source and analyzer cells. Readers were cautioned that trapping polymers in an analysis cell is a time dependent event, and that substantial variations in the calculated molecular weight could arise. Kahr and Wilkins[18] investigated nonpolar hydrocarbon polymers by LD-FTMS using silver cationization as a method for obtaining useful spectra of polystyrene, polyisoprene, polybutadiene, and polyethylene. Using a 7-T FTMS-2000 system, they were able to obtain completely resolved polymer spectra ranging in mass from 400 to 6000 Da with better than 12 ppm mass accuracy. Results were compared with manufacturer's GPC data.

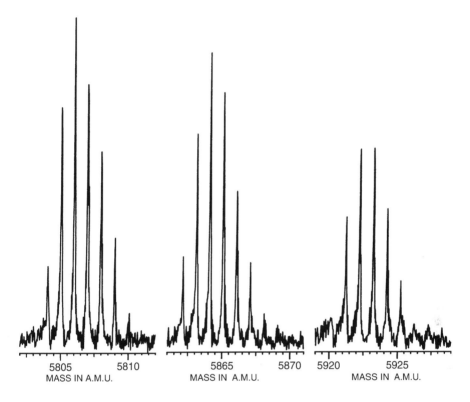

FIGURE 9.3
A portion of the high-resolution laser desorption FT mass spectrum of poly(propylene glycol) 4000 between m/z 5500 and m/z 6000. Average resolving power is 60,000. (Reproduced from ref. 16 with permission of the copyright holder)

9.3   Laser Ablation

Laser ablation uses higher laser power densities and smaller spot sizes than direct laser desorption. The outcome is intense, focused ablation of the sample surface resulting in the formation of craters on the sample probe. Besides providing analytically useful "fingerprint" information about materials, it is of interest in the semiconductor and electronics industries. Its coupling with the high resolution and trapping capabilities of FTMS make it of particular interest to companies such as IBM, whose scientists produced a series of papers on the topic. Creasy and Brenna[19] used 266 nm radiation from a Nd:YAG with a pulse length of 10 ns to ablate polyimide and graphite surfaces. The microprobe optical system used allowed for illumination and viewing of the sample in the magnetic solenoid, and the use of a 75-mm lens inside the vacuum chamber provided a minimum spot size of 5–8 μm. The mass spectra from polyimide show high-mass, even-carbon-number cluster ions. By measuring the size of the craters formed, it is possible to estimate the amount of ablated material, plasma particle densities, and ultimately to compare with theories on the formation of clusters.

Creasy and Brenna[20] continued their previous work by examining a copolymer of ethylene and tetrafluoroethylene (ETFE), polyphenylene sulfide (PPS), and a diamond-like carbon film (DLC). Ablation of these samples gave comparison information on the formation of fullerenes in the presence of H, S, and F atoms. In the case of ETFE, a large C_{60^+} peak was observed in the spectra, which showed a strong dependence on the number of laser pulses and laser power density.

These workers also investigated eight industrially important high-mass polymers by laser microprobe FTICR. These polymers included PEG 8000, poly(phenylene sulfide) (MW 1.0×10^6), poly(vinyl acetate) (MW 6.4×10^4), poly(styrene) (MW 2.5×10^6), PMMA (MW 4.6×10^4), poly(vinyl chloride) (3.7×10^4), poly(acrylonitrile) (MW 2.3×10^4), and poly(dimethylsiloxane) (MW 4.4×10^4). Brenna and Creasy[21] posed the question of whether broadband UV laser microprobe FTICR could be used to identify these polymers, and they also wanted to compare spectra with the more widely used TOF laser microprobe mass spectrometry (LAMMA). Results include the observation of odd-mass ions, carbon clustering (Figure 9.4), stable subunit condensation,

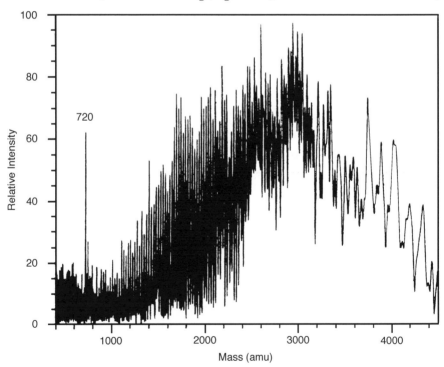

FIGURE 9.4

Positive ion mass spectrum of poly(phenylene sulfide) showing high-mass carbon cluster ions starting at C_{60} and extending past C_{400}. (Reproduced from ref. 21 with permission of the copyright holder)

and structurally significant negative ions in some cases. According to these workers, "data conclusively demonstrate disparities in ionic product distributions and therefore formation mechanisms for LAMMS based on TOF and FTICR. While positive-ion formation from both techniques is complex, FTICR spectra appear to be more sensitive to the original polymer structure than do literature TOF spectra."

The effect of gold particles on plasma-polymerized fluorocarbon films was examined by Creasy et al.[22] Laser ablation of gold-containing films produced ions that were similar to those formed in pyrolysis. This observation suggests that gold particles absorb energy, conductively heat the polymer, and affect the mechanism of the ablation.

9.4 Matrix-Assisted Laser Desorption/ Ionization (MALDI)

Recently, there has been significant interest in application of MALDI to many mass spectral techniques. The coupling of MALDI with FTMS is no exception. It is trivial, in most cases, to prepare polymer samples for MALDI analysis, requiring only a compatible solvent system that can dissolve both the analyte and matrix and a method for applying the solution to the sample probe tip. When a laser beam is impinged upon the probe tip, the matrix can preferentially absorb the laser energy and be ejected into the gas phase, taking along the analyte, which is typically co-crystallized in molar ratios ranging from 1:1 to 10000:1 (matrix:analyte). What makes MALDI work so well is the matching of the matrix absorption characteristics to the wavelength of the laser available. Because the ultraviolet was the first region of the spectrum to be examined, many combinations of matrices with popular UV laser systems (e.g., excimer, nitrogen, YAG) exist. Though the mechanisms are not fully understood, a small percentage of the analyte is ionized, allowing for its further manipulation and detection by mass spectrometry. Polymers almost exclusively undergo cation attachment in the gas phase. Undoped sample solutions make use of alkali salts left over from the original synthesis, or which arise as small impurities from solvents or matrix preparations. The most common observations are those of Na^+ or K^+ attachment to the polymer. In recent modeling studies, polar polymers have been shown to coordinate alkali salts to available oxygen sites along the polymer backbone. These theories are leading to study of polymer conformations in the gas phase.

Castoro et al.[23] were the first group to successfully couple internal MALDI generation for high-mass analysis by FTMS and show the extended mass range that is now available using this "soft" ionization technique. Using 355 nm radiation from an excimer-pumped dye laser for desorption/ionization and sinapinic acid as matrix, they successfully trapped MALDI ions from a PEG 10000 sample with masses as high as 14000 Da (Figure 9.5). They employed the gated trapping technique in which, prior to firing the laser, the source

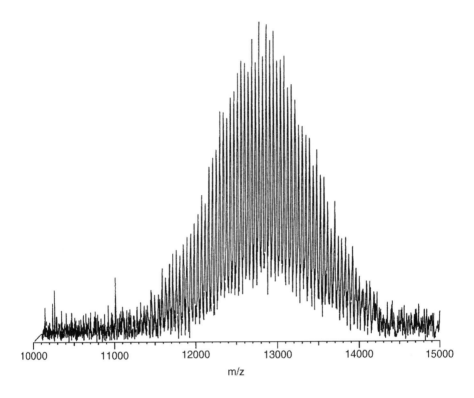

FIGURE 9.5
Matrix-assisted laser desorption/ionization FT mass spectrum of PEG 10000. (Reproduced from ref. 23 with permission of the copyright holder)

trapping plate is set to ground potential and the rear-most trapping plate is set to its highest value, typically 10 volts, setting up a decelerating field within the analysis cell. After firing the laser and waiting a fixed amount of time for ion transit (50 to 300 μs), the electrostatic trap is closed and the polymer ions remain trapped. This method has become the standard practice for trapping MALDI generated ions in FTMS instruments; MALDI-generated matrix and analyte ions are ejected with significant velocities and kinetic energies that scale with mass. The theory of gated trapping has been eloquently described by Knobeler and Wanczek.[24]

In some recent work, Easterling et al.[25] have built a 4.7 Tesla internal MALDI-FTICR instrument. The authors acknowledge that ion introduction into the homogeneous region of the magnetic field is simplified when ions are formed in the vicinity of the cell; this eliminates the need for additional ion lenses, multipole devices, or wire guides. Using other instrument modifications such as an open-ended cylindrical cell with capacitive coupling and an internal preamplifier, they demonstrate high performance analysis of singly charged molecular ions in the 1000 to 10000 Da range. Spectra are shown for PEG 8000 and PEG 4600. The authors propose that, because the linear flight space between end electrodes is tripled over that of most cubic cells, there is less mass discrimination in their case, using an open ended cylindrical cell over the conventional cubic cell design. The discrimination issue that affects overall molecular weight distribution calculations will be discussed further in the mass discrimination section.

The other method for introduction of ions into the magnetic field is the use of an external MALDI source. Proponents of external ion sources claim that external ion sources provide the necessary flexibility to interface a variety of techniques such as FAB, SIMS, and ESI. In addition, researchers believe external sources allow the tuning of instrumental parameters separately, thereby permitting optimization of the ion formation and detection events separately. An example of this approach was given by Heeren and Boon[26] who combined "in-source" pyrolysis for broadband screening of polymer additives with MALDI to characterize molecular weight distribution and perform end-group analysis. For the determination of an amine terminated poly(propyleneglycol), the authors also cited extensive discrimination using the external MALDI ion source. In their experiments, T_{gate} typically varied between 600–2000 μs, depending on the molecular weight range of interest.

White et al.[27] also developed a new external ion source 7 T FTICR utilizing an electrostatic ion guide. PEGs of number average molecular weights 1500, 2000, and 3400 were used as test samples to study time-of-flight effects. By adjusting the voltages of the ion guide, arrival times could be kept fairly consistent from 1.2 ms for PEG 1500 to 1.5 ms for PEG 3400. By applying quadrupolar axialization (discussed later in this chapter), the authors were able to measure PEG 3400 spanning a several hundred mass unit range with average resolving power of 400,000, due in part to the low (1×10^{-9} torr) system pressure.

Finally, the issue of polymer detection limits was addressed by Pastor et al.[28] Using PEG 2000 and 6000 standards, they determined a lower detection limit of 40 femtomoles (from 4 laser shots) and produced a spectrum with a signal-to-noise ratio of at least 5:1. Part of the added reproducibility of the MALDI events came from the use of an aerospray sample deposition technique that could produce highly uniform sample surfaces. Scanning electron microscopy was used to quantitate the laser spot sizes.

In an effort to prevent duplication of references, all further MALDI references will be included in other areas of this chapter to which they are appropriate.

9.5 Electrospray Ionization (ESI)

Although not a laser FTMS method, it is necessary to briefly consider the complementary technique, electrospray ionization, one of the most widely used alternatives for biological applications where high-mass analysis has been achieved through the addition of multiple charges. The introduction of electrospray ionization has revolutionized analysis of high-mass species. Because of its evaporative ionization process, molecular ion dissociation is minimized, thereby yielding multiple charging with correspondingly lower m/z values. However, very little work has employed the electrospray process with trapped ion techniques to analyze polymers. An obvious reason is the lack of control in producing charge states. As the number of trapped ions in an FTMS is spread over a larger m/z range, the effective sensitivity produced for any particular charge state is greatly reduced, as a consequence of decreased relative ion abundances. To make this point more explicitly, consider an example recently published by McLafferty and coworkers.[29] They analyzed a poly(ethylene glycol) 20,000 sample by ESI-FTMS. It was possible to resolve 5000 peaks from isotopic clusters representing 65 oligomers in 12 charge states. Analysis of data of this magnitude is a daunting task and clearly would not be the method of choice for polymer analysis, if simpler alternatives such as MALDI-FTMS are adequate. Additionally, most entry-level commercial FTMS systems would have a difficult time producing spectra of sufficient resolution to achieve this level of performance.

A more detailed account of ESI-FTMS of polymers was presented by O'Conner and McLafferty for poly(ethylene glycol)s of 4.5, 14 and 20 kDa.[30] Figure 9.6 shows their ESI/FT mass spectrum of PEG 14000 with resolving

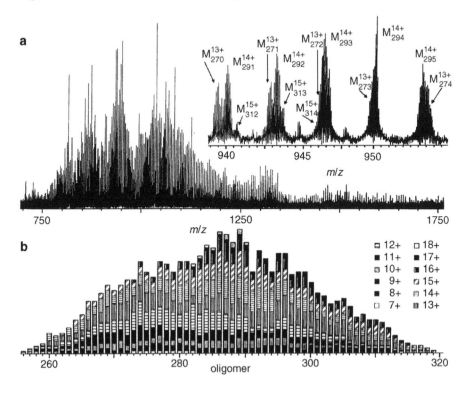

FIGURE 9.6
(a) Electrospray ionization FT mass spectrum of PEG 14000 with resolving power of 100,000. (b) M_r distribution from summed oligomer abundances. (Reproduced from ref. 30 with permission of the copyright holder)

power of 100,000. This is a highly complex spectrum which includes overlap of isotopic peaks from different charge states. One benefit of the higher resolving power of FTMS is identification of polymer impurities. The authors cite the example of the M_{294}^{14+} isotopic cluster. If one of the 294 $-CH_2CH_2O-$ groups (44 Da) is replaced with a propylene oxide group $-CH_2CH(CH_3)O-$ group (58 Da), then the peak cluster should be centered at 950.6 (extra m/z $14/14^+$). The abundances of these peaks are <5% of those of M_{294}^{14+}, demonstrating that the sample contains at most 0.02% ($1/294 \times 5\%$) monomer units containing an extra CH_2. As expected, data interpretation is very laborious. The algorithms employed never seemed to work individually, although combining attributes from several methods seemed to produce acceptable results, according to the authors.

9.6 Mass Discrimination

The fundamental premise of analyzing polymers using FTMS and the pulsed nature of a laser is that no mass discrimination exists that could bias the oligomer intensities and lead to incorrect calculation of molecular weight distributions. For accurate molecular weight distributions, no biasing effects should be present from the MALDI or cationization process, from ion transmission to the analysis cell, or from excitation or detection strategies. The first issue of "MALDI discrimination" is a hot topic of debate. Many people believe that for polydisperse polymers, oligomers at the start and end of the distribution may be ionized differently either through ion suppression effects as ions are being ejected into the gas phase or through preferential cationization. There is evidence that the choice of cation affects coordination with the polymer. There has been work done by time-of-flight mass spectrometry analyzing different choices of cations and their effects upon calculated molecular weight distributions. MALDI analysis of polymers as compared with GPC will continue to be an area of intense interest.

The most obvious form of mass discrimination occurs from ion transmission to the analysis cell and trapping. Because FTMS is a trapped ion technique, it is imperative that the analysis cell accommodate a representative portion of the polymer being analyzed. It logically follows that ions experiencing a time-of-flight effect may not be properly "sampled" if the trap is closed prematurely (biasing for lower mass species) or left open too long (biasing for higher mass species). Even with direct laser desorption studies, there was concern about accuracy in polymer characterization. Hogan and Laude[31] examined several factors that could lead to possible mass discrimination in LD-FTICR, including laser power density, trapping potential, and distance between the cell and desorption site. They showed, for example, that the number average molecular weight for PEG 600, PEG 1000, and PEG 1500 varied by 7, 10, and 12% as the desorption site was displaced over a 10-cm distance from the cell. By varying each of the instrumental parameters, they could achieve highly reproducible results that were within a few percent of the measured GPC values.

Trapping discrimination was also examined for MALDI-generated polymer ions in a cubic cell by Dey et al.[32] The use of a gated decelerating potential with a single delayed trapping time produces very noticeable discrimination when examining broad polymer distributions. Their work focused on examining the effect on poly(ethylene glycol) spectra as delay times following desorption/ionization and before applying static trapping voltages were systematically varied. Dey et al. proposed *post averaging* of time-domain transients to reconstruct the entire broadband polymer distribution. Using a 7 T FTMS-2000 system and a 10 μs sampling interval, the authors reported a reconstructed spectrum of a synthetic "polydisperse" sample made up of an equimolar mixture of PEG 1000, PEG 3000, PEG 6000, and PEG 8000. These encouraging results led to further research by Pastor and Wilkins[33] who concluded that there was a critical window of 2500 Da for the standard 2-inch cubic cell. A polymer that contains oligomers covering a range wider than 2500 Da will certainly produce discrimination if a single gated trapping time is used (see Figure 9.7). Polymers within this mass range can be properly

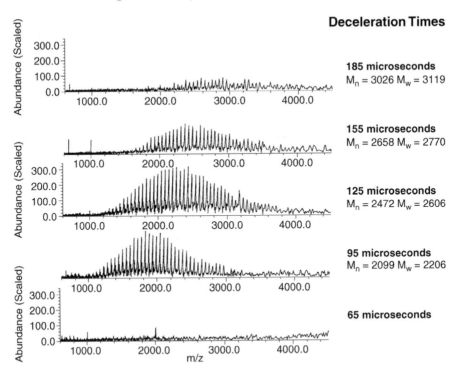

FIGURE 9.7
Spectra of hydroxyl-terminated polybutadiene 1350, taken at different gated trapping deceleration times following the laser pulse. Longer delays before applying static trapping voltages clearly show the time-of-flight effect of the ions entering the cell. (Reproduced from ref. 33 with permission of the copyright holder)

represented by optimizing the polymer's ion abundance over a small incremental time range. Thus, a single optimized trap time can be used to analyze polymers of narrow polydispersity. Employing a 3 T modified FTMS-1000 mass spectrometer equipped with a nitrogen laser, Pastor and Wilkins demonstrated characterization of both narrow and wide mass nonpolar polymers up to m/z 6000, including several polyisoprene, polybutadiene, and polystyrene samples.

Other groups have noted that the time-of-flight effect of the ions is a definite problem particularly when an external ion source is used. Easterling et al.[34] reported a reduced sensitivity to this discrimination effect by use of internal MALDI and an open-ended cylindrical analyzer cell of over three times the length of a standard cubic cell. They reported a mass shift of only 20 Da for a PEG 1000 polymer distribution over a 200 μs change in gated trapping time. Computer simulations showed a much improved field free region owing to the extended length of the flight path. Another method of constructing the "true" polymer distribution from its time segmented parts was proposed by O'Conner et al.[35] who advocate superimposing spectra on the same m/z axis rather than summing data transients. Their method is based on "acquiring a series of spectra at different trapping times and superimposing the spectra so that each oligomer has the intensity of its maximum intensity through the set of spectra." Use of a chromatographic polystyrene standard (MW 950) produced an error of 20% for the most probable polymer weight, M_p.

9.7   Quadrupolar Excitation/Ion Cooling

One of the latest innovations in FTMS has been the introduction of quadrupolar excitation/ion cooling (also known as quadrupolar axialization or ion axialization). Introduced as an ion manipulation technique, axialization can be used to compress ion clouds within the analysis cell, allowing for a tighter, more coherent motion during the excitation/detection events. A secondary benefit of applying axialization is its effect on nonselected ions, which are lost under high pressure through magnetron expansion of their orbits. The technique works by applying low sinusoidal in-phase voltages (typically 5 volts or less) to opposite pairs of cell plates. During the quadrupolar excitation event, magnetron motion is coupled to the cyclotron motion of the ions. In the presence of a collision gas, ions are cooled to the center of the cell both through cyclotron and axial relaxation. Magnetron relaxation, for ions not undergoing the applied frequency, causes a radial expansion of the remaining ions' orbits and eventual loss from the cell. This process can be used to accomplish highly specific mass selection. To date, axialization is an accepted technique for improving signal-to-noise ratios, mass resolving power, ion capture efficiency, and ion remeasurement efficiency. The technique has been reviewed by Guan et al.[36]

Showing interest in manipulating polymers, Pastor et al.[37] used quadrupolar excitation in a 7 T FTMS-2000 equipped with a dual cubic cell. First, the source cell was used to apply the axialization to the oligomers of interest. Following this, ions were transferred to the analyzer cell for detection. Using PEG 6000 as a test polymer, the authors demonstrated that the mass selection could be very precisely controlled by varying the applied amplitude of the single frequency quadrupolar excitation. Figure 9.8 shows

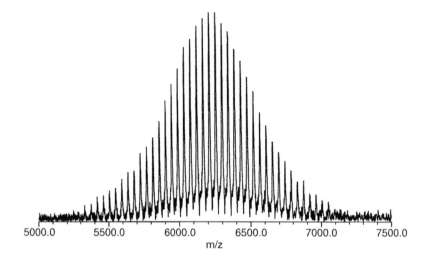

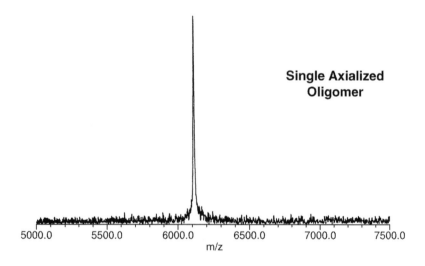

FIGURE 9.8

(a) MALDI FT mass spectra for PEG 6000 showing full distribution using source cell detection. (b) Mass spectrum of a selected oligomer using quadrupolar axialization to isolate and transfer to the analyzer cell. (Reproduced from ref. 37 with permission of the copyright holder)

the selection of a single oligomer from the polymer PEG 6000. This demonstrated an extremely effective way of reducing ion populations. It is a well-known fact that examining an entire polymer distribution can lead to reduced resolving powers at high trapping voltages, used to improve signal-to-noise, due to space charge interactions. A common solution is to reduce the overall number of trapped ions, leading to longer-lived transients with fewer interactions from other ions and, ultimately, better resolving power. In the past, to arrive at the same mass selection, chirp ejection pulses were applied to the polymer or the ions were subjected to stored waveform inverse Fourier transform (SWIFT).[38,39] However, both frequency sweep and SWIFT ion ejection techniques affect the radius of the analyte ion(s) of interest, and they usually require several applications of tailored excitation to cleanly select the peak(s) of interest. On the other hand, axialization performs the selection in a single step. Pastor et al. established that axialization could be applied to polymers with masses as high as 13,000 Da, the highest mass tested.

In a further development of axialization, Marto et al.[40] demonstrated broadband axialization of polymers using an external MALDI source and an electrostatic ion guide (EIG) in a 3 T FTICR. Broadband axialization can improve ion trapping efficiency of injected ions from an external source. The benefit is reduced off-axis ion displacement. It can also be used to select a subset of the polymer distribution, also achievable with single frequency axialization, but without the oligomer abundance distortions. Thus, the spectrum over the selected mass range should closely match the relative abundances of the original distribution. To achieve the broadband effect, Marto et al. used repeated SWIFT excitations controlled from a Macintosh II personal computer and a short C language program. For PEG 2000, an axialization range of 1950 to 2120 Da was selected, corresponding to selection of 5 oligomeric species.

Pitsenberger et al.[41] examined alternative excitation mechanisms for ion axialization and remeasurement in a 4.7 T FTMS using internal MALDI. Specifically, they looked at repetitive chirp, filtered noise, and high-amplitude single frequency excitation for ion axialization. Because the instrument employs an internally mounted preamplifier on the detection electrodes, said to improve overall S/N and "reduce the adverse effects of distributed capacitance on the image current," the authors instead applied two-plate axialization, a modified version of the axialization technique. For frequency chirp broadband axialization, a mass-to-charge range of 1500 to 2500 was selected for PEG 2000 with remeasurement efficiencies exceeding 99.5% using the open-ended cylindrical cell. The filtered white noise experiments yielded similar efficiencies but for a smaller mass range, 400 Da, due to inadequacies in the active bandpass filter used, which peaked at m/z 2000. Finally, the high-amplitude single frequency experiments, when combined with capacitive coupling of the excitation signal to the trapping plates, produced less axial ejection of ions and a wider retained range of ions. However, the distributions of axialization-selected oligomers still suffer from the uneven power applied during axialization, leading to distributions that center around the applied frequency. These authors also obtained spectra of PEG 4600, PEG 6000, and PEG 8000 that were collected using frequency burst excitation for ion axialization. These spectra appear remarkably similar to the PEG 4600 and PEG 8000 shown in the prior Easterling et al.[25] reference that were said to result from "single laser shots followed by the *amplified trapping*...."

Justification for using two-plate quadrupolar excitation has been reported by several groups including Jackson et al.[42] who demonstrated the qualitatively similar results of a mixture of PEG 1000 and PEG 1500 undergoing both two-plate and four-plate ion axialization. The authors note that despite the complex dynamics and interacting resonances produced from a two-plate excitation, the results show similar efficiencies in their ability to cool ions.

The effect of capacitive coupling on quadrupolar excitation (but using two plate axialization) was further demonstrated by Pitsenberger et al.[43] These authors were able to remeasure the 41-mer of PEG for up to 200 remeasurement cycles with 100% efficiency in a capacitively coupled cylindrical ion trap. Plots were shown demonstrating that the predominant loss of ions occurs through axial ejection during excitation. Lowering the excitation radius to prevent radial ejection, along with raising the trapping voltages to prevent axial ejection of low masses, makes the 100% remeasurement possible. Broadband remeasurement of a 600 m/z range for PEG 2000 produced a 99.5% efficiency after 50 remeasurements.

Finally, the ability to manipulate polymer distributions and reduce ion populations was further demonstrated by low-voltage on-resonance ion selection (LOIS).[44] This newly introduced technique uses the same high pressure collision gas as quadrupolar axialization but does not require the hardware switching of the latter. Thus, it is an easily implemented technique on most Fourier transform mass spectrometers and can offer some of the same advantages as ion axialization. Pastor and Wilkins showed high selectivity of several poly(ethylene glycol) samples ranging in mass from 1000 to 6000 Da. In some examples, oligomers were selected in various combinations, including alternating oligomers, and groups to demonstrate that any combination of ions could be mass selected. LOIS has also shown promise in ion remeasurement studies.[45] The theory is currently being investigated.

9.8   Direct Applications

Applications of specialty polymers are numerous. Of particular analytical interest are copolymers. Here, recent literature on laser desorption FTMS analysis of non-PEGs, PEOs, PSs, and PMMAs will be discussed briefly. These articles, which have appeared over the last 10 years, provide examples of the variety of polymer systems that can be analyzed by FTMS.

Srzic et al.[46] studied natural polymers of humic acid and lignins by LD-FTMS. Humic substances are commonly found in soil and water, and formed by the chemical and biological degradation of plants and animals. The resulting products can associate into complex organic structures. Lignin is a major cell wall component in wood and is composed of substituted phenylpropane units connected through ether links. Spectral peaks up to m/z 700 were observed for positive ions of humic substances collected from lake sediments. Samples of lignin from birch and spruce were more interesting because they showed the 444 Da repeat of a trimer building block and masses up to 3000 Da in negative ion mode.

LD-FTICR-MS was compared with high performance liquid chromatography (HPLC) for analysis of triton polymers, used as commercial surfactants. Liang et al.[47] found that octylphenol ethoxylates gave molecular weight distributions up to 3500 Da, with minimal fragmentation. The Triton X family of surfactants are commonly used in industrial cleaners and detergents, manufacturing processes, and agricultural applications. The authors show direct comparison with HPLC/UV chromatograms and LC/quad mass spectra (Figure 9.9). Overall, LD-FTICR-MS overcomes problems with fragmentation of higher molecular weight species, reduces analysis time, and provides fully resolved oligomer information.

An interesting method was presented for the analysis of perfluorinated polyethers (PFPE) by Cromwell et al.[48] The fluorinated polymers are commonly used as lubricants because of their low vapor pressures, chemical inertness, and thermal stability. Samples were examined in a cubic trap FTMS-2000 system using a 4.3 T magnetic field. Both a Kr/F excimer laser and a Nd/YAG laser were used for internal LD. The two-step procedure involved using the 248-nm excimer laser for desorption of the polymer and, immediately following, a second more tightly focused 532 nm laser ablation pulse to ablate metal cations from the surface beneath the polymer sample. Thus, cationization could be controlled, and the dynamics of the process are discussed. Mass spectra extended up to 10,000 Da are shown.

A variety of electrochemical polymerization studies have appeared. Elliott et al.[49] used LD-FTMS for the analysis of 4-methyl-4'-vinyl-2,2'-bipyridine-containing metal complexes. Mass spectral analysis in conjunction with thin layer chromatography (TLC) demonstrated that normal "polyvinyl-type" chains are formed through chain propagation. O'Malley et al.[50] used LD-FTMS to study chemically and electrochemically prepared poly(2-vinylthiophene). They identified three forms of poly(2-vinylthiophene): (i) free radical-initiated, (ii) a form arising from anodic oxidation of 2-vinylthiophene, and (iii) an insoluble film on the anode. Using high resolution FTMS, these workers drew conclusions about different mechanisms involved in the formation of each of the polymers.

Campana and co-workers[51] wrote a general tutorial on polymer analysis providing an excellent introduction to LD-FTMS.

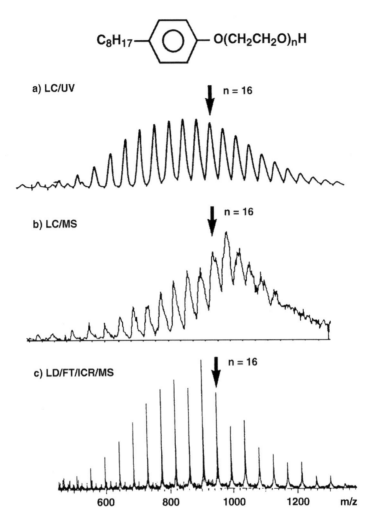

FIGURE 9.9
(a) LC/UV chromatogram of Triton sample, (b) LC/quadrupole mass spectral data of Triton sample, (c) LD-FTICR mass spectrum of Triton sample taken from a single time-domain data set. (Reproduced from ref. 47 with permission of the copyright holder)

9.9 Polymer Additives

It would be remiss not to discuss polymer additive analysis. Such materials are used as antioxidants, UV absorbers, and stabilizers, and dyes. An early paper dealt with LD-FTMS analysis of poly(methyl methacrylate) dyes such as 12-H-phthaloperine-12-one (orange dye), 1-(methylamino)anthraquinone (red dye), and 1,8-bis(phenylthio)anthraquinone. Hsu and Marshall[52] showed that LD/FTICR could determine these dyes in PMMA at least an order of magnitude better sensitivity than current methods using attenuated total reflectance infrared (ATR/IR), showing detection limits down to 0.1% vs. infrared's 1–2% by weight. The authors did comparisons of both untreated and redissolved PMMA and concluded that pretreatment of the polymer was not necessary to identify its dye component.

Probably the most comprehensive paper on LD-FTMS analysis of polymer additives was authored by Asamoto et al.[53] They examined about 30 additives commonly used with polyethylene and compared them with polyethylene extracts, originating from several sources including a wash bottle, a garbage can, and a tarpaulin. The materials were extracted for 8 hours using a Soxhlet apparatus and 150 ml of diethyl ether. The spectra showed the potential for LD-FTMS of these commercially available additives sold under the following trade names: Irganox (hindered phenolic amide), Naugard (long chain thiodipropionate ester), Ultranox (hindered phenol), Polygard (tris(nonylphenyl) phosphite), and Tinuvin (generally benzotriazole derivatives). Their results from the polyethylene extracts showed the presence of Irganox 3114 (783 Da), Naugard 524 (646 Da), and Naugard DSTDP (682 Da) along with several unknown additives that were also present in the spectra.

With the move toward higher mass additives to reduce volatility, Johlman et al.[54] examined additives with masses between 500 and 1300 Daltons and compared the results with spectra obtained employing FAB ionization. Using direct laser desorption FTMS in a 3 T system, spectra were found to be "superior" to the FAB spectra produced using both sector and triple-quad mass spectrometers, particularly in terms of reduced fragmentation.

Phosphite polymer stabilizers were analyzed by combining LD/EI/FTICR. Used to control molecular weight and color in melt processing, these antioxidants are difficult to analyze by extraction-LC, X-ray fluorescence, UV, or FTIR for a variety of reasons. Xiang et al.[55] examined these additives individually and in mixed polymers that were prepared by hot pressing to form thin films that could be attached to the probe tip. Additives examined included Ultranox 626 (604 Da), XR-2502 (636 Da), and Weston 618 (732 Da).

Examination of flame-retarding additives was done by Heeren et al.[56] using a direct temperature controlled pyrolysis external ion source and a 7 T FTICR. Samples were taken from common household appliances such as TV set housings, computer casings, and others and were pulverized to powder form. Direct heating of the filament probe with dried sample produced spectra with two distinct regions, corresponding to evaporation of nonbonded additives and pyrolysis of the polymer matrix. The unknown polymer blends were found to contain brominated biphenyls, brominated diphenyl ethers, tetrabromoBisphenol-A and its butylated isomers, polystyrene, and antimony oxides.

Laser desorption FTICR is also a valuable technique for the characterization of industrial materials. Simonsick and Ross[57] analyzed novel dispersants (used in emulsion polymerization), fluorinated surfactants, and natural oils with masses in the range of 500–3000 Da. Specifically, poly(ethylene oxide)

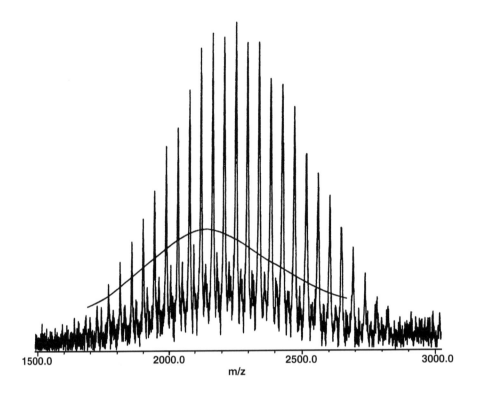

FIGURE 9.10
LD-FTICR-MS spectrum of StyrEomer-2100, a poly(ethylene oxide) methyl ether oligomer. The solid line represents the distribution from residual methoxy-capped poly(ethylene
oxide). (Reproduced from ref. 57 with permission of the copyright holder)

methyl ether styrene oligomers and tetrafluoroethylene-ethylene oxide copolymers were examined. Figure 9.10 is the spectrum of StyrEomer 2100, a mac-
romonomer that has shown stability for emulsion or dispersion
polymerization. The spectrum shows the presence of residual methoxy-
capped poly(ethylene oxide), which could cause detrimental end-use properties
such as increased water sensitivity in coatings.

9.10 Structure Analysis

Being a trapped ion technique, FTMS has always had the potential to allow
MS/MS studies to investigate polymer structure. However, to date, there are
relatively few literature reports of polymer structure studies. It is often found
that direct dissociation of polymers is a difficult problem. Current methods
for high-mass ion activation have met with limited success. For example, there
has been little work done by surface-induced dissociation (SID), and collision-
induced dissociation (CID) is known not to work well for singly charged ions

with masses above 3000 Da. Also, the amount of available energy that can be imparted to an ion is limited by the magnetic field strength and the dimensions of the analysis cell (the maximum kinetic energy, $KE = q^2B^2r^2/2m$). Thus, it is much easier to use an alternative mass spectrometry technique such as a sector instrument to achieve the desired energy necessary for polymer dissociation. FTMS researchers must rely on information obtained from fragmentation induced during the desorption, or they can use ionization methods that normally degrade the polymer to its lower mass constituents. Work continues in this area, particularly in the fields of sustained off-resonance irradiation collision-induced dissociation (SORI-CID), and quadrupolar axialization, which is used to improve the collection efficiency of ions. An earlier study by Lin et al.[58] showed the potential knowledge that could be gained from the collision-induced decomposition of polymers. Using a standard 3 Tesla FTMS 2000 mass spectrometer, they examined several poly(ethylene glycol)s and ethoxylated alcohols (EA) to study intramolecular hydrogen bonding. The results of CID analysis led to proposed mechanisms for the proton affinity behavior of the EAs, and the entropies of protonation for several species were calculated. Figure 9.11 shows the collision-induced decomposition spectra of $(C_{12}H_{25}(OC_2H_4)_7OH)H^+$ at energies of 2.9 and 7.3 eV. Pastor and Wilkins[59] have demonstrated that low mass polymer fragmentation up to m/z 3000 is achievable using the technique of SORI-CID. Both polar poly(ethylene glycol) and nonpolar polymers such as polyisoprene and polystyrene show distinct fragmentation patterns that can be used to help identify the end-groups present. These preliminary results establish that polymer dissociation in FTMS is worthy of further study to better control the parameters that affect energy deposition, ion selection, and product ion efficiency.

Another area of structure determination is end-group analysis. End-groups can influence the physical and mechanical properties of the polymer and suggest to polymer chemists reaction mechanisms that may not have been evident by alternative polymer analysis methods such as gel permeation chromatography (GPC), light scattering, and osmometry. Determining polymer end-group functionality traditionally has been accomplished by analyzing the derivatized polymer in parallel with its underivatized version. Mass spectrometry is utilized to get mass information, which together with knowledge of how the polymer was synthesized, leads to deducing the identity of end-groups. Using high resolution FTMS, De Koster et al.[60] have used graphical methods to determine the end-groups of several derivatized polyethylene glycols in an external MALDI 7 T FTICR mass spectrometer. By fitting the equation of a straight line to $m_{actual}(n) = m_{end\ group} + m_{cation} + n(m_{monomer})$, they determined the sum of the end-group and cationized masses to be the y-intercept of the straight line. PEG end-groups could be determined within a deviation of 30 millimass units for a mass of 1000 Da and within 5–200 millimass units for the 4000 Da range. Although useful, this approach is limited in its abilities to find end-groups for unknown polymer samples, particularly those whose end-groups contain the same number of elements but are distributed differently on the two ends of the polymer. For example, the oligomers

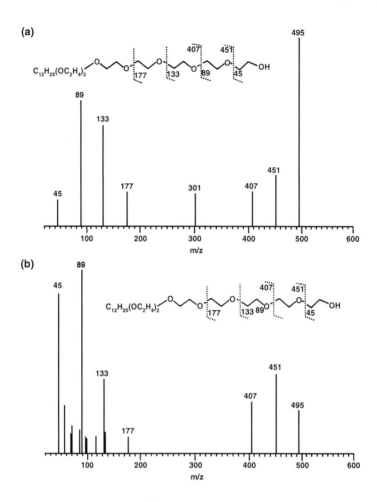

FIGURE 9.11
Collision-induced decomposition spectra of $(C_{12}H_{25}(OC_2H_4)_7OH)H^+$ at energies of (a) 2.9 eV and (b) 7.3 eV. (Reproduced from ref. 58 with permission of the copyright holder)

$CH_3O-(C_2H_4O)_n-CH_3$ and $CH_3CH_2O-(C_2H_4O)_n-H$, appear at the same mass but have elements positioned in different locations, changing the polymer.

In further work by the same group, van Rooij et al.[61] have used statistical methods to calculate the end-group functionality of methylated, propylated, and acetylated polyethylene glycol and polyvinyl pyrrolidone polymer samples. Using both a regression method and an averaging method, they could determine end-group masses within an accuracy of 3 millimass units for the molecular weight range from 500 to 1400 and within 20 millimass units for the molecular weight range from 3400 to 5000. The mathematical treatment of the experimental data adds new data reduction complexity. One drawback is the need to obtain an entire mass spectrum of the polymer. The authors state that because the error in the averaging method scales with the error in the experimental data, the method is only expected to work for masses up to 10,000 Da, provided that the spectra are unit mass resolved. This poses a problem since achievement of high resolving power for polymers usually requires limiting ion populations to reduce the space charge effects that ultimately degrade FTMS performance. The limited number of ions necessary for achieving unit resolving power with current instrument designs may not produce acceptable signal levels for analysis. Perhaps elaboration of remeasurement techniques for polymer analysis could provide a viable alternative approach.

Currently, it is the high resolution afforded by FTMS that makes it worthy of further investigation in the area of polymer analysis. Many subtleties in the preparation of the polymer can be illuminated by a simple high resolution mass spectrum. Towards this end, many researchers are striving to find new and innovative ways to address common polymer synthesis questions. A great amount of information can be extracted from a seemingly simple mass spectrum. For example, using high resolving power on a poly(oxyethylene)-poly(oxypropylene) polymer sample, van Rooij et al.[62] were able to determine block length distributions on the polymer system. The ultimate goal is full characterization of any polymer system using FTMS. As one can see, we are making great strides toward this end.

9.11 Concluding Remarks

It is clear that laser desorption Fourier transform mass spectrometry has significant potential as a polymer analysis tool. It works especially well for polar polymers and additives with molecular weights lower than 10000 Da. With proper attention to analytical details, it is a highly accurate and quite specific way to characterize these substances and can provide unrivaled mass resolution. As is obvious from this chapter, the application of LD-FTMS to polymer analytical problems is an active area of investigation in many laboratories. With the continuing proliferation of commercial FTMS instruments, LD-FTMS applications of polymer analysis should be expected to expand for the foreseeable future. There is no question that this method provides a versatile alternative to the numerous less specific analytical methods employed to date.

References

1. Dienes, T., Pastor, S. J., Schürch, S., Scott, J. R., Yao, J., Cui, S., and Wilkins, C. L., "Fourier Transform Mass Spectrometry—Advancing Years (1992-Mid. 1996)," *Mass Spectrom. Rev.*, 15, 163–211, 1996.

2. McCrery, D. A., Ledford, E. B. J., and Gross, M. L., "Laser Desorption Fourier Transform Mass Spectrometry," *Anal. Chem.*, 54, 1435–1437, 1982.

3. Wilkins, C. L., Weil, D. A., Yang, C. L. C., and Ijames, C. F., "High Mass Analysis by Laser Desorption Fourier Transform Mass Spectrometry," *Anal. Chem.*, 57, 520–524, 1985.

4. Brown, R. S., Weil, D. A., and Wilkins, C. L., "Laser Desorption-Fourier Tranform Mass Spectrometry for the Characterization of Polymers," *Macromol.*, 19, 1255–1260, 1986.

5. Brown, R. S. and Wilkins, C. L. In *Fourier Transform Mass Spectrometry: Evolution, Innovation, and Application,* Buchanan, M. V., Ed., American Chemical Society, 1987, pp 127–139.

6. Brown, C. E., Kovacic, P., Wilkie, C. A., Cody, R. B. J., and Kinsinger, J. A., "Laser Desorption/Fourier-Transform Mass-Spectral Analysis of Molecular Weight Distribution and End-Group Composition of Poly(p-phenylene)s Synthesized by Various Routes," *J. Polym. Sci. Polym. Lett. Ed.*, 23, 453–463, 1985.

7. Brown, C. E., Kovacic, P., Wilkie, C. A., Hein, R. E., Yaniger, S. I., and Cody, R. B. J., "Polynuclear and Halogenated Structures in Polyphenylenes Synthesized from Benzene, Biphenyl, and p-Terphenyl under Various Conditions: Characterization by Laser Desorption/Fourier Transform Mass Spectrometry," *J. Polym. Sci. Polym. Chem. Ed.*, 24, 255–267, 1986.

8. Brown, C. E., Kovacic, P., Cody, R. B. J., Hein, R. E., and Kinsinger, J. A., "Laser Desorption/Fourier Transform Mass Spectral Analysis of Heterocyclic Aromatic Polymers," *J. Polym. Sci. Polym. Lett. Ed.*, 24, 519–528, 1986.

9. Brown, C. E., Kovacic, P., Wilkie, C. A., Cody, R. B., Hein, R. E., and Kinsinger, J. A., "Laser Desorption/Fourier Transform Mass Spectral Analysis of Various Conducting Polymers," *Synth. Met.*, 15, 265–279, 1986.

10. Brown, C. E., Kovacic, P., Welch, K. J., Cody, R. B., Hein, R. E., and Kinsinger, J. A., "Laser Desorption/Fourier Transform Mass Spectra of Poly(phenylene sulfide), Polyaniline, Poly(vinyl phenol), Polypyrene, and Related Oligomers: Evidence for Carbon Clusters and Feasibility of Physical Dimension Measurement," *J. Polym. Sci. Polym. Chem. Ed.*, 26, 131–148, 1988.

11. So, H. and Wilkins, C. L., "On the First Observation of Carbon Aggregate Ions >C600+ by Laser Desorption Fourier Transform Mass Spectrometry," *J. Physe. Chem.*, 93, 1187–1189, 1989.

12. Schulten, H.-R. and Lattimer, R. P., "Applications of Mass Spectrometry to Polymers," *Mass Spectrom. Rev.*, 3, 231–315, 1984.

13. Lattimer, R. P., Harris, R. E., and Schulten, H.-R., "Applications of Mass Spectrometry to Synthetic Polymers," *Rubber Chem. Technol.*, 58, 577–621, 1985.

14. Nuwaysir, L. M. and Wilkins, C. L., "Laser Desorption Fourier Transform Mass Spectrometry of Polymers: Comparison with Secondary Ion and Fast Atom Bombardment Mass Spectrometry," *Anal. Chem.*, 60, 279–282, 1988.

15. Doherty, S. J. and Busch, K. L., "Secondary-Ion Mass Spectra and Fast-Atom-Bombardment Mass Spectra of Liquid Polymers: Alkoxylated Pyrazoles and Hydrazines," *Anal. Chim. Acta*, 187, 117–127, 1986.

16. Ijames, C. F. and Wilkins, C. L., "First Demonstration of High Resolution Laser Desorption Mass Spectrometry of High Mass Organic Ions," *J. Am. Chem. Soc.*, 110, 2687–2688, 1988.

17. Nuwaysir, L. M., Wilkins, C. L., and Simonsick, W. J. J., "Analysis of Copolymers by Laser Desorption Fourier Transform Mass Spectrometry," *J. Am. Soc. Mass Spectrom.*, 1, 66–71, 1990.

18. Kahr, M. S. and Wilkins, C. L., "Silver Nitrate Chemical Ionization for Analysis of Hydrocarbon Polymers by Laser Desorption Fourier Transform Mass Spectrometry," *J. Am. Soc. Mass Spectrom.*, 4, 453–460, 1993.

19. Creasy, W. R. and Brenna, J. T., "Large Carbon Cluster Ion Formation by Laser Ablation of Polyimide and Graphite," *Chem. Phys.*, 126, 453–468, 1988.

20. Creasy, W. R. and Brenna, J. T., "Formation of High Mass Carbon Cluster Ions from Laser Ablation of Polymers and Thin Carbon Films," *J. Chem. Phys.*, 92, 2269–2279, 1990.

21. Brenna, J. T. and Creasy, W. R., "High-Molecular-Weight Polymer Analysis by Laser Microprobe Fourier Transform Ion Cyclotron Resonance Mass Spectrometry," *Appl. Spectrosc.*, 45, 80–91, 1991.

22. Creasy, W. R., Zimmerman, J. A., Jacob, W., and Kay, E., "Pyrolysis and Laser Ablation of Plasma-Polymerized Fluorocarbon Films: Effects of Gold Particles," *J. Appl. Phys.*, 6, 2462–2471, 1992.

23. Castoro, J. A., Köster, C., and Wilkins, C., "Matrix-Assisted Laser Desorption/ Ionization of High-Mass Molecules by Fourier-Transform Mass Spectrometry," *Rapid. Commun. Mass Spectrom.*, 6, 239–241, 1992.

24. Knobeler, M. and Wanczek, K. P., "Theoretical Investigation of Improved Ion Trapping in Matrix-assisted Laser Desorption/Ionization Fourier Transform Ion Cyclotron Resonance Mass Spectrometry: Independence of Ion Initial Velocity," *Int. J. Mass Spectrom. Ion Proc.*, 163, 47–68, 1997.

25. Easterling, M. L., Pitsenberger, C. C., Kulkarni, S. S., Taylor, P. K., and Amster, I. J., "A 4.7 Tesla Internal MALDI-FTICR Instrument for High Mass Studies: Performance and Methods," *Int. J. Mass Spectrom. Ion Proc.*, 157/158, 97–113, 1996.

26. Heeren, R. M. A. and Boon, J. J., "Rapid Microscale Analyses with an External Ion Source Fourier Transform Ion Cyclotron Resonance Mass Spectrometer," *Int. J. Mass Spectrom. Ion Proc.*, 157/158, 391–403, 1996.

27. White, F. M., Marto, J. A., and Marshall, A. G., "An External Source 7 T Fourier Transform Ion Cyclotron Resonance Mass Spectrometer with Electrostatic Ion Guide," *Rapid Commun. Mass Spectrom.*, 10, 1845–1849, 1996.

28. Pastor, S. J., Wood, S. H., and Wilkins, C. L., "Poly(ethylene glycol) Limits of Detection Using Internal Matrix-assisted Laser Desorption/Ionization Fourier Transform Mass Spectrometry," *J. Mass Spectrom.*, 33, 473–479, 1998.

29. McLafferty, F. W., "High-Resolution Tandem FT Mass Spectrometry Above 10 kDa," *Acc. Chem. Res.*, 27, 379–386, 1994.

30. O'Conner, P. B. and McLafferty, F. W., "Oligomer Characterization of 4–23 kDa Polymers by Electrospray Fourier Transform Mass Spectrometry," *J. Am. Chem. Soc.*, 117, 12826–12831, 1995.

31. Hogan, J. D. and Laude, D. A. J., "Mass Discrimination in Laser Desorption/ Fourier Transform Ion Cyclotron Resonance Mass Spectrometry Cation-Attachment Spectra of Polymers," *Anal. Chem.*, 64, 763–769, 1992.

32. Dey, M., Castoro, J. A., and Wilkins, C. L., "Determination of Molecular Weight Distributions of Polymers by MALDI-FTMS," *Anal. Chem.*, 67, 1575–1579, 1995.

33. Pastor, S. J. and Wilkins, C. L., "Analysis of Hydrocarbon Polymers by Matrix-Assisted Laser Desorption/Ionization-Fourier Transform Mass Spectrometry," *J. Am. Soc. Mass Spectrom.*, 8, 225–233, 1997.

34. Easterling, M. L., Mize, T. H., and Amster, I. J., "MALDI FTMS Analysis of Polymers: Improved Performance Using an Open Ended Cylindrical Analyzer Cell," *Int. J. Mass Spectrom. Ion Proc.*, 169/170, 387–400, 1997.

35. O'Conner, P. B., Duursma, M. C., van Rooij, G. J., Heeren, R. M. A., and Boon, J. J., "Correction of Time-of-Flight Shifted Polymeric Molecular Weight Distributions in Matrix-Assisted Laser Desorption/Ionization Fourier Transform Mass Spectrometry," *Anal. Chem.*, 69, 2751–2755, 1997.

36. Guan, S., Kim, H. S., Marshall, A. G., Wahl, M. C., Wood, T. D., and Xiang, X., "Shrink-Wrapping an Ion Cloud for High-Performance Fourier Transform Ion Cyclotron Resonance Mass Spectrometry," *Chem. Rev.*, 94, 2161–2182, 1994.

37. Pastor, S. J., Castoro, J. A., and Wilkins, C. L., "High-Mass Analysis Using Quadrupolar Excitation/Ion Cooling in a Fourier Transform Mass Spectrometer," *Anal. Chem.*, 67, 379–384, 1995.

38. Marshall, A. G., Wang, T.-C. L., and Ricca, T. L., "Tailored Excitation for Fourier Transform Ion Cyclotron Resonance Mass Spectrometry," *J. Am. Chem. Soc.*, 107, 7893–7897, 1985.

39. Guan, S., and Marshall, A. G., "Stored Waveform Inverse Fourier Transfrom (SWIFT) Ion Excitation in Trapped-ion Mass Spectrometry: Theory and Applications," *Int. J. Mass Spectrom. Ion Proc.*, 157/158, 5–37, 1996.

40. Marto, J. A., Guan, S., and Marshall, A. G., "Wide-Mass-Range Axialization for High-Resolution Mass Spectrometry of Externally Generated Ions," *Rapid. Commun. Mass Spectrom.*, 8, 615–620, 1994.

41. Pitsenberger, C. C., Easterling, M. L., and Amster, I. J., "Efficient Ion Remeasurement Using Broadband Quadrupolar Excitation FTICR Mass Spectrometry," *Anal. Chem.*, 68, 3732–3739, 1996.

42. Jackson, G. S., Hendrickson, C. L., Reinhold, B. B., and Marshall, A. G., "Two-plate vs. Four-plate Quadrupolar Excitation for FTICR Mass Spectrometry," *Int. J. Mass Spectrom. Ion Proc.*, 165/166, 327–338, 1997.

43. Pitsenberger, C. C., Easterling, M. L., and Amster, I. J., "Effects of Capacitive Coupling on Ion Remeasurement Using Quadrupolar Excitation in High-Resolution FTICR Spectrometry," *Anal. Chem.*, 68, 4409–4413, 1996.

44. Pastor, S. J. and Wilkins, C. L., "Low-voltage On-resonance Ion Selection and Storage: An Alternative to Quadrupolar Axialization for FTMS," *Anal. Chem.*, 70, 213–217, 1998.

45. Pastor, S. J., Dienes, T., Yao, J., and Wilkins, C. L., "Investigation of Low-voltage On-resonance Ion Selection for Fourier Transform Mass Spectrometry," *J. Am. Soc. Mass Spectrom.*, 9, 931–937, 1998.

46. Srzic, D., Martinovic, S., Pasa Tolic, L., Kezele, N., Kazazic, S., Senkovic, L., Shevchenko, S. M., and Klasinc, L., "Laser Desorption Fourier Transform Mass Spectrometry of Natural Polymers," *Rapid Commun. Mass Spectrom.*, 10, 580–582, 1996.

47. Liang, Z., Marshall, A. G., and Westmoreland, D. G., "Determination of Molecular Weight Distributions of tert-Octylphenol Ethoxylate Surfactant Polymers by Laser Desorption Fourier Transform Ion Cyclotron Resonance Mass Spectrometry and High-Performance Liquid Chromatography," *Anal. Chem.*, 63, 815–818, 1991.

48. Cromwell, E. F., Reihs, K., de Vries, M. S., Ghaderi, S., Wendt, H. R., and Hunziker, H. E., "Transition-Metal Cationization of Laser-Desorbed Perfluorinated Polyethers with FTICR Mass Spectrometry," *J. Phys. Chem.*, 97, 4720–4728, 1993.

49. Elliott, C. M., Baldy, C. J., Nuwaysir, L. M., and Wilkins, C. L., "Electrochemical Polymerization of 4-Methyl-4'-vinyl-2,2'-bipyridine-Containing Metal Complexes: Polymer Structure and Mechanism of Formation," *Inorg. Chem.*, 29, 389–392, 1990.

50. O-Malley, R. M., Randazzo, M. E., Weinzierl, J. E., Fernandez, J. E., Nuwaysir, L. M., Castoro, J. A., and Wilkins, C. L., "Laser Desorption Mass Spectrometry of Chemically and Electrochemically Prepared Poly(2-vinylthiophene)," *Macromol.*, 27, 5107–5113, 1994.

51. Campana, J. E., Sheng, L. S., Shew, S. L., and Winger, B. E., "Polymer Analysis by Photons, Sprays, and Mass Spectrometry," *Trends in Anal. Chem.*, 13, 239–247, 1994.

52. Hsu, A. T. and Marshall, A. G., "Identification of Dyes in Solid Poly(methyl methacrylate) by Means of Laser Desorption Fourier Transform Ion Cyclotron Resonance Mass Spectrometry," *Anal. Chem.*, 60, 932–937, 1988.

53. Asamoto, B., Young, J. R., and Citerin, R. J., "Laser Desorption/Fourier Transform Ion Cyclotron Resonance Mass Spectrometry of Polymer Additives," *Anal. Chem.*, 62, 61–70, 1990.

54. Johlman, C. L., Wilkins, C. L., Hogan, J. D., Donovan, T. L., Laude, D. A. J., and Youssefi, M. J., "Laser Desorption/Ionization Fourier Transform Mass Spectrometry and Fast Atom Bombardment Spectra of Nonvolatile Polymer Additives," *Anal. Chem.*, 62, 1167–1172, 1990.

55. Xiang, X., Dahlgren, J., Enlow, W. P., and Marshall, A. G., "Analysis of Phosphite Polymer Stabilizers by Laser Desorption/Electron Ionization Fourier Transform Ion Cyclotron Resonance Mass Spectrometry," *Anal. Chem.*, 64, 2862–2865, 1992.

56. Heeren, R. M. A., de Koster, C. G., and Boon, J. J., "Direct Temperature Resolved HRMS of Fire-Retarded Polymers by In-Source PyMS on an External Ion Source Fourier Transform Ion Cyclotron Resonance Mass Spectrometer," *Anal. Chem.*, 67, 3965–3970, 1995.

57. Simonsick, W. J. J. and Ross, C. W. I., "The Characterization of Novel Dispersants, Fluorinated Surfactants, and Modified Natural Oils by Laser Desorption Fourier Transform Ion Cyclotron Resonance Mass Spectrometry (LD-FTICR-MS)," *Int. J. Mass Spectrom. Ion Proc.*, 157/158, 379–390, 1996.

58. Lin, H.-Y., Rockwood, A., Munson, M. S. B., and Ridge, D. P., "Proton Affinity and Collision-Induced Decomposition of Ethoxylated Alcohols: Effects of Intramolecular Hydrogen Bonding on Polymer Ion Collision-Induced Decomposition," *Anal. Chem.*, 65, 2119–2124, 1993.

59. Pastor, S. J. and Wilkins, C. L., "Sustained Off-resonance Irradiation and Collision-induced Dissociation for Structural Analysis of Polymers by MALDI-FTMS," *Int. J. Mass Spectrom. Ion Proc.*, 175, 81–92, 1998.

60. de Koster, C. G., Duursma, M. C., van Rooij, G. J., Heeren, R. M. A., and Boon, J. J., "Endgroup Analysis of Polyethylene Glycol Polymers by Matrix-Assisted Laser Desorption/Ionization Fourier-Transform Ion Cyclotron Resonance Mass Spectrometry," *Rapid Commun. Mass Spectrom.*, 9, 957–962, 1995.

61. van Rooij, G. J., Duursma, M. C., Heeren, R. M. A., Boon, J. J., and de Koster, C. J., "High Resolution End Group Determination of Low Molecular Weight Polymers by Matrix-Assisted Laser Desorption Ionization on an External Ion Source Fourier Transform Ion Cyclotron Resonance Mass Spectrometer," *J. Am. Soc. Mass Spectrom.*, 7, 449–457, 1996.

62. van Rooij, G. J., Duursma, M. C., de Koster, C. G., Heeren, R. M. A., Boon, J. J., Schuyl, P. J. W., and van der Hage, E. R. E., "Determination of Block Length Distributions of Poly(oxypropylene) and Poly(oxyethylene) Block Copolymers by MALDI-FTICR Mass Spectrometry," *Anal. Chem.*, 70, 843–850, 1998.

10

Matrix-Assisted Laser Desorption Ionization/ Mass Spectrometry of Polymers (MALDI-MS)

Giorgio Montaudo, Maurizio S. Montaudo, and Filippo Samperi

CONTENTS

10.1   Introduction

The introduction of Matrix-Assisted Laser Desorption/Ionization (MALDI)[1–5] and Electrospray Ionization (ESI)[6,7] (Chapter 1) has dramatically increased the mass range for molar mass analyses by mass spectrometry. In principle, both techniques are able to produce intact quasi-molecular ions of polymers with high molar mass (>100,000 Da).

However, ESI mass spectrometry occupies only a small, but valuable, segment in the analysis of synthetic polymers. It is better suited for the analysis of nearly monodisperse biopolymers, because of complications arising from the formation of multiply charged ions. Furthermore, for synthetic polymers the entanglement of chains prevents obtaining ESI spectra at high masses, also posing a limit to its utilization.[8]

Instead, MALDI-TOF allows desorption and ionization of very large molecules, even in complex mixtures. It gives information on the mass of individual oligomers from which repeat units, end-groups, the presence of rings, molar mass distributions, and other imformation can be derived.

The method plays an important role in polymer analysis also in view of the recurrent difficulties encountered by the conventional methods of characterization, such as size exclusion chromatography (SEC) or nuclear magnetic resonance (NMR), e.g., lack of calibration standards, low sensitivity, and the like.

The ionization process in MALDI-TOF proceeds through the capture of a proton or a metal ion (usually lithium, sodium, potassium), which forms a charged adduct with the molecular species. MALDI-TOF is applicable to the measurement of molar mass distributions in synthetic polymers because it usually produces singly charged quasi-molecular ions with negligible ion fragmentation. In the mass range up to 50 kDa, single oligomers can appear mass-resolved in MALDI-TOF spectra, enabling the determination of repeat units and end-group composition.

Hillenkamp and Karas developed MALDI-TOF mass spectrometry in 1988 for the analysis of large biomolecules (e.g., proteins),[3–5] but it was not demonstrated until 1992 that synthetic polymers also can be analyzed beyond molar masses of 10^5 Da.[9–12] This delay was probably due to the fact that methods of sample preparation for biopolymers, which use water-based solvents, are not applicable to most synthetic polymers.

Furthermore, synthetic polymers exhibit a marked polydispersity, which results in a lower signal-to-noise ratio. This is due to the fact that the signal intensity (which adds up to a single signal in the case of a biomolecule,[13–15]) is spread out over a number of oligomers having different degrees of polymerization. During the last few years, MALDI-TOF mass spectrometry has become an essential tool for the characterization of synthetic polymers.[16–26]

10.2   Fundamentals of the Method

10.2.1   The MALDI Process

MALDI makes use of short, intense pulses of laser light to induce the formation of intact gaseous ions (Chapter 1). Analyte molecules are not directly exposed to laser light, but are homogeneously embedded in a large excess of "matrix," which consists of small organic molecules. The matrix molecules exhibit a strong absorption at the laser wavelength used (usually the 337 nm of a nitrogen laser), and the matrix allows for an efficient and controllable energy transfer.

The high energy density obtained in the solid or liquid matrices, even at moderate laser irradiances, induces an instantaneous vaporization of a microvolume (called "plume"), and a mixture of ionized matrix and analyte molecules is released. The packet of ions generated in the process is accelerated to a fixed energy by an electric potential, ranging from 15 kV up to 30–35 kV. Depending on their mass-to-charge ratio (m/z), the ions have different velocities when they leave the acceleration zone and enter field-free flight tube (drift-tube). The ions impact onto an ion-detector, and the time interval between the pulse of laser light and the impact of each ion on the detector is measured. This produces signals whose intensities are proportional to the number of ions arriving at the electron multiplier (molar response). The MALDI-TOF mass spectrum is then obtained by recording the detector signal as a function of time.

According to Eq. 10.1, the square of the flight time is proportional to the m/z ratio,

$$m/z = 2 V t^2/l^2 \tag{10.1}$$

where m is the mass of the ion, z is the number of charges, V is the accelerating voltage, t is the ion flight time, l is the length of the flight tube. Since the V and l values are known, the m/z ratio can be calculated solely from Eq. 10.1.

In practice, however, exact values for the mass scale are obtained from the empirical Eq. 10.2, because of uncertainties in the determination of the flight time.

This uncertainty is due to a short delay in ion formation after the laser pulse, so that the real starting time of the ions is not identical to the time of the laser pulse (which provides the starting signal for the measurement of flight time).

$$m/z = at^2 + b \tag{10.2}$$

The constants a and b in Eq. 10.2 are measured by the flight times of two ions with known masses, which are used to mass calibrate the MALDI-TOF spectra.[13]

10.2.2 MALDI Matrices

An ideal matrix should have the following properties: high electronic absorption at the employed laser wavelength, good vacuum stability, low vapor pressure, good solubility in the solvents that can also dissolve the analyte, and good miscibility with the analyte in the solid state.

In the analysis of biopolymers, where the ionization usually takes place via proton transfer, the matrix also plays the role of the proton source.[3–15,27–30] In the analysis of synthetic polymers, ionization is usually achieved by cation attachment,[25] which is largely a matrix-independent process.

The most commonly used matrices for synthetic polymers are DHB (2,5-dihydroxybenzoic acid), HABA (2,(-4-hydroxyphenylazo)benzoic acid), IAA (3-β-indoleacrylc acid), dithranol (1,8,9-trihydroxyanthracene), sinapinic acid (3,5-dimethoxy-4-hydroxy cinnamic acid), all trans-retinoic acid, and 5-chlorosalicylic acid. Exhaustive lists of useful matrices are available in the literature.[18,21]

Many of these matrices for UV-MALDI can be applied in IR-MALDI which uses the Nd-YAG laser (3.27 μm, equivalent to 3050 cm^{-1}) or CO_2 laser (10.6 μm).[21,31–33] Excitation of the matrix by a 3.27 μm IR-laser is feasible.[33]

IR-MALDI spectra show a lower mass resolution compared to that obtained by UV-MALDI.[33] However, the IR-MALDI technique may be used for the analysis of halogenated polymers, which often show extensive fragmentation in UV-MALDI.[34]

The selection of a good matrix is still a trial and error process and the search of useful matrix compounds has been an active area in MALDI research.[28–30,35–41] Generally the choice and discovery of new matrix materials have been achieved more or less empirically and can be completely unrelated to the analyte in terms of structure or physical properties. The presence of specific functional groups (e.g., OH, COOH) is very important for the cationization process.[40]

In principle, in the desorption/ionization process, the amount of pulsed laser energy transferred to the analyte via matrix will depend on the laser power, on the nature of the matrix and sample, and on the dispersion of analyte molelules within the matrix. There exists a threshold irradiance that is matrix-dependent, below which ionization is not observed. Above this level ion production increases nonlinearly.

The mechanisms of ion formation in MALDI-MS are still not fully understood. Two main possibilities exist: ions may either be "pre-formed" in the solid state and simply be liberated upon laser irradiation, or they may be formed by ion-molecule reactions initiated by the laser pulse.[42,43]

10.2.3 Cationization in Synthetic Polymers

In contrast to biopolymers, the ionization of synthetic polymers usually occurs by cationization rather than protonation. In relatively polar polymers, sodium and/or potassium adduct ions are observed in the MALDI spectrum even if they are not added to the matrix/analyte mixture.[44,45] These cations are present as impurities in matrix, reagents, solvents, glassware, and other sources, and polymers having relatively high cation affinities do not necessarily require a high cation concentration in the MALDI sample. Most of the synthetic polymers having heteroatoms (e.g., polyesters, polyamides, polyethers, polyacrylates, and the like) are cationized by the addition of sodium or potassium salts.[46] Use of the delayed ion extraction technique (see Section 10.2.5) allows more time for cation attachment, and a substantial increase in signal intensity of cationized polymers may be obtained.[47]

Unsaturated hydrocarbon polymers such as polystyrene (PS), polybutadiene (PB), and polyisoprene (PI) can be successfully ionized after the addition of appropriate transition-metal salts (e.g., Ag^+X^-, $Cu^{++}X^-$), which interact with the double bonds of these polymers.[48] Saturated hydrocarbon polymers such as polyethylene and polypropylene are at present not very amenable to MALDI analysis because of the extremely low binding energy of the cation-macromolecule.[49] For these polymers, field desorption MS or field ionization MS might be alternative techniques (Chapter 8). Several groups have studied the cationization process, but systematic approaches regarding cation affinity are generally lacking, and this remains an interesting research opportunity for the future.

Protonation of polymers and proteins, under MALDI conditions, were studied by using a deuterium labeling method.[50] It was demonstrated that exchangeable protons from either carboxylic acid or hydroxyl groups of the matrix are important, but not essential, for the protonation of the analyte. In matrices with nonexchangeable protons, such as nitroanthracene, the exchangeable protons from solvent (e.g., acetone) were shown to contribute to the formation of protonated quasi-molecular ions. A more detailed knowledge of the proton affinities of the common MALDI matrices should play a critical role in understanding why some matrices favor proton transfer and are "hotter" than others.[51]

Some studies[51–55] deal with the conformation and energetics of the process, and compare the theoretical findings with experimentally observed MALDI data. Actually, different cations (such as lithium, sodium, potassium, and cesium) efficiently wrap the polymer around them.[52]

For aliphatic polyethers, the structure of the inner coordination sphere of the oxygen atoms around the cation was found to be cation dependent.[52]

It was also observed that larger cations are able to form stable conformations with a large oligomer, since more oxygen atoms are available for coordination with the cation.[52]

Cationization of polystyrene (PS) was investigated using silver, copper, zinc, cobalt, palladium, platinum, iron, chromium and aluminum salts. It was found that silver, copper, and palladium yield efficient cationization of PS oligomers, and it was argued that cationization occurred by gas phase ion-molecule reactions rather than pre-formed ions from the condensed phase.[53,54]

Polystyrene metal cations are more likely to be produced if the laser desorption process reduces the metals to the +1 state. This explains why salts of transition metals other than silver, copper, or palladium often fail to cationize polystyrenes under the MALDI condition used.[54,55]

Polystyrene has been found to have a higher affinity for attachment for Ag^+, Al^+, Cr^+, and Cu^+ than to K^+, whereas PEG shows a higher affinity for K^+ than for Al^+, Cr^+, and Cu^+. These preferential attachments were discussed in terms of the hard and soft acid principle.[55]

Cationization by means of divalent and trivalent metal salts has been demonstrated in MALDI experiments for different polymers, namely cyclic polyethylene oxide and α-cyclodextrin.[56] In all these cases analyte singly charged quasi-molecular ions are generated. The latter can be formed by adduction of the corresponding monovalent metal ion, by adduction of the divalent/trivalent metal ion with simultaneous deprotonation of the analyte molecules, or by simultaneous adduction of the divalent/trivalent metal ion and an appropriate number of deprotonated matrix moieties.[56]

The role of the counter anion in MALDI mass spectrometry analysis of PMMA has also been investigated.[57] In the presence of different sodium halides, decreasing ion yields were observed in the order I>Br>Cl. This result has been interpreted in terms of differing abundances of cations made available for gas-phase cationization by laser irradiation of the matrix/analyte/salt sample, caused by different lattice energies of the salts.[57]

10.2.4   Sample Preparation

Sample preparation is crucial in MALDI-MS and greatly influences the quality of the spectra. Detection sensitivity, selectivity, mass resolution, and other performance indicators are strongly dependent on the sample preparation.

In a typical preparation approach, appropriate amounts of matrix and polymer dissolved in compatible (preferably identical) solvents are mixed to yield a matrix/analyte molar ratio in the range 1000:1 to 10^6:1.

The optmimum ratio increases with increasing molar mass of the polymer sample. The metal ions required for an enhanced analyte cationization are added as their organic or inorganic salts, depending on their solubility, and mixed into the matrix/analyte solution.

After the selection of the MALDI matrix, cationization salt and solvent, several options are available for transferring the mixture onto the MALDI target.

The oldest procedure is the so-called dried droplet method.[3] In this method the three solutions are mixed, and 0.5–1 μl of the mixture is applied to the target and air-dried at room temperature. Under these conditions crystallization is relatively slow, thereby increasing the risk of segregation between sample and matrix and cationization salt.

More recently, a fast crystallization method was introduced, in which the target is put in a vacuum chamber in order to promote crystallization within a few seconds.[59] As a result, small crystals with little segregation are obtained, giving enhanced reproducibility and signal intensity, and higher resolution. Accelerated drying using a stream of high-purity nitrogen gas is an alternative means for reaching similar goals.[60] Improved homogeneity and better sensitivity were also obtained by spin coating techniques.[61]

In the analysis of polymers such as aromatic polyesters and aromatic dendrimers, improved results were obtained using the thin-layer method, where a matrix solution is prepared and allowed to crystallize and next the sample is added and dried.[62]

A promising technique of sample preparation for synthetic polymers is the electrospray sample deposition, where the matrix and the analyte solutions are sprayed onto the target surface under the influence of a high-voltage electric field.[63] The potential of this method was tested for PS and other polymers, using both the dried droplet (i.e., one-layer) and the thin-layer (i.e., two-layer) techniques.[62] Electrospray deposition yielded much higher signal intensities and much better shot-to-shot and spot-to-spot reproducibility, slightly favoring the one-layer method. The improved results were ascribed to the small and evenly sized crystals thus formed.[64]

Recently, microscopy and surface analysis techniques such as scanning electron microscopy (SEM), confocal fluorescence microscopic imaging, X-ray photoelectron spectroscopy (XPS), and time-of-flight secondary ion mass spectrometry (TOF-SIMS) have been successfully applied to study the homogeneity of the final MALDI sample preparation.[62,65,66] The microscopy technique showed better homogeneity of the sample preparation in the case of fast crystallization methods.[65] Evidence for segregation of salts from the matrix crystals was obtained by TOF-SIMS imaging by slow drying of a PEG1500 sample with DHB as matrix and sodium cations dissolved in methanol/water, or by fast drying of PMMA with DHB and sodium dissolved in acetone.[66] Upon changing from the dried droplet method to the electrospray deposition, homogeneous cocrystallization of analyte and matrix occurred, and very homogenous chemical images were obtained.[66] An improvement of the miscibility of PMMA in common matrices was obtained by addition of surfactant.[67]

A simple air spray deposition method was used with some PEG samples.[61] This method gave excellent spot-spot reproducibility, compared with the classical dry droplet method, and increased the usefulness of MALDI-TOF for end-group analysis of PEGs.[61]

A solid/solid sample preparation method, consisting of mixing analyte and matrix without any solubilization procedure, was reported for the MALDI analysis of some polyamides (e.g., poly(examethylene terephthalamide) and Ny_{12})

that are insoluble or poorly insoluble in common organic solvents.[68] Sample homogeneity was achieved simply by pressing a pellet from a finely ground powder of the polymer and matrix. The solid/solid method was also applied in the MALDI-TOF analysis of insoluble higher mass polycyclic aromatic hydrocarbons (PHAs).[69] A new matrix, namely 7,7,8,8-tetracyanoquinodimethane, was used in this case and very good signal-to-noise ratio, resolution, and spot-to-spot reproducibilty was obtained.[69]

Negative ion MALDI analysis of polymeric acids such as sulphonated polystyrene and poly(acrylic acid) requires desalting and conversion into the hydrogen form, thereby putting an additional demand on sample preparation.[70] The addition of ion exchange beads to the sample/mixture might prove adequate for the purpose of desalting, but one should be aware of the different solubility of the polymer in its neutralized form vs. the original salt form. Precipitation of the neutralized polymer onto the ion exchange beads might therefore occur. More sophisticated desalting options, such as membrane desalting and on-probe desalting using self-assembled monolayers, are used in the MALDI analysis of proteins[69] and can also apply to acidic synthetic polyelectrolytes.[71]

Synthetic polymers show a distribution of different chain lengths and endgroups. Appropriate methods of sample preparation are required for each class of polymer to avoid selective cationization and discrimination effects, which may affect polymer distributions and cause errors in the calculation of average molar mass values. Homogeneous mixtures of matrix and polymer in the condensed state are a prerequisite to avoid mass discrimination.

Accurate analysis of synthetic polymers requires some instrumental considerations, discussed in detail in the following section.

10.2.5 Instrumentation

Time-of-flight (TOF) mass spectrometers are ideally suited for use with the MALDI technique because of their theoretically unlimited mass range, high ion transmission, and for the pulsed nature of the laser used in this method.[72]

Mass resolution is the ability of an instrument to separate the signals from ions of similar mass, expressed as the mass of a given ion divided by the full width at half maximum of the peak (fwhm). Resolution in MALDI-TOF MS is mainly restricted by the ionization process, rather than by instrumental limits, because the ions have a certain time-span of formation, a spatial distribution, and a kinetic energy spread.[5,13,73,74]

In MALDI-TOF, the packets of ions produced by laser irradiation of matrix/analyte mixture and accelerated by a fixed electric potential (V) can be detected in two different ways: the linear and reflection (or reflectron) mode.[13] The two detection modes are complementary, since a higher resolution is obtained in the reflection mode and a higher sensitivity at high molar mass is achieved in the linear mode.

High resolution in the reflection mode is achieved by a reflecting field at the end of the flight tube with somewhat higher potential and the same polarity as the accelerating voltage. Ions with the same m/z but different velocities (which results in peak broadening in the linear mode) can be time-focused in the reflector.[72,74]

In the reflection mode the resolution is typically around 1000–2500, whereas in linear mode the resolution is around 500–800. The linear mode has a very high sensitivity and requires only a short lifetime (1 μs) for the detection of the molecular ions even in the case of a metastable decay on the flight path.

In the linear mode of a TOF instrument, however, the fragments from molecules that decompose after the acceleration zone have nearly the same velocity as the intact molecular ions. Thus these *metastable* ions are detected at the same flight time, but with a slight increase in peak width. In the reflection mode of a TOF instrument, ions that undergo fragmentation cannot be detected at their correct molar mass.[76]

With a special technique, namely "Post-source Decay" (PSD), however, the reflection mode can be used to analyze metastables.[77] By stepping down the reflection potential, all the fragments of a selected ion can be analyzed. The precursor ion is selected after the acceleration zone by a gating system, which deflects all unwanted ions perpendicular to the normal flight path.[77]

Analysis of metastables is quite efficient in the case of peptides, where information on the sequence of amino acids is readily obtained.[78] In the case of synthetic polymers, investigation of metastables can also be useful to deduce structure information, and the sequence of comonomers may also be obtained in the characterization of copolymers by the PSD method.[78]

Both detection methods, the linear mode and the reflection mode, have advantages and drawbacks and ought to be chosen according to the information desired. Generally, the linear mode is used for measuring molar mass distributions, especially for polymers with high molar masses (more than 10000 Da), whereas the reflection mode is used preferentially for determining the exact molar mass of individual oligomers, necessary for end-group determination or identification of side products, or impurities.

An algorithm that capitalizes on the correlation between the positions and velocities of desorbed ions has been developed to increase the mass resolution of linear TOF instruments.[74,79,80] It is applicable to low- and high-mass ions and may also improve the performance of reflectron mass analyzers.

Complete characterization for each ionization event can be obtained by the curved-field reflectron that, in addition to greatly enhancing signal-to-noise levels from multiple acquisition, provides the means to rapidly acquire product- and metastable-ion spectra.[81]

A significant improvement in mass resolution, respect to the above described "continuous extraction" procedure, was obtained by the introduction of "time lag focusing" or "delayed extraction" (DE).[82–88] This principle can be used in both linear and reflection mode yielding an improvement in resolution by a factor of above ten in both cases.

Whereas the reflection field compensates for the initial energy spread of the created ions, delayed extraction can also compensate for the time spread in ion formation. In this system the ions formed by MALDI are produced in a weak electrical field, and subsequently extracted by application of a high voltage pulse after a predetermined time delay of typically a few hundred nanoseconds (50–1000 ns) after a laser pulse. Thus, the initial desorption and ion formation process is separated from the ion acceleration step and, therefore, the time window required for ion formation does not contribute to peak broadening.

Another possible advantage of this experimental procedure is that during the delay most of the neutral species created during desorption are allowed to dissipate and are pumped away so that ion-neutral collisions are minimized. With this technique it is possible to achieve mass resolution of up to 10000, i.e., unit mass resolution can be obtained up to 7000 Da.[85]

The analytical capability of delayed extraction has been demonstrated in the MALDI-TOF analysis of several polymers such as PEG and PS.[84] Figure 10.1 compares 3 MALDI mass spectra obtained for a polyethylene glycol with narrow distribution, $D \leq 1.07$, obtained in: (a) linear mode, (b) reflection mode, (c) in reflection mode with delayed extraction. A higher mass resolution is evident when the delayed extraction procedure is used.[20]

In order to obtain mass spectra with good mass resolution and mass accuracy over a large mass range, a compromise between delay time and grid voltage values is necessary, especially when blends of polymers with different molar masses and or chemical structures are examined.[88]

Coupling a MALDI source with a Fourier Transform Ion Cyclotron Resonance Mass Spectrometer (FTICR-MS), has been reported to be a potentially advantageous method to accurately determine end-groups, molar mass distributions, and chemical composition of biological and synthetic polymers.[89–99] FTICR-MS provides high mass-resolving power, precise mass measurement, the possibility of isolating ions and manipulating them (multistage mass spectrometry), simultaneous detection of ions over a wide mass range, and ion measurement[23] (Chapter 9).

10.2.6 Ion Detection

The high-intensity background peaks resulting from the matrix often complicate MALDI-TOF mass spectra. This background results in limitations in both resolution and detector response, reducing the effectiveness of the MALDI technique for high molar mass polymers. It has been shown that signal-to-background ratio and the resolution improve as the laser power is reduced to the "threshold" irradiance. The threshold level depends on the detector used for measurement.[100]

Microchannel plate (MCP) detectors are ideally suited for high-sensitivity applications due to the rapid response coupled with a very high gain.[100] However, these same characteristics make the MCP detectors susceptible to saturation resulting from the low-mass ion component in the MALDI spectrum.

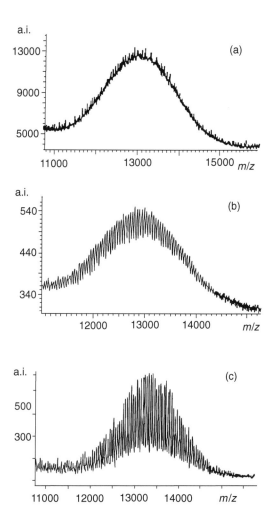

FIGURE 10.1
Positive-ion MALDI-TOF spectra of a polyethylene glycol SEC-standard sample obtained in:
(a) linear mode; (b) reflection mode; and (c) delayed extraction mode. (Reprinted with permission from Ref. 20, Copyright 1999 John Wiley & Sons Ltd)

Because a large component of the ion flux in MALDI is composed of low-mass ions ($m/z < 1000$), the saturation of the detector occurs frequently, resulting in a reduced capacity to detect high-mass ions. Low-mass ions can be deflected away from the ion optical axis using a pulsed deflector plate situated close to the entrance into the flight region.

Rapid bipolar pulsing of an electrostatic particle guide (EPG) permits selective elimination of unwanted background peaks that would normally saturate the detector, thus enhancing the dynamic range of the detector.[101] The EPG is an isolated wire electrode that spans the length of the flight tube, creating a potential field in the center which effectively "guides" ions to the detectors. Ions that are accelerated slightly perpendicular to the ion optical axis are captured in the potential field and transported to the detector, resulting in a dramatic improvement in sensitivity.

A cryogenically cooled superconducting tunnel junction (STJ) detector might solve poor detection efficiency exhibited by MCPs for large ions.[102] The cryogenic detectors comprise a class of particle and x-ray detectors that do not rely on ionization but register the impact of individual particles, charged or not, by responding to the deposited energy. The energy response capability of the detector introduces a way to detect isochronous ions of different charge.[102]

Hybrid detectors are different from dual MCPs in that they possess a single microchannel plate. The second amplification stage is often achieved using a scintillator which surmounts the entrance window of a photomultiplier. Sometimes a venetian-blind steel dynode is found in the initial part of a hybrid detector in order to perform ion-to-electron conversion.

10.2.7   MALDI Calibration

Mass calibration is a fundamental operation in TOF mass spectrometry. The time-of-flight data recorded in MALDI-TOF analysis are converted into mass-to-charge (m/z) values for ions using Eq. 10.2, Section 10.2.1. The external calibration procedure is usually adopted. It consists in recording the spectrum of a compound of known mass and computing the set of calibrating constants in Eq. 10.2. The set of constants are then used to calibrate the spectra of unknown compounds, recorded using the same instrumental conditions. Polymer samples with known structure and end-groups may be used. The accuracy in mass determination by this procedure is claimed to be 0.1%.[13] A more accurate calibration method is the internal calibration procedure that has been used to calibrate mass spectra of proteins.[5]

In the case of polymeric materials, internal standards are difficult to find, since each synthetic polymer class would need structurally identical species as internal standards, possibly monodisperse, and with mass values, to avoid overlap with the masses displayed by the polymer sample.

A precise calibration method for MALDI-TOF spectra of polymers, named "self-calibration," has been developed.[103,104] In this method, the flight time of a peak series (usually 20 or more peaks are needed for accurate results) are recorded and then given as input to a computer program, called CALTOF,[103] which yields as output the calibration constants a and b of Eq. 10.2.

The presence of multiply charged molecular ions in the MALDI-TOF spectrum can improve the accuracy of mass determination by averaging the values measured for different charge states. The gain, however, can be very little, if any, because of the significant spread in the molar mass data derived from the multiply charged ions.[105]

Furthermore, ions with different charge states lie on different calibration curves, depending on their initial energies and energy deficits. To overcome this problem and to improve the mass calibration in linear TOF instruments, a 3-point calibration procedure has been proposed, which takes into account the initial energies of the molecular ions and uses multiply charged ions in the calibration.[105]

The latter method gives 2 or 3 times better accuracy of mass determination than the conventional 2-point calibration.[104] In this method, multiply charged ions appear to lie on the same calibration curve.

In a recent work, improved calibration of MALDI-TOF mass spectra was obtained by optimization of the electrostatic field generated by the ions.[106] In contrast to conventional methods, where the relationship between ion flight time and mass is an arbitrary polynomial equation, this method is based on the physics of ion motion. Parameters needed to describe the physics are numerically optimized using a simple algorithm. Once these parameters are established, unknown masses can be determined from their times-of-flight.[106] This calibration method gives well-behaved results, since non-linear parameters (due to extraction delay, desorption velocity, and other events) are properly taken into account in the time-of-flight calculation.[106]

10.3   Applications of MALDI-TOF-MS to Synthetic Polymers

The amount of structural and compositional information that can be gathered from MALDI analysis of synthetic polymeric materials is substantial.[18,21–27,107–131] It includes repeat units and end-group identification, structural analysis of linear and cyclic oligomers, tracking of polymerization kinetics, and the estimate of composition and sequence for complex copolymer systems.

Another important application of MALDI-TOF mass spectrometry is the measurement of molar mass of synthetic polymeric materials up to 10^6 Da and beyond. The conventional MM characterization techniques (viscosity, SEC, light scattering) are indirect ways of measurement of molar masses, and the prospect of direct measurements of high MM and MMD has stirred much expectation among polymer scientists.

10.3.1   Determination of Molar Masses

Molar mass (MM) and molar mass distribution (MMD) values in macromolecular compounds can be calculated from their mass spectra, as discussed in Chapter 2. Most industrial polymers have average MM in the range of several tens of thousands and, often being polydisperse, their higher mass tails usually reach MM values of several hundred thousands and even higher.

Determinations of MM in relatively low MM polymer samples (up to a mass limit of about 10,000 Da) have appeared in the last decades, by a number of soft ionization MS techniques as static SIMS-TOF,[132] LD-FTMS,[133,134] DCI,[135] FD,[2] FAB,[1,2] ESI.[8] The ESI-TOF technique, which for proteins is capable of providing high MM measurements, in the case of synthetic polymers yields spectra only below 5000 Da, due to chain entanglement.[8]

Before the introduction of MALDI, mass spectrometry was not a widely applicable method for the determination of MM of synthetic polymers. Since MALDI-TOF allows desorption and ionization of very large molecules, it is now possible to perform the direct identification of mass-resolved polymer chains and the measurement of MM in numerous high polymers. MM values up to 1.5 million Daltons have been measured for PS monodisperse standards.[74] MALDI-TOF is therefore unique for the MM and MMD estimation in synthetic polymers by MS techniques.

Comparison of MALDI-TOF spectra of relatively low-mass polymers with those obtained by other ionization techniques (ESI,[38,125,126] SIMS,[125,127] FAB[127]), yields MM estimates in general agreement. MALDI-TOF was originally developed for proteins (exactly monodisperse polymers),[3–5,12–15,27,31] and its extension to the characterization of synthetic polymers was not straightforward. Contrary to the case of proteins, synthetic polymers may show a wide range of molar mass distributions, according to the synthetic method used in their preparation.

The determination of MM by SEC (which is an indirect method and needs appropriate standards for the calibration of the SEC traces) is very popular, and average values of MM and of MMD for all kinds of polymer samples can be routinely extracted from SEC measurements (Chapter 2). Therefore it was a logical choice to compare the MM values obtained from MALDI with those from SEC.

In the MALDI process, ions strike the detector and produce a current that is a function of the number of ions, so that the detector response is proportional to molar fractions. Therefore, in a mass spectrum the intensity of the peaks is proportional to the molar amount of each species, and the masses are displayed on a linear scale.

In contrast, the intensity of the SEC response is proportional to the weight amount of each species, and the masses are displayed on a logarithmic scale in SEC traces. The two traces are therefore not directly comparable, and this has to be taken into account when handling data from the two techniques.[113,114,128,131] Furthermore, SEC traces are continuous, i.e., the oligomers contributing to the intensity of the detector signal are not mass resolved, whereas MALDI-TOF spectra are typically mass-resolved up to about 50–60 kDa. At higher masses MALDI-TOF yields continuous traces, which is analogous to SEC.

The difference between SEC and MALDI traces does not constitute a problem in computing average molar masses, because the average MM of polymers are calculated from a summation of the abundance of each oligomer, and the same equations apply both to the SEC and MALDI-TOF techniques (Chapter 2).

The most serious problems encountered in achieving correct estimates of the molar masses of polymers from their MALDI-TOF spectra do have an instrumental origin. For the quantitative analysis of mass spectra of polymers, the number of charged adducts revealed by the MS detector must reflect the number of polymeric chains actually existing in the sample. This requires that the ionization yield of the various oligomer species present in polymers must be independent of chain size, and that the MALDI-TOF detector ought to show a constant response to ions as the mass of the oligomer increases.

However, the detectors currently in use in commercial MALDI-TOF instruments (see Section 10.2.6)[118] do not meet this crucial condition. Mass discrimination effects at high masses are observed, and therefore MALDI spectra of unfractionated polymer samples may not produce the correct MM estimate.[107]

Average MM estimates by MALDI-TOF for narrow disperse polymers were found to be in good agreement with conventional methods of MM determination,[25,34,37,38,44,64,74,107,113–128,131,136–138] but several authors have reported that for broad distribution polymers MALDI yields false MM values. In the initial systematic attempt to explore the accuracy of MM estimates obtained with MALDI-TOF, several polymethylmethacrylate (PMMA), polystyrene (PS), polyethyleneglycol (PEG) samples with a varying degree of MMD were analyzed.[107] Measurements of MALDI-TOF spectra in linear mode were used to estimate the MM and MMD of these polymer samples.[107] The results, reported in Table 10.1, show that the molar mass estimates provided by MALDI-TOF measurements agree with the values obtained by conventional techniques (such as SEC), only in the case of polymer samples with very narrow MMD ($M_w/M_n < 1.10$).

In the PMMA samples analyzed (see Table 10.1), when the polydispersion index reaches values around 1.10 the M_p (SEC) and M_p (MALDI) values may differ up to about 20%. At higher dispersions, MALDI spectra failed to yield reliable MM values, and the MM measured were much lower than those obtained by conventional methods.[107]

Also a number of condensation polymers such as Nylon 6, polycarbonate, and polyesters were studied (see Table 10.2). These polymers usually possess broad MM distributions, the value of the ratio M_w/M_n being usually around 2. From the inspection of Table 10.2 it can be seen that MALDI underestimates both M_w and M_n in the case of condensation polymers and that the ratio M_w/M_n derived from MALDI spectra of polydisperse polymers is strongly underestimated, and the MM distribution is much narrower. This evidence indicates that lighter molecules are preferentially detected in MALDI-TOF instruments, causing the underestimation of the presence of larger molecules and limiting the use of MALDI for MM and MMD determinations to "monodisperse" samples.

A great part of the MALDI work published to date has been concerned with low-mass polymers (<5–10 kDa), thus the value of the conclusions reached in

TABLE 10.1

Molecular Weight Distribution Data for Polymethylmethacrylates (PMMA), Polyethyleneglycols (PEG), and Polystyrenes (PS)

Sample	M_p(GPC)[a]	M_p(MALDI)[a]	$\Delta\%$[b]	M_w/M_n (GPC)[c]	M_w/M_n (MALDI)[d]
PMMA2400	2400	2100	12.5	1.08	1.10
PMMA3100	3100	2700	12.9	1.09	1.11
PMMA4700	4700	4200	10.6	1.10	1.08
PMMA6540	6540	5200	20.5	1.09	1.11
PMMA9400	9400	7500	20.2	1.10	1.08
PMMA12700	12700	10400	18.1	1.08	1.10
PMMA17000	17000	15000	11.8	1.06	1.03
PMMA29400	29400	27000	8.2	1.06	1.01
PMMA48600	48600	47000	2.1	1.05	1.01
PMMA95000	95000	90000	5.3	1.04	1.01
PMMA_W1[e]	33000	2200	94.0	2.50	1.15
PS5050	5050	5100	1.0	1.05	1.02
PS7000	7000	7020	0.3	1.04	1.02
PS9680	9680	9600	0.8	1.02	1.01
PS11600	11600	11300	2.6	1.03	1.02
PS22000	22000	20800	5.5	1.03	1.01
PS30300	30300	28000	7.6	1.03	1.02
PS52000	52000	46000	11.5	1.03	1.01
PS_W2[e]	9000	2000	77.0	2.00	1.06
PEG4100	4100	3900	4.9	1.05	1.01
PEG7100	7100	7420	4.3	1.03	1.02
PEG8650	8650	8610	0.5	1.03	1.02
PEG12600	12600	12790	1.5	1.04	1.01
PEG23600	23600	23710	0.1	1.06	1.02

[a] Most probable molecular weight.
[b] Percent difference between M_p(GPC) and M_p(MALDI).
[c] Calculated from the GPC curve using the formulas $\bar{M}_n = (\Sigma m_i N_i)/(\Sigma N_i)$, $\bar{M}_w = (\Sigma m_i^2 N_i)/(\Sigma m_i N_i)$.
[d] Calculated from the MALDI-TOF MS using the formulas $\bar{M}_n = (\Sigma m_i N_i)/(\Sigma N_i)$, $\bar{M}_w = (\Sigma m_i^2 N_i)/(\Sigma m_i N_i)$.
[e] Synthesized by free-radical polymerization.

these studies on the determination of MM and MMD is somewhat limited.[34,137] However, MALDI-TOF studies dealing with the MM determination of higher mass range polymers have appeared[44,74,107,123,124,131] confirming the above results.

Great caution is therefore needed in estimating MM and MMD of unfractionated polymers by MALDI-TOF, with accurate estimates being limited to nearly monodisperse polymer samples.[107] Since the great majority of synthetic polymers show a marked polydispersion, this conclusion has raised some concern about the general applicability of the MALDI-TOF method for the determination of MM in polymers. This concern, in turn, has stimulated several studies devised to test these results and to ascertain the reasons why MALDI spectra have shown such a limitation.[25,37,38,63,113–127,137]

TABLE 10.2

Molecular Weight Distribution Data for Nylon 6, Polycarbonates, and Other Polymers[107]

Sample	$\overline{M}_w{}^a$	$\overline{M}_w$ (MALDI)[b]	$\overline{M}_n$ (NMR)[c]	$\overline{M}_n$ (MALDI)[f]	$\Delta\%^d$	$\overline{M}_w/\overline{M}_n{}^e$	$\overline{M}_w/\overline{M}_n{}^b$
Polybutyleneadipate		3020	4000[f]	2090	47.8	—	1.45
Polycarbonate	24200	7930	17000[f]	6470	61.9	1.42	1.23
Polycaprolactone		6740	10000[f]	4690	53.1	—	1.44
Nylon 6 diamino	7000	2250	3000[g]	2020	48.5	2.33	1.11
Nylon 6 monoamino	15000	3080	6200[g]	2320	62.6	2.42	1.32
Nylon 6 hydrolysis	11000	2070	6800[g]	1850	72.8	1.62	1.12
Nylon 6 dicarboxyl	19000	2770	6400[g]	2120	66.7	2.97	1.30

[a] Calculated from intrinsic viscosity using appropriate k and a values.

[b] Calculated from the MALDI-TOF mass spectrum.

[c] Calculated from proton NMR or from end-group analysis.

[d] Percent difference between M_n(GPC) and M_n(MALDI).

[e] Calculated using viscosity and NMR data.

[f] Calculated from proton NMR.

[g] Calculated from end-group analysis.

10.3.1.1 Polymers with a Narrow MM Distribution

Samples with a narrow MM distribution are obtained in various ways. The most used method is anionic polymerization,[139] and this synthetic route can polymerize a large number of monomers. This includes styrene and substituted styrenes (p-methyl, 2,4,6 trimethyl, p-methoxy, α-methyl), nitrogen-containing compounds (vinylpiridine), acrylonitrile and substituted acrylonitriles (methacrylonitrile), a large number of alkylmethacrylates, heterocylic monomers such as ethylene oxide, propylene oxide, isobutylene oxide, ethylene sulfide and substituted ethylene sulfides, heterocylic dimers as glycolide and lactide, propiolactone and higher lactones, hexamethylcyclotrisiloxane and octamethylcyclotetrasiloxane along with dienes (butadiene, isoprene, piperylene, phenylbutadiene).

The theory predicts that the MM distribution of samples obtained by anionic polymerization follows the Poisson distribution, which is extremely narrow, as discussed in Chapter 2 of this book. Monodisperse polyelectrolytes such as poly(styrenesulfonic acid), poly(styrenecarboxlic acid), and poly(acrylic acid) can be synthesized from samples polymerized by anionic polymerization.[139] Fractionation also affords samples with a narrow MM distribution, but it requires substantial quantities of solvents and therefore it is less frequently adopted for samples that do not dissolve in water.[140]

Controlled radical (or "living") polymerization is a relatively new field which is rapidly developing. Samples obtained by this technique display a narrow MM distribution with a polydispersity index lower than 1.2, and often lower than 1.1.[139]

A large number of authors have recorded MALDI-TOF mass spectra of samples obtained by anionic polymerization and compared the values for

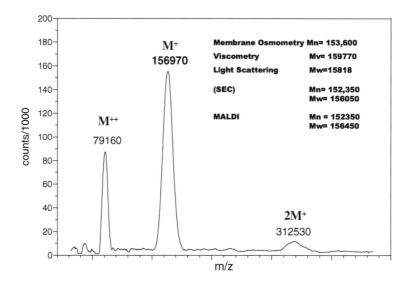

FIGURE 10.2
MALDI-TOF mass spectrum of a polystyrene sample.

M_n and M_w obtained by MALDI with the values obtained by osmometry, viscometry, light scattering, and SEC. In general there is good agreement with MALDI-TOF MM estimates, due to the fact that mass discrimination effects are negligible and instrumental limitations become less severe. Figure 10.2 shows the MALDI-TOF mass spectrum of a polystyrene sample obtained by anionic polymerization (initiator n-butyl lithium, termination by methanol).[117] The peak due to singly charged ions is at 157 kDa, whereas the peak due to doubly charged ions is at 79 kDa and the peak due to dimeric ions is at 312 kDa. The figure also reports the values for M_n and M_w obtained by conventional methods. It can be seen that the values for M_n and M_w obtained by MALDI agree with the latter.

Figure 10.3a shows the MALDI-TOF mass spectrum of a PMMA sample obtained by anionic polymerization (initiator cumyl lithium, termination by methanol).[107] It can be seen that the peak due to singly charged ions is at 48 kDa and that the peak due to dimeric ions is at 94 kDa. The figure also gives the values for M_n and M_w obtained by viscometry and SEC, which agree with the values obtained by MALDI.

Figure 10.3b reports the MALDI-TOF mass spectrum of a PMMA sample obtained with the same initiator, but using a higher $[M]/[I]$ ratio.[107] It can be seen that the spectrum is mass-resolved and that the values for M_n and M_w, namely $M_n = 8600$ and $M_w = 9600$, are in excellent agreement with the values obtained by viscometry, light scattering, and SEC.

Figure 10.4 reports the MALDI-TOF mass spectrum of a PEG sample obtained by ring-opening polymerization (initiator potassium ethanolate, termination by ethanol).[141] The comparison between the values for M_n and M_w

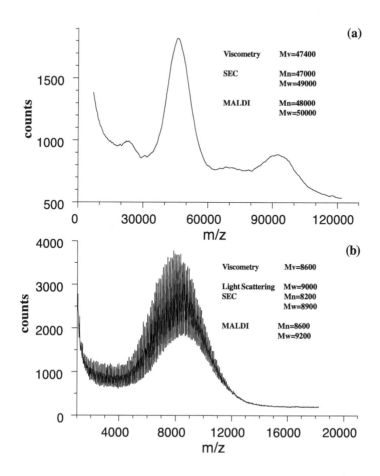

FIGURE 10.3
MALDI-TOF mass spectrum of PMMA that possesses a narrow MMD: (a) sample at 49 kDa, (b) sample at 9.4 kDa.

obtained by MALDI, namely $M_n = 7000$ and $M_w = 7200$, and the values obtained by osmometry, viscometry, light scattering, and SEC is satisfactory.

Lactones can be polymerized by ring-opening polymerization. MALDI-TOF spectra of anionic poly(lactic acid) (PLA) display negligible mass discrimination effects. Values of MM measured by MALDI turn out to be in good agreement with those obtained by conventional methods.[142]

Polybutadiene (PB) and polyisoprene (PI) samples obtained by anionic polymerization are employed as primary MM standard for SEC calibration. Yalcin et al. has recorded MALDI-TOF spectra of anionic PB and PI.[143] As an example, Figure 10.5 shows the MALDI-TOF mass spectrum of a polyisoprene sample with $M_n = 19300$. Yalcin et al. also analyzed PB6000, PI10500, and PB22000, and compared the MALDI method with conventional methods for measuring M_n and M_w values, finding a reasonable agreement.[143]

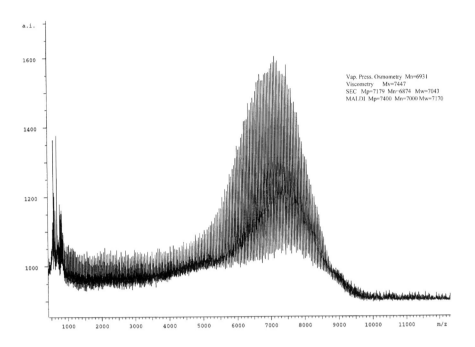

Vap. Press. Osmometry Mn=6931
Viscometry   Mv=7447
SEC Mp=7179 Mn=6874 Mw=7043
MALDI Mp=7400 Mn=7000 Mw=7170

FIGURE 10.4
MALDI-TOF mass spectrum of PEG7100.

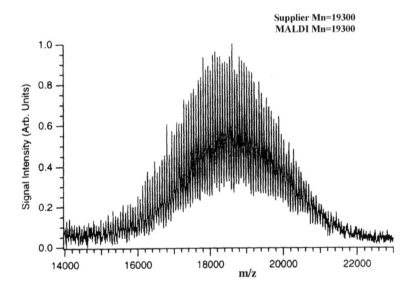

Supplier Mn=19300
MALDI Mn=19300

FIGURE 10.5
MALDI-TOF mass spectrum of polyisoprene 19300. (Adaped by permission from Ref. 143)

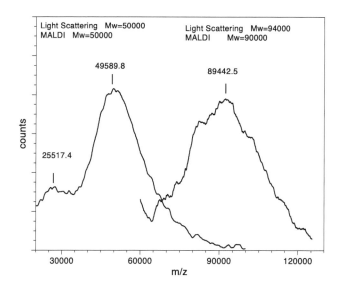

FIGURE 10.6

MALDI-TOF mass spectrum of SEC fractions of dextran. (Reprinted with permission from Ref. 140, Copyright 1995 John Wiley & Sons Ltd)

Garozzo et al. injected in a preparative SEC apparatus a polydisperse polymer (dextran) and analyzed the collected fractions by light scattering and by MALDI-TOF.[140] Figure 10.6 shows the MALDI-TOF mass spectrum of two dextran samples obtained by this method. The analysis of the spectra yielded $M_w = 50000$ for the sample on the left and $M_w = 90000$ for the sample on the right, and these values compare well with the light scattering values ($M_w = 50000$; $M_w = 94000$).

Figure 10.7 shows the MALDI-TOF mass spectrum of a sample obtained by ring-opening polymerization of a disubstituted cyclopropane.[144] Figure 10.7 also reports the values for M_n and M_w obtained by various conventional techniques, among which is NMR (NMR gives reliable results in the case of low MM polymers, as the present one) and the analysis of the MALDI spectrum.

Block copolymers produced by anionic polymerization do possess a very narrow MM distribution (Chapter 2). Wilczek-Vera et al. recorded the MALDI-TOF mass spectrum of a block copolymer containing units of α-methylstyrene and units of styrene.[145] MALDI peak intensities were inserted in the equations which define M_n and M_w and the result was $M_n = 4273$, $M_w = 4411$, in fair agreement with the molar mass averages obtained by SEC ($M_n = 4190$, $M_w = 4510$).[145]

Figure 10.8 reports the MALDI-TOF mass spectrum of a St-MMA block copolymer.[146] The peak due to singly charged ions is at 27000 Da, whereas the peak due to dimeric ions is at 54000 Da. The SEC analysis yielded $M_n = 26000$,

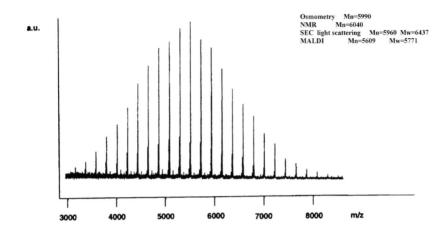

FIGURE 10.7
MALDI-TOF mass spectrum of a polymer obtained from a disubstituted cyclopropane. (Reprinted with permission from Ref. 144, Copyright 2000 American Chemical Society)

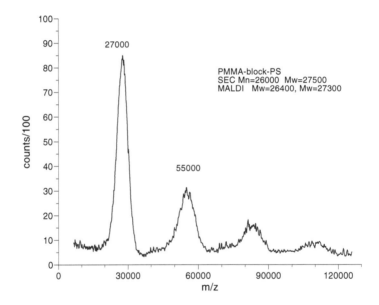

FIGURE 10.8
MALDI-TOF mass spectrum of a PMMA-block-PS copolymer sample obtained by anionic polymerization.

and the composition (determined by ^{1}H-NMR) turned out to be almost equimolar.[146] This implies that the styrene and MMA blocks have both $M_n = 13000$.

It can be concluded that the MALDI method for measuring M_n and M_w values in polymers and copolymers gives a good agreement with conventional methods.

10.3.1.2 Polymers with a Broad MM Distribution

When the sample possesses a broad MM distribution, in order to calculate the MM it is necessary to consider the intensities of peaks spanning over a wide range of mass numbers. This constitutes a serious problem in MALDI-TOF MS, since mass discrimination is likely to occur.

Before analyzing the effect of mass discrimination in the MM determination, it is interesting here to consider two special cases where MALDI-TOF measurements of wide distribution polymers are useful not for estimating the MM averages, but for detecting the MMD shape.

The first case occurs when a light-sensitive molecule initiates the polymerization process and the light source is pulsed, such as in the rotating sector polymerization[147] or in pulsed-laser polymerization (PLP).[148] In this case, the resulting polymer follows a peculiar MMD, since as the chain length (n) increases, the molar fraction of chains of size n, denoted by $I(n)$, grows and falls. More specifically, it falls exponentially,[147] then it grows, reaches a maximum, and then falls again. $I(n)$ possesses a point of inflection, at mass M_{inf}, which is given by:

$$M_{inf} = \Phi[I(n)] \qquad (10.3)$$

where Φ is an expression which involves the second derivative of $I(n)$ with respect to "n." By plotting the oligomer's abundance against its mass, and taking the second derivative, the point at which the second derivative changes from negative to positive is the point of inflection. M_{inf}, is important in assessing the rate constant of termination and the rate constant of chain transfer.[44, 147-152]

Figure 10.9 reports the MALDI-TOF mass spectrum of a polymer obtained by pulsed-laser polymerization that displays a series of peaks ranging from 3000 Da to 12000 Da.[148] The MMD resulting from the MS was plotted, and the point of inflection of the MMD (Eq. 10.3) was determined. The data allow one to determine the rate constant of termination and the rate constant of chain transfer.[148] The method based on MS has now become a standard.

The second case occurs when the polymeric sample possesses a large fraction of cyclic oligomers. The presence of cyclic oligomers is often noticed in polymers, although the abundance of cyclics decreases drastically as the mass grows, and cyclic oligomers with $n > 50$ are very rare. Several MS techniques have been used to detect cyclics in polymeric samples (Chapters 2 and 7).[153] MALDI can be used as well, since the mass range of cyclic oligomers is relatively limited, and therefore mass discrimination is not severe. Figure 10.10 reports the MALDI-TOF mass spectrum of a polylactic acid (PLA) sample.[154] In the region 1300–2600 Da, peaks due to cyclics ($\circ$) are stronger than peaks due to linear ($\times$) oligomers, whereas in the region 2600–4000 Da peaks due to cyclics are weaker. Remarkably, cyclic oligomers up to mass 4000 are indeed seen, corresponding to chains with a size up to $n = 70$ and even higher, showing an unusually extended range of PLA cyclics.[154]

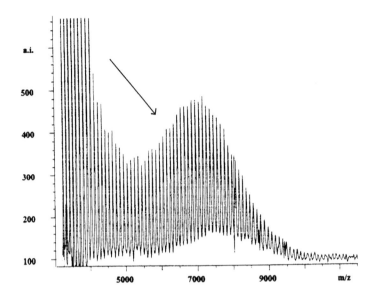

FIGURE 10.9
MALDI-TOF mass spectrum of a PMMA obtained by pulsed laser polymerization. The arrow indicates the position of the inflection point M_{inf}. (Reprinted with permission from Ref. 11, Copyright 1993 American Chemical Society)

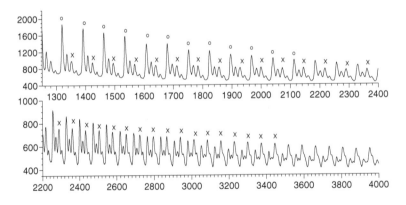

FIGURE 10.10
MALDI-TOF mass spectrum of poly(lactic acid). In the region 1300–2600 peaks due to cyclic are stronger than peaks due to linear, whereas in the region 2600–4000 peaks due to cyclic are weaker.

10.3.2   Effect of Mass Discrimination in the MM Determination

All the quantitative applications of MS to polymers, regardless of the ionization method utilized to produce ions, are based on the assumption that there is a quantitative correspondence between the relative abundance of the species to be analyzed and the relative intensity of MS peaks (see Chapter 2).

This assumption implies that there are no mass discrimination effects, i.e., that ion fragmentation, ionization yield, ion trasmittance, and detector response are independent of oligomer molar mass. Notably, in the case of the determination of composition and sequence in copolymers by MS techniques (Chapter 2), where one needs only to consider the relative intensity of the co-oligomers of the same length, the assumption that the relative peak intensities reflect the relative co-oligomers abundance has been proven valid.[145,155-167]

However, if we consider an extended mass range, these conditions are certainly not met by any MS technique of current use. The most notable case where it is necessary to consider the intensities of peaks spanning over a wide range of mass numbers is the determination of molar masses in polydisperse polymers (see Chapter 2).

Up to a decade ago, when the existing soft ionization MS techniques reached masses up to 5,000–10,000 Da, molar mass estimates produced by these MS techniques were reported to be in agreement with the results obtained by SEC analysis. The advent of MALDI-TOF boosted to 10^6 Da and beyond the range of masses detectable, and showed the existence of mass discrimination effects that rendered inaccurate the determination of molar masses in polydisperse polymers by MS.[18,21,116,117,120,131] Even when the detectors currently employed in MALDI instruments are equipped with a post acceleration device, the detection of high masses is still only partially efficient.[12] This was an important disadvantage of the technique, but it was crucial for the development of MALDI applications to synthetic polymers.

In order to overcome mass discrimination problems in MALDI, several strategies have been devised. An obvious way to proceed is to couple SEC fractionation with MALDI analysis of the fractions, in order to reduce the polydispersion of the original sample. This hyphenated method has been successful and it is now routinely applied in its off-line version, whereas several attempts are under way to develop a reliable on-line SEC/MALDI procedure (Section 10.3.3).

Another obvious approach is to try to design an ideal detector, capable of showing a constant response independent of mass value. Attempts in this direction have been reported,[18,21,120] and should these "ideal" detectors become available, it would be possible to calculate reliable MM and MMD directly from a single MALDI spectrum.

Still another attempt to cope with the problem of mass discrimination in MALDI has been reported. In this, an off-line correction of the detector response is made, eliminating spurious components in the signal and generating a new spectral baseline from which the molar mass of the polymer can be calculated.[123] The method utilizes the MALDI spectrum in continuous extraction to get the full ion yield from the detector. The results obtained up to the present are encouraging.[123,124]

Figure 10.11 reports the principal steps and the results obtained when the above procedure is applied to a BPA-polycarbonate (PC) sample. The upper part reports the MALDI-TOF mass spectrum of the PC sample. It displays a series of well-resolved peaks (see inset) in the region between 4000 and

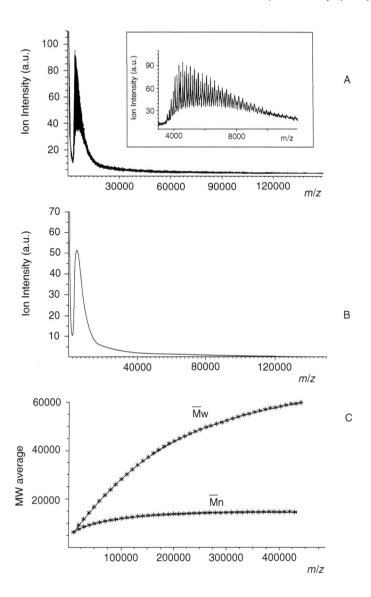

FIGURE 10.11
The procedure for the determination of M_n and M_w from the MALDI spectrum. MALDI-TOF
mass spectrum of a polycarbonate (a), the spectrum after smoothing and baseline subtraction
(b), the result of a calculation which yields M_n and M_w (c). (Reprinted with permission from Ref.
124, Copyright 1997 American Chemical Society)

8000 Da, followed by a quick decrease of the signal in the region between
8000 and 30000 Da and by a gentle decrease above 30000 Da. The middle
part of the picture reports the result obtained when one applies a strong
smoothing and selects a new level for the baseline. The last part of the picture

reports the result of M_n and of M_w as a function of the m/z value selected as end line. The correct estimate of MM is where the asymptote is reached.[123]

A further procedure for the correction of the decreasing detection response in MALDI-TOF spectra with increasing ion mass, is based on the use of PMMA standards of known MM to calibrate the detector response. Figure 10.12a shows the MALDI-TOF spectrum of an equimolar mixture of four PMMA primary standards with different MM. Although the mixture is equimolar, the areas under the four bell-shaped peaks are not the same, due to the fact that lighter molecules are preferentially desorbed and the response of the detector is not constant.

Figure 10.12b reports the ion yield, measured by taking the ratios among the areas under the four bell-shaped structures in the mass spectrum. It also

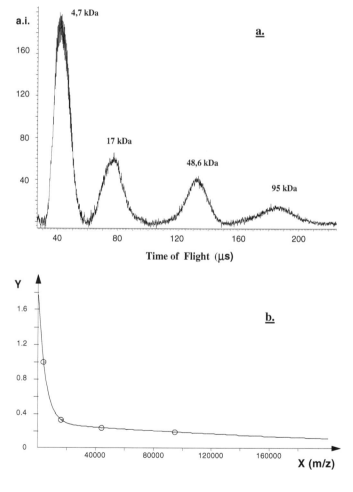

FIGURE 10.12
MALDI-TOF mass spectrum of a mixture of PMMA standards (a) and the function used for the correction (b). (Reprinted with permission from Ref. 131, Copyright 1998 John Wiley & Sons Ltd)

reports the function used to take into account the decreasing detection response in MALDI-TOF. From these data, an equation describing the MALDI-TOF response to increasing MM has been derived, and it can be used to correct the experimental spectra.[131]

10.3.3 Coupling Size Exclusion Chromatography and MALDI-TOF

From the early times of development of MS methods for polymer analysis, the value of coupling MS with SEC or LC was recognized, and a number of SEC/MS and LC/MS fractionations applied to polymers have appeared, using a variety of ionization techniques (Chapter 4).[168–175] These studies showed the unique value of SEC/MS and LC/MS in the structural characterization of homopolymers, copolymers, and complex polymer systems. With the advent of MALDI-TOF, a large number of investigations focused on MALDI as a SEC detector for the determination of end-groups and molar masses in narrow SEC fractions of synthetic polymers.[25,108–112,130,140,176–192,193,194]

Attempts to measure MM by MALDI of polydisperse polymer samples yielded poor results. Instead, correct MM estimates could be obtained when MALDI-TOF was used as a detector for SEC, collecting the SEC fractions and analyzing them with MALDI.

The earliest SEC/MALDI studies on MM determination[108,109,140] were prompted from the finding[107] that MM estimates derived from MALDI-TOF measurements are accurate only in the case of monodisperse polymers. The SEC fractionation of polydisperse polymers provided nearly monodisperse fractions, and MM measurements by MALDI on the samples collected could be used to calibrate the molar mass vs. retention volume in SEC plots. Thus, average MM of polydisperse polymers could be determined. At the same time, it was realized that the MALDI analysis of the fractions obtained in the SEC procedure allowed the identification of polymer end-groups as a function of the MM.[108,111,112,140,181] Both off-line and on-line SEC/MALDI-TOF methods have been described. Numerous applications to synthetic polymers have been published using the off-line method, which is typically operated by injecting about 0.5 mg of polymer.[176–189] The amount of sample present in each fraction (about 5 μg, on average) exceeds many times the quantity needed for a MALDI-TOF spectrum.

The method consists in fractionating the polymer by SEC and collecting a number of fractions (Figure 10.13). Selected fractions are then analyzed by MALDI-TOF, and the mass spectra of these nearly monodisperse samples allow the computation of reliable values of M_n and M_w corresponding to each fraction. When the M_n of each fraction is plotted vs. elution volume, a very reliable calibration for the chromatogram is obtained. The SEC trace and the calibration equation are fed into SEC software, which gives as an output M_n and M_w of the polymer. A further advantage of collecting SEC fractions of a polymer sample is that one can use them not only for performing the SEC/MALDI coupling, but also for SEC/NMR, or any other suitable

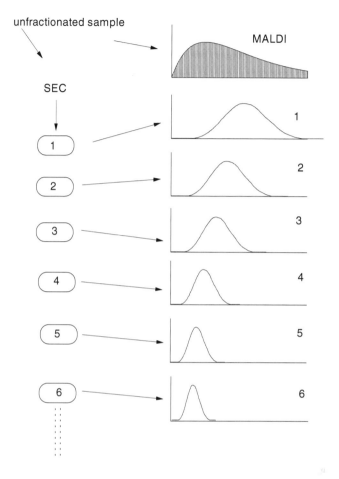

FIGURE 10.13
Scheme of the SEC-MALDI method.

coupling, to obtain complementary structural information.[189] Determination of bivariate distribution in PMMA/PBA and in PS/PAA copolymers was achieved by performing SEC/MALDI and SEC/NMR experiments.[189]

The off-line SEC/MALDI coupling is time-consuming and, despite the advantage illustrated above, the on-line coupling would be the ideal procedure. A number of methods and practical devices have been described to achieve SEC/MALDI on-line. The methods can be classified as "continuous flow MALDI" and "aerosol MALDI," respectively, according to the device selected for the coupling.[177,178]

Direct deposition was achieved by post-column mixing of the SEC solvent with matrix solution and then depositing it as a continuous track onto a rotating sample holder disk.[192] In another approach,[61] the SEC solvent was spray-deposited onto a rotating matrix-coated substrate, and the resulting polymer trail was characterized directly by MALDI. The coupling of SEC to a robotic interface also has been described, where the matrix is coaxially added to the SEC effluent and spotted dropwise on the MALDI target.[184]

In an aerosol SEC/MALDI experiment, the effluent from the SEC column was combined with a matrix solution and sprayed directly into a MALDI-TOF spectrometer.[177] Ions were formed by irradiation of the aerosol particles with a pulsed UV laser. The ions were separated in a two-stage TOF apparatus, and averaged mass spectra were stored throughout the SEC/MALDI experiment. The matter of coupling MALDI with SEC has been recently reviewed, and it was concluded that off-line SEC/MALDI is routine today, whereas considerable work is still necessary in order to make on-line SEC/MALDI a viable alternative to the off-line method.[178]

Perhaps, one may expect that once the exact experimental conditions for on-line SEC/MALDI separation have been established for a specific polymer system, the latter can be routinely repeated as needed, for instance in industrial practice. Conversely, the off-line SEC/MALDI method will always find its utility in research work, and when one is confronted with new polymeric materials.

SEC/ESI also has been developed[175] (see Chapter 4), although the ESI technique is mass limited for synthetic polymers, due to the effect of molecular entanglement[8] processes in high MM polymers during electrospray experiments.

Soon after the first reports on the use of SEC/MALDI to the characterization of polydisperse polymers,[108,112] several papers have appeared[176–189] confirming the application of SEC/MALDI to the calculation of the average MM of polydisperse samples of polymers. Progress in the area also has been monitored by some reviews.[185,188]

MALDI-TOF was coupled to SEC and to other techniques for the structural characterization of poly(dimethylsiloxane)-co-poly(hydromethylsiloxane), a copolymer used as a precursor of functionalized silicone grafts.[162]

Pasch et. al.[176] reported a SEC/MALDI investigation of polymers whose molar masses hardly exceeded 2000 Da. The mass values estimated for each fraction were not used to calculate average MM values of the whole polymers. The same authors have used preparative SEC to identify by MALDI-TOF the cyclic PMMA produced by group transfer polymerization.[195]

The accuracy of the SEC/MALDI procedure has been demonstrated.[112] A series of narrow distributed PMMA samples were injected into a SEC apparatus, and the elution volumes were recorded. A PMMA sample with a broad MMD ($M_n = 13000$ $M_w = 33000$) obtained by free radical polymerization was injected in an SEC apparatus, and SEC fractions were collected.[112] The MALDI-TOF mass spectra of the SEC fractions were recorded and the respective MM could be computed. Figure 10.14 reports the SEC calibration line obtained using the MALDI-TOF mass spectra of the SEC fractions and also the SEC calibration line obtained using the PMMA standards.[111,112] The two lines are virtually superimposed, thus confirming the accuracy of this procedure, which makes use of MALDI-TOF mass spectra of SEC fractions. The M_n and

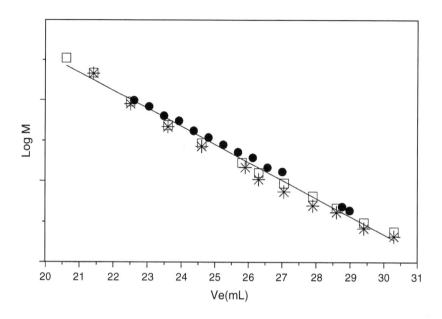

FIGURE 10.14

SEC calibration plots of the PMMA samples: □ M_w of the PMMA GPC standards as indicated by the supplier, ✳ M_w of the PMMA GPC standards obtained by MALDI-TOF spectra, ● M_w of the PMMA W1 fractions obtained by MALDI-TOF spectra.

M_w values for the whole sample, obtained by the SEC-MALDI method, turned out to be $M_n = 13000$ and $M_w = 36000$, which is within 5% of values given by the supplier.[111,112]

SEC-MALDI also proves useful in cases where the production of high polymers is accompanied by the formation of sizeable amounts of cyclic oligomers. The presence of a mixture of linear and cyclic chains in low molar mass polymers causes difficulties in establishing an appropriate SEC calibration curve for such samples.[196,197] The theory predicts that cycles are eluted at later times with respect to linear chains.[197] In fact, for a given polymer there is a small but defined difference between the hydrodynamic volumes of linear and cyclic oligomers of the same molar mass, and this difference is reflected in the elution volumes (V_e) of the linear and cyclic species of the same molar mass.[197]

It has been shown by SEC-light scattering experiments that cyclic oligomers of PDMS are eluted at slightly higher volumes with respect to linear oligomers of the same MM, owing to the smaller hydrodynamic volume. The ratio (M_{cycle}/M_{linear}) was found to be 1.24, independent of MM values.[197]

The experiment was repeated by MALDI-TOF. A commercial PDMS sample with a wide MMD (M_w/M_n about 2.5) was injected in a SEC apparatus and SEC fractions were collected.[111,112] Figure 10.15 reports the SEC trace along with the MALDI-TOF mass spectra of four SEC-fractions. The spectra possess an excellent S/N ratio, they display bell-shaped structures, and M_n

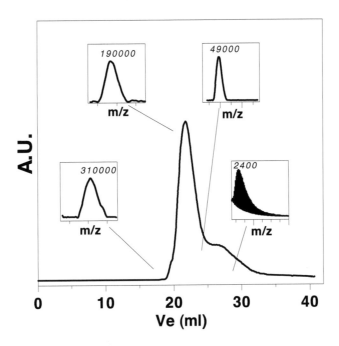

FIGURE 10.15

SEC trace of poly(dimethyl siloxane) (PDMS) sample in THF as solvent. The insets display the MALDI-TOF mass spectra of some selected fractions.

and M_w can be accurately computed. Another sample possessing a relevant fraction of cyclics was also analyzed. The MALDI-TOF mass spectra of the SEC fractions were of excellent quality.[112]

Figure 10.16 reports the SEC calibration line obtained using the MALDI-TOF mass spectra for the sample made of linear chains and also for the sample made of cyclics. It can be seen that the two lines are not superimposed, and that cycles are eluted later. Figure 10.17 reports the MALDI-TOF spectra of four SEC fractions collected from a PDMS sample, and it can be seen that the average molar mass for the earliest fraction is remarkably high: M_n = 417,000 and M_w = 424,000.[112]

The off-line SEC/MALDI procedure has been applied to two polycarbonate samples PC1 and PC2, which contained both linear and cyclic oligomers.[182,183] The results showed that the MALDI spectra of the SEC fractions allow the net detection of linear and cyclic oligomers contained in these samples, and also the simultaneous determination of the average molar masses of the linear and cyclic oligomers.[182,183] As expected, two parallel SEC calibration lines were obtained for linear and cyclic PC chains, and the ratio (M_{cycle}/M_{linear}) was found to be constant (about 1.20). The mass spectra obtained allowed calibration of the SEC curves against absolute MM, and thereafter calculation of the MM averages of the original samples from the SEC trace.[182,183]

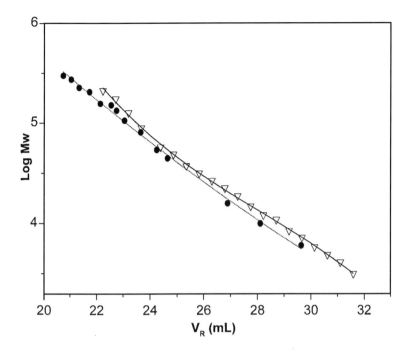

FIGURE 10.16
SEC calibration lines for poly(dimethyl siloxanes), MALDI-TOF average molecular weights vs. retention volume V_R of each SEC fractions. ● linear and ▽ cyclic macromolecules.

Applications of SEC/MALDI also proved important in the field of copolymer characterization. The calibration of SEC traces of copolymers is a difficult problem, and it requires some additional effort compared to the case of homopolymers. In fact, copolymer chains having the same MM may possess a different comonomer composition. A change in the comonomer composition may cause a variation in the overall dimensions of the copolymer chains. As a consequence, although having the same mass, their hydrodynamic volumes may be different and the elution volume of isobaric molecules will be subject to change as the copolymer composition varies. This poses a serious problem for the calibration of SEC traces.

Runyon et al. proposed a method to compute M_n and M_w of a copolymer sample which is based on the calibration lines obtained for homopolymer A and homopolymer B.[197]

The method consists of two steps. In the first step, after constructing the calibration lines for homopolymer A and homopolymer B, one records the SEC trace of the copolymer using a RI detector in series with a UV detector. Comparing the two detector responses, one measures the changes in w_A and w_B, the weight fraction of A units and of B units, as the elution volume grows. In the second step, M_n and M_w are computed from the SEC trace by using

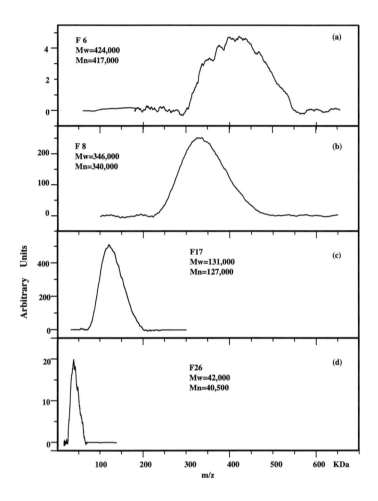

FIGURE 10.17
MALDI-TOF mass spectra, recorded in linear mode, of four selected poly(dimethyl siloxane) fractions: (a) fraction 6; (b) fraction 8; (c) fraction 17; and (d) fraction 26.

an interpolation formula:

$$\log(M_C) = w_A \log(M_A) + w_B \log(M_B) \tag{10.4}$$

where M_C is the molar mass of the copolymer, M_A and M_B are the molar masses of the two homopolymers, w_A and w_B are the weight fractions of A units and of B units. When the weight fraction of the two units in the copolymer is comparable ($w_A = 0.5$), the copolymer line falls in the middle, and one can draw it directly on the graph of the two homopolymer lines.

The method of Runyon et al. is based on the assumption that the relationship between the elution volumes of the copolymer and of the two "parent"

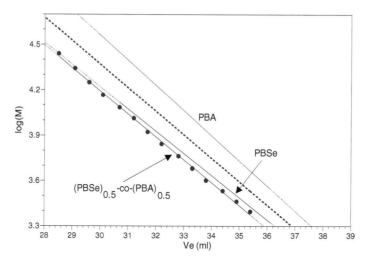

FIGURE 10.18

Calibration curves for PBSe, for PBA, and for the PBSe-PBA copolymer. (Reprinted with permission from Ref. 181, Copyright 1998 American Chemical Society)

homopolymers is linear.[197] This assumption doesn't hold, however. The relationship is more complex,[140] and the M_n and M_w values calculated by Eq. 10.4 are erroneous.

SEC/MALDI can be used to compute M_n and M_w of a copolymer sample and to overcome this limitation.[181] In one example, a random copolymer containing equimolar amounts of butylene adipate units and of butylene sebacate units, with a wide MMD (M_w/M_n about 2), was injected in a SEC apparatus, and SEC fractions were collected.[181] The MALDI-TOF mass spectra of the whole sample displayed peaks up to 14 kDa, whereas the MALDI-TOF mass spectra of the SEC fractions displayed peaks up to 75 kDa.[181] The spectra were characterized by an excellent S/N ratio, and the MM values for each fraction were determined.

Figure 10.18 reports the SEC calibration line for the random copolymer (the experimental points are marked as circles), along with the calibration lines for PBA (the butylene-adipate homopolymer) and for PBSe (the butylene-sebacate homopolymer).[181] The genuine SEC calibration line for the random copolymer was used to compute molar mass averages, which turned out to be $M_n = 8500$, $M_w = 12800$.[181] It can be noted that the SEC calibration line for the equimolar copolymer, according to Eq. 10.4, should lie between the two lines of the homopolymers. Instead, it is almost coincident with the calibration line of PBSe, probably because of the high hydrodynamic volume of the copolyester.[181]

A number of biodegradable copolyesters were analyzed by SEC/MALDI, the genuine SEC calibration lines were measured, and accurate MM averages were obtained.[181]

Nielen et al. used SEC/MALDI for the calculation of the average MM of a polydisperse MMA/MAA copolymer. The MM averages obtained by SEC/

MALDI (M_n = 18500 M_w = 36300) were in good agreement with manufacturer's data (M_n = 15000 and M_w = 34000).[179]

10.3.4 Coupling HPLC and MALDI-TOF

High-performance liquid chromatography (HPLC) can be used to separate macromolecules having the same backbone but different end-groups.[198] Combining HPLC with MALDI analysis, some materials have been characterized.[176,199–201] For instance the polymerization reaction of styrene using a carboxyl-functional initiator yields a mixture of chains with zero, one, and two carboxyl groups, and HPLC can separate the three components mixture.[199] MALDI-TOF spectra of the HPLC fractions were used to verify the efficiency of the separation.[199]

HPLC can be used also to separate macromolecules of different sizes. Spickermann et al. used HPLC to separate the high mass fraction of a polystyrene sample.[191] The MALDI-TOF spectrum of the high mass fraction turned out to be more informative than the MALDI-TOF spectrum of the whole sample, since the former displayed some peaks which were absent (or below the detection limit) in the latter spectrum.[191]

In another application various HPLC fractions of a poly(decamethylene adipate) sample were collected, and the HPLC chromatographic trace displayed strong peaks separated by deep valleys, suggesting that chains of different length were eluted at different times.[176] However, when the supposed homogeneous fractions were analyzed by MALDI-TOF, it was evident that oligomers with two, three, and even four repeat units were present in a single HPLC fraction.[176]

In the technique of liquid chromatography at the critical condition (LCCC), macromolecules of different sizes are eluted at the same time (Figure 10.19).[202] This peculiar elution behavior is achieved making use of columns in which the macromolecules are at the adsorption-elution transition, and Figure 10.19 reports the calibration curve for LCCC along with the curves for adsorption and size exclusion chromatography, respectively.[203–205]

LCCC can be used to separate macromolecules with different end-groups. Wachsen et al. used LCCC to separate the products obtained by partially degrading poly(lactic acid).[205] The MALDI-TOF spectra of the LCCC fractions turned out to display a smaller number of peaks compared to the MALDI-TOF spectrum of the whole sample, and this feature made the peak assignment procedure for MS peaks simpler and more reliable.[190,205]

LCCC is a powerful tool for the analysis of block copolymers. Lee et al. analyzed a copolymer containing blocks of poly(ethylene oxide) and blocks of poly(lactide) by LCCC, and they were able to determine the conditions in which chains with lactic blocks of different size are eluted at almost the same time.[201] The LCCC chromatogram shows a series of narrow peaks, and the MALDI-TOF spectra of the LCCC fractions confirm that each structure contains exclusively one type of oligomer.

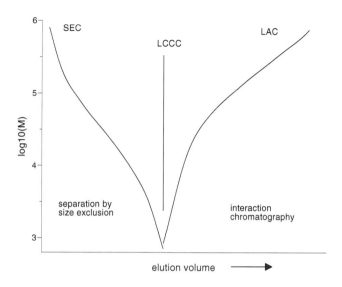

FIGURE 10.19
Calibration curves for Adsorption Chromatography, SEC, and for Chromatography at the Critical Conditions.

Pasch used an LCCC apparatus to separate a copolymer containing blocks of poly(ethylene oxide) and blocks of poly(propylene oxide).[202]

10.4 Structure and End-Group Identification

Synthetic polymers are characterized not only by differences in their molar mass distributions, they also exhibit "chemical distribution" (differing functional groups, sequences of monomers, and sequence length) and structural heterogeneities (linear, cyclic, grafted, and branched portions). Mass accuracy and mass resolution obtainable in MALDI-TOF MS often allows the direct structural determination of the species contained in each polymer sample and also of the end-groups that are present. Chemical structure and end-groups functionality of polyamides such as Ny_{12},[22] Ny_6,[206] $Ny_{6,T}$,[68] Ny_{10},[68] and $Ny_{6,10}$,[68] have been characterized by MALDI-TOF mass spectrometry. A detailed study of the end-groups present in Ny_6 samples obtained under different conditions may illustrate the capability of the method.[206]

The MALDI-TOF mass spectra of four Nylon 6 (Ny_6) samples, each terminated by different end-groups (i.e.: diamino-, amino-methyl-, dicarboxyl-, and amino-carboxyl) are shown in Figure 10.20.[206] These spectra permit a complete characterization of the molecular species present in each Ny_6 sample. The mass spectrum of the Ny_6-diamino terminated (Figure 10.20a) shows

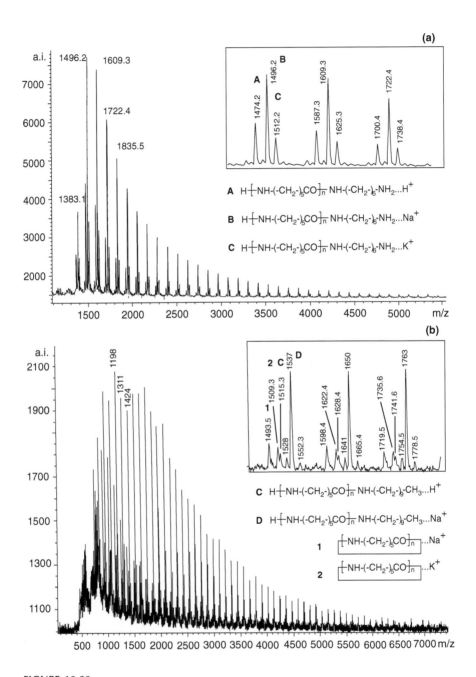

FIGURE 10.20
Positive-ions MALDI-TOF mass spectra of Ny6 samples: (a) reacted with diaminohexamethyl-ene; (b) reacted with decylamine; (c) reacted with adipic acid; (d) hydrolyzed with water in methansulfonic acid. (Continued)

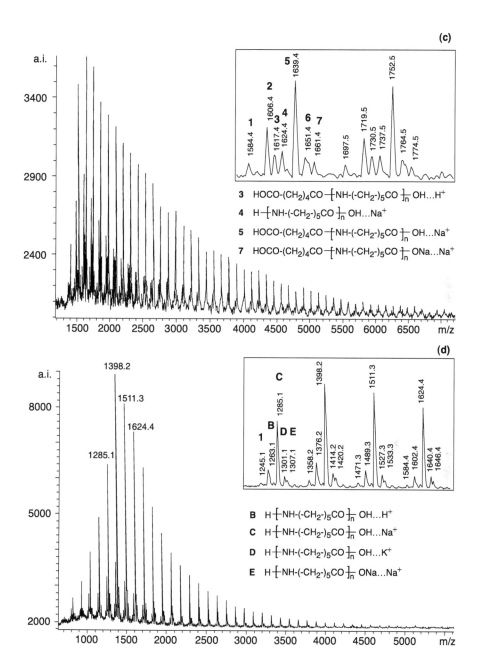

FIGURE 10.20
(*Continued*)

clusters of three intense peaks due to the protonated, sodiated, and potassiated quasi-molecular ions of the expected diamino-terminated Ny_6 oligomers. The protonated peaks (A) disappeared when tetrahydrofuran (THF) was used as a solvent instead of trifluoroethanol (TFE) solvent. The MALDI-TOF mass spectrum of the Ny_6 amino-teminated sample (Figure 10.20b) shows intense peaks due to sodiated ions (C) of the expected oligomers bearing NH_2 and $C_{10}H_{21}$ as terminal groups. The corresponding protonated ions (D) appear with low intensity. In the region 900–4000 Da, the MALDI spectrum also shows two mass series of low intensity that can be assigned to sodiated and potassiated ions of Ny_6 cyclic oligomers.

The MALDI-TOF mass spectrum of the Ny_6-dicarboxyl terminated (Figure 10.20c) shows a series of intense peaks ranging from 1000 to 7000 Da, assigned to Ny_6 oligomers bearing two COOH end-groups. Other peak series of lower intensity appear in the region 1000–4500 Da in Figure 10.20c. Peaks labeled as 1 and 2 correspond to protonated and sodiated Ny_6 cyclic oligomers, respectively. Peaks numbered 4 are due to sodiated ions of the oligomers with NH_2 and COOH end-groups.

The MALDI spectrum of Ny_6 hydrolyzed in aqueous methansulfonic acid (Figure 10.20d) is dominated by a series of intense peaks corresponding to sodiated ions of the Ny_6 oligomers with NH_2/COOH end-groups (peaks 4). Protonated (peaks 5) and potassiated (peaks 6) ions of these oligomers appear with low intensity.

The spectrum also displays other low-intensity peak series corresponding to protonated cyclic oligomers (peaks 1) and sodiated ions of the sodium salt of the NH_2/COOH-terminated Ny_6 chains (peaks E).[206]

Sometimes, the mass spectra of reaction mixtures are very complex, with partly superimposed peaks, and the end-group assignment is difficult. Doping the MALDI solutions with different alkali salts may provide a way of alleviating the problem.

The end-groups of a sample of poly(methylphenylsilane) (PMPS), obtained by the Wurtz coupling reaction of diorganodichlorosilane followed by fractionation with methanol, were characterized by comparing the MALDI spectra obtained with and without the addition of a specific alkaline salt (LiCl, KCl, or NaI) to the solution containing the sample and matrix.[207]

The mass shifts observed in the MALDI spectra of the salt-doped solutions with respect to the undoped solutions allowed the unambiguous assignments of the isobar peaks corresponding to PMPS oligomers. Six species of linear oligomers of PMPS bearing different types of end-groups ($-Si-(CH_3)_3$, $-O-Si-(CH_3)_3$, $-O-CH_3$, $-H$), were identified together with a sizeable amount of cyclic oligomers.[207]

The end-groups of several PEG samples were identified through their MALDI-TOF spectra, recorded in linear or reflection mode and in delayed extraction mode.[26,84,107,141,208–212] PEG samples used as primary standards for SEC were found to possess H/OH and OH/OH end-groups,[107,141,208] whereas several PEG samples used as surfactants were found to have a nonylphenol end-group.[26,84]

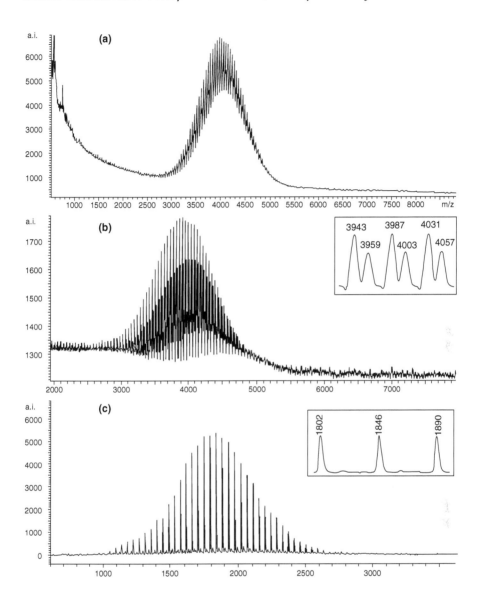

FIGURE 10.21

Positive ion MALDI-TOF spectra for sample PEG 4100 in the linear mode (a), for sample PEG 4100 in the reflection mode (b), and for sample PEG 2000 in the reflection mode (c). (Reprinted with permission from Ref. 141, Copyright 1995 American Chemical Society)

Figure 10.21 shows the MALDI-TOF spectra of two PEG samples. The PEG 4100 sample was synthesized by anionic ring-opening polymerization of ethylene oxide (EO) initiated with $C_2H_5O^-K^+$ in the presence C_2H_5OH, whereas the polymer PEG 2000 was prepared from EO and H_2O in the presence of an acid catalyst.[141]

The MALDI-TOF mass spectra of PEG 4100 recorded in linear and reflection modes are reported in Figures 10.21a and b, respectively. The spectrum recorded in reflection mode (Figure 10.21b) shows a higher resolution as compared to that in linear mode (Figure 10.21a), so that two mass series can be discerned (see inset in Figure 10.21b). The two mass series correspond to sodium- and potassium-cationized H-$(CH_2CH_2O)_n$-H oligomers, respectively.[141]

The MALDI-TOF mass spectrum of PEG 2000 (Figure 10.21c), recorded in reflection mode, shows one series of mass peaks that corresponds to sodium cationized HO-$(CH_2CH_2O)_n$-H oligomers. From this information it is often possible to deduce the type of initiation and termination mechanisms operating in the synthesis of PEG samples.[141]

The end-reactive groups of several types of heterotelechelic PEGs such as NH_2-PEG-OH, CHO-PEG-OH, and CHO-PEG-OCOC(CH_3)=CH_2, were also characterized by MALDI-TOF analysis.[209] A PEG sample end-labeled with pyrenebutyrate groups was characterized using HABA as a matrix, doped with a trace of either sodium chloride or potassium chloride.[210] Inspection of the MALDI-TOF spectra revealed that $80 \pm 2\%$ of the chains were doubly labeled, whereas the remaining chains contained only a single pyrene group.

Ethoxylate end-groups of a PEG sample used as a surfactant additive were detected by MALDI-MS analysis of fractions obtained by liquid chromatography at the critical condition (LCCC).[176] Three fractions were collected by LCCC and, as reported in Figure 10.22, the MALDI-TOF spectra show a peak-to-peak mass increment of 44 Da, confirming that all fractions consist of ethylene oxide-based oligomer chains. Accurate mass assignments of MALDI peaks demonstrated that end-groups of this PEG sample are the α-tridecyl-$\overline{\omega}$-hydroxy (C_{13}) and the α-pentadecyl-$\overline{\omega}$-hydroxy (C_{15}) alcohols, respectively.[176]

The hydroxy end-group functionalization of some PEG samples, carried out by reaction with ethyl iodide, has been investigated by MALDI-TOF analysis.[208] The decreasing amount of unsubstituted polymer and the formation of the expected end-groups was determined as a function of the reaction time. It was also observed that the ionization probabilities of functionalized and unfunctionalized polymers are different.[161]

MALDI-TOF MS has been useful for the identification of the end-chain of two heterotelechelic PEG spacers, containing a carboxyl group on one side and an allylic or hydroxyl group on the other terminus.[211] These polymers were synthesized by appropriate hydroxy end-group modification of PEG diols.

The high mass resolution that can be obtained by delayed extraction MALDI-TOF has permitted the characterization of two poly(ethylene glycol) derivatives of pharmaceutical significance, PEG-bis(ephedrine) and PEG-bis(acetaminophenone).[84] MALDI analysis confirmed that these polymers have ephedrine or acetaminophenone linked at both ends. End-group characterization, by MALDI-TOF, of derivatized and underivatized samples of octylphenol-teminated PEG revealed that the addition of carboxylic ends substantially enhances the detectability of these ethoxylated polymers.[212]

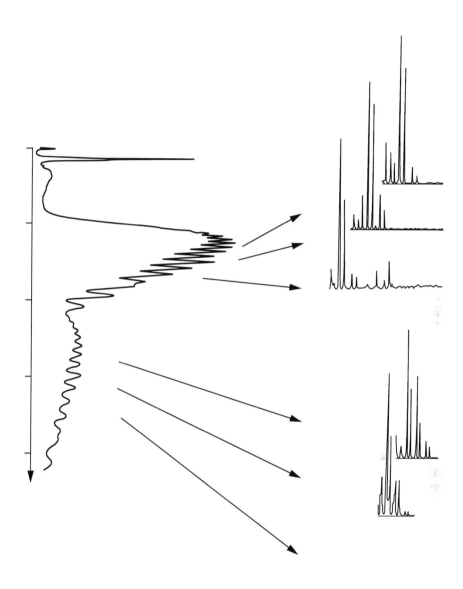

FIGURE 10.22

Separation of a technical polyethylene oxide (PEG) by liquid chromatography at the critical point of adsorption and analysis of fraction by MALDI-MS. Peak assignment indicates degree of polymerization (n). Column: Nucleosil 100 RP-18 (126 × 4 mm I.D.); eluent: acetonitrile-water (70:30 v/v). (Reprinted with permission from Ref. 176)

Some fluorescence-labeled poly(ethylene glycol) samples containing pyrene units at the chains ends, were also characterized.[213] These polymers are particularly useful for studying end-to-end cyclization kinetics.[213]

End-group characterization of PEG samples has also been performed by MALDI-FT-MS.[91,92] With respect to MALDI-TOF data, a higher resolution (30000 at 1000 Da and 6000 at 4300 Da) and higher mass accuracy (better than 15 ppm) were obtained for PEG 1000 and PEG 4000, respectively.[91,92]

Several papers report the use of MALDI-TOF for the end-group determination of PMMA synthesized by different methods: free-radical polymerization,[150,152,214] screened anionic polymerization (SAP),[215] group transfer polymerization (GTP),[195,215] catalytic chain polymerization (CCP), and emulsion polymerization.[216] All spectra consist of an envelope of different homologues, each separated by approximately 100 mass units, the repeat unit in PMMA.

The GTP, performed in the presence of trimethylsilyldimethylketene acetal as initiator in conjunction with tetrabutyl ammonium acetate catalyst, showed the formation of cyclic oligomers in addition to the expected linear oligomers.[195,215] In contrast, PMMA prepared by radical polymerization does not have this functional heterogeneity.[149,150] MALDI-TOF spectra of low molar mass PMMA, obtained by the polymerization of MMA using zirconocene as initiator, reveal that the process is not a living polymerization and that a back-biting cyclization process is involved.[217]

The mode of termination for thermally initiated free radical polymerization of styrene and MMA with AIBN as initiator has been evaluated by MALDI-TOF, and for low molar mass samples excellent agreement between MALDI-MS and SEC data was obtained.[150]

Termination can occur via either disproportionation or combination, resulting in chains with one or two initiator fragments, respectively. The ratio of the termination modes, determined for PMMA and PS, was in excellent agreement with literature data.[150]

In the case of PS, even subtle properties of the reaction could be revealed, e.g., a Diels-Alder rearrangement product due to thermal initiation as well as a chain scission product due to the MALDI conditions.[150]

MALDI end-group analysis was investigated for the determination of the chain transfer coefficients (Cs) for free-radical polymerization of PMMA with AIBN as initiator and 2-methyl-2-propanethiol as a chain transfer agent in toluene.[152] However, the values of Cs obtained were not consistent with those obtained with standard methods, and the relative intensities of the peak series relative to the oligomers with different end-groups were found to be dependent on the selection of cation (Li$^+$ or Na$^+$).[152]

The ester exchange reactivity of ester end-groups with respect to the backbone esters groups in a PMMA sample, prepared by free-radical polymerization of MMA in the presence of mercaptans as chain transfer agent, was monitored by MALDI-TOF.[214] The results indicate that a mixture was obtained containing a high amount of mono ester-exchanged PMMA and a low amount of both di ester-exchanged and unreacted PMMA.[214]

Delayed extraction MALDI-TOF has been employed to identify unusual end-groups in several PMMA samples with average molar masses between approximately 3500 and 12500 Da.[86] DE-MALDI-TOF spectra of these polymers, acquired using dithranol (DITH) as a matrix, 1,1,1,3,3,3-hexafluoro-iso-propanol (HFIP) as a solvent, and sodium iodide as cationizing agent, are shown in Figure 10.23. As a consequence of the improved signal-to-noise ratio and the good mass accuracy obtained, the different end-groups (Figure 10.23) of the PMMA samples analyzed were characterized and their structures were also confirmed by NMR analysis.[86]

The spectrum of the PMMA sample in Figure 10.23a presents a series of peaks corresponding to the sodiated ions of oligomers with structure 1. The expanded regions of spectra in Figure 10.23b,c, d, reveal two series of low intensity peaks that can be assigned to cyclic PMMA chains and to linear oligomers with hydrogen end-groups, respectively (structures 5 and 6).[86]

The DE-MALDI-TOF spectrum (Figure 10.23e) of a PMMA sample of higher molar mass (M_n = 7900, M_w = 8680), shows two intense series of peaks that have been assigned to PMMA chains with both hydrogen end-groups (species 7) and either a hydrogen end-group or a cyclic structure (species 8).[86]

MALDI-TOF analysis of a PMMA sample synthesized by anionic polymerization of MMA, initiated with tert-butyl lithium (tBuLi) in presence of aluminum alkyls at –78°C in toluene, yielded interesting information.[218] The MALDI spectra showed a series of peaks corresponding to the expected PMMA macromolecular chains with tBu/H end-groups, and another series corresponding to PMMA oligomers with the same end-groups and one tert-butyl isoprenyl ketone (tBVK) unit incorporated in the polymer chains.[218] The tBVK is supposed to be produced by the attack of the initiator on the carbonyl group of the monomer, in the initial step of the polymerization. A molar ratio Al/Li > 2 was necessary to avoid the formation of this unit.[218]

Free-radical polymerization in the presence of functional alkyl mercaptans yields a variety of semitelechelic (ST) polymers with different functional groups, and MALDI-TOF has been used to characterize the end-chains of ST-poly[N-(2-hydroxypropyl)methacrylamide] (ST-PHPMA) prepared using AIBN as initiator.[219] The analysis revealed that the end-groups of each polymer reflected the structure of the respective alkyl mercaptan used.[219]

Also in the case of relatively low molar mass polystyrene (PS), the mass-resolved molecular ions in the MALDI-TOF spectra have provided information concerning the end-groups present.[9,38,150,180,220–222]

The end-groups of two functionalized PS polymers, one terminated with a dimethylphenylsilyl [–Si–(CH$_3$)$_2$C$_6$H$_5$] end-group (PS-135), and the other with perfluoalkylsilyl [–Si(CH$_3$)$_2$(CH$_2$)$_2$C$_6$F$_{13}$] (PS-405) have been characterized by MALDI-TOF MS.[180] In this study, the effect of end-groups on desorption, ionization, and detection probabilities in MALDI-TOF MS analysis was also investigated. In fact, the relative ion yields of the two functionalized polystyrenes (PS-135 and PS-405) were compared relative to the standard proton-terminated polymer. Reporting the intensity ratios of the functionalized to unfunctionalized oligomers at each degree of polymerization,

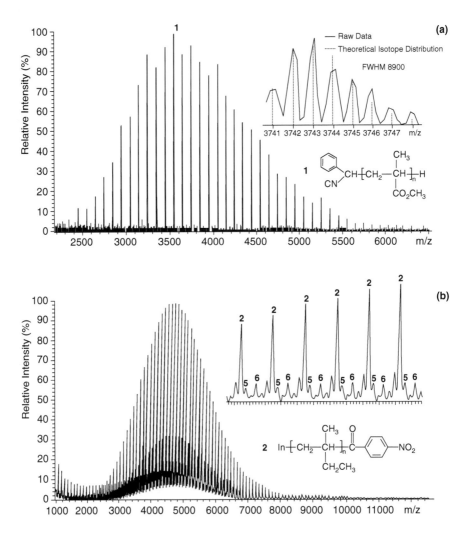

FIGURE 10.23

DE-MALDI-TOF mass spectra of: (a) PMMA E obtained in reflection mode (the inset shows the expansion of the sodium-molecular ion peak region for the 36-mer of PMMA E and the theoretical isotope pattern); (b) PMMA F obtained in linear mode (the inset shows the expansion of the region m/z 3575–4200); (c) PMMA G obtained in linear mode (the inset shows the expansion of the region m/z 3580–4275); (d) PMMA H obtained in linear mode (the inset shows the expansion of the region m/z 3600–4200); and (e) PMMA I obtained in linear mode (the inset shows the expansion of the region m/z 3550–4075). (Reprinted with permission from Ref. 86, Copyright 1997 John Wiley & Sons Ltd) (Continued)

the effect of end-group on ion yield could be established.[180] In fact, a value of 0.5 is expected if the ion yield of functionalized and unfunctionalized PS are the same. Agreement with this value was obtained for the oligomers with degree of polymerization $n = 8$–12, whereas a deviation was observed for

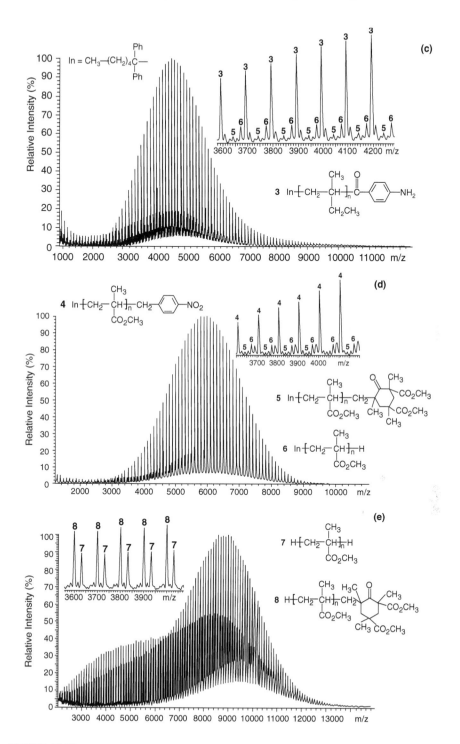

FIGURE 10.23
(*Continued*)

lower mass oligomers ($n < 8$) and for higher oligomer numbers ($n > 12$). The higher ion yield of the latter oligomer has probably resulted from improved interaction with the matrix, whereas for lower mass oligomers the end-group modification may allow it to ionize more easily than the protonated oligomers.[180]

MALDI-TOF spectra were obtained of carboxy-terminated oligostyrene samples, prepared by radical polymerization using a dicarboxylated initiator and efficiently fractionated by HPLC. The data show that in addition to the carboxy end-groups, further end-groups are formed due to contamination of the initiator or else to side reactions during the polymerization process.[199]

MALDI-TOF of PS samples prepared by TEMPO-mediated (TEMPO: 2,2,6,6-tetramethylpiperidine-N-oxyl) living free radical polymerization, have been performed using dithranol/silver trifluoroacetate or DHB as matrix.[222] The TEMPO-capped PS chains were observed only using the DHB matrix. With dithranol/silver trifluoroacetate these polymers undergo gas phase fragmentation during the MALDI analysis, and polystyrene chains with a methylene end-group were detected.[222] This fragmentation, only minor for conventionally prepared PS polymers, is enhanced when the chains contain TEMPO-based alkoxyamine end-groups.

The rupture of PS chains under MALDI conditions, observed in some studies, was hypothesized to occur near the chain ends, and causes some error in the MM determination.[150,222]

Oligomers and chemical composition of a series of aliphatic and aromatic polyesters have been characterized by MALDI-TOF mass spectrometry.[103,108,136,176,223–230] The analysis of several aliphatic polyesters revealed asymmetric oligomer distributions, hetero-terminated linear chains, and cyclic oligomers.[229] In order to obtain the structural identification of higher molar mass species or of insoluble portions of the starting polymer, an acidolysis was performed by following the progress of the reaction through the MALDI spectra.[229] The MALDI ionization efficiency appeared to be higher for carboxyl-terminated oligomers, causing a large disparity in the percentage of hydroxyl and carboxyl end-groups detected in the aliphatic polyesters.[229] However, a study on the MALDI-TOF spectra of some low-mass polyesters with hydroxyl and ester end-groups reported that these end functionalities do not influence the desorption/ionization process.[227]

Pertinent structure and end-groups in poly(ethyleneterephthalate) (PET) samples originated by different synthetic methods were determined by MALDI-TOF.[223] In this work, cyclic PET oligomers and a low amount of cyclic oligomers having a unit of diethylene glycol in the chain were revealed in the spectra of the oligomer mixtures extracted from technical PET. The presence of linear PET oligomers with methyl ester as end-groups, which happen to be isobaric with the cyclics, was excluded by a parallel NMR investigation.[223]

The MALDI-TOF spectrum of opportunely synthesized oligo(ethylene terephthalate diol)s (PET-OH) exhibits an intense series of peaks due to sodiated linear oligomers with two hydroxyl end-groups. The spectrum also shows some minor signals that are assigned to cyclic oligomers, terephthalic acid-terminated oligomers, and to oligomers containing impurities of diethylene glycol units.[223]

Polyester species terminated with hydroxyl or carboxyl groups were characterized in the MALDI-TOF analysis of polymers synthesized by enzymatic transesterification of adipic acid esters with 1,4 butanediol.[136]

The positive ions MALDI-TOF spectrum of poly(butylene adipate diol)s, made by bulk polymerization of terephthaloyl chloride with a molar excess of 1,4 butanediol, shows only one series of peaks, extending over mass range 1000–6000 Da. These may be assigned to dihydroxyl-terminated oligomers.[103]

Carboxylic, hydroxyl, adipoyl chloride, and sodium adipate end-groups of the high molar mass poly(butylene adipate) (PBA), synthesized by poly-condensation of an equimolar mixture of adipoyl chloride and 1,4-butanediol, were characterized by MALDI-TOF. Narrow dispersed fractions with low molar masses (<6000 Da) were collected by SEC fractionation, prior to MALDI analysis.[108]

Linear macromolecules having two hydroxyl end-groups were exclusively detected in polyesters obtained by bulk polymerization of spiro compounds such as O,O'-phthalid-3-ylidenecatechol and 4-methyl-O,O'-phthalid-3-ylidenecatechol.[224]

MALDI-TOF analysis showed that imidazoles used as initiators in the strictly alternating copolymerization of phenylglycidyl ether (PGE) and phthalic anhydride (PA) remain chemically bound to the polyester chains during the whole reaction.[230]

Positive ion MALDI-TOF mass spectrometry of macromolecules prepared by ring-opening polymerization of ε-caprolactone with tin(II) octoate [Sn(Oct)$_2$], in the presence of R'OH (water or alcohol) as co-initiator, revealed that there is a sizeable population of macromolecules having Sn atoms in the chains, either linear and/or cyclic ones:

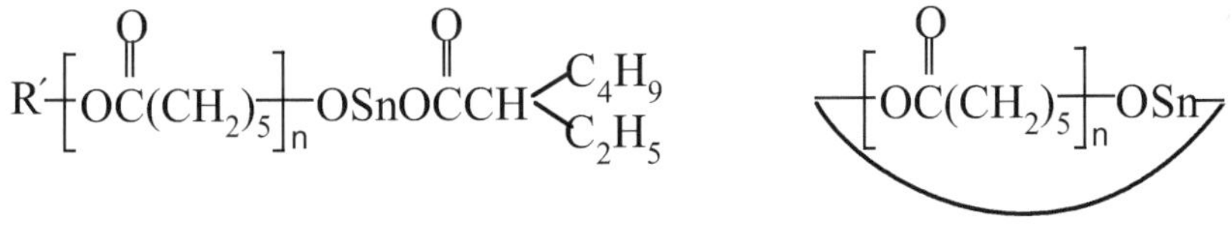

According to the kinetics, polymerization proceeds by the "active chain end" mechanism, i.e., adding monomers to the Sn-alkoxide linkage present at the chain ends.[231] The presence of the Sn atoms in the form shown above indicates that the polymerization is proceeding with Sn-alkoxides as active species.[231]

The MALDI mass spectra of four polylactides obtained by ring-opening polymerization of lactide, initiated by aluminum alkoxide derived from a Schiff's base, show well-resolved signals that can be assigned to both even-membered and odd-membered polylactide oligomers.[142] The unexpected odd-member, on the basis of the lactide ring-opening polymerization, may be

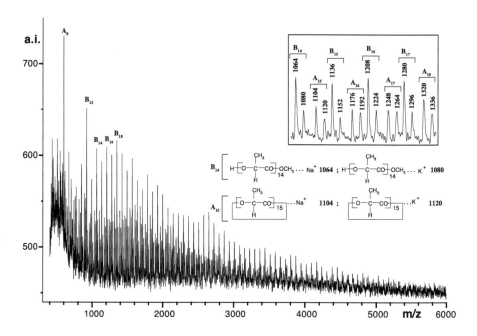

FIGURE 10.24
Positive ions MALDI-TOF mass spectrum of the low molar mass methanol soluble oligomers of a high conversion (98%) poly(D,L)lactide sample, recorded in reflection mode using HABA as a matrix. (Reprinted with permission from Ref. 142, Copyright 1996 American Chemical Society)

formed by ester exchange reactions which occur parallel to the polymerization process, causing a random cleavage of the polylactide chain.[142] Linear oligomers terminated by methoxy ester and hydroxy end-groups were observed in these polylactides. Open-chain oligomers with carboxylic end-groups were also observed; these may be formed by hydrolysis of ester groups. Cyclic oligomers were also found in the MALDI-TOF spectrum of a low molar mass fraction of a polylactide at high monomer conversion (98.0%) (Figure 10.24). This indicates that ester exchange reactions occur also in polylactide by intramolecular end-biting reactions.[142]

The end-groups present in poly(bisphenol A carbonate) (PC) have been identified by MALDI-TOF methods.[22,179,182,184] MALDI spectra of two polydisperse PC samples (PC1 and PC2), recorded in reflection mode with a mass resolution of about 3000 fwhm, show well-resolved peaks up to 16000 Da (Figure 10.25).[182] The inset expansions in Figure 10.25 show the presence of peaks belonging to six mass series, assigned in Table 10.3, which are desorbed as sodiated ions accompanied by less intense potassiated adducts.

Cyclic oligomers (species of type D, Table 10.3) appear as the most abundant species at low masses in both samples, whereas their peak intensities decrease rapidly with size. At high mass, linear oligomers with two phenylcarbonate end-groups (species A, Table 10.3) and with two p-tert-butylphenylcarbonate end-groups (species A', Table 10.3) predominate in the

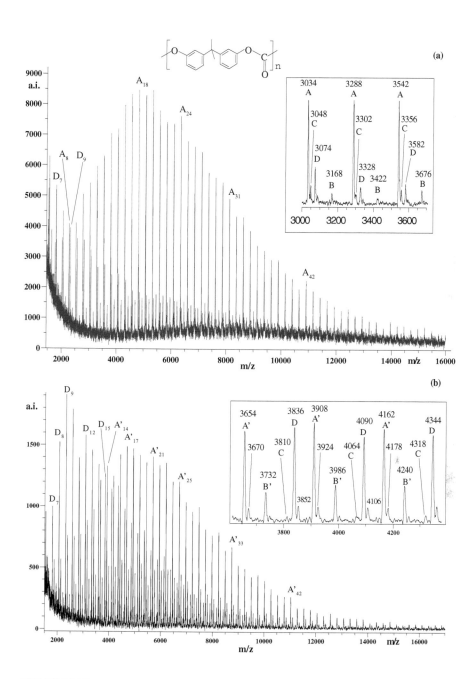

FIGURE 10.25
Positive ions MALDI-TOF MS spectra of unfractionated PC1 (a) and PC2 samples (b).

TABLE 10.3

Structural Assignments of the Peaks Appearing in the Inset Sections of the MALDI Spectra of Polydisperse PC1 and PC2 Samples Reported in Figure 10.25(a),(b)

Mass Series	Oligomers Structures	n	M Na⁺	M K⁺
A	(oligomer structure)	11	3034	
		12	3288	
		13	3542	
A′	(oligomer structure)	13	3654	3670
		14	3908	3924
		15	4162	4178
B	(oligomer structure)	12	3168	
		13	3422	
		14	3676	
B′	(oligomer structure)	14	3732	3748
		15	3986	4002
		16	4240	4256
C	(oligomer structure)	11	3048	
		12	3302	
		13	3556	
		14	3810	
		15	4064	
		16	4318	
D	(oligomer structure)	12	3074	3090
		13	3328	3344
		14	3582	3598
		15	3836	3852
		16	4090	4106
		17	4344	4350

spectra of PC1 and PC2 samples. Oligomers with two phenol groups (species C, Table 10.3) appear with very low intensity.[182]

The chemical composition of these PC samples was studied by MALDI-TOF analysis of the narrow distribution fractions obtained by SEC fractionation. Remarkably, MALDI spectra of the SEC fractions at high MM collected at lower elution volume display well-resolved oligomer peaks up to 70000 Da. The mass of these peaks confirms that they are mainly linear PC chains corresponding to species A and A′ (Table 10.3).[182]

MALDI spectra of the SEC fractions reported in Figure 10.26 are dominated by sodiated-ion peaks corresponding to linear poly(bisphenol A carbonate) species A (Table 10.3).[182] In contrast, MALDI spectra of lower MM SEC

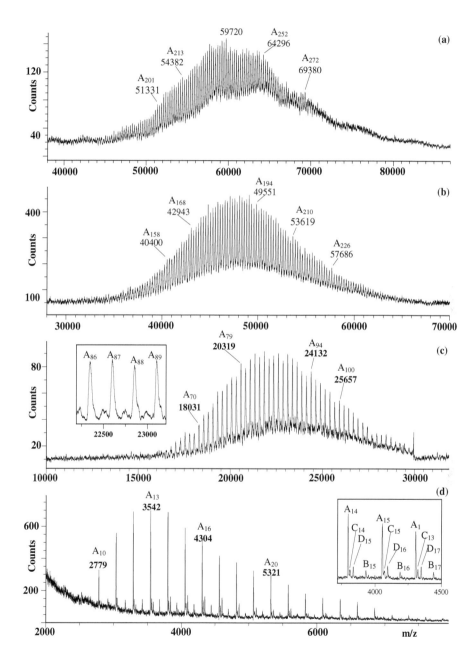

FIGURE 10.26
Positive ions MALDI-TOF mass spectra of the SEC fractions from sample PC1 collected at: (a) 27.8 mL, (b) 28.54 mL, (c) 30.2 mL, and (d) 35.3 mL.

fractions showed that cyclic oligomers are eluted at later times with respect to the isobaric linear oligomers.[182]

MALDI-TOF analysis also revealed that transesterification reactions between propylene carbonate and diols containing six or more carbon atoms in the molecule, carried out in presence of a coordination catalyst, lead to almost pure oligocarbonate diols.[232] When ethylene carbonate (EtC) was used, the resultant polymer contained a different combination of ethylene carbonate and ethylene oxide units. The transesterification with EtC, carried out in the presence of catalysts, such as K_2CO_3 and CsF, leads to oligomers with high amounts of oxyethylene fragments.[232]

Polythiophenes and regioregular, head-tail poly(3-alkylthiophenes) synthesized by different methods, have been characterized by MALDI-TOF.[233,234] More than one type of end-group, depending on the synthetic method employed, were characterized.[233] These polymers appear as molecular ions (M^+) in the MALDI spectra recorded using dithranol matrix, and cationized macromolecular chains were not detected.[223]

The structure of linear polysulfides has been characterized using 9-nitroanthracene/silver trifluoroacetate as the matrix/cation system.[235]

MALDI-TOF MS was also performed to characterize some well-defined mono- and ditelechelic polyphosphazenes.[236] The MALDI spectrum of non-telechelic polymers shows one series of peaks corresponding to sodiated ions of the expected $CF_3CH_2O-[CF_3(CH_2O)2-P=N]_n-P-(OCH_2CF_3)_4$ oligomers.[236] Oligomers having either $H_2C=CHCH_2N(H)-$ and $-P-(OCH_2CF_3)_4$ end-groups and both $H_2C=CHCH_2N(H)$ end functional groups were characterized in the MALDI spectrum of di-telechelic polyphosphazene.[236]

Diydroxy telechelic polyisobutylene (OH–PIB–OH), which is of great importance for the preparation of functionalized polymers and polymer networks, was characterized by MALDI-TOF using a DITH/CF_3COOAg matrix system.[237] Despite the apolar nature of PIB, the good quality MALDI spectra can be explained by the fact that PIB chains have two hydroxyl end-groups. This permits the formation of stable $[HO–PIB–OH +Ag]^+$ ions.[237] A good agreement was also found between the calculated isotope distributions and the distributions determined by means of MALDI.[237]

Structural confirmation of the end-groups and the repeat unit of some polydienes with average molar mass of about 24,000 Da, was also obtained by DE-MALDI-TOF using all-trans-retinoic acid as the matrix and copper(II) nitrate as the cationization agent.[143] In addition to the [polymer +Cu]$^+$principal distribution, a second distribution generated by the incorporation of the copper salt with the polymer chains was also observed.[143]

10.4.1   Post Source Decay

Since MALDI is a soft ionization method, fragmentation is generally not pronounced in the mass spectra. However, under specific instrumental and experimental conditions, different types of fragments can be observed in MALDI-TOF spectra.[21]

At higher laser fluences, fragmentation might occur in the ion-source region, sometimes referred to as "prompt fragmentation."[47] In addition, in delayed extraction ion sources, fast metastable fragmentation may occur during the delay time before extraction, i.e., on the ~100–200 ns time scale, referred to as "in-source decay" (ISD).[47,211] The increment in ISD fragmentation with laser energy at a fixed delay time is caused by a greater number of collisions in the expanding plume, leading to a greater amount of collisional activation in the source.

Prompt and ISD fragmentation yield fragment ions in the mass spectra in both the linear and reflection mode. In contrast, metastable fragmentation in the field-free region, referred to as "post-source decay" (PSD), occurs on the ~10 μs time scale and yields fragment ions in the mass spectrum from the reflection detector only. With the aid of a suitable precursor ion selector in the field-free region of the linear flight path and a scanning reflectron voltage, a PSD product (or daughter) ion spectrum can be obtained.[238]

Despite the low internal energy and low fragmentation efficiency obtained, MALDI-PSD is very useful for controlled fragmentation such as required, for example, in peptide sequencing,[238–240] isomeric oligosaccharide sequencing,[241] and end-group identification in synthetic polymers. In addition, a pulsed collision gas cell (CID) can be installed in the field-free region of MALDI TOF systems, thus yielding additional fragments in polymer analysis (PSD/CID). In contrast to the large number of papers on MALDI PSD (and PSD/CID) of biopolymers,[238–241] little can be found on synthetic polymers in the MALDI-TOF literature.[242–250] It may be noted that obtaining PSD spectra is a tedious, time-consuming process.

ISD fragmentation was observed in MALDI analysis of poly(propylene glycol) (PPG) and poly(ethylene glycol) when higher laser energy (~120 μJ) was used.[47]

MALDI analysis of hyperbranched polyesteramides showed rapid dissociation of higher oligomers into lower oligomers, and PSD/CID of mass selected ions was used to discriminate between fragment ions and the isobaric cyclic oligomers.[249] The results, as well those obtained by NMR, titration, and FD-MS, indicated that simple MALDI-TOF spectra yielded wrong conclusions about the chemical composition and functionality type distribution for the hyperbranched samples.[249]

Structural information on dendritic polyurethanes was obtained by PSD experiments, which allowed the detection of fragment ions arising from the cleavage of ester or amide bonds.[250]

Studies concerning the structural determination of some synthetic homopolymers by MALDI, combined with post-source decay or collisional-induced dissociation fragment ion analysis, show that the masses of individual end-groups can be determined from ion peaks generated by cleavage of the polymer backbone.[242–248]

PSD-MALDI and CID-MALDI investigations of polystyrene,[244,245] poly(alkyl methacrylates)s,[243,246,247] poly(ethylene glycol),[243] and poly(ethylene terephthalate),[243] have been carried using MALDI hybrid sector-orthogonal acceleration

TOF instruments. Mechanisms are proposed for the formation of the ion peak series observed in the spectra, and the masses of the end-groups of these polymers can be inferred from the data.[243-247] The fragment ions spectra obtained for PMMA and PS polymers are dependent on the mass-to-charge ratio of the precursor ion, the mechanism of cation attachment, the nature of the cation, and the mobilty of the polymer chain.[247]

Copper (Cu^{++}) and Silver (Ag^+) ions are commonly used for the MALDI-CID characterization of PS samples, since the corresponding clusters give greater sensitivity.[245,247] In the MALDI-CID spectra of poly(alkyl methacrylates),[246,247] and polystyrene,[247] distributions of fragment ions were found to be consistent with those originating from precursor ions.

PSD MALDI was used in order to differentiate between linear and branched PEG.[251] Linear PEG oligomers showed a fragmentation process occurring by cleavage on both sides of oxygen atoms, whereas branched PEG showed a dominant cleavage of ethylene oxide.[251]

MALDI-CID spectra of PEGs symmetrically terminated with butanoyl, benzoyl, and acetyl groups, have been performed using the novel hybrid magnetic-sector/time-of-flight instrumentation (MAG-TOF).[248] The authors reported a fragmentation mechanism of the alkali metal PEG adduct, which is in accord either with that proposed on the basis of the MALDI-CID spectra of PEG obtained with the hybrid sector-orthogonal acceleration TOF instrument,[171] or with that proposed on the basis of LSIMS mass spectrometry.[252] PSD fragmentation patterns were also reported for mass selected oligomers ion of polyisobutylene.[252]

Linear and cyclic structures of polycarbonates from bisphenol-A, from bisphenol-Z, and from 4,4'-dihydroxydiphenyl-3,3-pentane, were determined by PSD MALDI-TOF experiments.[242] It was found that the fragmentation behavior of these polycarbonates depends on the substituents bound to the central carbon atom of the bisphenol unit. The cleavage of the polymer backbone is suppressed, and the spectra are dominated by the loss of side groups. PSD MALDI spectra of these polycarbonates could be acquired only from Li adduct ions, because the sodium or potassium cationized adduct ions yield only dissociation of the cation from the polymer.[242]

PSD MALDI spectrum of the lithiated trimer (*m/z* 1096.2) from poly(bisphenol-A carbonate) (PC) end-capped with 4-tert-butylphenol at both ends, is shown in Figure 10.27. It shows one series of fragment ion peaks (*m/z* 904.3, 649.6, 395.5, and 141.1) that corresponds to the oligomer containing only one end-group, while retaining the lithium cation. The mechanisms hypothesized for the fragmentation of the PC chains are reported in Scheme 10.1. Fragmentation pathway A had been previously proposed to explain fragmentation of PC by electron impact and by SIMS measurements.[242]

PSD MALDI of a PC sample containing a mixture of differently terminated chains was also possible by analysis of the SEC fractions containing PC oligomers below 2000 Da. The intensity of the signals was high enough to allow the acquisition of a PSD spectrum of selected ions present in each SEC fraction.[242]

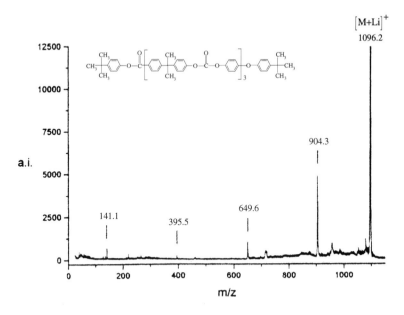

FIGURE 10.27
PSD-MALDI-TOF mass spectrum of the trimer from poly(bisphenol-A carbonate) (PC). (Reprinted with permission from Ref. 242)

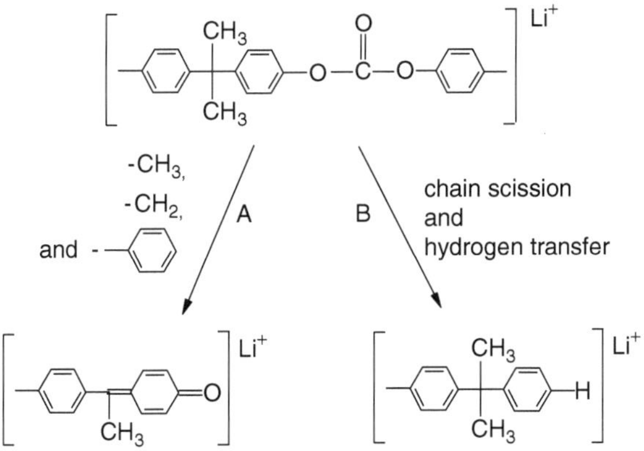

SCHEME 10.1
Possible mechanism for the fragmentation of poly(bisphenol A carbonate).

10.5   Copolymer Characterization

MALDI can be used for the analysis of copolymers.[145,155–167,254–258] The methodology for measuring MMD and MM averages is the same as for homopolymers. The mass numbers can be assigned to chemical structures, and endgroups may be identified along with the number of different repeat units.

As discussed in Chapter 2, mass spectra of copolymers are usually more complex as compared to homopolymers. Furthermore, the number of peaks increases with chain size. In the case of two different monomers (an AB copolymer), the number of peaks increases linearly, whereas in the case of three different monomers (an ABC terpolymer), the number increases quadratically with chain size. As a consequence, a higher mass resolution is required (as compared to homopolymers) in order to detect as separate peaks the signals due to the individual oligomers in the copolymer.

For instance, chains of PMMA of different length possess masses which are 100 Daltons apart and this implies that, in order to record a mass-resolved spectrum of a PMMA sample at 10 kDa, the resolution must be higher than 100. In the case of copolymers, peaks due to the different co-oligomers may appear very close to each other, as in the MMA/St copolymers (with the two repeat units differing by 4 mass units). As a consequence, in order to obtain a mass-resolved spectrum of a St/MMA copolymer at 10 kDa the resolution must be higher than 2500.

When the determination of the composition is not required, the assignment of mass spectral peaks can be directly carried out to establish what kind of oligomers is present. In a straightforward application, Vitalini et al.[256] recorded the negative ion MALDI-TOF mass spectrum of a copolymer containing units of bisphenol (referred to as A) and units of a derivatized fullerene (referred to as B). The mass spectrum (see Figure 10.28) covers a wide mass range (1600–14000 Da). The peaks are due to cyclic ions of the type A_nB where n extends from 1 to 15.

Yoshida et al.[258] recorded the MALDI-TOF spectra of a copolymer containing units of a substituted siloxane (referred to as A) and of dimethylsiloxane (referred to as B). The molar fraction of A in the copolymer (c_A) is known to be 0.2. The inspection of the assignments revealed that the number of A units varied in the range $n = 0$–2, whereas the number of B units varied in the range $n = 9$–24.

Nielen et al.[179] recorded the MALDI mass spectrum of a copolymer containing units of methacrylic acid (A) and methylmethacrylate (B). They observed peaks of the type B_n (pure MMA homosequences) and peaks of the type AB_{n-1} (with various types of end-groups). Peaks with two or more A units are absent, and this is consistent with the fact that the copolymer is known to possess only 16 mole % of A.

For block copolymers, the assignment of the MALDI peaks provides useful structure information. A_mB_n means a block of "m" consecutive units of A followed by a block of "n" consecutive units of B. Francke et al.[160] used a monomer with a triple bond to synthetize a poly(para-phenylene

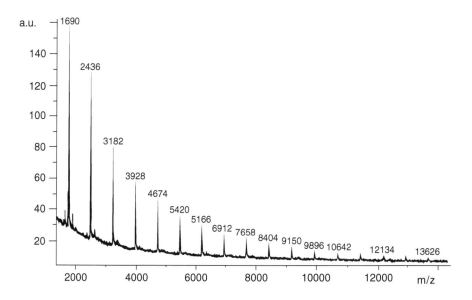

FIGURE 10.28

MALDI-TOF mass spectrum of a copolymer containing units of BPA-CH2 and units of a derivatized fullerene. (Reprinted with permission from Ref. 256, Copyright 1996 American Chemical Society)

ethylene)-block-poly(ethylene oxide) block copolymer. The MALDI-TOF spectrum of the copolymer display peaks due to ions corresponding to $A_m B_n$, where n takes the values 4,5,6 and m ranges from 10 to 20.

Mormann et al.[158] recorded the MALDI mass spectrum of a copolymer containing methylmethacrylate (MMA) and phenyl acrylate (PA). The copolymer spectrum displays peaks due to ions containing a relatively large number of MMA units (in the range $n = 15–53$), and only one, two, and three PA units. This result is incompatible with an exactly alternating structure, initially hypothesized.

Schriemer et al.[85] recorded the MALDI spectrum of a copolymer containing propylene oxide (A) and ethylene oxide (B). The most intense peaks in the mass spectrum are A_2B_{39}, A_2B_{40}, A_2B_{41}, A_2B_{42}, A_2B_{43}, and A_2B_{44}, implying that the copolymer is very rich in B units.

10.5.1 Determination of Copolymer Composition and Sequence

As discussed in Chapter 2, mass spectral peak intensities can be used to determine copolymer composition and sequence and various methods have been developed for this purpose. Montaudo et al.[163,259] investigated two methods suitable of performing the calculation.

The first method is the chain statistics approach, which compares the observed mass spectral intensities with the intensities derived from a specific statistical model for the sequence distribution. The second method is the direct method, which does not make use of statistics and employs instead a combination of mass spectral intensities.[163,259]

Wilczek-Vera et al.[155] applied the direct method to the MALDI-TOF mass spectrum of a block copolymer containing units of α-methylstyrene (A) and units of styrene (B). The resulting molar fraction of A units in the copolymer (c_A) was $c_A = 0.29$, which is consistent with the value $c_A = 0.31$ derived from NMR. The direct method was used by the authors to compute $\langle n_A \rangle$ and $\langle n_B \rangle$ (the number-average length of like monomers) and the result was $\langle n_A \rangle = 20.4$ and $\langle n_B \rangle = 7.3$. This implies that styrene blocks are much longer than α-methylstyrene blocks.

Raeder et al.[260] analyzed the MALDI-TOF mass spectrum of a copolymer containing units of styrene (A) and styrenesulfonic acid (B). They measured the abundances of the MALDI peaks and inserted the intensities into the key equation of the direct approach (see Chapter. 2). They found that $c_A = 0.06$, which implies that the average degree of sulfonation is 94%.

In Figure 10.29 another spectrum is reported, namely the MALDI-TOF spectrum of a copolymer containing units of α-methylstyrene (A) and units of methylmethacrylate (B).[130] The sample is a reference material denoted SRM1487, and it was obtained by anionic polymerization using a bifunctional initiator, namely the dimer of α-methylstyrene (sodium salt). The most intense peaks are between

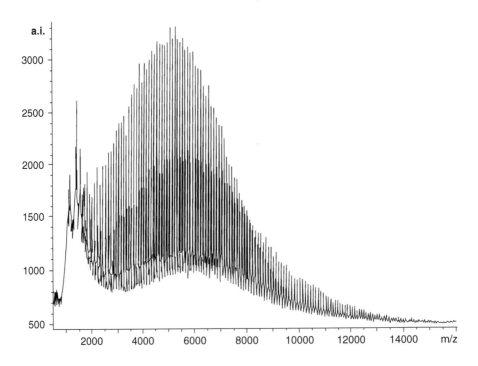

FIGURE 10.29
MALDI-TOF mass spectrum of a copolymer containing units of alfamethylstyrene and methylmethacrylate. (Reprinted with permission from Ref. 130, Copyright 1997 John Wiley & Sons Ltd)

4000 and 6000 Da. The copolymer is rich in A units, and peaks in the region 10000–14000 Da are well resolved. The MALDI intensities were used by the authors to determine the average number of A units in the chain, $\lambda(s)$, using the direct approach. The authors displayed the result of the calculation of $\lambda(s)$ along with the average number of A units in the chain, determined by a two detector SEC device. Excellent agreement was found.[130]

Chain statistics makes use of model distributions, and the most popular models are the Bernoulli model and the first-order Markoff model (see Chapter 2). The agreement factor (AF) is sometimes computed (see Chapter 2).

Figure 10.30a reports the first MALDI-TOF spectrum of a copolymer containing units of β-hydroxybutyrate (B) and β-hydroxyvalerate (V).[165] The copolymer

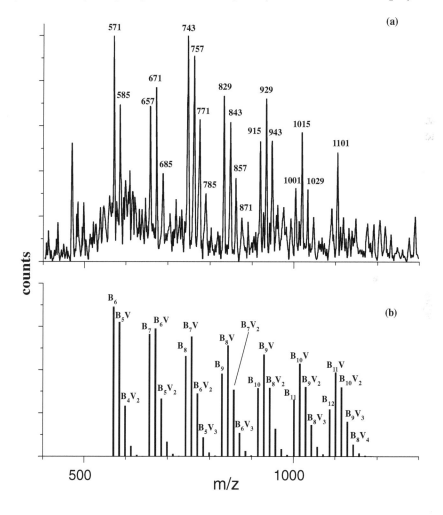

FIGURE 10.30
Positive-ion MALDI-TOF mass spectra of a copolymer containing units of hydroxybutirate and units of hydroxyvalerate. (a) Experimental mass spectrum and (b) calculated mass spectrum.

was obtained by partial methanolysis of a high molar mass copolyester (M_w = 750,000), and peaks are due to ions of the type HO–CH_2R–(OCH_2R)n–OCH_3...Na^+, where "n" is the degree of polymerization and R is a methyl or an ethyl group. Chain statistics was applied, and experimental MALDI peak intensities were given as an input to the MACO computer program (see Chapter 2).[165] The result was c_V = 0.13 (c_V is the molar fraction of V units), which is in excellent agreement with composition data obtained by other techniques. The agreement factor turned out to be AF = 0.11. Figure 10.30b reports the calculated mass spectrum, generated using the MACO program.[165]

Suddaby et al[157] recorded the MALDI-TOF mass spectrum of a series of five copolymers containing units of methyl methacrylate (referred to as MMA) and units of butyl methacrylate (referred to as BMA). They applied the direct method to plot the variation of the copolymer composition as the length of the chain increases. Although the method can detect subtle compositional heterogeneities, the five copolymers turned out to possess an extraordinarily homogeneous composition. Thereafter chain statistics was applied, and this yielded an estimate of the reactivity ratios (see Chapter 2). The calculations gave r_{MMA} = 1.09, r_{BMA} = 0.77, that differ from the literature values and also from the reactivity ratios determined by ^{1}H-NMR (r_{MMA} = 0.75, r_{BMA} = 0.98).[157] The authors also found a bias in the MALDI spectra toward chains rich in MMA, and concluded that MMA units favor ionization over the more hydrophobic BMA units, under the MALDI conditions used.[157]

Servaty et al. applied chain statistics to the MALDI-TOF mass spectrum of a copolymer sample containing units of hydromethylsiloxane (A) and units of dimethylsiloxane (B). The authors generated theoretical MALDI intensities using chain statistics.[162] They also determined the weight fraction of the chains that possess one functionalizable unit (hydromethylsiloxane unit), two functionalizable units, three functionalizable units, and so on. They found that the molar fraction of the chains without a functionalizable unit is by no means negligible (it accounts for 25% of the chains).

Figure 10.31 reports the MALDI-TOF spectrum of a copolymer containing units of butylene adipate (A) and units of butylene sebacate (B).[112] The synthetic route makes use of butandiol and the dimethyl esters of adipic and sebacic acids. The resulting copolymer was injected into the SEC apparatus, and the fraction eluting at 27 mL was collected. The mass spectral peaks are due to ions of the type: H_3CO–[CO(CH_2)$_4$COO(CH_2)$_4$O]$_m$–[CO(CH_2)$_8$COO(CH_2)$_4$O]$_n$–OCH_3...$Li+$, where *m* and *n* are the number of A and B units. The inset reports an expansion of a region of the spectrum. The peaks in the inset are due to chains with 11, 12, and 13 repeat units, and the base peak in the spectrum is A_6B_6. Chain statistics was applied and the result was c_A = 0.46 (c_A is the molar fraction of adipate units), which compares well with c_A = 0.47 obtained from NMR analysis.[112] The agreement factor turned out to be AF = 0.09, which is satisfactory. Table 10.4 reports the experimental and the calculated mass spectral intensities generated using the MACO program (see Chapter 2).

The MALDI-TOF mass spectrum of a copolymer containing units of butylene adipate and units of butylene succinate was recorded.[108] Contrary to

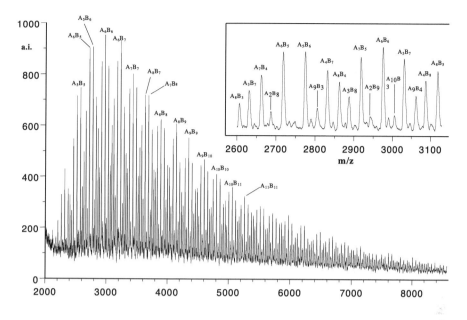

FIGURE 10.31

Positive-ion MALDI-TOF mass spectrum of a copolymer containing butylene adipate (A) and butylene sebacate (B) units.

the previous cases, the peaks in the spectrum are due to two types of ions (two mass series are present), and this complicates the peaks assignments. Despite this additional complexity, chain statistics was applied, and a reasonable agreement between the experimental and the calculated mass spectral intensities was found.[108]

Sometimes an extremely accurate determination of the sequence distribution of a copolymer sample is required. In this case, the method based on chain statistics should be preferred over the direct method. As an example, Montaudo[167] recorded the MALDI-TOF mass spectrum of a copolymer containing units of lactic acid residue (L) and units of glycine (G) obtained by ring-opening copolymerization of dilactide and 6-methyl-2,5-morpholinedione (the cyclic glycine-lactic acid dimer). The synthetic scheme yields a quasi-random copolymer with a peculiar sequence distribution, since it deviates from the standard one in which the two repeat units (L and G) are found at random along the copolymer chain. The deviation is due to chains having sequences containing two consecutive G units, which are not produced. This deviation is small and is therefore difficult to detect. Figure 10.32 reports a portion of the MALDI-TOF spectrum of the copolymer in the region 550–800. The peak at 588 is the most intense, and it is the sum of two contributions, namely $H-L_5G_3-OH$ K^+ and $H-L_5$ $(L')G_2-OH$ Na^+, where species L' possesses a ^{13}C atom. Analogously, the peak at m/z 603 is the sum of $H-L_6G_2-OH$ K^+ and $H-L_6(L')G_1-OH$ Na^+. Peaks at 617, 645, 660, 675, 690, 702, 717, 732, 474, 762 also have a double assignment.[167]

TABLE 10.4

Experimental and Calculated Relative Amounts of PBA/PBSe Lithiated Oligomers[a] Observed in the MALDI Spectrum Reported in Figure 10.31

Oligomers[b]	m/z[c]	$I_{exp.}$[d]	$I_{calc.}$[e]
10-mers			
A_7B_3	2407	290	398
A_6B_4	2463	570	653
A_5B_5	2519	780	766
A_4B_6	2575	810	642
A_3B_7	2631	500	376
A_2B_8	2687	240	147
AB_9	2743	50	34
B_{10}	2799	10	3
11-mers			
A_8B_3	2607	330	377
A_7B_4	2663	710	709
A_6B_5	2719	1010	970
A_5B_6	2775	1020	975
A_4B_7	2831	770	715
A_3B_8	2887	430	372
A_2B_9	2943	160	131
AB_{10}	2999	40	27
12-mers			
A_9B_3	2807	280	273
A_8B_4	2863	620	566
A_7B_5	2919	940	901
A_6B_6	2975	1090	1057
A_5B_7	3031	930	930
A_4B_8	3087	620	606
A_3B_9	3143	280	284
A_2B_{10}	3199	100	91
13-mers			
$A_{10}B_3$	3007	180	167
A_9B_4	3063	410	393
A_8B_5	3119	740	692
A_7B_6	3175	960	927
A_6B_7	3231	1010	952
A_5B_8	3287	720	744
A_4B_9	3343	410	436
A_3B_{10}	3399	160	186
A_2B_{11}	3455	40	54

AF = 9%[f]

[a] The lithiated molecular ions of the oligomers observed correspond to the following structure:
$H_3CO-[-CO(CH_2)_4COO(CH_2)_4O-]_m -[-CO(CH_2)_8COO(CH_2)_4O-]_n-OCH_3...Li^+$.

[b] A = butylene adipate, B = butylene sebacate.

[c] Observed m/z values after calibration procedure.

[d] Intensity of the MLi^+ ions in the MALDI-TOF spectrum.

[e] Intensities calculated using MACO4 program[166].

[f] AF = agreement factor between experimental and calculated MALDI-TOF spectrum.

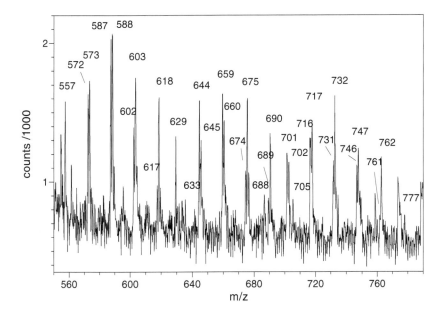

FIGURE 10.32
Expansion of the positive ion MALDI-TOF mass spectrum of a copolymer containing units of glycine and units of lactic, in the region of octamers, nonamers, and decamers.

Despite these difficulties, chain statistics was applied successfully. The MACO program evaluated the probability of finding two consecutive glycine units (GG), and it yielded a vanishingly low value, which is correct since the synthetic method prevents GG sequences. The result for copolymer composition was $c_G = 0.23$ (c_G is the molar fraction of G units), which compares well with $c_G = 0.24$, obtained from NMR analysis.[167]

Montaudo and Samperi[163] recorded the MALDI-TOF mass spectrum of a copolymer containing units of butylene adipate and units of butylene terephthalate. Chain statistics was applied and a reasonable agreement between the experimental and the calculated mass spectral intensities was found.

The MALDI-TOF mass spectrum of a copolymer containing units of butylene succinate, butylene adipate, and butylene sebacate has been recorded, and chain statistics was applied with success.[164]

10.6 Molecular Association

Applications of MALDI have been extended by exploring some peculiar aspects of the MALDI-TOF response to the phenomenon of molecular association in poly(bisphenol-A carbonate) (PC) samples.[183] Chain self-association was observed when a 10 mg/ml solution was injected onto SEC columns, using chloroform ($CHCl_3$) or tetrahydrofuran (THF) as eluent. The presence of association in PC was revealed by the difficulty of obtaining SEC fractions with the usually narrow MMD.[183] In fact SEC fractionation produced fractions containing PC chains of different size, and MALDI analysis showed that PC chains terminated with hydroxyl (OH) groups undergo self-association by hydrogen bonding. Chain association was found to produce aggregates with high hydrodynamic volume that were eluted through SEC columns at the same volume as that for higher molar mass chains.

The chain aggregates were broken when the SEC fractions, containing a heterogeneous mixture of PC chains of different size, were diluted in the HABA or IAA matrix used for the MALDI sample preparation. These matrices contain the carboxylic acid units, which are able to break the hydrogen bonds responsible for the formation of the chain aggregates.

The MALDI spectra of the PC fractions (Figure 10.33) show bimodal distributions of peaks, one at the expected high masses (in agreement with SEC calibration line measured in the absence of self-association), and the other at very low masses. The low mass peaks are due to PC chains terminated with OH groups (species B and C in Table 10.3), whereas the ions at high mass corresponding to PC chains capped at both ends (species A in Table 10.3).

Figures 10.33(a)-(d) report the MALDI spectra of PC fractions collected at the same elution volume (31.3 ml) in four different SEC runs. As shown in the spectra, a higher sample dilution, or the addition of a polar solvent such as ethanol to the $CHCl_3$ used as eluent in SEC analysis, is able to suppress self-association in PC samples. Figure 10.33a shows the MALDI spectrum of the fraction obtained by injecting a PC sample containing 850 ppm of OH end-groups, onto the SEC columns at a concentration of 2.5 mg/ml. Since the spectrum shows a narrow distribution of peaks centered at 15 kDa, it was concluded that this level of sample dilution (2.5 mg/ml) is able to suppress self-association.

When the same sample was injected into SEC apparatus at a sample concentration of 20.0 mg/ml, the MALDI spectrum showed a bimodal distribution of peaks (Figure 10.33b). Self-association can be suppressed (Figure 10.33c) using a $CHCl_3$/ethanol (95/5 v/v) mixture as eluent in SEC, even if the sample was injected into the SEC columns at a concentration of 20.0 mg/mL. The MALDI spectrum of an SEC fraction collected after injection of a 20 mg/mL solution of a PC sample with a reduced OH end-group concentration (100 ppm; Figure 10.33d) showed a very narrow distribution of peaks centered at about 15 kDa.[183]

The formation of aggregates of macrocyclic amphiphiles has been reported.[261] It was found that the amount of aggregates increases with increasing laser power, and it was assumed that the aggregate clusters are formed during the vaporization of the matrix/analyte molecule.[261]

In a recent investigation, the intermolecular association of a series of donor- and acceptor-substituted poly(p-phenylenevinylene)s was observed by SEC analysis, and MALDI-TOF was then used for the determination of the MM of these polymers.[262]

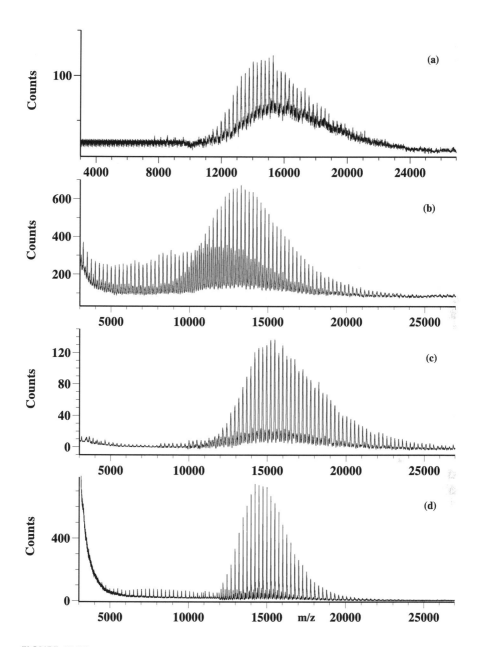

FIGURE 10.33

MALDI-TOF spectra of PC fractions collected at the same elution volume (31.3 mL) in four different SEC runs: **(a)** sample PC1 injected at a concentration of 2.5 mg/mL in $CHCl_3$; **(b)** sample PC1 injected at a concentration of 20 mg/mL in $CHCl_3$; **(c)** sample PC1 injected at a concentration of 20 mg/mL in $CHCl_3/C_2H_5OH$ 95/5 v/v; (d) sample PC3 injected at a concentration of 2.5 mg/mL in $CHCl_3$.

The relative stability constants of the crown ether complexes (18C6) with sodium, potassium, rubidium, and cesium in glycerol solution were measured by MALDI-TOF analysis.[263] The stability constants measured by MALDI-TOF were found to be in good agreement with those available in the literature, and the authors concluded that (under the conditions used) the MALDI results quantitatively described the relative concentrations of the complexes in the initial solution.[263]

10.7 Polymer Degradation

Recent studies have shown that MALDI-MS is a useful method to analyze pyrolysis products from polar polymers. MALDI has been used to examine pyrolyzates from poly(bisphenol-A carbonate),[264] segmented polyurethane,[265] PEG,[266] poly(tetrahydrofuran) (PTHF),[267] and poly(propylene glycol).[269] MALDI-MS was found to have advantage because, with respect to traditional mass spectral techniques such as GC/MS (Chapter 4) and DPMS (Chapter 5), pyrolyzates with much higher MM can be studied. Further development has recently resulted from comparing DPMS and MALDI-TOF in the study of the thermal degradation processes of polymers.[264,265]

In such experiments, one may partially degrade a polymeric sample, keeping it under inert atmosphere (e.g., nitrogen, argon) at a certain temperature, and then obtain the MALDI spectrum of the sample to observe the thermally induced changes. The spectrum will consist of a mixture of undegraded and degraded chains. The MALDI analysis may prove incomplete, since only the degradation products most thermally stable will survive the heating at atmospheric pressure. DPMS data, being taken on-line and in a continuously evacuated system that provides very short transport times from the hot zone, may complement the MALDI data by supplying information on less thermally stable pyrolysis products and on possible intermediates.[264,265]

The isothermal pyrolysis of BPA-polycarbonate (PC) was studied by heating it at a fixed temperature (between 350°C and 450°C) under a nitrogen stream, and the MALDI-TOF spectra of this thermally treated PC were analyzed.[264] This showed that a rearrangement of the carbonate group leads to the formation of xanthone units in PC chains of sizeable molar mass. The xanthones (aromatic units) are considered to be precursors of graphite-like structures in the char residue that is produced at temperatures higher than 450°C under inert atmosphere.[264] The PC pyrolysis products formed at 400°C at different heating times are observed in MALDI spectra in the mass range 2900–3100 reported in Figures 10.34a–d.[264] The MALDI data agree with those obtained by DPMS, and the structure of the PC pyrolyzates is in accord with the proposed mechanism of thermal degradation (Scheme 5.8, Chapter 5).[264] Furthermore, kinetic studies of PC isothermal pyrolysis were possible by

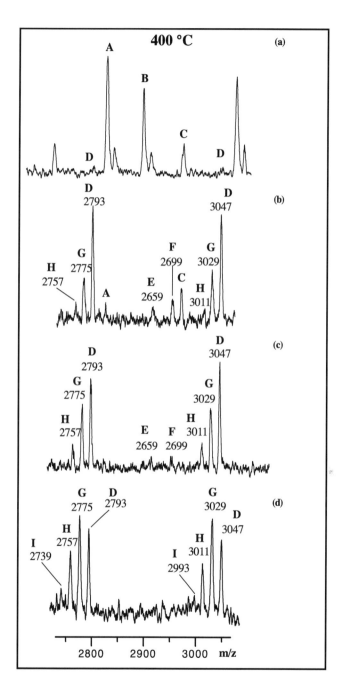

FIGURE 10.34

MALDI-TOF mass spectra of the THF soluble fraction extracted from the pyrolysis residue of PC obtained after heating at 400°C for: (a) 0 min, (b) 15 min, (c) 30 min, and (d) 60 min. (Reprinted with permissiom from Ref. 264, Copyright 1999 American Chemical Society)

monitoring the diverse molecular species present in the MALDI spectra of the partially degraded samples.[264]

The pyrolysis product of a segmented polyurethane consisting of 4,4'-methylenebis-(phenylisocyanate) (MDI), poly(butylene adipate) (PBA), and 1,4-butanediol (BDO) have been studied by MALDI-TOF.[265] Several series of oligomeric pyrolysis products were observed over the range 800–10,000 Da. Dissociation of the urethane linkage to yield products with isocyanate and hydroxyl end-groups was observed at the lowest temperatures (ca. 250°C). Linear polyester oligomers with hydroxyl and/or vinyl end-groups were detected. Cyclic polyester oligomers were also observed. At higher temperatures (>300°C) the nitrogen-containing pyrolysis products are no longer present in the residue. Dehydration of the linear and cyclic polyester pyrolyzates occurs at these temperatures, producing olefinic end-groups.[265]

The low-temperature (150–300°C) inert atmosphere degradation of PEG (2000 Da) was studied by MALDI-MS.[266] Pyrolysis ensued as low as 150°C, and the initial pyrolyzed products were found to have hydroxyl and ethyl ether end-groups formed via C-O homolytic cleavage followed by hydrogen abstraction. At higher temperatures, the abundance of the ether end-groups increases as more C-C cleavage occurs. Vinyl ether end-groups increase at higher temperatures (250–300°C), due to dehydration of hydroxyl end-groups.[266] The pyrolysis products were also characterized by chemical ionization mass spectrometry (CI-MS), and the assignment of their structures was aided by tandem mass spectrometry (CI-MS/MS) and by deuteration of hydroxyl end-groups in the pyrolyzate.[266]

The same methods have also been applied to identify the oligomeric pyrolysis products of PTHF.[267] Eleven series of oligomeric pyrolyzates were characterized by MALDI-MS. Pyrolysis of PTHF ensues about 175°C in inert atmosphere, and the initial pyrolysis products all have at least one hydroxyl end-group, retained from the original low molar mass polymer. The other end-group is ethyl ether, propyl ether, butyl ether, or aldehyde. MALDI spectra of the pyrolysis products at higher temperatures (250–350°C) showed that there is an increasing tendency to form products with a combination of alkyl ether and/or aldehyde end-groups. The amount of pyrolysis products containing the hydroxyl end-group diminishes at the higher temperatures, and butenyl ether end-groups are observed to an appreciable extent. The latter functionality is apparently formed mainly via dehydration of oligomers terminated with OH groups. A free radical mechanism was proposed to explain the main degradation products of PTHF sample.[267]

Structural information on copolymers may be obtained by combining chemical or thermal degradation methods with MALDI-MS analysis.[270–272] Oligomers formed by thermal and chemical degradation of poly[(R)-3-hydroxybutanoate] (PHB) were characterized by MALDI-TOF.[270] The molar mass distributions determined by MALDI analysis of the partially degraded material are similar to those caluated by SEC.[270]

Selective chemical degradation combined with MALDI analysis was used for the characterization of polyether and polyester polyurethane (PUR) soft blocks.[271] Ethanolamine and phenylisocyanate were used for the recovery of PTHF and polyester soft blocks, respectively; accurate MM of these blocks were determined by the SEC/MALDI method.[271]

Highly crosslinked networks synthesized by photopolymerization of dimethacrylated sebacic acid were degraded in a phosphate saline solution, and the poly(methacrylic acid) degradation product, purified from sebacic acid, was characterized by MALDI-TOF.[273]

The MM of the poly(methacrylic acid) degradation product was evaluated as a function of the network formation (i.e., double-bond conversion), rate of initiation, and monomer size. MALDI-TOF results, supported by ^{1}H-NMR, showed that distribution of the kinetic chain lengths was relatively narrow, with average lengths shorter than calculated from experimentally measured rate data. This indicated the influence of diffusion-controlled kinetics as well as chain transfer. It was also found that the average kinetic chain length shifted to lower values with increasing initiation rate and double bond conversion.[273]

MALDI-TOF analysis of an ozonolysis degradation product has been used to distinguish a random styrene-butadiene copolymer from a block styrene-butadiene copolymer (ABA), containing 45 wt% and 38% styrene, respectively.[272] Several acrylonitrile-butadiene copolymers were also characterized using ozonolysis/MALDI-MS.[272] The composition calculated from the oligomer distributions detected by MALDI-TOF was close to the reported composition for these copolymers (typically within 5 wt%). The discrepancy in the values was explained, in part, by a compositional bias resulting from the ozonolysis process. This bias arises when only one of the comonomers reacts efficiently with ozone.[291]

Oxidative and hydrolytic degradation processes occurring in PET were studied by means of MALDI-TOF,[225,226] and the thermal oxidation of a poly (ethylene glycol) containing a 1,3-disubstituted phenolic group in the chain was shown by MALDI-TOF to give carbonization and the release of a wide range of volatile degradation products.[201]

The formation of different oligomers during the hydrolytic degradation of a commercial PET sample was observed by MALDI-MS. An ester scission process was found to generate acid terminated oligomers H–[GT]$_m$–OH and T–[GT]$_m$–OH and ethylene glycol terminated oligomers H–[GT]$_m$–G, where G is an ethylene glycol unit and T is a terephthalic acid unit. The scission of ester bonds during the chemical treatment led to a strong decrease in the number of cyclic oligomers ([GT]$_m$). The presence of diacid-terminated species demonstrated a high degree of degradation.[226,227]

In a recent work,[275] MALDI-TOF has been used to monitor the thermal oxidation products of a Nylon 6 sample, at 250°C in air. The MALDI spectra (Figure 10.35) of the thermo-oxidation products of Ny6 samples terminated with carboxylic end-groups (species A and B, Table 10.5) showed the presence of polymer chains containing aldehydes, amides, methyl and N-formamide terminal groups (Table 10.5), generated in the primary oxidation process. Oligomers terminated with carboxyl end-groups (species G, Table 10.5) might be formed by the further oxidation of aldehyde-containing compounds.

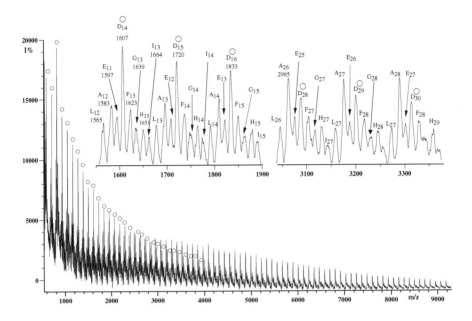

FIGURE 10.35
Positive ions MALDI-TOF mass spectrum of an Ny6 sample heated at 250°C for 30 min in air.
(Reprinted with permission from Ref. 275, Copyright 2001, John Wiley & Sons.)

The formation of azomethynes, from further reaction of aldehydes with amino-
terminated Ny6 chains, is also supported by the appearance of specific peaks
in the MALDI spectra (species I, Figure 10.35, Table 10.5).[275]

10.8 Selected Polymers

MALDI has been used to analyze a variety of polymers, and listings of MALDI
studies on synthetic polymers have appeared.[18,21] In this section, a survey
of the MALDI literature on selected polymers is reported.

10.8.1 Polystyrene

A host of papers on MALDI-TOF of PS have appeared.[9,10,22,26,38,39,41–43,47,50,54,55,62,66,74,84,109,113,115,117,122,150–152,180,191,199,220–222,244–246,276,277] Although the overwhelming majority of the PS samples analyzed possess a narrow MM distribution, PS samples
with a broad MM distribution,[107] along with macrocyclic PS samples,[194,277] have
also been investigated.

The first report of MALDI on polystyrene samples is the seminal paper by
Bahr et al.,[9] where the matrix used was 2-nitrophenyl octyl ether, which is
a liquid. Liquid matrices can be best used with instruments equipped with
horizontal sample plates. In the case of vertical plates, the thickness of the
matrix layer at a specific spot on the probe tip varies with time (due to gravity).

TABLE 10.5

Structural Assignments of Peaks Displayed in the Inset Expansions in Figure 10.35 and Corresponding to Sodiated Molecular Ions

Species	Structure	$M+Na^+(n)$
A	$HOOC\text{-}(CH_2)_8\text{-}CO\text{---}[NH\text{-}(CH_2)_5\text{-}CO]_n\text{---}OH$	1583 (12), 1696 (13) 1809 (14), 1922 (15)
B	$HOOC\text{-}(CH_2)_8\text{-}CO\text{---}[NH\text{-}(CH_2)_5\text{-}CO]_n\text{---}O^-\ Na^+$	1605 (12), 1718 (13) 1831 (14), 1944 (15)
C	$H\text{---}[NH\text{-}(CH_2)_5\text{-}CO]_n\text{---}OH$	1625 (14), 1738 (15) 1851 (16), 1964 (17)
D	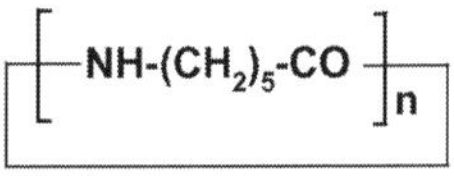 $[\text{---}NH\text{-}(CH_2)_5\text{-}CO\text{---}]_n$ (cyclic)	1607 (14), 1720 (15) 1833 (16), 2965 (26) 3078 (27), 3192 (28)
E	$CH_3\text{-}(CH_2)_3\text{-}CO\text{---}[NH\text{-}(CH_2)_5\text{-}CO]_n\text{---}NH_2$ $CH_3\text{-}(CH_2)_3\text{-}CO\text{---}[NH\text{-}(CH_2)_5\text{-}CO]_n\text{---}OH$	1597 (11), 1710 (12) 1823 (13), 1936 (14) 2953 (25), 3066 (26)
F	$H\text{-}CO\text{-}(CH_2)_4\text{-}CO\text{---}[NH\text{-}(CH_2)_5\text{-}CO]_n\text{---}NH_2$ $H\text{-}CO\text{-}(CH_2)_4\text{-}CO\text{---}[NH\text{-}(CH_2)_5\text{-}CO]_n\text{---}OH$	1623 (13), 1736 (14) 1849 (15), 3207 (27) 3302 (28), 3433 (29)
G	$HO\text{-}CO\text{-}(CH_2)_4\text{-}CO\text{---}[NH\text{-}(CH_2)_5\text{-}CO]_n\text{---}NH_2$ $HO\text{-}CO\text{-}(CH_2)_4\text{-}CO\text{---}[NH\text{-}(CH_2)_5\text{-}CO]_n\text{---}OH$	1639 (13), 1749 (14) 1865 (15), 3223 (27) 3318 (28), 3449 (29)
H	$H\text{-}CO\text{-}(CH_2)_4\text{-}CO\text{---}[NH\text{-}(CH_2)_5\text{-}CO]_n\text{---}NH\text{-}CHO$	1651 (13), 1764 (14) 1877 (15), 3235 (27) 3348 (28), 3461 (29)
	$H\text{---}[NH\text{-}(CH_2)_5\text{-}CO]_n\text{---}NH\text{-}CHO$	1651 (14), 1764 (15) 1877 (16), 3235 (28) 3348 (29), 3461 (30)

(Continued)

TABLE 10.5

Structural Assignments of Peaks Displayed in the Inset Expansions in Figure 10.35 and Corresponding to Sodiated Molecular Ions (continued)

Species	Structure	M+Na$^+$(n)
I	CH$_2$=N-CH$_2$-(CH$_2$)$_4$-CO$-$[$-$NH-(CH$_2$)$_5$-CO$-$]$_n$$-$NH-CHO	1664 (13), 1777 (14) 1890 (15), 3248 (27),
L	H-CO-N=CH-(CH$_2$)$_4$-CO$-$[$-$NH-(CH$_2$)$_5$-CO$-$]$_n$$-$NH-CHO	1565 (12), 1678 (13) 1791 (14), 3149 (26) 3262 (27), 3375 (28)

As a consequence, the distance between the sample and the acceleration plate varies, causing continuous variation in the MALDI conditions.

For this reason, the usual practice nowadays is to record MALDI spectra of PS samples using solid matrices. A PS sample with a MM of about 1.5 million was recorded using all-trans retinoic acid.[74]

Polymeric samples often contain impurities or additives (such as antioxidants, plasticizers, surfactants), and some of these compounds (for example, surfactants) are known to act as a "poison," depressing the efficiency of the MALDI process. In order to enhance the number of ions produced by MALDI, it may be necessary to purify the analyte from these contaminants prior to analysis.[74] Some MALDI matrices, such as all-trans retinoic acid,[74] are particularly sensitive to impurities, whereas for other matrices (like HABA[220] and dithranol[38]) the loss of efficency is small. Hence the latter matrices may be preferable when purification is a problem.

Though several groups have observed attachment of alkali metal cation to polystyrene,[220,278] most MALDI practitioners add silver or copper salts to produce good metal-polystyrene cation signals, dithranol as a matrix.[54,119,121,244,278] Good spectra were also obtained using palladium salts, but other transition metal salts were unfavorable under the conditions used.[279] It is noteworthy that silver salts were found to be the best cationization reagents in all the cases studied.

Some authors have also reported that polystyrene cationization in MALDI depends strongly on the ability of MALDI to produce the metal in the +1 oxidation state. It has been proposed that the cationization may proceed through gas-phase metal attachment reactions under the condition used.[54,279]

Medium-quality spectra of PS were obtained using IAA[138] or 9-nitroanthracene,[113] whereas 4-(phenylazo)-resorcinol[39] and DHB yield low-quality spectra above 50 kDa.

10.8.2 Polymethylmethacrylate

PMMA has been extensively studied by MALDI-TOF.[9,44,137,214,281] Bahr et al.[9] reported the first MALDI spectrum of PMMA, using DHB as a matrix. This matrix is not optimal, since it does not contain chemical groups similar to the groups present in the polymer repeat unit. All-trans indoleacrylic-acid (IAA) is chemically similar to PMMA, and Danis et al.[44] used it to record MALDI spectra of a PMMA sample which has high molar mass, namely $M_w = 260,000$. Following this line of reasoning, the fact that IAA also gave good results in the case of poly (butyl methacrylate) is not surprising.[138]

10.8.3 Aliphatic Polyethers

A class of polymers that has received considerable attention by researchers in the field of MALDI is the aliphatic polyethers,[80,141,210,213,284–289] with particular reference to poly(ethylene glycol), poly(propylene glycol), and poly(tetramethylene glycol), abbreviated as PEG, PPG, and PTMG. It is well-known from studies with other ionization techniques (e.g., FAB, DCI, FD, LD, see Chapters 5, 6, 7, 8) that these polymers show a strong tendency to readily form ions in the gas phase. Some workers have analyzed PEG samples using HABA to determine their MM, and obtained spectra with an excellent signal-to-noise ratio.[141]

Very often the goal of the MALDI-TOF analysis is to record the mass spectrum at the highest possible resolution, with less concern for the signal-to-noise ratio. In this case, one must select a matrix that is not ejected at high initial velocity, since a high initial velocity causes a broadening of mass spectral peaks in MALDI-TOF and degrades the resolution. Thus, DHB is preferred to HABA, since it is known that HABA is ejected at a speed almost double that of DHB.[27] A better resolution was recorded in some cases using a HABA/KI system.[30]

Some authors[80,260] analyzed also PPG using DHB. Several liquid matrices were used in MALDI-TOF analysis of PEG, and excellent agreement was found between the molar mass distribution determined with liquid and solid matrices.[287]

The preferred conformation of PEG 600 in the gas phase has been obtained by the coupling of a MALDI source to an ion chromatograph. Drift time in the ion chromatograph provide a measure of the size/conformation of the ion in the gas phase. The PEG chains were detected as sodium-cationized ions in the MALDI spectrum, and mobilities of $PEG_9.Na^+$ to $PEG_{19}.Na^+$ oligomers were reported.[52] Detailed modeling of $PEG_9.Na^+$ with molecular mechanics methods indicated that the lowest energy structure has the Na^+ ion "solvated" by the polymer chain with seven oxygen atoms as nearest neighbors. The excellent agreement between the model and experiment supports the deduced gas-phase structures of the sodiated-PEG chain.[52]

Though MALDI is considered a soft ionization technique, some authors proposed that many signals observed in the low-mass range (less than m/z 1000–2500) of PEG 6000, are due to the pyrolysis of neutral high-mass oligomers under MALDI conditions.[47,289] In fact, together with the expected ions of PEG chains having end-groups correlated with the polymerization procedure, were also detected ions such as monomethylated PEG, dimethylated PEG, monovinyl PEG, and divinyl PEG, formed according to the hypothesized Scheme 10.2.[289]

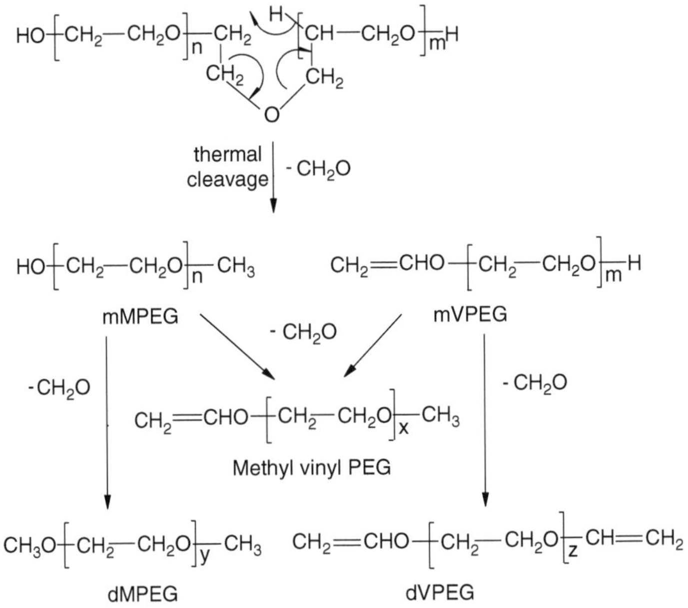

$$HO-[CH_2-CH_2O]_n-CH_2 \quad CH-CH_2O-[_m H$$

thermal cleavage | - CH₂O

$$HO-[CH_2-CH_2O]_n-CH_3 \qquad CH_2=CHO-[CH_2-CH_2O]_m-H$$

mMPEG   mVPEG

- CH₂O

-CH₂O

$$CH_2=CHO-[CH_2-CH_2O]_x-CH_3$$

Methyl vinyl PEG

-CH₂O

$$CH_3O-[CH_2-CH_2O]_y-CH_3 \qquad CH_2=CHO-[CH_2-CH_2O]_z-CH=CH_2$$

dMPEG   dVPEG

SCHEME 10.2
Fragmentation pathway of the high-mass PEG oligomers under MALDI conditions.

10.8.4 Polyesters

Numerous MALDI-TOF spectra of aliphatic and aromatic polyesters have appeared.[103,104,108,126,176,179,181,182,190,225–231,243,270,290] Aliphatic polyesters are easily dissolved in THF or chloroform, and DHB,[270] HABA,[103] and IAA[230,265] give excellent MALDI-TOF spectra.

On the other hand, aromatic polyesters such as PET and PBT do not dissolve in THF or chloroform, and different solvents (such as tetrachloroethane, trifluoracetic acid and chloroform/trifluoracetic mixtures) must be used. Neverthe-less, PET, PBT, and poly(1,2-dihydroxybenzene phthalate) can be analyzed using HABA[223,224] or dithranol.[20]

Some authors have used DHB and IAA matrices for MALDI-TOF analysis of some aliphatic and aromatic polyesters and have observed that molar mass distribution values exhibit a certain variability with both laser fluence and matrix.[290] A certain degree of selectivity in the cationization process by Na⁺ and K⁺, according to the structure and the MM of the oligomers, was also observed.[290]

Molar masses estimated by MALDI-TOF were close to those calculated by ESI for a series of polyester paints[126] and for some low-molar mass polyesters.[227] However, the results significantly underestimated the MMD of the polyester paint studed.[126] Differences in the relative abundances of branched and cyclic species in ESI vs. MALDI mass spectra were also noted.[126]

10.8.5 Polycarbonates

The MALDI-TOF analysis of poly(bisphenol-A carbonate)[102,179,181,182] and of other carbonates[256,257] is straightforward, since these polymers desorb well. As a consequence, the choice of the matrix is not critical. However, when the scope of the analysis is to obtain mass-resolved spectra at 60 kDa and above, only one matrix has been used, namely HABA.[182,183]

MALDI-TOF spectra of commercial polycarbonates display peaks due to open-chain macromolecules along with peaks due to cyclics. Polycarbonates containing porphyrin or fullerene units in the main chain fullerene are easily analyzed by MALDI-TOF.[257]

10.8.6 Polyamides

MALDI-TOF studies of polyamides are rare, although these polymers desorb very well.[22,68,206] Most studies are on Nylon 6 and HABA has been the best matrix. Nylon 6 does not dissolve in THF, acetone, or chloroform, but it does dissolve in trifluoroethanol and formic acid.

10.8.7 Polyelectrolytes

Polyelectrolytes require desalting and conversion to the "hydrogen" form prior to MALDI analysis. Sinapinic acid is often used as a matrix for polylectrolytes such as poly(styrenesulfonic acid)[260] and poly(acrylic acid).[291] Figure 10.36 reports the MALDI-TOF mass spectrum (recorded in the negative-ion mode) of poly(styrenesulfonic acid), which possesses a narrow MM distribution.[260] The most intense peaks are those at 130,000 and 200,000 due respectively to

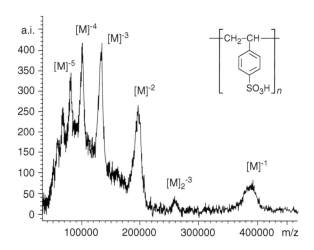

FIGURE 10.36
MALDI-TOF mass spectrum of a poly(styrenesulfonic acid) sample. (Reprinted with permission from Ref. 70, Copyright 1995 American Chemical Society)

triply and doubly charged ions, and the peak at 390,000 (due to ions bearing a single charge) is much less intense. The degree of sulfonation of polystyrenes was calculated from the difference in MMD deduced by MALDI-TOF analysis of the sulfonated and parent PS samples.[70]

Polyacrylamides are water-soluble polymers. In the early days of MALDI of polymers, they attracted the attention of researchers in the field, since one can use sample preparation recipes developed for proteins. This avoids the tedious operation of performing various attempts with different matrices.

DHB was used as a matrix to analyze poly(N–N diethylacrylamide) (PDEAA) prepared by group transfer polymerization,[292] and also PDEAA prepared by anionic polymerization.[293] HABA was used in the analysis of a substituted polyacrylamide obtained by free radical polymerization.[219]

The MALDI-TOF mass spectrum of a polymer based on an acrylamide with an α-helical peptide structure in the side chain was shown to yield a series of peaks up to 14000 Da.[294]

10.8.8 Epoxy Resins and Polysilanes

Epoxy resins and polymers containing silicon atoms in the backbone have been extensively studied by MALDI-TOF in order to determine their molar mass distribution and chemical functionality.[37,38,49,109,111,112,123,125,286–288,295]

MALDI-TOF was used to deduce the three-dimensional structure of complex silsesquioxane polymers.[295] These polymers have silicon coordinated with three bridging oxygen atoms in the form of $[RSiO_{3/2}]$ and can form a wide variety of complex three-dimensional structures. Specifically, the branched-linear silsesquioxane may react with itself to form intramolecular closed loops. Four distinct levels of molecular structure were observed from the MALDI-TOF spectrum.

The mass separation of the major cluster of peaks was found to be 108.25 Da as expected from the synthesis. The mass separation of 18 Da between peaks within a major cluster indicated the loss of H_2O and the formation of intramolecular closed loops. The data indicated also that the percentage of closed loops did not vary with oligomer size, and that the molecular structure was intermediate between branched-linear and a simple ladder structure, with no evidence that fully condensed structures had been formed in significant amounts.[295]

10.8.9 Dendrimers and Hyperbranched Polymers

Dendrimers are monodisperse polymers with precisely controlled macromolecular architecture. Although dendrimer synthesis is complex, the lack of chain entanglements results in a lower viscosity with respect to linear polymers of the same MM.[296]

The polymerization of A_xB monomers yields highly branched polymers (hyperbranched), containing a multitude of end-groups. Hyperbranched polymers are similar to dendrimers in the sense that they are less prone than linear polymers to form entanglements and to undergo crystallization.[297]

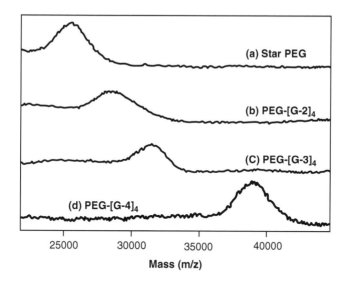

FIGURE 10.37
MALDI-TOF mass spectra of a dendrimer sample at various stages of its growth. The dendrimer grows on a star-shaped nucleus made of PEG. (Reprinted with permission from Ref. 298, Copyright 1999 American Chemical Society)

MALDI can be used to follow dendrimer growth. Some papers report the characterization of dendrimers grown on a star-shaped nucleus made of PEG[252] and of some aromatic polyester dendrimers. [299,300] Figure 10.37 shows the MALDI-TOF mass spectra of a dendrimer which grows on a star-shaped nucleus made of PEG, at various stages of its growth.[298] Initially, the molar mass is around 25000, then it changes to 28000 and to 33000, and eventually it reaches 39000.

Polyether dendrimers,[301] phosphorus-containing dendrimers,[302] and polyurethane dendritic wedges[303] also have been characterized by MALDI-TOF.

The MALDI-TOF mass spectrum of a linear block copolymer made of PEG and polyether dendrimer blocks indicated that the number of ions produced decreases very rapidly (almost exponentially) as the laser irradiance is turned down, until a threshold is reached where no ions are produced.[301] It was also found that the best resolution is achieved with a laser power 40% above the threshold. This is unusual, since normally the mass-spectral resolution falls steadily as laser power grows.[301]

The average molar mass and the structure of some hyperbranched polymers with phenol terminal groups[304] and of some hyperbranched polyglycerols[305] have been characterized by MALDI-TOF analysis.

The role of cyclization in hyperbranched polymers was also investigated.[306] Cyclization in hyperbranched pentafluorophenyl-terminated poly(benzyl ether)s was indicated by the presence of ions at 20 Da less than the masses of acyclic species, owing to the loss of the HF chain ends during polymerization.[306]

10.8.10   Hydrocarbon Polymers and Carbon-Rich Polymers

Hydrocarbon polymers, in particular nonaromatic polyolefins, are materials that are extremely difficult to analyze by MALDI mass spectrometry, primarily because it is difficult for them to form cation attachment ions.

A few papers report the characterization of polydienes, such as polybutadienes,[39,143,221] and polyisoprene, by MALDI-TOF.[143,221] Polybutadienes of narrow polydispersity with masses up to 300,000 Da, and polyisoprenes of narrow polydispersity with masses up to 150,000 Da, were well-characterized using all-trans-retinoic acid as the matrix and copper(II) nitrate as the cationization agent.[143] Intense signals due to Cu^+ attachment to the polymer chains were observed, while much weaker spectra were obtained using siver nitrate as cationization salt.[143] Oligomers with average molar masses up to 5000 Da have been well-characterized by MALDI-FTMS.[93]

Information about the degradation pathways occurring in polyolefins exposed to vacuum-ultraviolet radiation (VUV) were obtained by MALDI-TOF analysis of oligomers formed by VUV photolysis of linear C36-alkane.[307] Both molar masses and end-groups of degradation product were studied.

MALDI-TOF analysis was also applied to the characterization of carbon-rich polymers such as polyparaphenylene,[129] poly(4'-vinylhexaphenylbenzene),[308] and large polycyclic aromatic hydrocarbons (PAHs).[69,318] The mechanical mixing of the analyte with matrix powders was used as the sample preparation method, since the PAHs are insoluble.[69] In this case, 7,7,8,8,-tetracyanoquinodimethane, a new matrix with promising properties, was used.

Carbon-rich polymers can be obtained by free radical polymerization of 4'-(4-vinylphenylene)hexaphenylbenzene, and the MALDI-TOF analysis can be carried out using dithranol as the matrix.[308]

10.8.11   Fluorinated Polymers

A few papers report the MALDI-TOF characterization of fluorinated polymers.[289,309] Fluorinated polymers are insoluble or only sparingly soluble in common organic solvent systems, and modified MALDI sample preparation methods were applied to overcome this problem.[289] It was found that the signal intensity of the fluorinated oligomers changes remarkably with sample preparation methodology.[289] Positive and negative ion MALDI spectra of the fluorinated polymers were recorded using DHB as a matrix. The presence of fluorinated groups offers the advantage of inductive stabilization of anionic charge sites for improved signals in the negative ion mode.[309]

10.8.12   Polymer Mixtures

Mixtures of narrow distribution polymers were used in several studies about mass discrimination problems in MALDI-TOF.[25,26,88]

In a recent work it was found that, under similar sample preparation and MALDI instrument conditions, different ion detection systems produce a different response in polymer mixtures.[116] This is due to several factors:

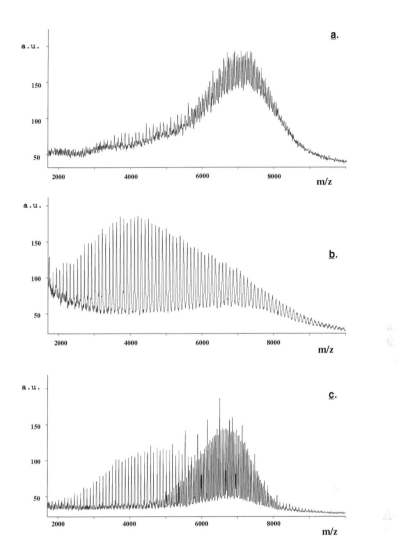

FIGURE 10.38

Positive-ion DE-MALDI-TOF mass spectra of a PEG/PMMA blend recorded adopting different delay times grid voltage % values. (a) 4600 ns and 88%; (b) 100 ns and 96%; and (c) 2800 ns and 96%. (Reprinted with permission from Ref. 88, Copyright 1999 John Wiley & Sons Ltd)

(i) the type of detection system, (ii) saturation effects, and (iii) signal-to-noise limitations. Gating the matrix ion signal, adjusting the operating parameters, and working under nonsaturating conditions may minimize complications from the first two factors. Correcting for the third factor is more difficult, given the limits of the technology.[116]

MALDI applications concerning mixtures of different polymers are scarce, probably because of significant differences in cationization efficiencies and mass discrimination effects. Figures 10.38a–c reports the DE-MALDI-TOF spectra of an equimolar mixture of two narrow dispersed PEG and PMMA samples having similar molar masses.[88] These mass spectra were recorded under various experimental conditions, and their examination reveals the presence of mass discrimination effects.[88] In fact, the spectrum in Figure 10.38a shows a signal centered at about m/z 7000, consisting of a series of peaks separated by 44 Da, due to the PEG component. A series of less intense peaks differentiated by 100 Da, due to PMMA, also appears in the spectrum at lower mass values. In contrast, the spectrum in Figure 10.38b is composed almost exclusively of this second family of peaks, as if the material consisted only of PMMA. The effective composition of the blend is shown instead by the spectra in Figure 10.38c.[88]

10.8.13   Miscellaneous

In addition to the polymers described above, the MALDI technique has also been employed for the characterization of several synthetic polymeric materials. The literature reports the MALDI characterization of polyacrylonitrile (PAN),[276] poly(ether sulfone) (PES),[124] poly(dimethyl phenylene oxide) (PDMPO),[124] and functionalized poly(p-phenylene)s.[200] Analysis by MALDI-TOF of aromatic polyethers (such as PEEK), can be found,[279] and several macrocyclic samples were also characterized by MALDI-TOF.[310–317]

References

1. Burlingame, A. L., Boyd, R. K., and Gaskell, S. J., Mass Spectrometry, *Anal. Chem.*, 70, 647R, 1998.
2. Koenig, J. L., *Spectroscopy of Polymers*, 2a Ed., Elsevier, 1999.
3. Karas, M. and Hillenkamp F., Laser Desorption Ionization of proteins with Molecular Masses exceeding 1000 Daltons, *Anal. Chem.*, 60, 2299, 1988.
4. Hillenkamp, F., Laser Desorption Mass Spectrometry: Mechanism, Techniques and Applications, *Advances In Mass Spectrometry*, Vol. II A, p. 354, 1988.
5. Hillenkamp, F., Karas, M., Beavis, R. C., and Chait, B. T., Matrix-Assisted Laser Desorption/Ionization Mass Spectrometry of Biopolymers, *Anal. Chem.*, 63, 1193 A, 1991.
6. Fenn, J. B., Mann, M., Meng, C. K., Wong, S. F., and Whitehouse, C. M., Electrospray ionization—principles and practice, *Mass Spectrom. Rev.*, 9, 37, 1990.
7. Campana, J. E., Castoro, J. A., Sheng, L.-S, Shew, S. L., Stout, H., and Winger, B. E., Characterization of Polymers by Laser Desorption Ionization and Electrospray Mass Spectrometry, *40th International Society for the Advancement of Material and Process Engineering (SAMPE) Symposium and Exhibition*, Anaheim, CA, 1995.
8. Festag, R., Alexandratos, S. D., Joy, D. C., Wunderlich, B., Annis, B., and Cook, K. D., Effects of Molecular Entanglements During Electrospray of High Molecular Weight Polymers, *J. Am. Soc. Mass Spectrom.*, 9, 299, 1998.

9. Bahr, U., Deppe, A., Karas, M., Hillenkamp, F., and Giessman, U., Mass Spectrometry of Synthetic Polymers by UV-Matrix-Assisted Laser Desorption/Ionization, *Anal. Chem.*, 64, 2866, 1992.

10. Karas, M., Bahr, U., Deppe, A., Stahl, B., and Hillenkamp, F., Molar Mass Determination of Macromolecular Compounds by Matrix-assisted Laser Desorption Ionization Mass Spectrometry, *Makromol. Chem., Macromol. Symp.*, 61, 397, 1992.

11. Danis, P. O., Karr, D. E., Westmoreland, D. G., Piton, M. C., Christie, D. I., Clay, P. A., Kable, S. H., and Gilbert, R. G., Measurements of Propagation Rate Coefficients Using Pulsed-laser Polymerization and Matrix-assisted Laser Desorption/Ionization Mass Spectrometry *Macromolecules*, 26, 6684, 1993.

12. Spengler, B., Kirsch, D., Kaufmann, R., Karas, M., Hillenkamp, F., and Giessmann, U., The Detection of Large Molecules in Matrix-assisted UV-laser Desorption, *Rapid Commun. Mass Spectrom.*, 4, 301, 1990.

13. Cotter, R. J., Time-of-Flight Mass Spectrometry for the Structural Analysis of Biological Molecules, *Anal. Chem.*, 64, 1027 A, 1992.

14. Bahr, U., Karas, M., and Hillenkamp, F., Analysis of biopolymers by Matrix-assisted Laser Desorption/Ionization (MALDI) Mass Spectrometry, *Fresenius J. Anal. Chem.*, 348, 783, 1994.

15. Hillenkamp, F., Laser Desorption Time-of-Flight Mass Spectrometry, in *Proceedings of the International Conference on Biological Mass Spectrometry, Kyoto, 1992,* Matsuo T., Ed., Sanei Publishing Co., Kyoto, Japan, p. 22, 1992.

16. Creel, H. S., Prospects for the Analysis of High Molar Mass Polymers Using MALDI Mass Spectrometry, *Trends Polym. Sci.*, 1, 336, 1993.

17. Montaudo, G., Mass Spectrometry of Synthetic Polymers, Mere Advances or Revolution?, *Trends Polym. Sci.*, 4, 81, 1996.

18. Rader, H. I. and Schrepp, W., MALDI-TOF mass spectrometry in the Analysis of Synthetic Polymers, *Acta Polym.*, 49, 272, 1998.

19. Wu, K. J. and Odom, R.W., Characterizing Synthetic Polymers by MALDI MS, *Anal. Chem.*, 69, A456, 1998.

20. Scamporrino, E. and Vitalini, D., Recent Advances in Mass Spectrometry of Polymers in *Modern Techniques for Polymer Characterization*, Pethric, R. A. and Dawkins, J. V., Eds., Wiley, 1999.

21. Nielen, M. W. F., Maldi Time-of-flight Mass Spectrometry of Synthetic Polymers, *Mass Spectrom. Rev.*, 18, 309, 1999.

22. Pasch, H. and Ghahary, R., Analysis of Complex Polymers by MALDI-TOF Mass Spectrometry, *Macromol. Symp.*, 152, 267, 2000.

23. Scrivens, J. H., The Characterization of Synthetic Polymer Systems, *Adv. Mass Spectrom.*, 13, 447, 1995.

24. Cottrell, J. S., Koerner, M., and Gerhards, R., The Characterization of Synthetic Polymers by Matrix-assisted Laser Desorption/Ionization Mass Spectrometry, *Rapid Comm. Mass. Spectrom.*, 9, 1662, 1995.

25. Spickermann, J., Martin, K., Räder, H. J., Müllen, K., Schlaad, H., Müller, A. H. E., and Krüger, R.-P., Quantitative Analysis of Broad Molecular Weight Distributions Obtained by Matrix-assisted Laser Desorption/Ionization Time-of-flight Mass Spectrometry, *Eur. Mass Spectrom.*, 2, 161, 1996.

26. Thomson, B., Suddaby, K., Rudin, A., and Lajoie, G., Characterisation of Low Molecular Weight Polymers Using Matrix Assisted Laser Desorption Time-of-Flight Mass Spectrometry, *Eur. Polym. J.*, 32, 239, 1996.

27. Ehring, H., Karas, M., and Hillenkamp, F., Role of Photoionization and Photochemistry in Ionization Processes of Organic Molecules and Relevance for Matrix-assisted Laser Desorption Ionization Mass Spectrometry, *Org. Mass Spec.*, 27, 472, 1992.

28. Beavis, R. C., Chaudhary, T., and Chait, B. T., α-Cyano-4-hydroxycinnamic Acid as Matrix for Matrix-assisted Laser Desorption Mass Spectrometry, *Org. Mass Spectrom.*, 27, 156, 1992.

29. Beavis, R. C., Matrix-assisted Ultraviolet Laser Desorption: Evolution and Principles, *Organic Mass Spectrometry*, 27, 653, 1992.

30. Juhasz, P., Costello, C. E., and Biemann, K., Matrix-Assisted Laser Desorption Mass Spectrometry with 2-(4-Hydroxyphenylazo)benzoic Acid Matrix, *J. Am. Soc. Mass Spectrom.*, 4, 399, 1993.

31. Overberg, A., Karas, M., and Hillenkamp, F., Matrix-assisted Laser Desorption of Large Biomolecules with a TEA-CO_2-Laser, *Rapid Commun. Mass Spectrom.*, 5, 129, 1991.

32. Cramer, R., Hillenkamp, H., and Haglund, R. F. Jr., Infrared Matrix-Assisted Laser Desorption and Ionization by Using a Tunable Mid-Infrared Free-Electron Laser, *J. Am. Soc. Mass Spectrom.*, 7, 1187, 1996.

33. Weidner, St., Kühn, G., and Friedrich, J., Infrared-matrix-assisted Laser Desorption/Ionization and Infrared-Laser Desorption/Ionization Investigations of Synthetic Polymers, *Rapid Comm. Mass. Spectrom.*, 12, 1373, 1998.

34. McEwen, C. N., Peacock, P. M., and Guan, Z. Complication in the Analysis of Synthetic Polymers by MALDI, *Proc. 46th ASMS Conf. Mass Spectrom.*, Allied Topics, Orlando, FL, 409, 1998.

35. Fitzgerald, M. C., Parr, R. G., and Smith, L. M., Basic Matrices for the Matrix-Assisted Laser/Desorption Mass Spectrometry of Proteins and Oligonucleotides, *Anal. Chem.*, 65, 3204, 1993.

36. Russel, M. R., Lamb, J. H., and Lim, C. K., 5,10,15,20-*meso*-Tetra(hydroxyphenyl)chlorin as a Matrix for the Analysis of Low Molecular Weight Compounds by Matrix-assisted Laser/Desorption Time-of-flight Mass Spectrometry, *Rapid Comm. Mass Spectrom.*, 9, 968, 1995.

37. Tang, X., Dreifuss, P. A., and Vertes, A., New Matrices and Accelerating Voltage Effects in Matrix-assisted Laser Desorption/Ionization of Synthetic Polymers, *Rapid Comm. Mass. Spectrom.*, 9, 1141, 1995.

38. Belu, A. M., DeSimone, J. M., Linton, R. W., Lange, G. W., and Friedman, R. M., Evaluation of Matrix-Assisted Laser Desorption/Ionization for Polymer Characterization, *J. Am. Soc. Mass Spectrom.*, 7, 11, 1996.

39. Liu, D. H. M., and Sclunegger, U. P., Matrix-Assisted Laser Desorption/Ionization of Synthetic Polymers with Azo Compound Matrices, *Rapid Comm. Mass. Spectrom.*, 10, 483, 1996.

40. Xu, N., Huang, Z. H., Watson, J. T., and Gage, D. A., Mercaptobenzothiazoles: A New Class of Matrices for Laser Desorption Ionization Mass Spectrometry, *J. Am. Soc. Mass Spectrom.*, 8, 116, 1997.

41. Macha, S. F., Limbach, P. A., and Savickas, P. J., Application of Nonpolar Matrices for the Analysis of Low Molecular Weight Nonpolar Synthetic Polymers by Matrix-Assisted Laser Desorption/Ionization Time-of-Flight Mass Spectrometry, *J. Am. Soc. Mass Spectrom.*, 11, 731, 2000.

42. Lehmann, E., Knochenmuss, R., and Zenobi, R., Ionization Mechanism in Matrix-assisted Laser Description/Ionization Mass Spectrometry: Contribution of Pre-formed Ions, *Rapid. Commun. Mass Spectrom.*, 11, 1483, 1997.

43. Knochenmuss, R., Lehmann, E., and Zenobi, R., Polymer Cationization in Matrix-assisted Laser Desorption/Ionization, *Eur. Mass. Spectrom.*, 4, 421, 1998.
44. Danis, P. O. and Karr, D. E., A Facile Sample Preparation for the Analysis of Synthetic Organic Polymers by Matrix-assisted Laser Desorption/Ionization, *Org. Mass Spec.* 28, 923, 1993.
45. Lloyd, P. M., Scrivener, E., Maloney, D. R., Haddleton, D. M., and Derrick, P. J., Cation Attachment to Synthetic Polymers in Matrix-Assisted Laser Desorption/ Ionization Mass Spectrometry, *Polymer Preprint*, 847, 1996.
46. King, R. C., Goldschmidt, R., Xiong, Y., and Owens, K. G., Mechanistic Studies of the Cationization of Synthetic Polymers by Alkaly Metals in the Matrix-assisted Laser Desorption/Ionization Experiment. *43rd ASMS Conf. Mass Spectrom.*, Atlanta, GA, 1237, 1995.
47. Mowat, I. A., Donovan, R. J., and Maier, R. R. J., Enhanced Cationization of Polymers Using Delayed Ion Extraction with Matrix-Assisted Laser Desorption/ Ionization, *Rapid Comm. Mass. Spectrom.*, 11, 89, 1997.
48. Mowat, I. A. and Donovan, R. J., Metal-ion Attachment to Nonpolar Polymers During Laser Desorption/Ionization at 337 nm, *Rapid Comm. Mass. Spectrom.*, 9, 82, 1995.
49. Reinhold, M., Meier, R. J., and de Koster, C. G., How Feasible is Matrix-assisted Laser/Desorption/Ionization Time-of-flight Mass Spectrometry Analysis of Polyolefins?, *Rapid Comm. Mass. Spectrom.*, 12, 1962, 1998.
50. Wong, C. K. L., So, M. P., and Dominic Chan, T.-W., Origins of the Proton in the Generation of Protonated Polymers and Peptides in Matrix-assisted Laser Desorption/Ionization, *Eur. Mass. Spectrom.*, 4, 223, 1998.
51. Burton, R. D., Watson, C. H., Eyler, J. R., Lang, G. L., Powell, D. H., and Avery, M. Y., Proton Affinities of Eight Matrices Used for Matrix-assisted Laser Desorption/Ionization, *Rapid. Commun. Mass Spectrom.*, 11, 443, 1997.
52. von Helden, G., Wyttenbach, T., and Bowers, M. T., Conformation of Macromolecules in the Gas Phase: Use of Matrix-Assisted Laser Desorption Methods in Ion Chromatography, *Science*, 267, 1483, 1995.
53. Hoberg, A.-M., Haddleton, D. M., Derrick, P. J., and Scrivens, J. H., Evidence for Cationization of Polymers in the Gas Phase During Matrix-assisted Laser/ Desorption Ionization, *Eur. Mass. Spectrom.*, 3, 471, 1997.
54. Rashidezadeh, H., Hung, K., and Guo, B., Probing Polystyrene Cationization in Matrix-assisted Laser/Desorption Ionization, *Eur. Mass. Spectrom.*, 4, 429, 1998.
55. Llenes, C. F. and O'Malley, R. M., Cation Attachment in the Analysis of Polystyrene and Polyethylene Glycol by Laser-desorption Time-of-flight Mass Spectrometry, *Rapid. Commun. Mass Spectrom.*, 6, 564, 1992.
56. Wong, C. K. L. and Dominic Chan, T.-W., Cationization Processes in Matrix-assisted Laser Desorption/Ionization Mass Spectrometry: Attachment of Divalent and Trivalent Metal Ions, *Rapid. Commun. Mass Spectrom.*, 11, 513, 1997.
57. Hoberg, A.-M., Haddleton, D. M., Derrick, P. J., Jackson, A. T., and Scrivens, J. H., The Effect of Counter Ions in Matrix-assisted Laser Desorption/Ionization of Poly(methyl methacrylate), *Eur. Mass. Spectrom.*, 4, 435, 1998.
58. Steenvoorde, R. J. J. M., Breuker, K., and Zenobi, R., The Gas-phase Basicities of Matrix-assisted Laser Desorption/Ionization Matrices, *Eur. Mass. Spectrom.*, 3, 339, 1997.

59. Nicola, A. J., Gusev, A. I., Proctor, A., Jackson, E. K., and Hercules, D. M., Application of the Fast-evaporating Sample Preparation Method for Improving Quantification of Angiotesin II by Matrix-assisted Laser Desorption/Ionization, *Rapid. Commun. Mass Spectrom.*, 9, 1164, 1995.

60. Gusev, A. I., Wilkinson, W. R., Proctor, A., and Hercules, D. M., Improvement of Signal Reproducibility and Matrix/Comatrix Effects in MALDI Analysis, *Anal. Chem.*, 67, 1034, 1995.

61. Haddleton, D. M., Waterson, C., and Derrick, P. J., Comment: A Simple, Low-cost, Air-spray Method for Improved Sample Preparation for Matrix-assisted Laser Desorption/Ionization Mass Spectrometry of Derivatised Poly(ethylene glycol), *Eur. Mass. Spectrom.*, 4, 203, 1998.

62. Hanton, S. D., Cornelio Clark, P. A., and Owens, K. G., Investigations of Matrix-Assisted Laser Desorption/Ionization Sample Preparation by Time-of-Flight Ion Mass Spectrometry, *J. Am. Soc. Mass Spectrom.*, 10, 104, 1999.

63. Axelsson, J., Scrivener, E., Haddleton, D. M., and Derrik, P. J., Mass Dicrimination Effects in an Ion Dectector and Other Causes for Shifts in Polymer Mass Distribution Measured by Matrix-Assisted Laser Desorption/Ionization Time-of-Flight Mass Spectrometry, *Macromolecules*, 29, 8875, 1996.

64. Axelsson, J., Hoberg, A.-M., Waterson, C., Myatt, P., Shield, G. L., Varney, J., Haddleton, D. M., and Derrik, P. J., Improved Reproducibility and Increased Signal Intensity in Matrix-Assisted Laser Desorption/Ionization as a Result of Electrospray Sample Preparation, *Rapid. Commun. Mass Spectrom.*, 11, 209, 1997.

65. Dai, Y., Whittal, R. M., and Li, L., Confocal Fluorescenze Microscopic Imaging for Investigating the Analyte Distribution in MALDI Matrices, *Anal. Chem.*, 71, 1087, 1999.

66. Yalcin, T., Dai, Y., and Li, L., Matrix-Assisted Laser Desorption/Ionization Time-of-Flight Mass Spectrometry for Polymer Analysis: Solvent Effect in Sample Preparation, *J. Am. Soc. Mass Spectrom.*, 9, 1303, 1998.

67. Kassis, C. M., De Simone, J. M., Linton, R. W., Lange, G. W., and Friedman, R. M., An Investigation into the Importance of Polymer-Matrix Miscibility Using Surfactant Modified Matrix-assisted Laser Desorption/Ionization Mass Spectrometry, *Rapid Comm. Mass. Spectrom.*, 11, 1462, 1997.

68. Skelton, R., Dubois, F., and Zenobi, R., A MALDI Sample Preparation Method Suitable for Insoluble Poymers, *Anal. Chem.*, 72, 1707, 2000.

69. Przybilla, L., Brand, J.-D., Yoshimura, K., Räder, J., and Müllen, K., MALDI-TOF Mass Spectrometry of Insoluble Giant Polycyclic Aromatic Hydrocarbons by a New Method of Sample Preparation, *Anal. Chem.*, 72, 4591, 2000.

70. Danis, P. O. and Karr, D. E., Analysis of Poly(styrenesulfonic acid) by Matrix-Assisted Laser Desorption Time-of-Flight Mass Spectrometry, *Macromolecules*, 28, 8548, 1995.

71. Brockman, A. H., Dodd, B., and Orlando, R., A Desalting Approach for MALDI-MS Using On-Probe Hydrophobic Self-Assembled Monolayers, *Anal. Chem.*, 69, 4716, 1997.

72. Mamyrin, B. A., Laser Assisted Reflectron Time-of-Flight Mass Spectrometry, *Int. J. Mass Spectrom. Ion Proc.*, 1, 131, 1994.

73. Ioanoviciu, D., Cuna, C., and Ardelean, P., Peak Profiles of Ions Originating From Solid Surfaces in Time-of-flight Mass Spectrometers Incorporating Reflectrons, *Rapid Comm. Mass. Spectrom.*, 7, 999, 1993.

74. Schriemer, D. C. and Li, L., Detection of High Molecular Weight Narrow Polydisperse Polymers up to 1.5 Milion Daltons by MALDI Mass Spectrometry, *Anal. Chem.*, 68, 2721, 1996.

75. Whittal, R. M. and Li, L., High-Resolution Matrix-Assisted Laser Desorption/Ionization in a Linear Time-of-Flight Mass Spectrometer, *Anal. Chem.*, 67, 1950, 1995.

76. Colby, S. M., King, T. B., and Reilly, J. P., Improving the Resolution of Matrix-assisted Laser Desorption/Ionization Time-of-flight Mass Spectrometry by Exploiting the Correlation between Ion Position and Velocity, *Rapid Comm. Mass. Spectrom.*, 8, 865, 1994.

77. Spengler, B., Kirsch, D., and Kaufmann, R., Fundamental Aspects of Postsource Decay in Matrix-Assisted Laser Desorption Mass Spectrometry. 1. Residual Gas Effects, *J. Phys. Chem.*, 96, 9678, 1992.

78. Kaufmann, R., Kirsch, D., Tourmann, J. L., Machold, J., Hucho, F., Utkin, Y., and Tsetlin, V., Matrix-Assisted Laser Desorption Ionization (MALDI) and Post Source Decay (PSD) Product Ion Mass Analysis Localize a Photolabel Crosslinked to the Delta-Subunit of Nachr Protein by Neurotoxin II, *Eur. Mass Spectrom.*, 1, 313, 1995.

79. Spengler, B., Kirsch, D., Kaufmann, R., and Jaeger, E., Peptide Sequencing by Matrix-assisted Laser-desorption Mass Spectrometry, *Rapid Comm. Mass. Spectrom.*, 6, 105, 1992.

80. Ioanoviciu, D., Complete Third-order Resolution Formulae for Time-of-flight Mass Spectrometers Incorporating Reflectrons, *Rapid Comm. Mass. Spectrom.*, 7, 1095, 1993.

81. Cornish, T. J. and Cotter, R. J., A Curved-field Reflectron for Improved Energy Focusing of Product Ions in Time-of-flight Mass Spectrometry, *Rapid Comm. Mass. Spectrom.*, 7, 1037, 1993.

82. Vestal, M. L., Juhasz, P., and Martin, S. A., Delayed Extraction Matrix-assisted Laser Desorption Time-of-flight Mass Spectrometry, *Rapid Comm. Mass. Spectrom.*, 9, 1044, 1995.

83. Brown, R. S. and Lennon, J. J., Mass Resolution Improvement by Incorporation of Pulsed Ion Extraction in a Matrix-Assisted Laser Desorption/Ionization Linear Time-of-Flight Mass Spectrometer, *Anal. Chem.*, 67, 1998, 1995.

84. Whittal, R. M., Schriemer, D. C., and Li, L., Time-Lag Focusing MALDI Time-of-Flight Mass Spectrometry for Polymer Characterization: Oligomer Resolution, Mass Accuracy, and Average Weight Information, *Anal. Chem.*, 69, 2734, 1997.

85. Schriemer, D. C., Whittal, R. M., and Li, L., Analysis of Structurally Complex Polymers by Time-Lag Focusing Matrix-Assisted Laser Desorption Ionization Time-of-Flight Mass Spectrometry, *Macromolecules*, 30, 1955, 1997.

86. Jackson, A. T., Yates, H. T., Lindsay, C. I., Didier, Y., Segal, J. A., Scrivens, J. H., and Brown, J., Utilizing Time-lag Focusing Matrix-assisted Laser Desorption/Ionization Mass Spectrometry for the End-Group Analysis of Synthetic Polymers, *Rapid Comm. Mass. Spectrom.*, 11, 520, 1997.

87. Larsen, B. S., Byrd, H. C. M., and McEwen, C. N., Instrumental Improvements for Characterization of Synthetic Polymers, *Polym. Preprints*, 1999.

88. Vitalini, D., Mineo, P., and Scamporrino, E., Effect of Combined Changes in Delayed Extraction and Potential Gradient on the Mass Resolution and Ion Discrimination in the Analysis of Polydisperse Poymers and Polymer Blends by Delayed Extraction Matrix-assisted Laser Desorption/Ionization Time-of-flight Mass Spectrometry, *Rapid Comm. Mass. Spectrom.*, 13, 2511, 1999.

89. Castoro, J. A., and Wilkins, C. L., High Resolution Matrix-Assisted Laser Desorption/Ionization of Small Proteins by Fourier Transform Mass Spectrometry, *Analytical Chemistry*, 65, 2621, 1993.

90. Sheng, L.-S, Covey, J. E., Shew, S. L., Winger, B. E., and Campana, J. E., Matrix-assisted Laser Desorption Ionization Fourier Mass Spectrometry, *Rapid Comm. Mass. Spectrom.*, 8, 498, 1994.

91. de Koster, C. G., Duursma, M. C., van Rooij, G. J., Heeren, R. M. A., and Boon, J. J., Endgroup Analysis of Polyethylene Glycol Polymers by Matrix-assisted Laser Desorption/Ionization Fourier-transform Ion Cyclotron Resonance Mass Spectrometry, *Rapid Comm. Mass Spectrom.*, 9, 957, 1995.

92. van Rooij, G. J., Duursma, M. C., Heeren, R. M. A., Boon, J. J., and de Koster C. G., High Resolution End Group Determination of Low Molecular Weight Polymers by Matrix-Assisted Laser Desorption Ionization on an External Source Fourier Transform Ion Cyclotron Resonance Mass Spectrometry, *J. Am. Soc. Mass Spectrom.*, 7, 449, 1996.

93. Pastor, S. J. and Wilkins, C. L., Analysis of Hydrocarbon Polymers by MALDI-FTMS, *J. Am. Soc. Mass Spectrom.*, 8, 225, 1997.

94. van der Hage, E. R. E., Duursma, M. C., Heeren, R. M. A., Boon, J. J., Nielen, M. W. F., Weber, A. J. M., de Koster, C. G., and de Vries, N. K., Structural Analysis of Polyoxyalkyleneamines by Matrix-Assisted Laser Desorption/Ionization on an External Ion Source FT-ICR-MS and NMR, *Macromolecules*, 30, 4302, 1997.

95. van Rooij, G. J., Duursma, M. C., de Koster, C. G., Heeren, R. M. A., Boon, J. J., Schuyl, P. J. W., and van der Hage, E. R. E., Determination of Block Length Distributions of Poly(oxypropylene) and Poly(oxyethylene) Block Copolymers by MALDI-FTICR Mass Spectrometry, *Anal. Chem.*, 70, 843, 1998.

96. O'Connor, P. B., Duursma, M. C., van Rooij, G. J., Heeren, R. M. A., and Boon, J. J., Correction of Time-of-Flight Shifted Polymeric Molecular Weight Distributions in Matrix-Assisted Laser Desorption/Ionization Fourier Transform Mass Spectrometry, *Anal. Chem.*, 69, 2751, 1997.

97. Pastor, S. J., Castoro, J. A., and Wilkins, C. L., High-Mass Analysis Using Quadrupolar Excitation/Ion Cooling in a Fourier Transform Mass Spectrometer, *Anal. Chem.*, 67, 379, 1995.

98. Dey, M., Castoro, J. A., and Wilkins, C. L., Determination of Molecular Weight Distributions of Polymers by MALDI-FTMS, *Anal. Chem.*, 67, 1575, 1995.

99. Castoro, J. A. and Wilkins, C. L., High Resolution Laser Desorption/Ionization Fourier Transform Mass Spectrometry, *Trends in Analytical Chemistry*, 13, 229, 1994.

100. Ens, W., Mao, Y., Mayer, F., and Standing, G., Properties of Matrix-assisted Laser Desorption. Measurements with a Time-to-Digital Converter, *Rapid Comm. Mass. Spectrom.*, 5, 117, 1991.

101. Hanson, C. D. and Just, C. L., Selective Background Suppression in MALDI-TOF Mass Spectrometry, *Anal. Chem.*, 66, 3676, 1994.

102. Benner, W. H., Horn, D. M., Jaklevic, J. M., Frank, M., Mears, C., Labov, S., and Barfknecht, A. T., Simultaneous Measurement of Flight Time and Energy of Large Matrix-assisted Laser Desorption Ionization Ions with a Superconducting Tunnel Junction Detector, *J. Am. Soc. Mass Spectrom.*, 8, 1094, 1997.

103. Montaudo, G., Montaudo M. S., Puglisi, C., and Samperi, F., Self-calibrating Property of Matrix-assisted Laser Desorption/Ionization Time-of-flight Spectra of Polymeric Materials, *Macromolecules*, 8, 981, 1994.

104. Montaudo, G., Montaudo, M. S., Puglisi, C., and Samperi, F., Determination of Absolute Mass Values in MALDI-TOF of Polymeric Materials by a Method of Self-Calibration of the Spectra, *Anal. Chem.*, 66, 4366, 1994.
105. Costa Vera, C., Zubarev, R., Ehting, H., Hakansson, P., and Sunqvist, B. U. R., A Three-point Calibration Procedure for Matrix-assisted Laser Desorption/ Ionization Mass Spectrometry Utilizing Multiply Charged Ions and Their Mean Initial Velocities, *Rapid Comm. Mass. Spectrom.*, 10, 1429, 1996.
106. Christian, N. P., Arnold, R. J., and Reily, J. P., Improved Calibration of Time-of-Flight Mass Spectra by Simplex Optimization of Electrostatic Ion Calculations, *Anal. Chem.*, 72, 3327, 2000.
107. Montaudo, G., Montaudo, M. S., Puglisi, C., and Samperi, F., Characterization of Polymers by Matrix-assisted Laser Desorption/Ionization Time-of-flight Mass Spectrometry: Molecular Weight Estimates in Samples of Varying Polydispersity, *Rapid Comm. Mass Spectrom.*, 9, 453, 1995.
108. Montaudo, G., Garozzo, D., Montaudo, M. S., Puglisi, C., and Samperi, F., Molecular and Structural Characterization of Polydisperse Polymers and Copolymers by Combining MALDI-TOF Mass Spectrometry with GPC Fractionation, *Macromolecules*, 28, 7983, 1995.
109. Montaudo, G., Montaudo, M. S., Puglisi, C., and Samperi, F., Molecular Weght Distribution of Poly(dimethylsiloxane) by Combining Matrix-assisted Laser Desorption/Ionization Time-of-flight Mass Spectrometry with Gel-permeation Chromatography Fractionation, *Rapid Comm. Mass Spectrom.*, 9, 1158, 1995.
110. Montaudo, G., MALDI-TOF as GPC Detector for MW & MWD Measurements in Polydisperse Polymers and Copolymers, *Polym. Preprints*, 290, 1996.
111. Montaudo, G., Montaudo, M. S., Puglisi, C., and Samperi, F., Molecular Weight Determination and Structural Analysis in Polydisperse Polymers by Hyphenated Gel Permeation Chromatography/Matrix-Assisted Laser Desorption Ionization-Time of Flight Mass Spectrometry, *Int. J. Polym. Anal. & Characterization*, 3, 177, 1997.
112. Montaudo, M. S., Puglisi, C., Samperi, F., and Montaudo, G., Application of Size Exclusion Chromatography Matrix-assisted Laser Desorption/Ionization Time-of-flight to the Determination of Molecular Masses in Polydisperse Polymers, *Rapid Comm. Mass Spectrom.*, 12, 519, 1998.
113. Lloyd, P. M., Suddaby, K. G., Varney, J. E., Scrivener, E., Derrick, P. J., and Haddleton, D. M., A Comparison Between Matrix-assisted Laser Desorption/ Ionization Time-of-flight Mass Spectrometry and Size Exclusion Chromatography in the Mass Characterisation of Synthetic Polymers with Narrow Molecular-mass Distributions: Poly(methyl methacrylate) and Poly(styrene), *Eur. Mass Spectrom.*, 1, 293, 1995.
114. Jackson, C., Larsen, B., and McEwen, C., Comparison of Most Probable Peak Values As Measured for Polymer Distributions by MALDI Mass Spectrometry and by Size Exclusion Chromatography, *Anal. Chem.*, 68, 1303, 1996.
115. Martin, K., Spickermann, J., Räder, H. J., and Müllen, K., Why Does Matrix-assisted Laser Desorption/Ionization Time-of-flight Mass Spectrometry Give Incorrect Results for Broad Polymer Distributions?, *Rapid Comm. Mass. Spectrom.*, 10, 1471, 1996.
116. Schriemer, D. C. and Li, L., Mass Discrimination in the Analysis of Polydisperse Polymers by MALDI Time-of-Flight Mass Spectrometry. 1. Sample Preparation and Desorption/Ionization Issues, *Anal Chem.*, 69, 4169, 1997.

117. Schriemer, D. C. and Li, L., Mass Discrimination in the Analysis of Polydisperse Polymers by MALDI Time-of-Flight Mass Spectrometry. 2. Instrumental Issues, *Anal Chem.*, 69, 4176, 1997.

118. Byrd, H. C. M. and McEwen, C. N., The Limitations of MALDI-TOF Mass Spectometry in the Analysis of Wide Polydisperse Polymers, *Anal Chem.*, 72, 4568, 2000.

119. Guo, B., Chen, H., and Liu, R. X., Observation of Varying Molecular Weight Distributions in Matrix-assisted Laser Desorption/Ionization Time-of-Flight Analysis of Polymethyl Methacrylate, *Rapid Comm. Mass. Spectrom.*, 11, 781, 1997.

120. Rashidzadeh, H. and Guo, B., Use of MALDI-TOF to Measure Molecular Weight Distributions of Polydisperse Poly(methyl methacrylate), *Anal. Chem.*, 70, 131, 1998.

121. Chen, H. and Guo, B., Use of Binary Solvent Systems in the MALDI-TOF Analysis of Poly(methyl methacrylate), *Anal. Chem.*, 69, 4399, 1997.

122. Zhu, H., Yalcin, T., and Li, L., Analysis of the Accuracy of Determining Average Molecular Weights of Narrow Polydispersity Polymers by Matrix-Assisted Laser Desorption/Ionization Time-of-Flight Mass Spectrometry, *J. Am. Soc. Mass Spectrom.*, 9, 275, 1998.

123. Montaudo, G., Scamporrino, E., Vitalini, D., and Mineo, P., Novel Procedure for Molecular Weight Averages Measurements of Polydisperse Polymers Directly from Matrix-assisted Laser Desorption/Ionization Time-of-flight Mass Spectra, *Rapid Comm. Mass Spectrom.*, 10, 1551, 1996.

124. Vitalini, D., Mineo, P., and Scamporrino, E., Further Application of a Procedure for Molecular Weight and Molecular Weight Distribution Measurement of Polydisperse Polymers from Their Matrix-Assisted Laser Desorption/Ionization Time-of-Flight Mass Spectra, *Macromolecules*, 30, 5285, 1997.

125. Yan, W., Ammon, D. M. Jr., Gardella, J. A., Maziarz, E. P. Jr., Hawkridge, A. M., Grobe III, G. L., and Wood, T. D., Quantitative Mass Spectrometry of Technical Polymers: A Comparison of Several Ionization Methods, *Eur. Mass Spectrom.*, 4, 467, 1998.

126. Hunt, S. M., Sheil, M. M., and Derrick, P. J., Comparison of Electrospray Ionisation Mass Spectrometry with Matrix-assisted Laser Desorption Ionisation Mass Spectrometry and Size Exclusion Chromatography for the Characterisation of Polyester Resins, *Eur. Mass Spectrom.*, 4, 475, 1998.

127. Parees, D. M. Hanton, S. D., Clark, P. A. C., and Willcox, D. A., Comparison of Mass Spectrometric Techniques for Generating Molecular Weight Information on a Class of Ethoxylated Oligomers, *J. Am. Soc. Mass Spectrom.*, 9, 282, 1998.

128. Lehrle, R. S. and Sarson, D. S., Polymer Molecular Weight Distributions: Results from Matrix-assisted Laser Description Ionization Compared with those from Gel-permeation Chromatography, *Rapid Comm. Mass Spectrom.*, 9, 91, 1995.

129. Räder, H. J., Spickermann, J., Kreyenschmidt, M., and Müllen, K., MALDI-TOF Mass Spectrometry in Polymer Analytics. 2a. Molecular Weight Analysis of Rigid-Rod Polymers, *Macromol. Chem. Phys.*, 197, 3285, 1996.

130. Guttman, C. M., Blair, W. R., and Danis, P. O., Mass Spectrometry and SEC of SRM 1487, a Low Molecular Weight Poly(methyl methacrylate) Standard, *J. Polym. Sci. B: Polym. Phys.*, 35, 2049, 1997.

131. Scamporrino, E., Maravigna, P., Vitalini, D., and Mineo, P., A New Procedure for Quantitative Correction of Matrix-assisted Laser Desorption/Ionization Time-of-flight Mass Spectrometric Response, *Rapid Comm. Mass. Spectrom.*, 12, 646, 1998.

132. Bletsos, I. V., Hercules, D. M., van Leyen, D., Hagenhoff, B., Niehuis, E., and Benninghoven, A., Molecular Weight Distributions of Polymers Using Time-of-Flight Secondary-Ion Mass Spectrometry, *Anal. Chem.*, 63, 1953, 1991.

133. Wilkins, C. L., Weil, D. A., Yang, C. L. C., and Ijames, C. F., High Mass Analysis by Laser Desorption Fourier Transform Mass Spectrometry, *Anal. Chem.*, 57, 520, 1985.

134. Brown, R. S., Weil, D. A., and Wilkins, C. L., Laser Desorption-Fourier Transform Mass Spectrometry for the Characterization of Polymers, *Macromolecules*, 19, 1255, 1986.

135. Vincenti, M., Pelizzetti, E., Guarini, A., and Costanzi, S., Determination of Molecular Weight Distributions of Polymers by Desorption Chemical Ionization Mass Spectrometry, *Anal. Chem.*, 64, 1879, 1992.

136. Chaudhary, A. K., Critchley, G., Diaf, A., Beckman, E. J., and Russell, A. J., Characterization of Synthetic Polymers Using Matrix-Assisted Laser Desorption/Ionization-Time of Flight Mass Spectrometry, *Macromolecules*, 29, 2213, 1996.

137. McEwen, C., Jackson, C., and Larsen, B., The Fundamentals of Characterizing Polymers using MALDI Mass Spectrometry, *Polym. Preprints*, 314, 1996.

138. Danis, P. O., Karr, D. E., Simonsick, W. J. Jr., and Wu, D. T., Matrix-Assisted Laser Desorption/Ionization Time-of-Flight Mass Spectrometry Characterization of Poly(butyl methacrylate) Synthesized by Group-Transfer Polymerization, *Macromolecules*, 28, 1229, 1995.

139. Hsieh, Quirk, *Anionic Polymerization*, Marcel Dekker, New York, 1996.

140. Garozzo, D., Impallomeni, G., Spina, E., Sturiale, L., and Zanetti, F., Matrix-assisted Laser Desorption/Ionization Mass Spectrometry of Polysaccharides, *Rapid Comm. Mass Spectrom.*, 9, 937, 1995.

141. Montaudo, G., Montaudo, M. S., Puglisi, C., and Samperi, F., Characterization of Polymers by Matrix-Assisted Laser Desorption Ionization-Time of Flight Mass Spectrometry. End-Group Determination and Molecular Weight Estimates in Poly(ethylene glycols), *Macromolecules*, 28, 4562, 1995.

142. Montaudo, G., Montaudo, M. S., Puglisi, C., Samperi, F., Spassky, N., LeBorgne, A., and Wisniewski, M., Evidence for Ester-Exchange Reactions and Cyclic Oligomer Formation in the Ring-Opening Polymerization of Lactide with Aluminum Complex Initiators, *Macromolecules*, 29, 6461, 1996.

143. Yalcin, T., Schriemer, D. C., and Li, L., Matrix-Assisted Laser Desorption Ionization Time-of-Flight Mass Spectrometry for the Analysis of Polydienes, *J. Am. Soc. Mass Spectrom.*, 8, 1220, 1997.

144. Penelle, J. and Xie, T., Ring Opening Polymerization of Diisopropyl Ciclopropane-1,1 Dicarboxylate under Living Anionic Conditions: A Kinetic and Mechanicistic Study, *Macromolecules*, 33, 4667, 2000.

145. Wilczek-Vera, G., Danis, P. O., and Eisenberg, A., Individual Block Length Distributions of Block Copolymers of Polystyrene-block-Poly(α-methylstyrene) by MALDI/TOF Mass Spectrometry, *Macromolecules*, 29, 4036, 1996.

146. Montaudo M. S., unpublished results.

147. Flory P., *Principles of Polymer Chemistry*, Cornell University Press, Ithaca, New York, 1971.

148. Coote, M. L., Zammit, M. D., and Davis, T. P., Determination of Free-radical Rate Coefficients Using Pulsed-laser Polymerization, *Trip*, 4, 189, 1996.

149. Zammit, M. D., Davis, T. P., and Haddleton, D. M., Determination of the Propagation Rate Coefficient (kp) and Termination Mode in the Free-Radical Polymerization of Methyl Methacrylate, Employing Matrix-Assisted Laser Desorption Ionization Time-of-Flight Mass Spectrometry for Molecular Weight Distribution Analysis, *Macromolecules*, 29, 492, 1996.

150. Zammit, M. D., Davis, T. P., Haddleton, D. M., and Suddaby, K. G., Evaluation of the Mode of Termination for a Thermally Initiated Free-Radical Polymerization via Matrix-Assisted Laser Desorption Ionization Time-of-Flight Mass Spectrometry, *Macromolecules*, 30, 1915, 1997.

151. Schweer, J., Sarnecki, J., Mayer-Posner, F., Müllen, K., Räder, H. J., and Spickermann, J., Pulsed-Laser Polymerization/Matrix-Assisted Laser Desorption/ Ionization Mass Spectrometry: An Approach Toward Free-Radical Propagation Rate Coefficients of Ultimate Accuracy?, *Macromolecules*, 29, 4536, 1996.

152. Kapfenstein, H. M. and Davis, T. P., Studies on the Application of Matrix-assisted Laser Desorption/Ionisation Time-of-flight Mass Spectrometry to the Determination of Chain Transfer Coefficients in Free Radical Polymerization, *Macromol. Chem. Phys.*, 199, 2403, 1998.

153. Montaudo, G., Mass Spectral Determination of Cyclic Oligomers Distributions in Polymerization and Degradation Reactions, *Macromolecules*, 24, 5289, 1991.

154. Spassky, N., Platzgraf, H., Simic, V., and Montaudo, M.S., Inter and Intramolecular Ester Exchange Reactions in the Ring Opening Polymerization of Lactides using Lanthanide Alkoxide Initators, *Macromol. Chem. Phys.*, 201, 2432, 2000.

155. Wilczek-Vera, G., Yu, Y., Waddell, K., Danis, P. O., and Eisenberg, A., Analysis of Diblock Copolymers of Poly(α-methylstyrene)-block-polystyrene by Mass Spectrometry, 32, 2180, 1999.

156. Schädler, V., Spickermann, J., Räder, H.-J., and Wiesner, U., Synthesis and Characterization of α,ω-Macrozwitterionic Block Copolymers of Styrene and Isoprene, *Macromolecules*, 29, 4865, 1996.

157. Suddaby, K. G., Hunt, K. H., and Haddleton, D. M., MALDI-TOF Mass Spectrometry in the Study of Statistical Copolymerization and Its Application in Examining the Free Radical Copolymerization of Methyl Methacrylate and *n*-Butyl Methacrylate, *Macromolecules*, 29, 8642, 1996.

158. Mormann, W., Walter, J., Pasch, H., and Rode, K., Copolymerization of MMA and Methyl α-Phenylacrylate by GTP, *Macromolecules*, 31, 249, 1998.

159. Kukulj, D., Heuts, J. P. A., and Davis, T. P., Copolymerization of Styrene and α-Methylstyrene in the Presence of a Catalytic Chain Transfer Agent, *Macromolecules*, 31, 6034, 1998.

160. Francke, V., Räder, H. J., Geerts, Y., and Müllen, K., Synthesis and Characterization of a Poly(*para*-phenyleneethylene)-*block*-poly(ethylene oxide) Rod-coil Block Copolymer, *Macromol. Rapid Commun.*, 19, 275, 1998.

161. Bednarek, M., Biedron, T., and Kubisa, P., Synthesis of Block Copolymers by Atom Transfer Radical Polymerization of *tert*-Butyl Acrylate with Poly(oxyethylene) Macroinitiators, *Macromol. Rapid Commun.*, 20, 59, 1999.

162. Servaty, S., Köhler, W., Meyer, W. H., Rosenauer, C., Spickermann, J., Räder, H.-J., and Wegner, G., MALDI-TOF-MS Copolymer Analysis: Characterization of a Poly(dimethylsiloxane)-co-Poly(hydromethylsiloxane) as a Precursor of a Functionalized Silicone Graft Copolymer, *Macromolecules*, 31, 2468, 1998.

163. Montaudo, M. S. and Samperi, F., Determination of Sequence and Composition in Poly(butyleneadipate-co-butyleneterephthalate) by Matrix-assisted Laser Desorption/Ionization Time-of-flight Mass Spectrometry, *Eur. Mass. Spectrom.*, 4, 459, 1999.

164. Montaudo, M. S., Puglisi, C., Samperi, F., and Montaudo. G., Structural Characterization of Multicomponent Copolyester by Mass Spectrometry, *Macromolecules*, 31, 8666, 1998.

165. Abate, R., Ballistreri, A., Montaudo, G., Garozzo, D., Impallomeni, G., Critchley, G., and Tanaka, K., Quantitative Applications of Matrix-assisted Laser Desorption/Ionization with Time-of-flight Mass Spectrometry: Determination of Copolymer Composition in Bacterial Copolyesters, *Rapid Comm. Mass Spectrom.*, 7, 1033, 1993.

166. Carroccio, S., Rizzarelli, P., and Puglisi, C., Matrix-assisted Laser Desorption/Ionisation Time-of-flight Characterisation of Biodegradable Aliphatic Copolyesters, *Rapid Comm. Mass Spectrom.*, 14, 1513, 2000.

167. Montaudo, M. S., Sequence Constraints in a Glycine-Lactic Acid Copolymer Determined by Matrix-assisted Laser Desorption/Ionization Mass Spectrometry, *Rapid Comm. Mass Spectrom.*, 13, 639, 1999.

168. Montaudo, G., Puglisi, C., Scamporrino, E., and Vitalini, D., Mass Spectrometric Analysis of the Thermal Degradation Products of Poly(o-,m-,p-phenylene sulfide) and of the Oligomers Produced in the Synthesis of these Polymers. *Macromolecules*, 19, 2157, 1986.

169. Montaudo, G., Puglisi, C., Scamporrino, E., and Vitalini, D., Separation and Characterization of Cyclic Sulfides Formed in the Polycondensation of Dibromoalkanes with Aliphatic Dithiols. *Macromolecules*, 19, 2869, 1986.

170. Hagenhoff, B., Benninghoven, A., Barthel, H., and Zoller, W., Supercritical Fluid Chromatography and Time-of-Flight Secondary Ion Mass Spectrometry of Poly(dimethylsiloxane) Oligomers in the Mass Range 1000-10000 Da, *Anal. Chem.*, 63, 2466, 1991.

171. Prokai, L. and Simonsick, W. J., Electrospray Ionization Mass Spectrometry Coupled with Size-Exclusion Chromatography. *Rapid Commun. Mass Spectrom.*, 7, 853, 1993.

172. Nielen, M. W. F. and Buijtenhuijs, F. A., Polymer Analysis by Liquid Chromatography/Electrospray Ionization TOF Mass Spectrometry, *Anal. Chem.*, 71, 1809, 1999.

173. Ballistreri, A., Garozzo, D., Giuffrida, M., Impallomeni, G., and Montaudo, G., Sequencing Bacterial Poly(β-hydroxybutyrate-co-valerate) by Partial Methanolysis, HPLC Fractionation and FAB Mass Spectrometry, *Macromolecules*, 22, 2107, 1989.

174. Ballistreri, A., Montaudo, G., Impallomeni, G., Lenz, R. W., Ulmer, H. W., and Fuller, R. C., Synthesis and Characterization of Polyesters Produced by Rhodospirillum Rubrum from Pentenoic Acid. *Macromolecules*, 23, 5059. 1990.

175. Prokai, L. and Simonsick, W. J., Electrospray Ionization Mass Spectrometry Coupled with Size-Exclusion Chromatography. *Rapid Commun. Mass Spectrom.*, 7, 853, 1993.

176. Pasch, H. and Rode, K., Use of Matrix-assisted Laser Desorption/Ionization Mass Spectrometry for Molar Mass-sensitive Detection in Liquid Chromatography of polymers, *J. Chromatogr.*, 699, 21, 1995.

177. Fei, X. and Murray, K., On-Line Coupling of Gel Permeation Chromatography with MALDI Mass Spectrometry, *Anal. Chem.*, 68, 3555, 1996.

178. Murray, K., Coupling Matrix-Assisted Laser Desorption/Ionization to Liquid Separations, *Mass Spectrom. Rev.*, 16, 283, 1997.

179. Nielen, M. W. and Malucha, S., Characterization of Polydisperse Synthetic Polymers by Size-exclusion Chromatography/Matrix-assisted Laser Desorption/Ionization Time-of-flight Mass Spectrometry, *Rapid Comm. Mass Spectrom.*, 11, 1194, 1997.

180. Danis, P. O., Saucy, D. A., and Huby, F. J., Application of MALDI/TOF Mass Spectrometry Coupled with Gel Permeation Chromatography, *Polym. Prep.*, 17, 311, 1996.

181. Montaudo, M. S., Puglisi, C., Samperi, F., and Montaudo, G., Molar Mass Distribution and Hydrodynamic Interactions in Random Copolyesters Investigated by Size Exclusion Chromatography/Matrix-Assisted Laser Desorption Ionization, *Macromolecules*, 31, 3839, 1998.

182. Puglisi, C., Samperi, F., Carroccio, S., and Montaudo, G., SEC/MALDI Analysis of Poly(Bisphenol A carbonate). 1. End-Groups and Molar Masses Determination, *Rapid Comm. Mass Spectrom.*, 13, 2260, 1999.

183. Puglisi, C., Samperi, F., Carroccio, S., and Montaudo, G., SEC/MALDI Analysis of Poly(Bisphenol A carbonate). 2. Self Association Due to Phenol End-Groups, *Rapid Comm. Mass Spectrom.*, 13, 2268, 1999.

184. Nielen, M. W. F., Polymer Analysis by Micro-Scale Size-Exclusion Chromatography/MALDI Time-of-Flight Mass Spectrometry with a Robotic Interface, *Anal. Chem.*, 70, 1563, 1998.

185. Barth, H. G., Boyes, B. E., and Jackson, C., Size Exclusion Chromatography, *Anal. Chem.*, 68, 445R, 1996.

186. Barth, H. G., Boyes, B. E., and Jackson, C., Size Exclusion Chromatography and Related Separation Techniques, *Anal. Chem.*, 70, 251R, 1998.

187. Kassis, C. E., DeSimone, J. M., Linton, R. W., Remsen, E. E., Lange, G. W., and Friedman, R. M., A Direct Deposition Method for Coupling Matrix-assisted Laser Desorption/Ionization Mass Spectrometry with Gel Permeation Chromatography for Polymer Characterization, *Rapid Comm. Mass Spectrom.*, 11, 1134, 1997.

188. Hanton, S. D. and Liu, M., GPC Separation of Polymer Samples for MALDI Analysis, *Anal Chem.*, 72, 4550, 2000.

189. Montaudo, M. S. and Montaudo, G, Bivariate Distribution in PMMA/PBA Copolymers by Combined SEC/NMR and SEC/MALDI Measurements, *Macromolecules*, 32, 7015, 1999.

190. Krüger, R.-P., Much, H., Schuls, G., and Wachsen, O., New Aspects of Determination of Polymer Heterogeneity by 2-Dimensional Orthogonal Liquid Chromatography and MALDI-TOF-MS, *Macromol. Symp.*, 110, 155, 1996.

191. Spickermann, J., Räder, H. J., Müllen, K., Müller, B., Gerle, M., Fischer, K., and Schmidt, M., MALDI-TOF Characterization of Macromonomers, *Macromol. Rapid. Commun.*, 17, 885, 1996.

192. Dwyer, J. and Botten, D., A Novel Sample Preparation Device for MALDI-MS, *Internat. Lab.*, 13A, 1997.

193. Pasch, H., Deffieux, A., Henze, I., Schappacher, M., and Rique-Lubert, L., Analysis of Macrocyclic Polystyrenes. 1. Liquid Chromatographic Investigations, *Macromolecules*, 29, 8776, 1996.

194. Pasch, H., Deffieux, A., Ghahary, R., Schappacher, M., and Rique-Lubert, L., Analysis of Macrocyclic Polystyrenes. 2. Mass Spectrometric Investigations, *Macromolecules*, 30, 98, 1997.

195. Pasch, H. and Gores, F., Matrix-assisted Laser Desorption/Ionization Mass Spectrometry of Synthetic Polymers: 2. Analysis of Poly(methyl methacrylate), *Polymer*, 36, 1999, 1995.

196. Semlyen, J. A., Ed., *Cyclic Polymers*, Elsevier, London, 1986, chap 3.

197. Runyon, J. R., Barnes, D. E., Rudd, J. F., and Tung, L. H., Multiple Detectors for Molecular Weight and Compositional Analysis of Copolymers by Gel Permeation Chromatography *J. Applied Polymer Sci.*, 13, 2359, 1969.

198. Uglea, C.V., *Liquid Chromatography of Oligomers*, Marcel Dekker, New York, 1996.

199. Braun, D., Henze, I., and Pasch, H., Functionality Type Analysis of Carboxy-terminated Oligostyrenes by Gradient High Performance Liquid Chromatography, *Macromol. Chem. Phys.*, 198, 3365, 1997.

200. Remmers, M., Müller, B., Martin, K., Räder, H.-J., and Köhler, W., Poly(p-phenylene)s. Synthesis, Optical Properties, and Quantitative Analysis with HPLC and MALDI-TOF Mass Spectrometry, *Macromolecules*, 32, 1073, 1999.

201. Lee, H., Lee, W., Chang, T., Choi, S., Lee, D., Ji, H., Nonidez, W. K., and Mays, J. W., Characterization of Poly(ethylene oxide)-block-poly(L-lactide) by HPLC and MALDI-TOF Mass Spectrometry, *Macromolecules*, 32, 4143, 1999.

202. Pasch, H., Liquid Chromatography at the Critical Point of Adsorption—A New Technique for Polymer Characterization, *Macromol. Symp.*, 110, 107, 1996.

203. Pasch, H., Hyphenated Techniques in Liquid Chromatography of Polymers, *Advan. Polym. Sci.*, 150, 1, 2000.

204. Kitayama, T., Janco, M., Ute, K., Niimi, R., Hatada, K., and Berek, D., Analysis of Poly(ethyl methacrylate)s by On-line Hyphenation of Liquid Chromatography at the Critical Adsorption Point and Nuclear Magnetic Resonance Spectroscopy, *Anal. Chem.*, 72, 1518, 2000.

205. Wachsen Reichert Kruger, R. P., Much, H., and Schultz, G., Thermal Decomposition of Biodegradable Polyesters. 3. Studies on the Mechanism Oligo-L-lactide, *Polym. Degrad. Stability*, 55, 225, 1997.

206. Montaudo, G., Montaudo, M. S., Puglisi, C., and Samperi, F., Characterization of End-Groups in Nylon 6 by MALDI-TOF Mass Spectrometry, *J. Polym. Sci. A: Polym. Chem.*, 34, 439, 1996.

207. Montaudo, G., Montaudo, M. S., Puglisi, C., Samperi, F., and Sepulchre, M., End-group Characterization of Poly(methylphenylsilane) by Alkali Metal Salts Doped MALDI-TOF Mass Spectra, *Macromol. Chem. Phys.*, 197, 2615, 1996.

208. Weidner, St. and Kühn, G., Chemical End-group Derivatization of Poly(ethyleneglycol)-Investigation by Matrix-assisted Laser Desorption/Ionisation Mass Spectrometry, *Rapid Comm. Mass. Spectrom.*, 10, 942, 1996.

209. Nagasaki, Y., Ogawa, R., Yamamoto, Kato, M., and Kataoka, K., Synthesis of Heterotelechelic Poly(ethylene glycol) Macromonomers. Preparation of Poly(ethylene glycol) Possessing a Methacryloyl Group at One End and a Formyl Group at the Other End, *Macromolecules*, 30, 6489, 1997.

210. Whittal, R. M., Li, L., Lee, S., and Winnik, M. A., Characterization of Pyrene End-labeled Poly(ethylene glycol) by High Resolution MALDI Time-of-flight Mass Spectrometry, *Macromol. Rapid Commun.*, 17, 59, 1996.

211. Völcker, N. H., Klee, D., Hanna, M., Höcker, H., Bou, J. J., Martinez de Ilarduya, A., and Munõz-Guerra, S., Synthesis of Heterotelechelic Poly(ethylene glycol)s and their Characterization by MALDI-TOF-MS, *Macromol. Chem. Phys.*, 200, 1363, 1999.

212. Barry, J. P., Carton, W. J., Pesci, K. M., Anselmo, R. T., Radtke, D. R., and Evans, J. V., Derivatization of Low Molecular Weight Polymers for Characterization by Matrix-assisted Laser Desorption/Ionization Time-of-flight Mass Spectrometry, *Rapid Comm. Mass. Spectrom.*, 11, 437, 1997.

213. Williams, J. B., Gusev, A. I., and Hercules, D. M., Use of Liquid Matrices for Matrix-Assisted Laser Desorption Ionization of Polyglycols and Poly(dimethylsiloxanes), *Macromolecules*, 29, 8144, 1996.

214. Esselbon, E., Fock, J., and Knebelkamp, A., Block Copolymers and Telechelic Oligomers by End-Group Reaction of Polymethacrylates, *Macromol. Symp.*, 91, 1996.

215. Maloney, D., Hunt, K. H., Lloyd, P. M., Muir, A. V. G., Richards, S. N., Derrick, P. J., and Haddleton, D. M., Polymethylmethacrylate End-group Analysis by Matrix-assisted Laser Desorption Ionisation Time-of-flight Mass Spectrometry (MALDI-TOF-MS), *J. Chem. Soc., Chem. Commun.*, 5, 561, 1995.

216. Thomson, B., Wang, Z., Paine, A., Lajoie, G., and Rudin, A., A Mass Spectrometry Investigation of the Water-Soluble Oligomers Remaining after the Emulsion Polymerization of Methyl Methacrylate, *J. Polym. Sci. Part A: Polym. Chem.*, 33, 2297, 1995.

217. Li, Y., Ward, D. G., Reddy, S. S., and Collins, S., Polymerization of Methyl Methacrylate Using Zirconocene Initiators-Polymerization Mechanisms and Applications, *Macromolecules*, 30, 1875, 1997.

218. Schlaad, H., Müller, A. H. E., Kolshorn, H., and Krüger, R.-P., Mechanism of Anionic Polymerization of (Meth)acrylates in the Presence of Aluminium Alkyls. 3. MALDI-TOF-MS Study on the Vinyl Ketone Formation in the Initiation Step of Methyl Methacrylate with *tert*-Butyl Lithium, *Polymer Bull.*, 35, 177, 1995.

219. Lu, Z. R., Kopeckovà, P., Wu, Z., and Kopecek, J., Synthesis of Semitelechelic Poly[N-(2-hydroxypropyl)methacrylamide] by Radical Polymerization in the Presence of Alkyl Mercaptans, *Macromol. Chem. Phys.*, 200, 2022, 1999.

220. Montaudo, G., Montaudo, M. S., Puglisi, C., and Samperi, F., 2-(4-Hydroxyphenylazo)-benzoic acid: A Solid Matrix for Matrix-assisted Laser Desorption/Ionization of Polystyrene, *Rapid. Commun. Mass Spectrom.*, 8, 1011, 1994.

221. Danis, P. O., Karr, D. E., Xiong, Y., and Owens, K. G., Methods for the Analysis of Hydrocarbon Polymers by Matrix-assisted Laser Desorption/Ionization Time-of-flight Mass Spectrometry, *Rapid. Commun. Mass Spectrom.*, 10, 862, 1996.

222. Dourges, M. A., Charleux, B., Vairon, J. P., Blais, J. C., Bolbach, G., and Tabet, J. C., MALDI-TOF Mass Spectrometry Analysis of TEMPO-Capped Polystyrene, *Macromolecules*, 32, 2495, 1999.

223. Weidner, St., Kühn, G., and Just, U., Characterization of Oligomers in Poly(ethylene terephthalate) by Matrix-assisted Laser Desorption/Ionisation Mass Spectrometry, *Rapid Comm. Mass. Spectrom.*, 9, 697, 1995.

224. Scamporrino, E., Vitalini, D., and Mineo, P., Synthesis and MALDI-TOF MS Characterization of High Molecular Weight Poly(1,2-dihydroxybenzene phthalates) Obtained by Uncatalyzed Bulk Polymerization of O,O'-Phthalid-3-ylidenecatechol or 4-Methyl-O,O'-phthalid-3-ylidenecatechol, *Macromolecules*, 29, 5520, 1996.

225. Weidner, St., Kühn, G., Friedrich, J., and Schröder, H., Plasmaoxidative and Chemical Degradation of Poly(ethylene terephthalate) Studied by Matrix-assisted Laser/Desorption Ionization Mass Spectrometry, *Rapid Comm. Mass. Spectrom.*, 10, 40, 1996.

226. Weidner, S., Kühn, G., Werthmann, B., Schroeder, H., Just, U., Borowski, R., Decker, R., Schwarz, B., Schmuecking, I., and Seifert, I., A New Approach of Characterizing the Hydrolytic Degradation of Poly(ethylene terephthalate) by MALDI-MS, *J. Polym. Sci. A: Polym. Chem.*, 35, 2183, 1997.

227. Guittard, J., Tessier, M., Blais, J. C., Bolbach, G., Rozes, L., Maréchal, E., and Tabet, J. C., Electrospray and Matrix-assisted Laser Desorption/Ionization Mass Spectrometry for the Characterization of Polyesters, *J. Mass Spectrom.*, 31, 1409, 1996.

228. Krause, J., Stoeckli, M., and Schlunegger, U. P., Studies on the Selection of New Matrices for Ultraviolet Matrix-assisted Laser Desorption/Ionization Time-of-flight Mass Spectrometry, *Rapid Commun. Mass Spectrom.*, 10, 1927, 1996.

229. Williams, J. B., Gusev, A. I., and Hercules, D. M., Characterization of Polyesters by Matrix-assisted Laser Desorption Ionization Mass Spectrometry, *Macromolecules*, 30, 3781, 1996.

230. Leukel, J., Burchard, W., Krüger, R. P., Much, H., and Schulz, G., Mechanism of the Anionic Copolymerization of Anhydride-cured Epoxies—Analyzed by Matrix-assisted Laser Desorption Ionization Time-of-flight Mass Spectrometry (MALDI-TOF-MS), *Macromol. Rapid Commun.*, 17, 359, 1996.

231. Kowalski, A., Duda, A., and Penczek, S., Mechanism of Cyclic Ester Polymerization Initiated with Tin(II) Octoate. 2. Macromolecules Fitted with Tin (II) Alkoxide Species Observed Directly in MALDI-TOF Spectra, *Macromolecules*, 33, 689, 2000.

232. Rokicki, G. and Kowalczyk, Synthesis of Oligocarbonate Diols and Their Characterization by MALDI-TOF Spectrometry, *Polymer*, 41, 9013, 2000.

233. Liu, J., Loewe, R. S., and McCullough, R. D., Employing MALDI-MS on Poly(alkylthiophenes): Analysis of Molecular Weights, Molecular Weight Distributions, End-Group Structures, and End-Group Modifications, *Macromolecules*, 32, 5777, 1999.

234. Visy, C., Lukkari, J., and Kankare, J., Electrochemically Polymerized Terthiophene Derivates Carrying Aromatic Substituents, *Macromolecules*, 27, 3322, 1994.

235. Mahon, A., Kemp, T. J., Varney, J. E., and Derrick, P. J., Ions Derived from Linear Polysulfide Oligomers Using Matrix-assisted Laser Desorption/Ionisation Time-of-flight Mass Spectrometry, *Polymer*, 39, 6213. 1998.

236. Allcock, H. R., Nelson, J. M., Prange, R., Crane, C. A., and de Denus, C. R., Synthesis of Telechelic Polyphosphazenes via the Ambient Temperature Living Cationic Polymerization of Amino Phosphoranimines, *Macromolecules*, 32, 5736, 1999.

237. Kéki, S., Deák, G., Mayer-Posner, F. J., and Zsuga, M., MALDI-TOF MS Characterization of Dihydroxy Telechelic Polyisobutylene, *Macromol. Rapid Commun.*, 21, 770, 2000.

238. Lee, H. and Lubman, D. M., Sequence-Specific Fragmentation Generated by Matrix-Assisted Laser Desorption/Ionization in a Quadrupole Ion Trap/Reflectron Time-of-Flight Device, *Anal. Chem.*, 67, 1400, 1995.

239. Katta, V., Chow, D. T., and Trohde, M., Application of In-Source Fragmentation of Protein Ions for Direct Sequence Analysis by Delayed Extraction MALDI-TOF Mass Spectrometry, *Anal. Chem.*, 70, 4410, 1998.

240. Kaufmann, R., Chaurand, P., Kirsch, D., and Spengler, B., Post-source Decay and Delayed Extraction in Matrix-assisted Laser Desorption/Ionization-Reflectron Time-of-Flight Mass Spectrometry. Are There Trade-offs?, *Rapid Comm. Mass. Spectrom.*, 10, 1199, 1996.

241. Garozzo, D., Nasello, V., Spina, E., and Sturiale, L., Discrimination of Isomeric Oligosaccharides and Sequencing of Unknowns by Post Source Decay Matrix-assisted Laser Desorption/Ionization Time-of-flight Mass Spectrometry, *Rapid Comm. Mass. Spectrom.*, 11, 1561, 1997.

242. Przybilla, L., Räder, H.-J, and Müllen, K., Post-source Decay Fragment Ion Analysis of Polycarbonates by Matrix-assisted Laser Desorption/Ionization Time-of-flight Mass Spectrometry, *Eur. Mass Spectrom.*, 5, 133, 1999.

243. Jackson, A. T., Yates, H. T., Scrivens, J. H., Critchley, G., Brown, J., Green, M. R., and Bateman, R. H., The Application of Matrix-assisted Laser Desorption/Ionization Combined with Collision-induced Dissociation to the Analysis of Synthetic Polymers, *Rapid Comm. Mass. Spectrom.*, 10, 1668, 1996.

244. Jackson, A. T., Yates, H. T., MacDonald, W. A., Scrivens, J. H., Critchley, G., Brown, J., Deery, M. J., Jennings, K. R., and Brookes, C., Time-Lag Focusing and Cation Attachment in the Analysis of Synthetic Polymers by Matrix-Assisted Laser Desorption Ionization-Time-of-Flight Mass Spectrometry, *J. Am. Soc. Mass Spectrom.*, 8, 132, 1997.

245. Jackson, A. T., Yates, H. T., Scrivens, J. H., Green, M. R., and Bateman, R. H., Matrix-Assisted Laser Desorption/Ionization-Collision Induced Dissociation (MALDI-CID) of Poy(Styrene), *J. Am. Soc. Mass Spectrom.*, 9, 269, 1998.

246. Jackson, A. T., Yates, H. T., Scrivens, J. H., Green, M. R., and Bateman, R. H., Utilizing Matrix-Assisted Laser Desorption/Ionization-Collision Induced Dissociation for the Generation of Structural Information from Poy(Alkyl Methacrylate)s, *J. Am. Soc. Mass Spectrom.*, 8, 1206, 1997.

247. Scrivens, J. H., Jackson, A. T., Yates, H. T., Green, M. R., Critchley, G., Brown, J., Bateman, R. H., Bowers, M. T., and Gidden, J., The Effect of the Variation of Cation in the Matrix-Assisted Laser Desorption/Ionization-Collision Induced Dissociation (MALDI-CID) Spectra of Oligomeric Systems, *Int. J. Mass Spectrom. Ion. Process.*, 165, 363, 1997.

248. Bottrill, A. R., Giannakopulos, A. E., Waterson, C., Haddleton, D. M., Lee, K. S., and Derrick, P. J., Determination of End-Groups of Synthetic Polymers by Matrix-Assisted Laser Desorption/Ionization: High-Energy Collision-Induced Dissociation, *Anal. Chem.*, 71, 3637, 1999.

249. Kwakkenbos, G., Muscat, D., van Benthem, R., and de Koster, C. G., Effect of In-source Decay on the Characterization of Hyperbranched Polyesteramides by MALDI-TOF-MS, *Proc. 47th ASMS Conf. Mass Spectrom. Allied Topics*, Dallas, TX, 1999.

250. Puapaiboon, U., Taylor, R. T., and Jai-nhuknann, J., Structural Confirmation of Polyurethane Dendritic Wedges and Dendrimers Using Post Source Decay Matrix-assisted Laser Desorption/Ionization Time-of-flight Mass Spectrometry, *Rapid Comm. Mass Spectrom.*, 13, 516, 1999.

251. Kowalski, P., Guttman, C., and Wallace, PSD Analysis of Polymers by MALDI TOF MS, *Proc. 46th ASMS Conf. Mass Spectrom. Allied Topics*, Orlando, FL, p. 1060, 1998.

252. Selby, T. L., Wesdemiotis, C., and Lattimer, R. P., Dissociation Characteristics of [M + X](+) Ions (X=H, Li, K) from Linear and Cyclic Polyglycols, *J. Am. Soc. Mass Spectrom.*, 5, 1081, 1994.

253. Varney, J. E., Derrick, P. J., Szilagyi, Z., and Vehey, K., Structural Elucidation of Synthetic Polymers Using Matrix-assisted Laser Desorption/Ionization Time-of-flight Mass Spectrometry Coupled with Post-source Decay, *Proc. 45th ASMS Conf. Mass Spectrom. Allied Topics*, Palm Springs, CA, p. 317, 1997.

254. Vitalini, D., Mineo, P., Di Bella, S., Fragalà I., Maravigna, P., and Scamporrino, E., Synthesis and Matrix-Assisted Laser Desorption Ionization-Time of Flight Characterization of an Exactly Alternating Copolycarbonate and Two Random Copolyethers Containing Schiff Base Copper(II) Complex Nonlinear Optical Units in the Main Chain, *Macromolecules*, 29, 4478, 1996.

255. Vitalini, D., Mineo, P., and Scamporrino, E., Synthesis and Characterization of Some Main Chain Porphyrin Copolyformals, Based on Bisphenol A and Long Linear Aliphatic Units, Having a Low Glass Transition Temperature, *Macromolecules*, 32, 60, 1999.

256. Vitalini, D., Mineo, P., and Scamporrino, E., Synthesis and Characterization of Some Copolyformals Containing Different Amounts of Fullerene Units, *Macromolecules*, 32, 4247, 1999.

257. Vitalini, D., Mineo, P., Iudicelli, V., Scamporrino, E., and Troina, G., Preparation of Functionalized Copolymers by Thermal Processes: Porphyrination and Fullerenation of a Commercial Polycarbonate, *Macromolecules*, 33, 7300, 2000.

258. Yoshida, S., Yamamoto, S., and Takamatsu, T., Detailed Structural Characterization of Modified Silicone Copolymers by Matrix-Assisted Laser Desorption/Ionization Time-of-flight Mass Spectrometry, *Rapid Comm. Mass Spectrom.*, 12, 535, 1998.

259. Montaudo, M. S. and Montaudo G., Further Studies on the Composition and Microstructure of Copolymers by Statistical Modeling of Their Mass Spectra, *Macromolecules*, 25, 4264, 1992.

260. Räder, H. J., Spickermann, J., and Müllen, K., MALDI-TOF Mass Spectrometry in Polymer Analytics: 1. Monitoring the Polymer-Analogous Sulfonation Reaction of Poly(Styrene), *Macromol. Chem. Phys.*, 196, 3967, 1995.

261. Höger, S., Spickermann, J., Morrison, D. L., Dziezok, P., and Räder, H. J., Aggregates of Shape Persistent Macrocyclic Amphiphiles Detected by MALDI-TOF Spectroscopy, *Macromolecules*, 30, 3110, 1997.

262. Klärner, G., Former, C., Martin, K., Räder, H.-J., and Müllen, K., Connective CC Double Bond Formation for the Synthesis of Donor- and Acceptor-Substituted Poly(*p*-phenylenevinylene)s, *Macromolecules*, 31, 3571, 1998.

263. Dubois, F., Knochenmuss, R., and Zenobi, R., Can Complexation Constants in Solution be Measured by Matrix-assisted Laser Desorption/Ionization Mass Spectrometry?, *Eur. Mass Spectrom.*, 5, 267, 1999.

264. Puglisi, C., Samperi, F., Carroccio, S., and Montaudo, G., MALDI-TOF Investigation of Polymer Degradation Pyrolysis of Poly(Bisphenol A carbonate), *Macromolecules*, 32, 8821, 1999.

265. Lattimer, R. P., Polce, M. J., and Wesdemiotis, C., MALDI-MS Analysis of Pyrolysis Products from a Segmented Polyurethane, *J. Anal. Appl. Pyrolysis*, 48, 15, 1998.

266. Lattimer, R. P., Mass Spectral Analysis of Low-temperature Pyrolysis Products from Poly(ethylene glycol), *J. Anal. Appl. Pyrolysis*, 56, 61, 2000.

267. Lattimer, R. P., Mass Spectral Analysis of Low-temperature Pyrolysis Products from Poly(tetrahydrofuran), *J. Anal. Appl. Pyrolysis*, 57, 57, 2001.

268. Burkoth, A. K. and Anseth, K., MALDI-TOF Characterization of Highly Cross-Linked, Degradable Polymer Networks, *Macromolecules*, 32, 1438, 1999.

269. Barton, Z., Kemp, T. J., Buzy, A., and Jennings, K. R., Mass Spectral Characterization of the Thermal Degradation of Poly(propylene oxide) by

Electrospray and Matrix-Assisted Laser Desorption Ionization, *Polymer*, 36, 4927, 1995.

270. Bürger, H. M., Müller, H.-M., Seebach, D., Börnsen, K. O., Schär, M., and Widmer, H. M., Matrix-Assisted Laser Desorption as a Mass Spectrometry Tool for the Analysis of Poly[(R)-3-hydroxybutanoates]. Comparison with Gel Permeation Chromatography, *Macromolecules*, 26, 4783, 1993.

271. Mehl, J. T., Murgasova, R., Dong, X., and Hercules, D. M., Characterization of Polyether and Polyester Polyurethane Soft Blocks Using MALDI Mass Spectrometry, *Anal. Chem.*, 72, 2490, 2000.

272. Zoller, D. L. and Johnston, M. V., Microstructure of Butadiene Copolymers Determined by Ozonolysis/MALDI Mass Spectrometry, *Macromolecules*, 33, 1664, 2000.

273. A. K. and Anseth, K. S., MALDI-TOF Characterization of Highly Cross-Linked, Degradable Polymer Networks, *Macromolecules*, 32, 1438, 1999.

274. Kemp, T. J., Berridge, R., Eason, M. D., and Haddleton, D. M., Spectroscopic, Physical and Product Studies of the Thermal Degradation of Poly(ethylene glycol) Containing a 1,3-Disubstituted Phenolic Group, *Polym. Deg. Stab.*, 64, 329, 1999.

275. Chionna, D., Puglisi, C., Samperi, F., Carroccio, S., Turturro, A., and Montaudo, G., Structure of the Species Produced in the Thermal Oxidative Degradation of Nylon 6 Determined by MALDI-TOF Mass Spectrometry, *Macromol. Rapid Comm.*, 22, 524, 2001.

276. Linnemayr, K., Vana, P., and Allmaier, G., Time-delayed Extraction Matrix-assisted Laser Desorption Ionization Time-of-flight Mass Spectrometry of Poly-acrylnitrile and Other Synthetic Polymers with the Matrix 4-Hydroxyben-zylidene Malononitrile, *Rapid Comm. Mass Spectrom.*, 12, 1344, 1998.

277. Pithawalla, Y. B., Gao, J., Yu, Z., and El-Shall, S., Even/Odd Alternation in Styrene Cluster Ions. Evidence for Multiple Cyclization During the Early Stages of Polymerization and the Inhibition Effect of Water, *Macromolecules*, 29, 8558, 1996.

278. Deery, M. J., Jennings, K. R., Jasieczek, C. B., Haddleton, D. M., Jackson, A. T., Yates, H. T., and Scrivens, J. H., A Study of Cation Attachment to Polystyrene by Means of Matrix-Assisted Laser Desorption/Ionization and Electrospray Ionization-Mass Spectrometry, *Rapid Commun. Mass Spectrom.*, 11, 57, 1997.

279. Rashidezadeh, H. and Guo, B., Investigation of Metal Attachment to Polystyrenes in Matrix-Assisted Laser Desorption Ionization, *J. Am. Soc. Mass Spectrom.*, 9, 724, 1998.

280. Shen, X., He, X., Chen, G., Zhou, P., and Huang, L., MALDI-TOF Mass Spectrometry Characterization of C_{60} End-capped Polystyrene Prepared by ATRP, *Macromol. Rapid Commun.*, 21, 1162, 2000.

281. Larsen, S. B., Simonsick, W. J. Jr., and McEwen, C. N., Fundamentals of the Applications of Matrix-Assisted Laser Desorption-Ionization Mass Spectrometry to Low Mass Poly(methylmethacrylate) Polymers, *J. Am. Soc. Mass Spectrom.*, 7, 287, 1996.

282. Lehrle, R. S. and Sarson, D. S., Degradation and Selective Desorption of Polymer Can Cause Uncertainty in MALDI (Matrix-assisted Laser Desorption Ionisation) Measurements, *Polym. Deg. and Stab.*, 51, 197, 1996.

283. Sakurada, N., Fukuo, T., Arakawa, R., Ute, K., and Hatada, K., Characterization of Poly(methyl methacrylate) by Matrix-assisted Laser Desorption Ionization Mass Spectrometry. A Comparison with Supercritical Fluid Chromatography and Gel Permeation Chromatography, *Rapid Comm. Mass Spectrom.*, 12, 1895, 1998.

284. Giannakopulos, A. E., Bahir, S., and Derrick, P. J., Comment: Reproducibility of Spectra and Threshold Fluence in Matrix-assisted Laser Desorption/Ionisation (MALDI) of Polymers, *Eur. Mass. Spectrom.*, 4, 127, 1998.

285. Lee, S., Winnik, M. A., Whittal, R. M., and Li, L., Synthesis of Symmetric Fluorescently Labeled Poly(ethylene glycols) Using Phosphoramidites of Pyrenebutanol and Their Characterization by MALDI Mass Spectrometry, *Macromolecules*, 29, 3060, 1996.

286. Pasch, H., Unvericht, R., and Resch, M., Analysis of Epoxy Resins by Matrix-assisted Laser Desorption/Ionization Mass Spectrometry, *Ang. Makromol. Chem.*, 212, 191, 1993.

287. Matsumoto, K., Shimazu, H., and Yamaoka, H., Synthesis and Characterization of Polysilabutane Having Oligo(Oxyethylene)phenyl Groups on the Silicon Atom, *J. Polym. Sci.: Polym. Chem.*, 36, 225, 1998.

288. Mandal, H. and Hay, A. S., Synthesis and Characterization by MALDI-TOF MS of Polycyclic Siloxanes Derived from *p*-Substituted Novolac Resins by MALDI-TOF MS, *J. Polym. Sci.: Polym. Chem.*, 36, 2429, 1998.

289. Marie A., Fournier, F., and Tabet, J. C., Characterization of Synthetic Polymers by MALDI-TOF/MS: Investigation into New Methods of Sample Target Preparation and Consequence on Mass Spectrum Finger Print, *Anal. Chem.*, 72, 5106, 2000.

290. Blais, J. C., Tessier, M., Bolbach, G., Remaud, B., Rozes, L., Guittard, J., Brunot, A., Maréchal, E., and Tabet, J. C., Matrix-assisted Laser Desorption Ionisation Time-of-flight Mass Spectrometry of Synthetic Polyesters, *Int. J. Mass Spectrom. Ion Processes*, 144, 131, 1995.

291. Danis, P. O., Karr, D. E., Mayer, F., Holle, A., and Watson, C. H., The Analysis of Water-soluble Polymers by Matrix-assisted Laser Desorption Time-of-flight Mass Spectrometry, *Org. Mass Spec.*, 27, 843, 1992.

292. Eggert, M. and Freitag, R., Poly-N,N-diethylacrylamide Prepared by Group Transfer Polymerization: Synthesis, Characterization, and Solution Properties, *J. Polym. Sci. Part A: Polym. Chem.*, 32, 803, 1994.

293. Freitag, R., Baltes, T., and Eggert, M., A Comparison of Thermoreactive Water-Soluble Poly-N,N-diethylacrylamide Prepared by Anionic and by Group Transfer Polymerization, *J. Polym. Sci. Part A: Polym. Chem.*, 32, 3019, 1994.

294. Murata, H., Sanda, F., and Endo, T., Synthesis and Radical Polymerization of a Novel Acrylamide Having an *α*-Helical Peptide Structure in the Side Chain, *J. Polym. Sci. Part A: Polym. Chem.*, 36, 1679, 1998.

295. Wallace, W. E., Guttman, C. M., and Antonucci, J. M., Molecular Structure of Silsesquioxanes Determined by Matrix-Assisted Laser Desorption/Ionization Time-of-Flight Mass Spectrometry, *J. Am. Soc. Mass Spectrom.*, 10, 224, 1999.

296. Hult, A., Johansson, M., and Malmstrom, E., Hyperbranched Polymers, *Advan. Polym. Sci.*, 143, 1, 1999.

297. Hawker, C., Dendridic and Hyperbranched Macromolecules. Precisely Controlled Macromolecular Architectures, *Advan. Polym. Sci.*, 147, 133, 1999.

298. Yu, D., Vladimirov, N., and Fréchet, J. M. J., MALDI-TOF in the Characterization of Dendritic-Linear Block Copolymers and Stars, *Macromolecules*, 32, 5186, 1999.

299. Sahota, H. S., Lloyd, P. M., Yeates, S. G., Derrick, P. J., Taylor, P. C., and Haddleton, D. M., Characterisation of Aromatic Polyester Dendrimers by Matrix-assisted Laser Desorption Ionisation Mass Spectrometry, *J. Chem. Soc., Chem. Commun.* 2445, 1994.

300. Mowat, I. A., Donovan, R. J., Feast, W. J., and Stainton, N. M., Matrix-assisted Laser Desorption/Ionization Mass Spectrometry of Aryl Ester Dendrimers, *Eur. Mass Spectrom.*, 4, 451, 1998.

301. Leon, J. W. and Fréchet, J. M. J., Analysis of Aromatic Polyether Dendrimers and Dendrimer-linear Block Copolymers by Matrix-assisted Laser Desorption Ionization Mass Spectrometry, *Polym. Bull.*, 35, 449, 1995.

302. Blais, J. C., Turrin, C. O., Caminade, A., and Majorat, J. P., MALDI-TOF Mass Spectrometry for the Characterization of Phosphorus-containing Dendrimers. Scope and Limitations, *Anal. Chem.*, 72, 5097, 2000.

303. Puapaiboon, U. and Taylor, R., Characterization and Monitoring Reaction of Polyurethane Dendritic Wedges and Dendrimers Using Matrix-assisted Laser Desorption/Ionization Time-of-flight Mass Spectrometry, *Rapid Comm. Mass Spectrom.*, 13, 508, 1999.

304. Martínez, C. A. and Hay, A. S., Preparation of Hyperbranched Macromolecules with Aryl Fluoride and Phenol Terminal Functionalities Using New Monomers and Cs_2CO_3 or $Mg(OH)_2$ as the Condensation Agent, *J. Polym Sci.: Polym. Chem.*, 35, 2015, 1997.

305. Sunder, Hanselmann, H., Frey, H., and Mülhaupt, R., Controlled Synthesis of Hyperbranched Polyglycerols by Ring-Opening Multibranching Polymerization, *Macromolecules*, 32, 4240, 1999.

306. Gooden, J. K., Gross, M. L., Mueller, A., Stefanescu, A. D., and Wooley, K. L., Cyclization in Hyperbranched Polymer Syntheses: Characterization by MALDI-TOF Mass Spectrometry, *J. Am. Chem. Soc.*, 120, 10180, 1998.

307. Weidner, S., Holländer, A., and Kühn, G., Matrix-assisted Laser Desorption/Ionization Mass Spectrometry of C_{36}-Alkane Following Degradation by Vacuum-ultraviolet Radiation, *Rapid Comm. Mass Spectrom.*, 11, 447, 1997.

308. Kübel, C., Chen, S.-L., and Müllen, K., Poly(4'-vinylhexaphenylbenzene)s: New Carbon-Rich Polymers, *Macromolecules*, 31, 6014, 1998.

309. Latourte, L., Blais, J.-C., and Tabet, J.-C., Desorption Behavior and Distributions of Fluorinated Polymers in MALDI and Electrospray Ionization Mass Spectrometry, *Anal. Chem.*, 69, 2742, 1997.

310. Kricheldorf, H. R. and Eggerstedt, S., MALDI-TOF Mass Spectrometry of Tin-initiated Macrocyclic Polylactones in Comparison to Classical Mass-spectroscopic Methods, *Macromol. Chem. Phys.*, 200, 1284, 1999.

311. Kricheldorf, H. R., Langanke, D., Spickermann, J., and Schmidt, M., Macrocycles. 10. Macrocyclic Poly(1,4-butanediol-esters)s by Polycondensation of 2-Stanna-1,3-dioxepane with Dicarboxylic Acid Chlorides, *Macromolecules*, 32, 3559, 1999.

312. Jedlinski, Z., Juzwa, M., Adamus, G., Kowalczuk, M., and Montaudo, M., Anionic Polymerization of Pentadecanolide. A New Route to a Potentially Biodegradable Aliphatic Polyester, *Macromol. Chem. Phys.*, 197, 2923, 1996.

313. Ding, Y. and Hay, A. S., Cyclic Aromatic Disulfide Oligomers: Synthesis and Characterization, *Macromolecules*, 29, 6386, 1996.

314. Wang, Y.-F. and Hay, A. S., A Facile Synthesis and the Polymerization of Macrocyclic 1,4-Phenylene Sulfid (PPS) Oligomers, *Macromolecules*, 29, 5050, 1996.

315. Ding, Y. and Hay, A. S., Synthesis of Macrocyclic Aryl Ethers Containing the Tetraphenylbenzene Moiety, *Macromolecules*, 29, 3090, 1996.

316. Xie, D., Ji, Q., and Gibson, H. W., Synthesis and Ring-Opening Polymerization of Single-Sized Aromatic Macrocycles for Poly(arylene ether)s, *Macromolecules*, 30, 4814, 1997.

317. Wang, Y.-F. and Hay, A. S., Macrocyclic Arylene Ether Ether Sulfide Oligomers: New Intermediates for the Synthesis of High-Performance Poly(arylene ether ether sulfide)s, *Macromolecules*, 30, 182, 1997.

318. Yoshimura, K., Przybilla, L., Ito, S., Brand, J. D., Wehmeir , M., Räder, H. J., and Müllen, K., Characterization of Large Synthetic Polycyclic Aromatic Hydrocarbons by MALDI- and LD-TOF Mass Spectrometry, *Macromol. Chem. Phys.*, 202, 215, 2001.

11

Two-Step Laser Desorption Mass
Spectrometry

Mattanjah S. de Vries and Heinrich E. Hunziker

CONTENTS

11.1 Introduction

The study of organic polymers by mass spectrometry has been an active field
of research in recent years. Both time-of-flight (TOF) and Fourier transform
ion cyclotron resonance mass spectrometry (FTMS) have been used in combination with a variety of ionization methods. These include secondary ion

mass spectrometry (SIMS),[1] matrix-assisted laser desorption and ionization (MALDI),[2] and neutral laser desorption, combined with cationization,[3-5] and photoionization. The challenge in such studies lies in getting the molecules into the gas phase and ionizing them without fragmentation. This is an important objective because potential new contributions of mass spectrometry to polymer analysis may result from the possibility of obtaining exact oligomeric distributions. Traditional analysis methods, such as nuclear magnetic resonance (NMR) and various forms of chromatography, generally provide average distributions or envelopes. Details of chain length distributions can easily be missed. For example bimodal distributions, branching variations, functional group distributions, or exact compositions of copolymers are all difficult to deduce from averaged measurements. Mass spectrometry therefore holds great promise for advanced polymer analysis, provided one can obtain parent mass distributions without fragmentation, without mass bias, and with exact mass measurements.

There is now an arsenal of techniques that have been shown to be capable of providing mass spectra of whole polymers. A major subgroup of these new methods uses pulsed lasers to vaporize and ionize the high molecular weight material.[6] Two important effects have been put to use in this approach to aid in the production of stable molecular ions: Chemical ionization and matrix-assisted desorption. Wilkins and others have shown many examples where cationization with alkali ions, added to the polymer as salts, yields oligomeric mass distributions.[4,7-9] Hillenkamp and coworkers have originated and developed the MALDI technique, in which a low molecular weight organic matrix serves the dual function of entraining large molecules into the vapor phase without decomposition, and ionizing them chemically, usually by proton transfer.[2,10] These techniques are discussed elsewhere in this book. One general drawback of cationization and MALDI is the need for sample preparation. Another limitation is that matrices and dopants are often material-specific; in fact not all materials lend themselves to these types of analysis. An example is the class of perfluorinated polyether polymers (PFPEs), which we will discuss below.

In this chapter we concern ourselves with the laser-induced vaporization of neat polymers. In this case, without ionic dopants or ions created by laser-induced breakdown, there is no chemical ionization and the gas phase polymer molecules are neutral. Post-ionization is thus necessary to convert them into ions. Since this is usually done by a second pulsed laser the procedure has also been called two-step laser mass spectrometry.[11-15]

11.2   Step 1: Laser Desorption

While thermal desorption of polymers generally causes thermal degradation, one can generalize that many polymers—as well as other complex molecules—can be lifted into the gas phase intact, provided the heating rate is fast enough. Heating rates of the order of 10^{11} degrees Kelvin/second are easily achieved with commercial nanosecond pulsed lasers and suffice in many cases. The extension of the method to shorter pulse lengths is not yet extensively investigated and may offer further improvements. Some theoretical models exist to explain the phenomenon with very limited systematic experimental data for comparison. Hall discusses laser desorption in terms of a competition between rates for desorption and thermal chemistry.[16] With a high pre-exponential factor, the former would prevail at higher temperatures. Therefore reaching higher temperatures sufficiently quickly would lead to preferential desorption. Zare and Levine discuss laser desorption in terms of a bottleneck model.[17] One assumes that most of the laser energy is absorbed by the substrate rather than by the adsorbate. At the nanosecond time scale, far from thermal equilibrium, the energy redistributes in such a way that the bond between the molecule and the surface is the first weak link that is encountered. Thus this bond acts as the "bottleneck" and is most likely to be broken, leading to intact desorption. Experimentally, Li et al. have varied the time scale for cooling after the laser pulse by preparing layered substrates with materials of different heat conductance.[18] Spectroscopic studies of desorbed material have been reported by Zenobi on aniline[11] and by D.H. Levy on indole[19] and on Trp-Gly.[20] These studies show, among other things, that these molecules come off the surface with internal temperatures above room temperature, but well below the calculated surface temperature. Much work still remains to be done in order to completely characterize the process.

In general, laser desorption produces neutrals in excess of ions by probably several orders of magnitude. In terms of sensitivity, on the other hand, several orders of magnitude can be lost in the ionization step. As mentioned before, both MALDI and electrospray produce ions directly in the ionization step. The advantage of desorbing neutrals and thus separating the ionization from the volatilization step will be illustrated in the examples below.

11.3 Step 2: Laser Ionization

In the techniques that we discuss here, the ionization step is always achieved separately with a second pulsed laser. In all cases we aim to produce a very soft ionization pathway in order to minimize fragmentation. Depending on the analytical challenge, ionization can either be as general as possible, or just the opposite, very selective for specific polymers. The most obvious way to achieve ionization with pulsed lasers is photoionization, which we discuss in Sections 11.3.3 and 11.3.4. However, some molecular species resist photoionization by virtue of a very high ionization potential and the absence of a suitable chromophore. Therefore in Sections 11.3.1 and 11.3.2, we will discuss five indirect approaches, in which the ionization laser merely generates the ionizing agent, either in the form of ions or of low energy electrons.

First: *Laser-generated cationization.* In Section 11.3.1 we show how one can employ a laser, focused on a transition metal target, in order to generate a metal plasma. The resulting metal ions subsequently cationize neutral polymers that were desorbed by an independent desorption laser at low fluence.

Second: *Photoelectron attachment.* In Section 11.3.2 we follow a similar idea, in which the ionizing laser produces photoelectrons from a suitable metal target. These very low energy electrons can then attach to independently desorbed neutral polymers in order to form negative ions.

Third: *Single-photon ionization.* Conceptually the simplest and most general form of photoionization is single-photon ionization by a photon with energy exceeding the vertical ionization potential. We have found that for larger molecules, such as polymers, this approach usually leads to fragmentation, unless the desorbed molecules are effectively cooled in order to minimize internal energy. We achieve this by jet cooling. In order for this approach to be effective for most polymers, photon energies in the vacuum ultraviolet region of the spectrum are required.

Fourth: *Two-photon ionization.* For polymers with a suitable chromophore, particularly aromatic groups, we can use two-photon ionization, which makes it possible to perform photoionization at commercial laser wavelengths. The opportunity to match a wavelength with a specific type of chromophore can provide a certain degree of selectivity of detection. This approach can also successfully be applied to ionize polymers with end-group chromophores.

Fifth: *Resonance-enhanced multiphoton ionization (REMPI).* In the most sophisticated form of photoionization we vary the wavelength of an excitation laser, such that a second photon (from the same or another laser) ionizes only when the first photon is resonant with a specific molecular level. Thus we obtain a combination of optical spectroscopy and mass spectrometry.

11.3.1   Laser-Generated Cationization

As a first approach to post-ionization of desorbed neutral polymers we discuss laser-generated cationization. This approach differs from other desorption/cationization techniques because the desorption step and the cationization step are completely separate. First polymers are gently laser desorbed at low fluence with one laser. Independently atomic metal ions are generated by a second pulsed laser that is tightly focused on a metal surface to create a plume of metal ions. Gas phase collisions above the sample surface subsequently produce the cationized complexes. Separation of the desorption and cationization processes allows independent optimization of the two laser/material interactions. This approach is especially useful in situations where other ionization methods fail, such as the example of PFPEs discussed here.[21]

Perfluorinated polyethers are widely used throughout industry as lubricants. Compared to their hydrocarbon analogues, they are more chemically inert and thermally stable, and have very low vapor pressures at room temperature.

This makes them very attractive for use in harsh environments and areas where polymer degradation is a concern. These polymers are commercially available with average molecular weights ranging from 2000 to 10000. A typical structure is as follows:

$$E_1-O-[R_3]-E_2 \qquad (11.1)$$

where the repeat group $R_i = [CF_2]_i-0$, $i = 1, 2, 3$, and E_i represent the end-groups.

11.3.1.1 Experimental Details

All mass spectra were obtained with a Fourier transform ion cyclotron resonance mass spectrometer (FTMS). The theory and applications of FTMS spectrometry are well-established and will not be discussed here.[22,23] In our apparatus the sample was located at one of the trap plates of the cell, inside a superconducting magnet. Two pulsed lasers were used, one for the PFPE desorption and another for metal ion formation. Typically, 2 to 4 mJ/pulse of 248 nm or 193-nm light (20 ns fwhm) from an excimer laser was softly focused to an elliptical spot with a 2-mm long axis for desorption. To create the metal ions, a 0.3-mJ pulse of 532-nm light (10 ns fwhm) from the doubled output of a Nd:YAG was tightly focused. The metal substrates were prepared from foils of various metals (typically 0.010 in. thick). Surface preparation was not critical because the experiment was not sensitive to PFPE/metal surface interactions since the thickness of the polymer films was on the order of 1 μm.

11.3.1.2 Molecular Weight Distributions

We have applied this technique to determine the chain length distribution and average molecular weight of PFPE samples of type P1, with nonfunctionalized end-groups:

$$E_1 = -CF_2-CF_3$$

$$E_2 = -CF_2-CF_2-CF_3.$$

This material is commercially produced under the brand names Demnum-S65® or Krytox®, the latter having branched repeat groups. A representative mass spectrum obtained by this method is shown in Figure 11.1. There are other methods that can be used to determine average molecular weights such as HPLC and ^{19}F-NMR. However, one gets limited information on the distribution of chain lengths or variations in the end-groups of the polymer molecules from these methods.

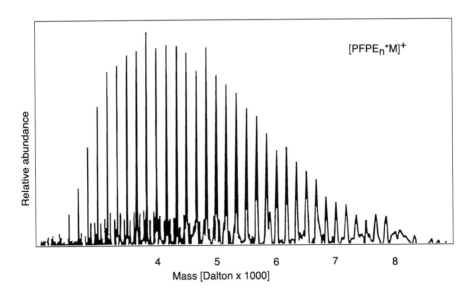

FIGURE 11.1
Mass spectrum of cationized Demnum S-65, having structure I and $\langle M \rangle$ =4600 Da. The PFPE
film was on an Au/Ni coated Cu sample. The desorbed polymer was cationized by Ni+.

The method for calculation of the average molecular weight of a polymer
sample using FTMS has been described previously.[4] Each peak in a mass
spectrum corresponds to a given chain length and set of end-groups. At the
low resolution used, each peak encompasses the distribution of isotopes
associated with that species. To account for the change in resolution of FTMS
with increasing mass and the change in the isotope distribution with an increasing number of C-atoms, it is necessary to measure the area under each peak
to obtain the relative amount of that species in a given sample. The assumption is made that the desorption efficiency remains constant for the different
length PFPEs in a given sample. This appears to hold true for higher molecular weight samples ($MW > 2000$) but not for those with molecular weights
under ~2000. It is also assumed that there is no bias due to nonuniform
excitation or detection of ions of different mass in the ICR cell.

The number average molecular weight, $\langle M \rangle$, is calculated according to

$$\langle M \rangle = \frac{\Sigma I_n M_n}{\Sigma I_n}$$

When using this equation with I_n equal to the area under each peak, we
found that $\langle M \rangle$ came out higher than the values obtained from ^{19}F-NMR measurements. However, if one makes the simple assumption that the cationization
cross section increases linearly with the polymer chain length

$$I_n = \text{area}(n)/n \qquad (11.2)$$

TABLE 11.1

Comparison of $\langle M \rangle$ Determined by FTMS and ^{19}F NMR

Sample	FTMS	^{19}F
Demnum S65	4600 ± 85	4560 ± 0
	3970 ± 10	4030 ± 50

a much better agreement results. The average molecular weights calculated this way for two different polymer samples and the values obtained for the same samples using ^{19}F NMR are presented in Table 11.1. The uncertainties in the $\langle M \rangle$ values determined by FTMS represent the standard deviation of five independent measurements. The agreement between the two methods is very good, attesting to the validity of using this method to determine chain length distributions of PFPE compounds.

11.3.2 Electron Attachment

As a variation to the laser-generated cationization technique, we can also generate electrons, rather than ions, by focusing a laser on a metal surface. With the photon energy exceeding the work function of the metal, photoelectrons are ejected with very low kinetic energy. Those very low energy photoelectrons can serve as a soft ionization source that potentially causes much less fragmentation than the traditional 70 eV electron impact ionization source. One can also use photoelectrons to generate negative ions by electron attachment, as in the following example.

The experimental details are identical to those in Section 11.3.1, except that the cell voltages on the end plates are adjusted in order to trap negative ions instead of positive ions. We used a single laser here to produce both neutral desorption and electrons, although the two processes could also be separated for individual optimization, as in Section 11.3.1.

Figure 11.2 shows a characteristic negative ion mass spectrum for a Demnum-S65® sample ($\langle n \rangle = 25$) on an Au surface. Only the mass peaks above 1000 Da are shown; all lower mass ions were ejected with a SWIFT waveform for this spectrum.[24] Fall-off of the peak intensities below 1200 Da is due to the limited resolution of the SWIFT waveform. The spectrum was taken with 0.5 mJ/pulse of 193-nm light focused to a ~100 μm diameter spot and averaged over six spots on the sample for 500 shots/spot. No positive ions were generated under these conditions.

The spectrum consists of sets of three peaks that repeat every 166 Da. This spacing is equivalent to the mass of the polymer repeat unit. Two of these sets are shown in the inset of Figure 11.2. The two major peaks are spaced by 50 Da, corresponding to a CF_2 group, while the minor peak is spaced 22 Da from the high mass peak in the group. Small ripples on the low mass side of the major peaks, which are spaced by a constant frequency, are artifacts due to "ringing." The intensity of each group grows roughly exponentially

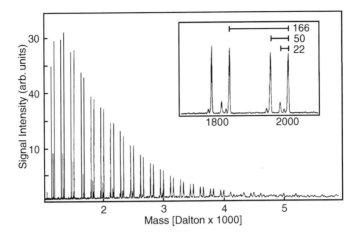

FIGURE 11.2
Negative ion spectrum obtained from desorption and electron attachment of a Demnum S-65 PFPE sample ($\langle M \rangle = 4600$ Da). An expanded view of two peak sets is shown in the insert (mass scale and spacings are in Da).

toward lower mass. Measurements of the lower masses verified that this trend continues down to $n = 0$. The exponent of this increase is found to depend on the laser energy and wavelength, with higher pulse energies and wavelengths producing a steeper increase.

The two major peaks in each group of the spectrum can be explained by a simple model involving two competing reaction channels. During or after desorption of a polymer molecule, a photoelectron from the metal surface attaches to the polymer causing dissociation via one of two pathways

$$E_1-(R_3)_n-E_2 + e^- \rightarrow E_1-(R_3)_x-CF_2CF_2CF_2O^-$$
$$+ {}^{\bullet}CF_2CF_2CF_2O-(R_3)_{n-x-2}-E_2$$
(11.3)

$$E_1-(R_3)_n-E_2 + e^- \rightarrow E_1-(R_3)_x-CF_2CF_2CF_2{}^{\bullet}$$
$$+ {}^-OCF_2CF_2CF_2O-(R_3)_{n-x-2}-E_2.$$
(11.4)

The negative fragment observed carries the $-CF_2CF_2CF_2O-$ group. The peaks with the 50 Da difference correspond to negative ion fragments with long or short end-groups (E_1 or E_2, respectively). The rough equivalency between the intensities of the two peaks indicates that the two channels are equally probable. This same fragmentation pathway has been observed for CF_3-O-CF_3 in the gas phase.[25] In this case electron attachment leads to the formation of $CF_3O^- + CF_3$.

The overall shape of the distribution can be understood by considering that each fragment can undergo further fragmentation. These secondary reactions can result from electron attachment to a neutral fragment or dissociation of a negatively charged fragment. The distribution can be modeled by successive random dissociations along the polymer chain. In other words, each $-CF_2-O-CF_2-$ center on a negatively charged polymer is equally reactive. When this process is repeated several times, it produces a fragment distribution rising exponentially toward the monomer. Increasing the laser energy increases the number of random breaks. As the number of breaks increases, the distribution gets steeper, being weighted toward shorter chains. This is consistent with experimental observations for higher laser pulse energies and shorter wavelengths.

The possible secondary reactions are important to understanding the negative ion spectrum. The $-CF_2O^-$ end-group may dissociate into $-CFO + F^-$. This would create a PFPE with an acid fluoride end-group which, upon electron attachment, could form a negative ion according to

$$E_1-(R_3)_m-CF_2CF_2CFO + e^- \rightarrow E_1-(R_3)_y-CF_2CF_2CF_2^{\bullet}$$
$$+ \ ^-OCF_2CF_2CF_2O-(R_3)_{m-y-2}-CF_2CF_2CFO$$

(11.5)

accounting for the presence of the third peak in each set at 22 Da below the main fragment peaks.

Negative ions corresponding to parent PFPE molecules were searched for extensively at many different laser fluences, at 248 nm as well as 193 nm desorption wavelengths, with various PFPE film thicknesses, and with different metal substrates. No such ions were ever observed. This suggests that the barrier to dissociation of the negatively charged polymers is very low. Therefore, after electron attachment, the rate of polymer dissociation is much greater than that of negative ion stabilization under the conditions of our experiment. We also observed that the extent of fragmentation decreased at longer desorption wavelengths and detected no negative ions at 532 nm under low laser fluence conditions. This can be attributed to insufficient photon energy to generate photoelectrons from the substrate.

11.3.3 VUV Ionization

Conceptually one of the most promising forms of soft ionization is photoionization. In principle it is possible to choose photon energies that just exceed the vertical ionization energy of the neutral in order to impart minimal excess energy and thus limit fragmentation. In practice there are two problems with this concept. First the ion is produced with excess energy due to the internal energy of the neutral. Even if the ionization step does not add much energy, the ionic dissociation energy can often be less than that of the neutral. As a result the ion often fragments, even though the neutral was stable. This problem is particularly severe for larger molecules, where the large number of degrees of freedom imply that the molecule can contain a large amount of internal energy even at moderate temperatures.[26-29] As we will see below, the key to minimizing fragmentation is cooling the neutrals prior to ioniza-tion. The second practical problem is that one cannot easily tune the ionization wavelength such that it fits any molecule to within a fraction of an electron volt. If, on the other hand, one wishes to employ a single wavelength to serve as a general photoionization source for most molecules, then it should be of the order of 10 eV. At that energy a large number of molecules can be ionized, notably including most fully saturated hydrocarbons.

11.3.3.1 Experimental Details

In order to overcome both problems outlined above, we will show examples of the combination of laser desorption, jet cooling, and vacuum ultraviolet (VUV) single-photon ionization at 125 nm (corresponding to 10 eV). Several previous studies have used VUV postionization without cooling.[30,31] The VUV pulses were generated by nonresonant third harmonic generation in phase-matched rare gas mixtures.[32] Because of its potential for achieving higher VUV flu-ences we used resonance-enhanced four wave mixing in Hg vapor instead, employing only a single, tunable dye laser at 625.70 nm to generate VUV at 125.14 nm.[32]

Figure 11.3 schematically shows the experimental setup. The approach consists of a pulsed nozzle, skimmer, and reflectron TOF mass analyzer. The 125 nm VUV beam was introduced along the jet axis and focused in the ionization volume. A doubled Nd:YAG laser (532 nm) was used for desorp-tion, with pulse energy densities of 10 to 100 mJ/cm^2 and a spot size of about 1 mm diameter. Vaporization was caused by substrate heating only. Optimal injection of the vapor into the jet expansion occurred when the desorption spot was about 1 mm in front of and 0.5 mm below the nozzle, and the desorption laser was fired near the peak of the jet gas pulse. The pulsed valve was operated with Ar or Xe at a backing pressure of 8 atmospheres.

For comparison we also used a setup without jet cooling in the form of a modified laser microprobe ion source coupled to a reflectron TOF mass analyzer.[15]

11.3.3.2 Poly(ethylene oxide)

We applied the jet/VUV technique to some readily available samples of small, aliphatic polymers. Figure 11.4 shows mass spectra of two different samples of poly(ethylene oxide). Panel B shows a pure reference sample of nominal $Mp = 1000$ Da, Panel A a laboratory detergent called Liqui-Nox (Alconox Inc.). Both spectra show an oligomeric distribution of parent ions. Using oligomer peak heights as a measure of abundance, we calculate $M_n = 840$ Da and $M_w = 880$ for the reference sample, whose masses are consistent with the formula $H-(O-CH_2-CH_2)_n-OH$. The spectrum of this sample also shows a distribution of low molecular weight fragments with masses $n \times 44$ Da whose abundance increases monotonically with decreasing weight. In addition there is a prominent odd-mass fragment at 71 Da. The polymer in Liqui-Nox is shorter, and its mass values also conform to the formula $n \times 44$ Da. Since this

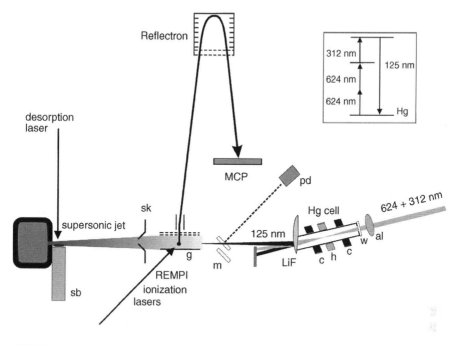

FIGURE 11.3

Schematic diagram of the experimental setup for laser desorption with jet cooling and postionization. Abbreviations: sb, sample bar; sk, skimmer; g, ion source grids; m, moveable mirror; pd, photodiode; mcp, multichannel plate detector; LiF, plano-convex lithium fluoride lens; c, annular cooler; h, annular heater; w, fused silica window; al, achromatic lens.

is a detergent, the composition is probably either $C_5H_{11}-(O-CH_2-CH_2)_n-OH$, or $C_9H_{19}-\phi-(O-CH_2-CH_2)_n-OH$.

The $\Delta m = 44$ fragment series fits none of the products found or intermediates postulated in thermal degradation studies,[33,34] but is consistent with earlier laser desorption work where the same series was observed in cationized form[9] and attributed to the random scission process

$$-CH_2-CH_2-O-CH_2- \rightarrow (CH_2-CH_2 + O-CH2) \rightarrow$$
$$-CH=CH_2{}^\bullet + {}^\bullet HO-CH_2-.$$

The vinyl-terminated product fits the observed fragment series. It must arise from pyrolysis because it is not a typical ether ion fragment[35] and can be detected by cationization.[9] The discrepancy with slow thermal degradation presumably arises because the vinyl product is formed in a high activation energy unimolecular reaction which predominates under the conditions of laser desorption. A possible mechanism is a concerted hydrogen transfer.[36] In that case the C–O bond scission shown in parentheses in the above reaction diagram would be simultaneous with a hydrogen transfer. The odd mass fragment observed at 71 Da could be the radical

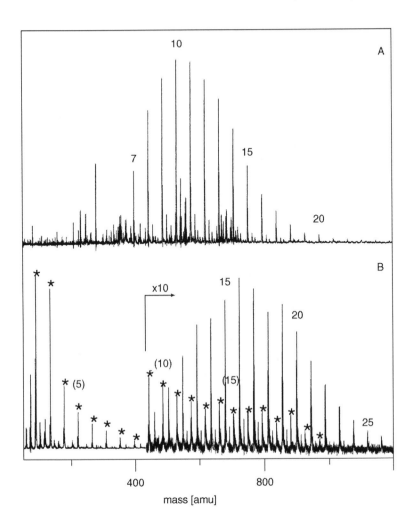

FIGURE 11.4

Jet/VUV mass spectra of poly(ethylene oxide) samples. A: Liqui-Nox laboratory detergent. B: Reference sample with nominal $M_p = 1000$; A fragment series in spectrum B is labeled with asterisks. The number, n, of ethylene oxide repeat units is given for some peaks (in parentheses for fragment series).

$CH_2-CH_2-O-CH=CH_2$, derived from the vinyl terminal by a simple C—O bond rupture.

11.3.3.3 Poly(isoprene)

Figure 11.5 shows the mass spectrum of a sample of poly(isoprene) with nominal $M_p = 1000$ Da. It corresponds to the structure $C_4H_9-(CH_2-C(CH_3) = CH-CH_2)_n-H$. From the peak heights one calculates $M_n = 820$ and $M_w = 870$. At the low mass end this spectrum shows peaks for isoprene monomer (68 Da) and dimer (136 Da).

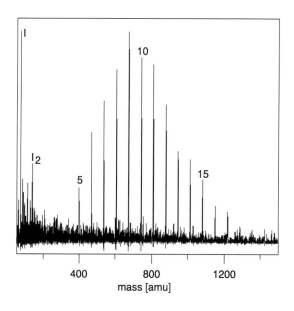

FIGURE 11.5
Jet/ VUV mass spectrum of poly(isoprene) with nominal $M_p = 1000$. The number, n, of isoprene repeat units is given for some peaks. I and I_2 refer to isoprene monomer and dimer.

11.3.3.4 Poly(dimethyl siloxane)

Jet/VUV mass spectra for three different samples of poly(dimethyl siloxane) with nominal M_w = 800, 2000, and 6000 Da are shown in Figure 11.6. In this case the masses observed are all fragments missing one methyl group. We did not observe the molecular ions corresponding to the structure $(CH_3)_3Si-(O-Si(CH_3)_2)_n-CH_3$, $M = 88 + n \times 74$ Da; instead the spectrum shows the M–15 Da ions. We also examined the cyclic tetramer and pentamer, $(O-Si(CH_3)_2)_m$, $m = 4$ and 5, and observed only the M–15 Da ions.

Cooling can only be effective if the ion has a stable ground state and fragmentation is thermally activated. This example reminds us that this is not always the case. The complete absence of a molecular ion for all linear and cyclic poly(dimethylsiloxane)s shows that their molecular ions as produced by VUV ionization are intrinsically unstable. Judging from the ionization potential of $(CH_3)_3Si-O-Si(CH_3)_3$, 9.64 ± 0.01 eV,[37] the excess energy for VUV ionization is small and similar to that of the paraffins, where cooling suppresses fragmentation completely.

For the linear polymer samples shown in Figure 11.6 the M_w (M_n) values determined from the mass spectra of the three samples are 950 (920), 1310 (1240), and 2150 (2010) Da, respectively. There is clearly a deficit of signal for higher oligomers, increasing with the M_w of the sample and accompanied by an increasing proportion of low mass fragments in the mass spectrum.

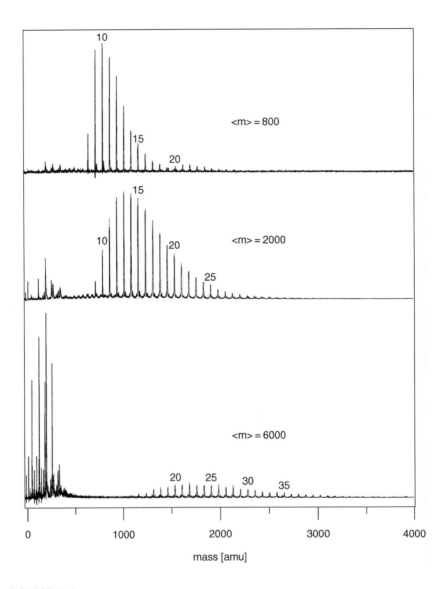

FIGURRE 11.6

Jet/VUV mass spectra of 3 samples of poly(dimethyl siloxane) whose nominal M_p values are indicated in the figure. Numbers of repeat units, n, are indicated for some of the oligomer peaks which are all (M-15) Da masses.

11.3.3.5   Polystyrene

In order to compare ionization efficiencies of two-photon and one-photon ionization we investigated an aromatic polymer that can readily be ionized by two-photon ionization at 193 nm, namely polystyrene. The sample composition corresponds to the structure C_4H_9—$(CH_2$—$CH\phi)_n$—H, where ϕ = phenyl. Figure 11.7 shows representative mass spectra of samples with nominal M_p

Polystyrene

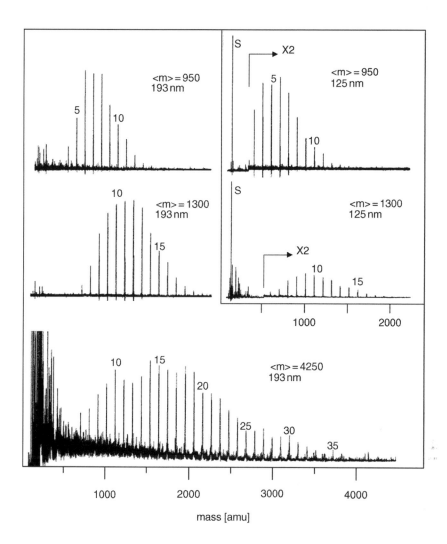

FIGURE 11.7

One-photon (125 nm) and two-photon (193 nm) ionization mass spectra of 3 samples of polystyrene whose nominal M_p values are indicated in the figure. Some peaks are labeled with the number, n, of styrene repeat units. The low mass peak designated S in the 125-nm ionization spectra is styrene. Note the vertical scale change (×2) in these traces.

values of 950, 1300, and 4250. For 193 nm ionization the values of M_w (M_n) determined from the mass spectra were, respectively, 910 (860), 1340 (1270), and 2120 (1830) Da. For 125 nm ionization, the values were 780 (700) and 1220 (1150) Da, and the spectrum of the highest molecular weight sample was too weak to be evaluated. Just as for poly(dimethyl siloxane), there is a discrimination against higher oligomers that increases with the average molecular weight of the sample and is accompanied by an increasing proportion of low mass fragments. A prominent fragment observed with 125 nm ionization is styrene; this was also detected by Feldmann et al. as a photoablation product of polystyrene using 118.4 nm ionization.[30] A comparison of the one-photon and two-photon ionization traces shows that the former exhibit a bias toward low mass when compared to the latter. Since the method of ionization is the only difference between these spectra, it follows that ionization efficiency falls off more rapidly with molecular weight for the one-photon process.

The onset of mass discrimination depends on the type of polymer. We consider four causes for this effect: (i) preferential fragmentation of high mass ions, (ii) competition between vaporization and thermal decomposition in the desorption step, (iii) reduced ionization efficiency for high mass oligomers, and (iv) mass dependence of the detector. Since we have no evidence for less efficient cooling of high mass neutrals we regard cause (i) as unlikely. Schlag and Levine have presented arguments and some evidence for cause (iii).[38] They proposed a mass dependence of the molecular ionization efficiency of the form $\exp(-M/M_0)^{3/2}$, where M is the molecular weight and M_0 an empirical constant. This may account for part of the mass bias we observe, but since the exponential falloff is steepest for the low masses, it cannot account for the increased distortion we observe for $M > 1000$ Da. Mass dependent detection can probably be ruled out at the ion kinetic energies that we are using, as also evidenced by the detection of much higher mass polymers described below. The discrimination against higher oligomers is probably due to cause (ii) which predominates in this mass range. For high mass molecules dissociation becomes faster than vaporization. Desorption changes into ablation, which gives rise to the increased proportion of small fragments detected with samples of high average molecular weight.

Notice that even for $M < 1000$ Da there appears to be relatively more efficient two-photon than one-photon ionization of the higher oligomers of polystyrene. Schlag et al. reported a similar effect for small, aromatic peptides.[39] They used identical total ionization energy, while in our case comparison is less direct because the energies are different (9.9 eV for one-photon, 12.9 eV for two-photon ionization). The effect may be due to increased oscillator strength (more chromophores) of higher oligomers in the first step of the two-photon process.

11.3.4 Two-Photon Ionization

In two-photon ionization the first photon excites polymers to an electronic state and a second photon ionizes the excited molecules. This approach has the following advantages over single-photon ionization. One can employ commercial lasers, avoiding the complication of sum frequency mixing or other schemes for VUV generation. Typical lasers for this purpose include the quadrupled Nd:YAG laser and excimer lasers. Two-photon processes in the absence of saturation of both steps depend quadratically on laser power, while single-photon processes depend linearly on power. On the other hand the mixing or doubling schemes required to generate VUV photons depend nonlinearly on the power of the pump lasers. Therefore on the one hand the single-photon scheme has higher ionization efficiency; however on the other hand in practice much higher laser fluences are available for two-photon ionization. The result is that sensitivities can be achieved in practice that can be higher by order of magnitudes.

Two-photon ionization requires a chromophore at a practical wavelength to absorb the first photon. This constitutes both a weakness and a strength of the method. On the one hand not all polymers can be two-photon ionized, on the other hand wavelength dependence introduces selectivity of detection.

11.3.4.1   *Photofrin*

Figure 11.8 shows a comparison of a number of mass spectral techniques, which were all applied to measure the oligomeric distribution of a small polymer, Photofrin. This is a complex mixture of nonmetalic porphyrins, linked primarily through ether bonds.[40,41] The compound was developed as a drug for use in photodynamic therapy for the treatment of solid tumors. Its characterization, important for drug approval, has proven to be a serious analytical challenge. Figure 11.8 shows the result of analysis by FAB (fast atom bombardment), UV- and IR-MALDI, electron spray ionization, and two-step laser mass spectrometry with jet cooling.[42] In the latter case ionization was performed at 193 nm. The major conclusion from this direct comparison is that all four techniques show a very similar oligomeric distribution. We note that another study, which employed two-step laser mass spectrometry without jet cooling produced almost exclusively monomers,[43] which once again points to the importance of cooling in order to reduce fragmentation with photoionization.

11.3.4.2   *Two Photon Ionization of End-Group Chromophores*

Figure 11.9 shows the mass spectrum of a monofunctionalized homopolymer with repeat unit R_3 and a functional end-group A, consisting of aromatic ester as follows:

$$A = -CF_2-(CO)-O-CH_2CH_2-O-C_6H_5,$$

The monofunctionalized polymer is of the form:

$$A-(R_3)_n-E_2. \tag{11.6}$$

Bifunctional polymers have the form:

$$A-O-(R_3)_n-A. \tag{11.7}$$

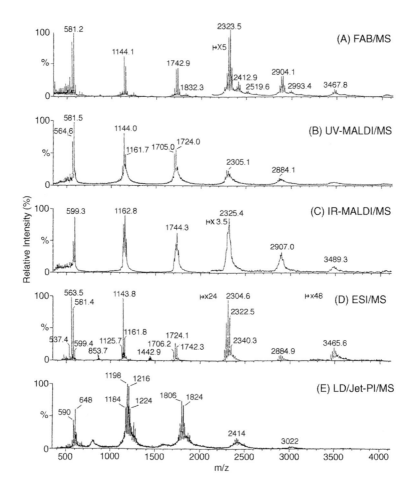

FIGURE 11.8
Mass spectra of desalted Photofrin, in the positive ionization modes, obtained by (A) FAB/MS, (B) UV-MALDI/MS, (C) IR-MALDI/MS, (D) ESI/MS (nozzle-skimmer voltage 100 V). (E) Mass spectrum of per-methyl ester of Photofrin, in the positive ionization mode, obtained by LD/Jet-PI(193 nm)/MS.

The monofunctional material is commercially available under the brand name Demnum-SP[®].[44] We performed two-photon ionization with 193 nm. All major peaks correspond to parent ions and are spaced apart by 166 Da, the mass of a repeat unit. They range from polymers with $n = 5$ ($m/z = 1114$ Da) to $n = 40$ ($m/z = 6924$ Da). Values of n are indicated with selected peaks. This mass spectrum qualitatively shows the distribution of chain lengths in the sample. The measured distribution may be affected by mass dependencies in the experiment, such as transmission, detector response, and entrainment efficiency. However the average of this distribution is consistent with NMR. Minor peaks in the spectrum are due to (i) bifunctional polymers that are present in this sample as an impurity and (ii) polymers missing CF_2 in one

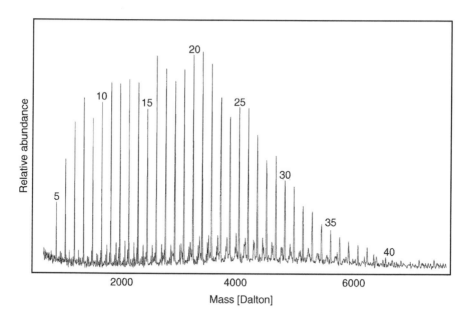

FIGURE 11.9
Laser desorption REMPI time of flight mass spectrum of Demnum-SP®. Ionization wavelength 193 nm. Major peaks represent parent masses, separated by the 166 Da mass of a repeat unit: $[CF_2]_3$–O. The numbers of units are indicated for some peaks.

of their repeat units or their end unit. We note that the presence of bifunctional polymers cannot be observed by either NMR or size exclusion chromatography, and in fact would lead to an erroneous assignment of the average chain length distribution in the case of NMR.

Figure 11.10a shows the mass spectrum of a co-polymer of the type:

$$P-O-[(R_1)_k(R_2)_l]-P,$$

where **P** represents an end-group containing a piperonyl chromophore:

$$P = -CF_2-CH_2-O-C_8H_7O_2.$$

This material is commercially available under the name AM-2001®. Ionization was performed with 193 nm. Every peak in this mass spectrum corresponds to a parent mass with one of the possible combinations of k and l. Information contained in this mass spectrum goes beyond the distribution of chain lengths. The ability to distinguish the abundances of individual (k, l) combinations provides for a much more refined characterization of co-polymers, as will be further discussed below.

From an analytical perspective there is the limitation that the molecule needs to have a chromophore in order to be detected by REMPI. We have extended the applicability of the technique by chemically attaching chromophores to

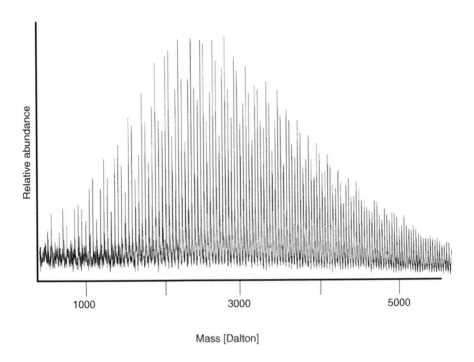

Mass [Dalton]

FIGURE 11.10
Laser desorption REMPI time of flight mass spectrum of (a) AM-2001®: P–O–[(CF$_2$–O)$_k$(C$_2$F$_4$–O)$_l$]–P, with P = –CF$_2$–CH$_2$–O–C$_8$H$_7$O$_2$, (b) Z-Dol® with esterified end-groups: A–O–[(CF$_2$–O)$_k$(C$_2$F$_4$–0)$_l$]–A with A = –CF$_2$–CH$_2$–O–CO–CH$_2$–C$_6$H$_5$. Ionization wavelength 193 nm.

a number of commercial PFPEs that do not have chromophores, particularly those with alcohol and acid end groups.

Figure 11.10b shows the mass spectrum of a co-polymer of the type:

$$\mathbf{A'} - O - [(\mathbf{R}_1)_k (\mathbf{R}_2)_l] - \mathbf{A'}$$

$$\mathbf{A'} = -CF_3 - CH_2 - O - CO - CH_2 - C_6H_5,$$

where **A'** takes the place of the alcohol end-group of the original polymer and was obtained by esterifying it with phenyl acetyl chloride. The alcohol polymer is commercially available under the name Z-Dol®. Ionization was performed at 193 nm. The difference between this material and that in Figure 11.10a is that in the case of AM-2001 the end-group is itself a chromophore, while in the case of Z-Dol the chromophore first had to be attached. Furthermore our analysis shows that the distribution of repeat units as a function of chain length is different in the two cases, as discussed below.

To demonstrate the increased level of detail that is available because parent molecules can be detected without fragmentation for individual (*k*, *l*) combinations, we have plotted the relative abundances in a different way. Figure 11.11 shows relative peak integrals plotted in a grid of *k* and *l*.

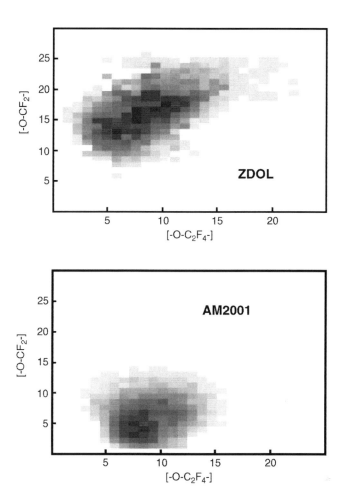

FIGURE 11.11
Relative abundances of copolymers of type A–O–[(CF$_2$–O)$_k$(C$_2$F$_4$–0)$_l$]–A, plotted as a function of k and l. (a) AM-2001®, (b) Z-Dol®, (c) model calculation for a random copolymer.

Figure 11.11a is for the AM-2001; Figure 11.11b is for the Z-Dol. For comparison Figure 11.11c shows a plot for a model distribution, which assumes a purely random copolymer. This takes the form of a binomial distribution, with $I_{k,l}$ denoting the relative abundance of a polymer with a repeat unit combination (k, l) as follows:

$$I_{k,l} = \left(\frac{k+l}{l}\right) p_1^k p_2^l$$

$$P_1 + P_2 = 1$$

P_1 and P_2 represent the relative probabilities of adding either an R_1 or an R_2, unit as the chain is being built up during the polymerization. The ratio $f = P_1/P_2$ is the only free parameter in the model. Furthermore we have restricted the overall chain length by multiplying the distribution with an envelope function that was taken from the actual molecular weight distribution. We cannot completely reproduce the actual distributions of Figures 11.11a or 11.11b with any single value of f. In the model the line of maximum abundance must pass through the origin for any f while for the actual polymers this is not the case. We find f to be different depending on the overall length of the polymer chain. For Z-Dol the smaller polymers tend to have relatively more single carbon repeat units, while the larger ones on average have somewhat more than two carbon repeat units. For AM2001 this trend appears to be inverted. This phenomenon must be related to the detailed kinetics of the polymerization reactions that formed these polymers.

Attachment of chromophore end-groups can also be used as a labeling technique. We have used it to investigate degradation by friction of the non-functionalized, commercial PFPE polymer, Demnum-S65. Friction tends to break this polymer randomly at the ether linkages, producing carbonyl fluoride and new C_3 end-groups. The effect is similar to that of electron attachment as discussed in Section 11.3.2 although the mechanism may be different. In the presence of water the carbonyl fluoride is transformed into carboxylic acid, which we selectively labeled with a phenoxy group. The resulting laser desorption REMPI mass spectrum of the labeled acid fragments is shown in Figure 11.12. It exhibits a molecular weight distribution that peaks around 1500 Da and averages about 2000 Da, while that of the original polymer peaks around 3800 Da and averages 4600 Da.[21] Fragmentation is evident in the pattern of major peaks that occur in pairs 50 Da apart: The labeled acid group appears with either the $-CF_2CF_2CF_3$ or the $-CF_2CF_3$ end-group of the original polymer, labeled □ and ◆, respectively. Peaks with the former end-group are stronger since additional C_3 end-groups are formed during fragmentation, and these products may fragment again. There is a third set of peaks 50 Da below those with the $-CF_2CF_3$ end-group, labeled ◇. Two possible explanations for this set are (1) loss of C_2F_4 from a radical intermediate of the fragmentation process or (2) preferential chain-breaking at R_2 repeat units that are present in minor amounts, as mentioned above.

11.3.5 REMPI

The next refinement in polymer detection by photoionization comes in the form of resonance enhanced multiphoton ionization (REMPI).[13,14,45,46] This is a form of two-photon ionization in which the first photon is tuned to a resonant transition in the molecule. By varying the wavelength one obtains an excitation spectrum, potentially providing vibrational spectroscopy of the excited electronic state of the polymer. Cooling is essential in order to obtain resolvable spectral features. We demonstrate the principle with the example of a PFPE with an aromatic chromophoric end-group of type A1.

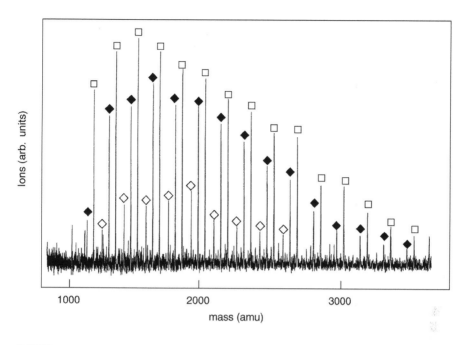

FIGURE 11.12

Laser desorption REMPI time of flight spectrum of degradation products of Demnum-S65 after frictional wear. The spectrum is obtained by esterification of fragments that have acid end-groups. Ionization wavelength 193 nm. The symbols differentiate three different series of fragments, as described in the text.

A REMPI spectrum results from monitoring a particular mass to charge ratio in the mass spectrum while tuning the wavelength of the ionization laser. Figure 11.13 shows REMPI spectra of a series of molecules of increasing size and complexity which were chosen to model the chromophore end of the polymer and converge to the structure of the PFPEs.[47] We assign the main features in these spectra to the respective S_{0-0} transitions for two reasons: (1) In each case the peak is close to the wavelength of the known S_{0-0} transitions of phenol and anisole; (2) no other spectral feature that can be assigned to the electronic origin is observed when scanning at least 1300 cm^{-1} to the red. Panel A shows the spectrum of phenol; panel B, anisole; and panel C, 2-phenoxyethanol. Panels D through G show 2-phenoxyethyl esters of perfluorinated carboxylic acids, with increasing length of the perfluoroalkyl chain. In H through J a series of esters with branched perfluorinated polyether chains is shown. Finally panels K through N show spectra of straight chain perfluorinated polyethers found in the PFPE sample with the shorter chain distribution. Generally, the spectra evolve smoothly from phenol to the polymers. The transition shifts to higher energy in going from phenol (A) to the first ester (D). In D, E, and F structure is observed that may be related to several electronic origins corresponding to different conformations of the molecule. In proceeding from the esters to the branched ethers (H–J)

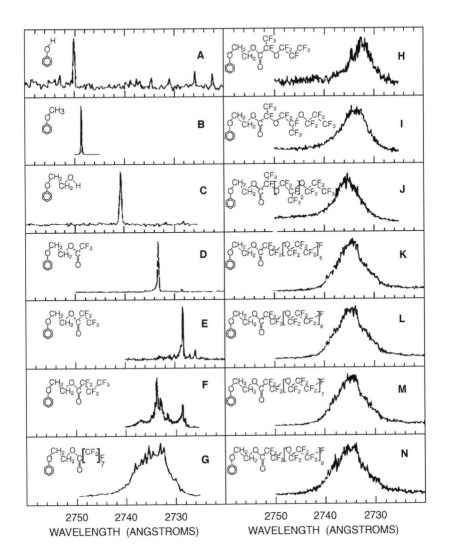

FIGURE 11.13

R2PI spectra of the series of model compounds and PFPEs illustrated in Figure 11.1: (A) phenol; (B) anisole; (C) 2-phenoxyethanol; (D–G) 2-phenoxyethyl esters of perfluorinated carboxylic acids: perfluoroacetate, perfluorobutyrate, perfluoropropionate, and perfluorooctanoate; (H–J) 2–phenoxyethyl esters with branched ether chains: $n = 1, 2,$ and 3; (K–N) type I PFPE with $n = 5, 6, 7,$ and 9.

the spectra shift back slightly toward lower energy. The straight chain polymers (K–N) show no further shift or broadening. It appears that the spectra have converged to a limiting value at three ether oxygens in the chain (J) and the branched and straight chain polyethers have similar spectra.

A comparison of the branched polyether in panel J with the straight chain polyether with two more additional repeat units in panel K suggests that there is no difference in the spectra of the branched and straight chain polymers. The progression through panels K–N, corresponding to increasing the number of repeat units, indicates that both the position and width of the spectra are insensitive to chain length in this range.

In general, the spectral shift has reached its limit at the perfluoroacetate (panel D). Although an anomalous shift does appear in the perfluoropropionate (panel E), the larger molecules seem to prefer a conformation closer to one that resembles the perfluoroacetate. The spectral width seems to have converged with the first branched ether (panel H). The PFPE spectrum (position and width) seems to have converged at three repeat units, assuming no difference in branched and straight chain spectra, as the data suggest.

Apparently, additional repeat units beyond three are too remote from the chromophore to induce additional shifts. The additional units also do not significantly affect the ionization efficiency: With increasing length of the ether chain, vibrations involving low frequency torsional and bending motions that remain populated in the beam must continue to increase in number and decrease in vibrational frequency. Transitions originating from these vibrational levels certainly add additional congestion to the spectrum. However, they must have small shifts relative to those already present in the spectrum and add unobservable broadening to the already broadened peaks.

We note that the increasing spectral bandwidth with increasing molecular size observed in Figure 11.13, should serve as a caution about comparing ionization efficiencies at fixed wavelength. An apparent decline in ionization efficiency may be due to sampling a smaller fraction of the molecular population as the molecules get larger and the broadening effects cited above become important.

11.3.5.1 Van der Waals Dimers of Polymers

When material is desorbed with high enough density into the early part of the supersonic expansion it is possible to form small van der Waals clusters. Figure 11.14 shows spectra of the dimers of type A1, obtained from the sample with the lower average molecular weight. Since the sample contained a distribution of chain lengths, each peak in the mass spectrum is due to a mixture of dimers. Each dimer mass can only be assigned in terms of the sum of two monomer chain lengths. For example, a dimer mass peak corresponding to a chain length of 16 can contain contributions of monomers with n and m repeat units in any combination for which $n + m = 16$. All the dimer spectra show a broadening and a characteristic redshift of about 110 cm^{-1} from the monomer wavelength.

For comparison with the polymer clusters, we measured the spectrum of the anisole dimer. The spectrum showed a sharp origin shifted 215 cm^{-1} from the anisole monomer. The trimer absorbed in the same region, but gave a much broader signal (about 100 cm^{-1} full width at half maximum) without any sharp structure.

We also obtained REMPI spectra of doubly functionalized PFPEs of type A2 for a number of chain lengths. These spectra exhibit broadening and

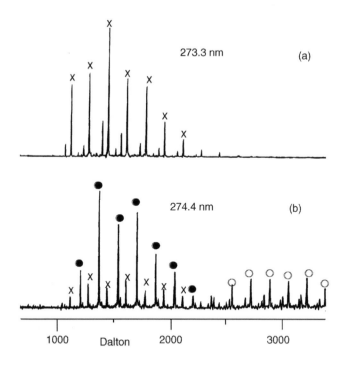

FIGURE 11.14

Mass spectra of the type I PFPE sample with $\langle n \rangle = 8$. (a) Ionization at 273.3 nm yields predominantly parent ions of the type I PFPE, which are indicated by the filled circles. (b) Ionization at 274.4 nm favors ionization of the type B polymers present as an impurity in the sample (crosses) and van der Waals dimers of the type I polymers (open circles). Peaks due to these two components of the mass spectrum are not visible in panel A. The parent masses of the type I polymer are again indicated by the filled circles.

redshifts remarkably similar to those of the dimer spectra. The similarity in the spectra of the dimers of the type A1 polymers and the isolated, doubly functionalized type A2 polymers leads to the conjecture that the chromophore environment in both cases may be similar.

Another indication that the two chromophores are not independent in the type A2 polymers can be seen in the mass spectra obtained at two different wavelengths. One way of achieving this similarity is to form an intramolecular complex in the type A2 polymers that resembles the intermolecular complex in the van der Waals dimer. This would involve similar interactions between pairs of chromophores in each case. If the type A2 polymers in fact had independent chromophores, they would appear with the type A1 polymers in the mass spectrum in panel A, which was taken with the ionization laser tuned to 273.3 nm, near the peak of the type A1 polymer absorption. Instead, they are most pronounced in the mass spectrum taken at 274.8 nm, the ionization wavelength that favors the van der Waals dimers.

We note that the possibility of dimers forming due to chromophore-chain interactions can be eliminated by the absence of dimers between type A1 polymers and nonfunctionalized polymers. Since the polymer mixture contains 28% of the nonfunctionalized polymer, strong interaction between the chromophore and the chain would result in a significant portion of the dimers involving one functionalized and one nonfunctionalized member. These were not observed. We conclude, therefore, that the van der Waals complex must be formed by intimate interaction between the chromophores.

If the intramolecular and intermolecular complexes are formed in the same way, as their wavelength spectra suggest, we expect that they both involve interaction of the chromophore ends. This scenario has two prerequisites. First, the chains must be flexible enough so that during the jet expansion they can efficiently bend to bring the chromophores together. In other words, the barriers to internal rotation must be low enough for the polymers to explore many conformations while they are being cooled, since the experiments suggest that the two ends of the molecule find each other with high efficiency. Second, the interaction between chromophores must be strong enough to effectively form the intramolecular complex once the chromophores are brought together.

We believe it is significant for the formation of the intramolecular dimer of the type A2 polymer that the chromophore dimer binding energy is substantially larger than the torsional barrier heights. High quality *ab initio* calculations for a model ether compound, 1,2-dimethoxyethane, predict barriers in the range of 500 to 800 cm^{-1}.[48] Analogous barriers for the perfluorinated compound are expected to be similar or lower.[48] When the polymers are laser-vaporized from the surface they must have an internal temperature equal to at least room temperature or higher. At this stage they have sufficient internal energy to surmount the barriers to internal rotation and freely explore many conformations. A certain fraction of these conformations bring the chromophore ends together, but at this point the molecules may be too energetic to form a complex.

As the internal energy of the molecules is reduced during the expansion, there comes a stage when there is still sufficient energy for the conformations to interconvert but the chromophores begin to get trapped in the dimer geometry if they happen to come together. If the dimer binding energy is larger than the critical barrier heights, there is still sufficient vibrational energy in the chain for conformational changes after the dimers have begun to form. As the internal energy continues to drop, the chain conformations are eventually frozen in, but only after almost all the chromophores have formed intramolecular dimers. Since the torsional barriers are likely larger than 400 cm^{-1}, we expect to see metastable conformations, in agreement with our observations for some of the model compounds.

Combinations of polymer chains of variable flexibility with different chromophores will provide an interesting arena for making predictions based on barriers to internal rotation and binding energy. Gas phase spectroscopic measurements, similar to those reported here, of the interaction between chromophores at remote positions on polymer chains may be excellent tests of those predictions. It may be possible to gauge the internal barrier heights with a sequence of chromophores of ranging dimer strength.

11.4   Summary

Two-step laser mass spectrometry, especially in combination with jet cooling, offers ways to obtain complete and accurate oligomeric distributions in many cases at a level of detail that is unavailable from conventional techniques. Additional information can often be obtained from the wavelength selectivity in the ionizing step, ultimately allowing for optical spectroscopy in combination with mass spectrometry. Thus the techniques described in this chapter extend the tools of polymer analysis and can provide new insights in polymer properties.

References

1. Lub, I., van Vroonhove, F.C.B.M., Bruninx, E., and Benninghoven, A., "Interaction of Nitrogen and Ammonia Plasmas with Polystyrene and Polycarbonate Studied by X-Ray Photoelectron-Spectroscopy, Neutron-Activation Analysis and Static Secondary Ion Mass-Spectrometry," *Polymer* **30**, 40–44, 1989.
2. Hillenkamp, F., Karas, M., Beavis, R.C., and Chait, B.T., "Matrix-Assisted Laser Desorption Ionization Mass-Spectrometry of Biopolymers," *Anal. Chem.* **63**, A1193–A1202, 1991.
3. van der Peyl, G.J.Q., Isa, K., Haverkamp, J., and Kistemaker, P.G., "Gas-Phase Ion-Molecule Reactions in Laser Desorption Mass-Spectrometry," *Org. Mass Spectrom.* **16**, 416–420, 1981.
4. Brown, R.S., Weil, D.A., and Wilkins, C.L., "Laser Desorption Fourier-Transform Mass-Spectrometry for the Characterization of Polymers," *Macromolecules* **19**, 1255–1260, 1986.
5. Llenes, C.F.O., and OMalley, R.M., "Cation Attachment in the Analysis of Polystyrene and Polyethylene-Glycol by Laser-Desorption Time-of-Flight Mass-Spectrometry," *Rapid Commun. Mass Spectr.* **6**, 564–570, 1992.
6. Posthumus, M.A., Kistemaker, P.G., Meuzelaar, H.L.Z., and de Brauw, ten Noever, "Laser Desorption-Mass Spectrometry of Polar Non-Volatile Bio-Organic Molecules," *Anal. Chem.* **50**, 985–919, 1978.
7. Wilkins, C.L., Weil, D.A., Yang, C.L.C., and Ijames, C.F., "High Mass Analysis by Laser Desorption Fourier-Transform Mass-Spectrometry," *Anal. Chem.* **57**, 520–524, 1985.
8. Mattern, D.E. and Hercules, D.M., "Laser Mass-Spectrometry of Polyglycols— Comparison with Other Mass-Spectral Techniques," *Anal. Chem.* **57**, 2041–2046, 1985.
9. Cotter, R.J., Honovich, J.P., Olthoff, J.K., and Lattimer, R.P., "Laser Desorption Time-of-Flight Mass-Spectrometry of Low-Molecular-Weight Polymers," *Macromolecules* **19**, 2996–3001, 1986.
10. Bahr, U., Deppe, A., Karas, M., and Hillenkamp, F., "Mass-Spectrometry of Synthetic-Polymers by UV Matrix-Assisted Laser Desorption Ionization," *Anal. Chem.* **64**, 2866–2869, 1992.

11. Zenobi, R., "In Situ Analysis of Surfaces and Mixtures by Laser Desorption Mass Spectrometry," *International Journal of Mass Spectrometry and Ion Processes.* **145** (1–2), 51–77, 21 July, 1995.

12. Arrowsmith, P., de Vries, M.S., Hunziker, H.E., and Wendt, H.R., "Pulsed Laser Desorption Near a Jet Orifice: Concentration Profiles of Entrained Perylene Vapor," *Applied Physics B* **46**(2), 165–173, 1988.

13. Meijer, G., de Vries, M.S., Hunziker, H.E., and Wendt, H.R., "Laser Desorption Jet-Cooling Spectroscopy of para-Amino Benzoic Acid Monomer, Dimer, and Clusters," *Journal of Chemical Physics.* **92** (12), 7625–7635, 1990.

14. Meijer, G., de Vries, M.S., Hunziker, H.E., and Wendt, H.R., "Laser Desorption Jet-Cooling of Organic Molecules," *Appl. Phys. B* **51**, 395–403, 1990.

15. de Vries, M.S., Elloway, D.J., Wendt, H.R., and Hunziker, H.E., "Photoionization Mass Spectrometer with a Microscope Laser Desorption Source," *Review of Scientific Instruments.* **63**, 3321–3325, 1992.

16. Hall, R.B., "Pulsed-Laser-Induced Desorption Studies of the Kinetics of Surface-Reactions," *J. Phys. Chem.* **91**, 1007–1015, 1987.

17. Zare, R.N. and Levine, R.D., "Mechanism for Bond-Selective Processes in Laser Desorption," *Chem. Phys. Lett.* **136**, 593–599, 1987.

18. Li, Y., McIver, R.T., and Hemminger, J., "Experimental-Determination of Thermal and Nonthermal Mechanisms for Laser Desorption from Thin Metal-Films," *J. Chem. Phys.* **93**, 4719–4723, 1990.

19. Elam, J.W. and Levy, D.H., "Ultraviolet Laser Desorption of Indole," *J. Chem. Phys.* **106**, 10368–10378, 1997.

20. Elam, J.W. and Levy, D.H., "Laser Ablation of Trp-Gly," *J. Phys. Chem.* **102**, 8113–8120, 1998.

21. Cromwell, E.F., Reihs, K., de Vries, M.S., Ghaderi, S., Wendt, H.R., and Hunziker, H.E., "Laser Desorption FTMS of Perfluorinated Polyethers by Transition Metal Cationization," *J. Phys. Chem.* **97**, 4720, 1993.

22. Marshall, A.G. and Grosshans, P.B., "Fourier-Transform Ion-Cyclotron Resonance Mass-Spectrometry—the Teenage Years," *Anal. Chem.* **63**, A215–A229, 1991.

23. Lehman, T.A. and Bursey, M.M., *Ion Cyclotron Resonance Spectrometry,* John Wiley & Sons, New York, 1976.

24. Chen, L., Wang, T.-C.L., Ricca, T.L., and Marshall, A.G., "Trapped Ion Mass-Spectrometry," *Anal. Chem.* **59**, 449–454, 1987.

25. Spyrou, S.M., Sauers, I., and Christophorou, L.G., *J. Chem. Phys.* **78**, 7200– 7216 (Electron-attachment to the perfluoroalkanes normal-cnf2n+2 (n = 1–6) and i-c4f10, 1983).

26. Danon, A., Amirav, A., Silberstein, J., Salman, Y., and Levine, R.D., "Internal Energy Effects on Mass Spectrometric Fragmentation," *J. Phys. Chem.* **93**, 49–55, 1989.

27. Amirav, A., "Electron Impact Mass Spectrometry of Cholesterol in Supersonic Molecular Beams," *J. Phys. Chem.* **94**, 5200–5202, 1990.

28. Danon, A., Amirav, A., Silberstein, J., Salman, Y., and Levine, R.D., "Internal Energy Effects on Mass Spectrometric Fragmentation," *J. Phys. Chem.* **93**, 49–55, 1993.

29. Nir, E., Hunziker, H.E., and de Vries, M.S., "Fragment Free Mass Spectrometric Analysis with Jet Cooling/VUV Photoionization," *Anal. Chem.* **71**, 1674–1678, 1999.

30. Feldmann, D., Kutzner, J., Laukemper, J., MacRobbert, S., and Welge, K.H., "Mass Spectroscopic Studies of the Arf-Laser Photoablation of Polystyrene," *Appl. Phys. B* **44**, 81–85, 1987.

31. Pallix, J.B., Schuehle, U., Becker, C.H., and Huestid, D.L., "Advantages of Single-Photon Ionization Over Multiphoton Ionization for Mass-Spectrometric Surface-Analysis of Bulk Organic Polymers," *Anal. Chem.* **61**, 805–811, 1989.

32. Hilbig, R. and Wallenstein, R., "Resonant Sum and Difference Frequency Mixing in Hg," *IEE J. Quant. Electr.* **QE-19**, 1759–1770, 1983.

33. Grassie, N. and Mendoza, G.A.P., "Thermal-Degradation of Polyether-Urethanes. 1. Thermal-Degradation of Poly(Ethylene Glycols) Used in the Preparation of Poly-urethanes," *Polym. Degrad. Stab.* **9**, 155–165, 1984.

34. Bortel, E. and Lamot, R., "Investigation on Degradation of High Molecular Poly(Ethylene Oxide) in Solid-State," *Makromol. Chem.* **178**, 2617–2628, 1977.

35. McLafferty, F.M., *Interpretation of Mass Spectra*, University Science Books, Mill Valley, California, 1980.

36. Allen, G., Aggarwall, S.L., and Russo, S., *Comprehensive Polymer Science, First Supplement*, Pergamon Press, Oxford/New York/Seoul/Tokyo, 1992.

37. Lias, S.G., Bartmess, J.F., Liebman, J.I., Holmes, R.D., Levin, R.D., and Mallard, W.G., "Gas-Phase Ion and Neutral Thermochemistry," *J. Phys. Chem. Ref. Data* **17**, 1–861, 1988.

38. Schlag E.W. and Levine, R.D., "Ionization, Charge Separation, Charge Recombination, and Electron-Transfer in Large Systems," *J. Phys. Chem.* **96**, 10608–10616, 1992.

39. Schlag, E.W., Grotemeyer, J., and Levine, R.D., "Do Largre Molecule Ionize?" *Chem. Phys. Lett* **190** (6), 521–527, 1992.

40. Dougherty, T.J., Boyle, D.G., Weishaupt, K.R., Henderson, B.A., Potter, W.R., Bellnier, D.A., and Wityk, K.E.K.E., "Photoradiation Therapy—Clinical and Drug Advances," in *Porphyrin Photosensitization*, edited by Kessel, D. and Dougherty, T.J., Plenum Press, New York, 1983, p. 3.

41. Byrne, C.J., Marshallsay, L.V., and Ward, A.D., "The Structure of the Active Material in Hematoporphyrin Derivative," *Photochemistry and Photobiology* **46**, 575–580, 1987.

42. Siegel, M.M., Tabei, K., Tsao, R., Pastel, M.J., Pandey, R.K., Berkenkamp, S., Hillenkamp, F., and de Vries, M.S., "Comparative Mass Spectrometric Analyses of Photofrin Oligomers by FAB/MS, UV- & IR-MALDI/MS, ESI/MS and LD/Jet-PI/MS," *J. Mass Spectrom.* **34**, 661–669, 1999.

43. Zhan, Q., Voumard, P., and Zenobi, R., "Chemical-Analysis of Cancer-Therapy Photosensitizers by 2-Step Laser Mass-Spectrometry," *Anal. Chem.* **66**, 3259–3266, 1994.

44. Demnum is a registered trademark of Daikin Industries.

45. Grotemeyer, J., Boesl, U., Walter, K., and Schlag, E.W., "Biomolecules in the Gas-Phase 2: Multiphoton Ionization Mass-Spectrometry of Angiotensin-i," *Or. Mass Spectrosc.* **21**, 595–597, 1986.

46. Li, L. and Lubman, D., "Pulsed Laser Desorption Method for Volatilizing Thermally Labile Molecules for Supersonic Jet Spectroscopy," *Rev. Sci. Instrum.* **59**, 557–561, 1988.

47. Anex, D.S., de Vries, M.S., Knebelkamp, A., Bargon, J., Wendt, H.R., and Hunziker, H.E., "Resonance-Enhanced Two-Photon Ionization Time-of-Flight Spectroscopy of Cold Perfluorinated Polyethers and Their External and Internal Van der Waals Dimers," *International Journal of Mass Spectrometry and Ion Processes.* **131**, 319–334, 1994.

48. Private communication, D.Y. Yoon.

Index

C

N

O

P

T

U

Printed and bound by CPI Group (UK) Ltd, Croydon, CR0 4YY

28/10/2024

01780251-0003